Indian National Science Academy

The **Indian National Science Academy** series is dedicated to publishing books on multidisciplinary topics, bridging the gaps between various scientific disciplines. This series offers readers a comprehensive understanding of diverse subjects, meticulously curated to include cutting-edge research and innovative perspectives from fields such as mathematics, statistics, physics, history of science, medical and life sciences, humanities, and social sciences. By integrating knowledge from these diverse fields, the series provides a platform for groundbreaking discoveries and fosters a collaborative approach to scientific inquiry.

It is an invaluable resource for students, researchers, educators and enthusiasts alike, offering valuable insights and practical applications that transcend traditional academic boundaries. From exploring the mathematical foundations of statistical models to understanding the social implications of medical advancements, this series is essential for anyone seeking to navigate the ever-evolving landscape of modern science. The series not only informs but also inspires new ways of thinking and problem-solving, making it a cornerstone for interdisciplinary learning and innovation.

Virendra M. Tiwari
Editor

Himalayan Natural Hazards

Springer

Editor
Virendra M. Tiwari
CSIR-North East Institute of Science
and Technology
Jorhat, Assam, India

ISSN 3059-3743 ISSN 3059-3751 (electronic)
Indian National Science Academy
ISBN 978-981-92-3783-8 ISBN 978-981-92-3784-5 (eBook)
https://doi.org/10.1007/978-981-92-3784-5

Mathematics Subject Classification: 86A05, 76M99, 86A40, 86-XX, 86-08, 86A15

Responsible Editor: Aishwarya Gupta

This Springer imprint is published by the registered company Springer Nature Singapore Pte Ltd.
The registered company address is: 152 Beach Road, #21-01/04 Gateway East, Singapore 189721, Singapore

Foreword

The Himalaya, one of the youngest and most dynamic mountain systems on Earth, currently exist at the crossroads of tectonic instability, climatic variability, ecological fragility, and escalating anthropogenic pressures. The region is susceptible to a range of natural hazards, including devastating earthquakes, landslides, glacial lake outburst floods, cloudbursts, and shifting hydrological patterns. Such phenomena serve to establish the Himalayan region as a natural laboratory for understanding the interactions among multiple hazards, whose effects extend beyond national borders and impact millions of individuals. Comprehending these evolving threats is thus not solely an academic endeavor but rather a critical scientific and societal imperative.

This volume, produced under the auspices of the Indian National Science Academy and published by Springer Nature, constitutes a timely and significant contribution to enhancing interdisciplinary understanding of Himalayan hazards and risk processes. By consolidating perspectives from various fields, including geology geophysics, climatology, geomorphology, hydrology, engineering, and disaster management, the book captures the complexity of the Himalayan system and underscores the necessity for integrated scientific methodologies in hazard assessment and resilience planning.

The chapters contained within this volume not only synthesize current scientific knowledge but also prioritize the importance of field observations, emerging technologies, hazard modelling, and community-centred mitigation strategies. Particularly noteworthy is the acknowledgement that Himalayan hazards are intricately linked to ongoing climate change, landscape evolution, and socio-economic vulnerabilities. Such integrative insights are essential for the formulation of sustainable policies and the establishment of informed disaster preparedness frameworks across the Himalayan region.

This book is intended to serve as a vital reference for researchers, policymakers, disaster management professionals, students, and others who seek to comprehend the evolving dynamics of the Himalayan environment. It is anticipated that the scientific discussions and case studies compiled within this volume will foster further research, enhance regional cooperation, and contribute significantly to the development of safer and more resilient Himalayan communities. The editors and contributors merit recognition for assembling a comprehensive and forward-looking work on one of the world's most fragile yet critically important mountain systems.

With best compliments,

New Delhi, India

Prof. Shekhar C. Mande

Preface

The Himalayan region, extending roughly 2500 km, is highly susceptible to a variety of natural hazards, including earthquakes, landslides, glacial lake outburst floods (GLOFs), and extreme meteorological events. This area experiences approximately 90% of the seismic activity within the Indian subcontinent, having witnessed at least four major earthquakes with a moment magnitude (Mw) of 8 or above, alongside more than 25 significant earthquakes (Mw $\geq$ 7) over the past century. The resulting fatalities from these seismic events are estimated to be around one million lives lost throughout this period. In addition to seismic threats, the region is burdened by numerous geological and hydrological challenges, including flash floods, debris flows, and GLOFs, compounded by extreme weather anomalies and forest fires, which can trigger cascading disasters. The frequency of these events is increasing, underscoring the critical need for enhanced understanding and preparedness. Consequently, it is imperative to systematically catalogue existing research, document prior mitigation interventions, and advance efforts to refine hazard assessments and develop robust mitigation strategies.

This book incorporates these multifaceted challenges through its diverse chapters. It offers an in-depth examination of Himalayan tectonics and the spectrum of geological and meteorological hazards, including interactions among multiple hazards and forest fire dynamics. Furthermore, the text evaluates the current effectiveness of early warning systems and the vulnerabilities of civil infrastructure to these hazards. Each chapter is crafted by domain experts with extensive professional experience, providing a consolidated entity of knowledge that has yet to be compiled in a single resource. This collection may provide a reference resource for experts and researchers in earth sciences—such as geologists, seismologists, meteorologists, and climate scientists—along with professionals involved in construction and risk management efforts in the Himalayan region to enhance their comprehension of the area's natural hazard characteristics, associated risks, and methods for effective mitigation.

It is not prudent to present a detailed precis of each chapter of the book, as these are eloquently articulated by the experts in the field; however, a succinct overview of the presentation is provided below.

Recent geophysical investigations in the Himalayan region have highlighted critical variations in crustal architecture and seismogenic processes, as well as heterogeneity in crustal thickness, seismic velocities, and gravitational anomalies. Crustal thickness exhibits a gradient, ranging from approximately 40 km beneath the Lesser Himalaya to 60 km beneath the Higher Himalaya. The decollement interface, which delineates the Himalayan crust from the Indian plate that is underthrusting, exhibits complex characteristics that significantly influence the patterns of seismic rupture. While advancements in seismic imaging have yielded valuable data, a denser network of observation points is essential for more granular mapping of the complex geological features. The magnetotelluric (MT) investigations across the Himalayan arc suggest high-resistivity blocks, underpinned by conductive pathways associated with partial melt zones in southwestern Tibet, within the Karakoram and Ladakh batholiths in the trans-Himalayan region. In Himachal Pradesh and Uttarakhand, the Main Himalayan Thrust (MHT) ramp exhibits high conductivity, likely attributable to fluid-saturated weak zones that extend into central Nepal, corroborated by seismological data. The Sikkim sector is characterized by two distinct lithospheric blocks separated by the Main Central Thrust Zone, whereas southern Tibet exhibits a prominent, deep, conductive zone indicative of partial melting. In western Nepal, a unique lithospheric architecture featuring an exceptionally thick sedimentary basin is revealed.

Analyses of historical and current earthquake data indicate that seismicity along the Himalayan arc varies due to the geometry and coupling of the Main Himalayan Thrust (MHT) and structural segmentation. The Sikkim Himalaya is identified as a distinct segment characterised by transverse tectonics, evident in strike-slip earthquake mechanisms and lower crustal-uppermost mantle earthquakes influenced by Indian plate convergence. Numerical simulations indicate that the convergence of the Indian plate acts as a catalyst for earthquakes, with high strain energy concentrations identified in both the eastern and western segments of the Himalaya. A high frequency of earthquakes in the Himalayan region is predominantly associated with the Main Himalayan Thrust (MHT), and a significant portion of Himalayan arc, often referred to as the central seismic gap, has remained devoid of major seismic events for centuries, underscoring the need for detailed investigations into the dynamics of strain accumulation. The study, leveraging recent GNSS data, provided estimates of strain rates and their intricate relationship with seismicity, and documented a robust correlation between strain and seismic activity.

In the geodynamically evolving Himalaya, landslide problems are critical, particularly in areas of the Main Central Thrust and Main Boundary Thrust zones. These zones experience repeated rock movement, leading to shearing and fragmentation. Key factors include steep bedrock slopes and triggers like rainfall, seismic activity, and erosion. The thrust zones also pose a risk of landslide damming, which can lead to historical lake formations and downstream effects. The studies presented in the book and reference therein emphasise the importance of landslide susceptibility mapping and risk assessment to enhance disaster risk reduction.

Analyses of precipitation from 1979 to 2020 reveal variable monthly patterns due to local topography and large-scale atmospheric dynamics in the Himalaya. Precipitation patterns correlate with high vertically integrated moisture transport

(VIMT), moist static energy (MSE), and cloud cover, which support deep convection. It is also found that the El Niño-Southern Oscillation (ENSO) significantly influences Himalayan precipitation, with El Niño years linked to drier conditions and La Niña years to wetter conditions. Amplified monsoon rainfall can drive landslides, underscoring the need for integrated forecasting. Snow avalanches are common in the high-altitude regions of the Himalaya. Data shows that new challenges, especially regarding wet snow avalanches, and demands future strategies for managing avalanche risks in the IHR. Growing glacial lakes and related floods pose significant concerns for the Indian Himalayan Region (IHR). The damage from recent GLOFs in the 21st century highlights the highest risk in the western Himalaya and significant threats in eastern regions like Sikkim. Predicting GLOFs remains challenging and current GLOF models struggle due to a lack of understanding of the complex mechanisms involved.

Climate change and human activities have heightened fire risks, shifting research focus from post-fire damage assessments to predictive modelling and real-time monitoring. The Himalayan region is experiencing warming faster than the global average, causing distress among residents due to various geological, ecological, and hydrological disasters. Climate change has led to excessive rainfall, resulting in landslides, flash floods, and glacier lake outburst floods (GLOFs), as well as other impacts such as forest fires and cascading effects from earthquakes. These issues are worsened by unplanned constructions and inadequate infrastructure, disrupting water, energy, food, health, and biodiversity sectors.

The seismic risk in the IHR involves site effects, vulnerable buildings, and fragile access networks, requiring an integrated, multi-scalar approach. A tiered risk assessment framework can guide land-use and operational strategies. Key actions include establishing a Himalayan Seismic Observatory, expanding strong-motion networks, and launching short-term safety initiatives for schools and hospitals. In the medium term, retrofitting at-risk buildings and implementing earthquake early warning systems are priorities. Long-term goals focus on comprehensive microzonation, standardised permitting, and enhanced infrastructure resilience, supported by financial mechanisms to reduce risk. Developing effective early warning systems (EWS) is vital to reducing their significant impacts. Seismic Network-based Early Warning for the Himalayan hazards could be a possibility. As risks from human activities grow in complexity, transitioning to Multi-Hazard Early Warning Systems (MHEWS) is vital for effective preparedness and resilience.

Jorhat, India Virendra M. Tiwari

Acknowledgements

I extend our gratitude to Prof. Ashutosh Sharma, Past President of the Indian National Science Academy (INSA), Prof. S. C. Mande, President of INSA, and the INSA Council for their endorsement of the INSA-Springer Series Book and for their unwavering support throughout this initiative. Prominent, respected scientists, Prof. H. K. Gupta, and Prof. Shailesh Nayak, have promptly provided a note for this publication. I express our sincere appreciation for their brief note and messages. I wish to express my sincere appreciation to all contributors and reviewers for their valuable feedback, which significantly enhanced the quality of the material presented within this book. Additionally, I would like to acknowledge my colleagues at CSIR-NEIST, Bijit Kumar Choudhury and Radhika Chettri, as well as Sudhanshu Agarwal at INSA, for their assistance at various stages of the publication process; their contributions have been invaluable.

Jorhat, India
April 2026

Virendra M. Tiwari

Message by Harsh K. Gupta

The Himalayan mountain belt is one of the most active and complex geodynamic regions on Earth, where ongoing tectonic deformation, climatic variability, and fragile ecosystems combine to create a landscape highly susceptible to natural hazards. Earthquakes, landslides, glacial hazards, riverine floods, and extreme weather events have repeatedly demonstrated the Himalayan terrain's vulnerability and the urgent need for integrated scientific understanding and possible mitigation of these natural hazards.

In this regard, *Himalayan Natural Hazards* is a commendable and timely contribution that brings together current scientific perspectives on the nature, processes, and impacts of hazards across the Himalayan belt. Produced by the Indian National Science Academy and published by Springer Nature, the volume focuses on the growing importance of interdisciplinary research in addressing the challenges posed by these natural hazards.

The book presents a rich compilation of studies encompassing seismology, tectonics, geomorphology, hydrology, climate science, remote sensing, and disaster risk management. By integrating theoretical developments with field observations and modern analytical techniques, the contributors have provided a valuable insight into both the causes and consequences of Himalayan hazards. The emphasis on hazard assessment, monitoring, mitigation, and resilience further enhances the practical relevance of this scientific compilation.

The emphases laid in this volume that Himalayan hazards cannot be viewed in isolation is a very important aspect. The interactions between geological processes, climatic influences, environmental degradation, and human activities demand holistic approaches that combine scientific innovation with sustainable policy frameworks. The discussions presented in this book, therefore, contribute not only to academic advancement but also to broader societal efforts to reduce disaster risk and strengthen resilience in vulnerable mountain communities.

This volume will undoubtedly serve as an important resource for researchers, students, professionals, planners, and policymakers engaged in Himalayan studies and natural hazard research. The editors and contributors deserve sincere appreciation for compiling such a comprehensive and insightful work on one of the world's most sensitive and strategically important mountain regions.

May 2026

Harsh K. Gupta
Insa Senior Scientist
Fellow Indian Academy of Sciences
Foreign Member Russian Academy of Sciences
Fellow International Science Council

Message by Shailesh Nayak

The Himalayan region is characterized by its exceptional biodiversity and ecological importance, but it is equally defined by its extreme geological and environmental vulnerabilities. As one of the globe's most tectonically active mountain ranges, the Himalaya is prone to a wide array of natural hazards, including seismic events, landslides, avalanches, glacial lake outburst floods, and severe meteorological phenomena. In recent decades, the interplay of climate change, rapid urbanization, infrastructure development, and environmental deterioration has exacerbated the susceptibility of this delicate ecosystem.

This book on *Himalayan Natural Hazards* represents a significant scientific contribution, particularly as the understanding of multi-hazard processes and their cascading impacts becomes crucial. It assembles a diverse range of expertise across disciplines such as geosciences, atmospheric sciences, hydrology, engineering, disaster risk reduction, and environmental studies, thus providing a comprehensive perspective on hazard dynamics specific to the Himalayan context.

A key strength of this volume is its interdisciplinary approach, which integrates empirical observations, sophisticated analytical techniques, remote sensing technologies, numerical modelling, and risk assessment methodologies. The chapters elucidate the intricate scientific challenges posed by Himalayan hazards while underscoring the urgent need for sustainable mitigation and adaptation strategies. Furthermore, there is a significant emphasis on translating scientific insights into actionable frameworks for disaster preparedness, resilience enhancement, and policy formulation.

This script is poised to be a vital reference for researchers, academics, policymakers, engineers, planners, and students in the fields of earth and environmental sciences. It not only chronicles cutting-edge scientific developments but also stimulates future research into the evolving hazards facing the Himalayan region.

The editors and contributors are to be commended for their efforts in assembling this comprehensive and impactful volume, which advances our understanding of one of the world's most dynamic and vulnerable mountain systems.

Shailesh Nayak

Contents

Part I
Geological Hazards

Chapter 1
The Role of Himalayan Deformation and Structural Architecture in Understanding Himalayan Natural Hazards

Malay Mukul, Enjamamul Haque, Manas Mukul, and Vinee Srivastava

Abstract The Himalayan mountain belt is an active fold-thrust belt characterised by a basal decollement, the Main Himalayan thrust. Slip along the basal decollement has been the crux of Himalayan deformation, resulting in the generation of Himalayan topography through the emplacement of in-sequence thrust sheets since ~ 55 million years ago. The Himalayan hinterland has been uplifted to more than 8 km above mean sea level and is the region where most Himalayan glaciers and associated frontal glacial lakes are located. Parts of this topography continue to be uplifted along the active Lesser Himalayan Duplex (LHD), which has evolved as the dominant active structure in the Himalaya. This Himalayan structural architecture is key to understanding the topographic development and associated natural hazards, such as earthquakes, landslides and glacial lake outburst floods, along the Himalayan mountain front. This chapter examines the relationships between Himalayan deformation, topography development, climatic conditions and associated natural hazards in the Himalaya, within the context of how Earth's surface processes and deformation kinematics interact. Understanding these connections and incorporating them into developmental activities holds the key to sustainable development in the Himalaya.

Keywords Himalayan natural hazards · Himalayan wedge · Seismic hazard · Landslides · Faults · GLOFs · Main Himalayan thrust

M. Mukul (✉)
Continental Deformation Laboratory, Department of Earth Sciences, Indian Institute of Technology Bombay, Mumbai, India

E. Haque · V. Srivastava
Department of Earth and Environmental Sciences, Indian Institute of Science Education and Research, Bhopal, India

M. Mukul
School of Computer Applications, Kalinga Institute of Industrial Technology, Bhubaneshwar, India

V. M. Tiwari (ed.), *Himalayan Natural Hazards*, Indian National Science Academy,
https://doi.org/10.1007/978-981-92-3784-5_1

1.1 Introduction

Natural hazards are 'those elements of the physical environment, harmful to man and caused by forces extraneous to him' (Burton & Kates, 1964). Natural hazards such as landslides, floods and earthquakes pose persistent and widespread threats across the Himalayan region, resulting in significant fatalities, displacement and infrastructure damage each year (Froude & Petley, 2018; Schuster & Highland, 2001). The hazards we face are closely tied to the dynamic evolution of the Himalaya, shaped by the interplay of tectonic forces that build the topography and climatic conditions that erode it (Burbank et al., 1996; Molnar & Lyon-Caen, 1988). Landslides, for instance, occur as a gravitational response to an imbalance between driving and resisting forces on slopes, controlled by both intrinsic factors such as rock cohesion and pore-space dynamics and extrinsic factors such as slope geometry, water infiltration and tectonic stress (Duncan & Wright, 2005; Iverson, 2000). Similarly, floods, particularly flash floods driven by Landslide Lake Outburst Floods (LLOFs) and Glacial Lake Outburst Floods (GLOFs), are intensified by steep terrain, heavy monsoonal precipitation and the destabilisation of glacial lakes due to climate warming. Earthquakes, on the other hand, result from the kinematics of slip transfer along the basal decollement of the Himalaya to accommodate the downward subduction of the Indian plate beneath the southern margin of Tibet (e.g. Bilham et al., 2001; Grandin et al., 2012; Lave & Avouac, 2000; Mukul et al., 2007, 2018). Earthquakes build and shape topography, cause widespread structural damage, disrupt critical infrastructure and pose severe risks to human life.

Unlike earthquake hazards that occur sporadically, flood and landslide processes are often persistent, interrelated and seasonally reinforced, especially in the Himalayan fold-thrust belt. Here, active tectonic deformation, high-relief topography, variable lithology and intense monsoonal precipitation are intertwined in ways that can produce a complex, hazard-prone landscape marked by chronic landslide instability, flood vulnerability and seismic risk (Kamp & Owen, 2011; Kargel et al., 2016). However, the slip along the basal decollement has been the crux of Himalayan deformation. It has generated and evolved the Himalayan topography by emplacement of in-sequence and out-of-sequence thrust sheets since ~ 55 Ma. We contend that this deformation-driven topography, associated climate and climate change are the primary controls behind Himalayan natural hazards. We first discuss the Himalayan structural architecture and then detail how it is connected with Himalayan natural hazards.

1.1.1 Himalayan Structural Architecture and Topographic Growth

The Himalaya is a thin-skinned style fold-thrust belt and is a wedge-shaped orogen characterised by a basal decollement (Fig. 1.1), the Main Himalayan thrust (MHT) (e.g. Boyer & Elliott, 1982; DeCelles et al., 1998; Mukul, 2000, 2010; Nelson et al., 1996; Schelling & Arita, 1991; Srivastava & Mitra, 1994; Thakur, 1993). In such orogens, slip along the basal decollement is the key to understanding the fold-thrust belt evolution and is best understood in the context of the symptomatic critical wedge theory, wherein the deforming, wedge-shaped orogen exhibits three stages of evolution over the geological timescale (DeCelles & Mitra, 1995; Mitra, 1997). The equilibrium state in the evolution of the orogen is the critical state, wherein slip along the basal decollement propagates all the way from the hinterland to the foreland, resulting in significant in-sequence thrusting and emplacement of the major thrust faults and associated thrust sheets in the Himalaya, such as the Main Frontal thrust (MFT), Main Boundary thrust (MBT), Ramgarh thrust (RT), Main Central thrust (MCT), South Tibetan Detachment system (STDS) and the Tethyan thrust belt. Among the thrust sheets, the STDS and MCT occupy a larger portion (> 60%) of the Himalayan fold-thrust belt's aerial extent in parts of India, Nepal, and Bhutan (Fig. 1.2a and b). From north to south, these thrusts delineate different zones in the Himalaya: (a) The Indus-Tsangpo Suture Zone (ITSZ) is the backstop to the Himalayan wedge, (b) The STDS carries the Tethyan Himalayan Sequence (THS), (c) The MCT sheet consists of the Greater Himalayan Sequence (GHS), (d) The MBT and RT carry the Lesser Himalayan Sequence (LHS) and (e) The MFT sheet includes the Siwalik foothills or Sub-Himalayan Sequence (SHS). Topographic growth during this stage is primarily at the front of the wedge through the lengthening of the wedge and the emplacement of a new thrust sheet. However, the emplacement of a new thrust sheet also results in the uplift of the existing thrust sheets, which eventually evolve into a topographic gradient that increases from foreland to hinterland. The Himalayan wedge exhibits a topographic gradient (Fig. 1.3) with two of the world's highest peaks (Mt. Everest, 8848.86 m, and Mt. Kanchenjunga, 8586 m) located in the central part of the orogen in the eastern Nepal Himalaya.

There are two non-equilibrium, transient stages in the evolution of a wedge-shaped orogen with a basal decollement. The first is the supercritical state, where the wedge lengthens by the emplacement of a new thrust sheet, accompanied by frontal topographic growth to attain the equilibrium critical state. In addition, the topography is reduced in the hinterland, typically by normal faulting (DeCelles & Mitra, 1995; Mitra, 1997; Mitra et al., 2026; Mukul et al., 2007). This state has also been postulated in the Himalaya by 'dominant' structures (e.g. the Kangmar antiform; Nelson et al., 1996) and normal faults related to the STDS (Mukul, 2000, 2010). The second transient state is the subcritical state, characterised by internal deformation within the deforming wedge through out-of-sequence deformation. A subcritical wedge, dominated by secular or focused erosion, is stalled and exhibits internal deformation by forming out-of-sequence thrusts, duplexes, back-thrusts and fault-related folding to attain equilibrium. The Lesser Himalayan Duplex (e.g. DeCelles et al., 1998; McQuarrie et al., 2008; Mitra et al., 2010; Srivastava & Mitra, 1994), out-of-sequence thrusts (e.g. Haque et al., 2026; Morley, 1988; Mukul et al., 2007; Mukherjee, 2015; Mukul et al., 2017) and channel flow (Beaumont et al., 2001; Grujic et al., 2002; Wobus et al., 2005) have been recognised as out-of-sequence deformation

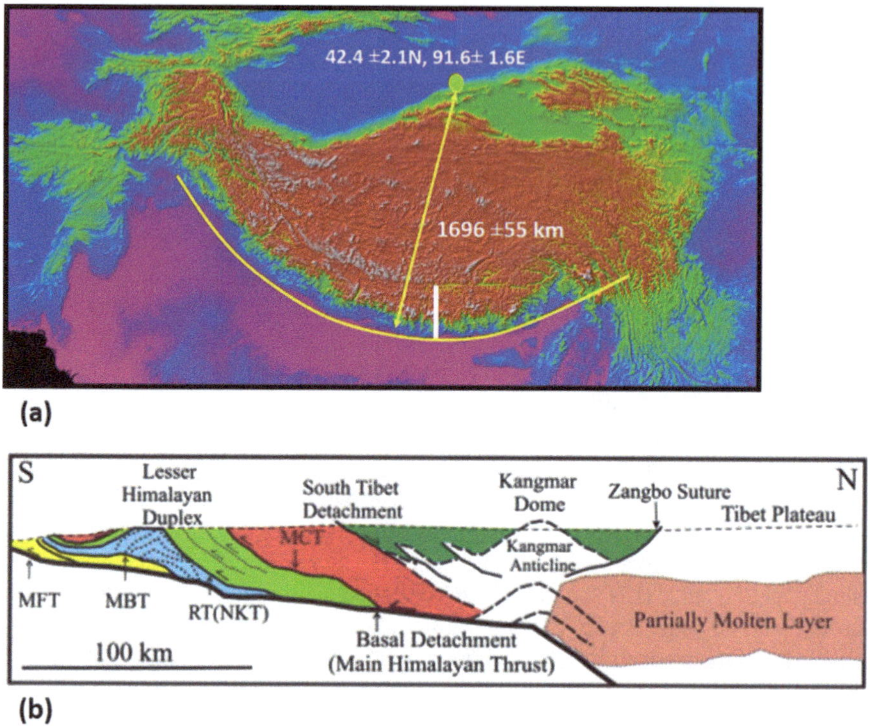

Fig. 1.1 Himalaya is **a**) arcuate in shape (Bendick & Bilham, 2001) in map view and **b**) a thin-skinned style, wedge-shaped mountain belt in cross-section, as shown in the Darjiling-Sikkim Himalaya (after Mukul, 2010; Nelson et al., 1996). The Main Himalayan thrust (MHT) is the basal decollement of the Himalayan wedge. Listric thrust faults such as the Main Central thrust (MCT), Ramgarh thrust (RT), Main Boundary thrust (MBT) and the Main Frontal thrust (MFT) sole into the MHT. The brittle-ductile transition is located ~ 100 km from the mountain front defined by the MFT (Mukul et al., 2018). The white line in **a** represents the line of the cross-section shown in **b**. The image in **a** was provided by and used with permission from Chris Duncan

features in the Himalaya. The out-of-sequence deformation builds topography within the wedge, north of the mountain front in the Himalayan thrust belt.

The symptomatic critical wedge model was initially developed for the North American Cordilleran fold-thrust belt (DeCelles & Mitra, 1995; Mitra, 1997), which is not an active fold-thrust belt. This approach was extended to the Himalayan fold-and-thrust belt, an active thrust belt, at younger neotectonic (Mukul, 2000; Mukul et al., 2007), seismotectonic and decadal time scales (Mukul, 2005, 2010; Mukul et al., 2018). The symptomatic critical wedge model at younger time scales provides a framework for linking Himalayan deformation, structural architecture and topographic growth to Himalayan Natural Hazards. This chapter systematically explores these connections and clearly describes our current understanding.

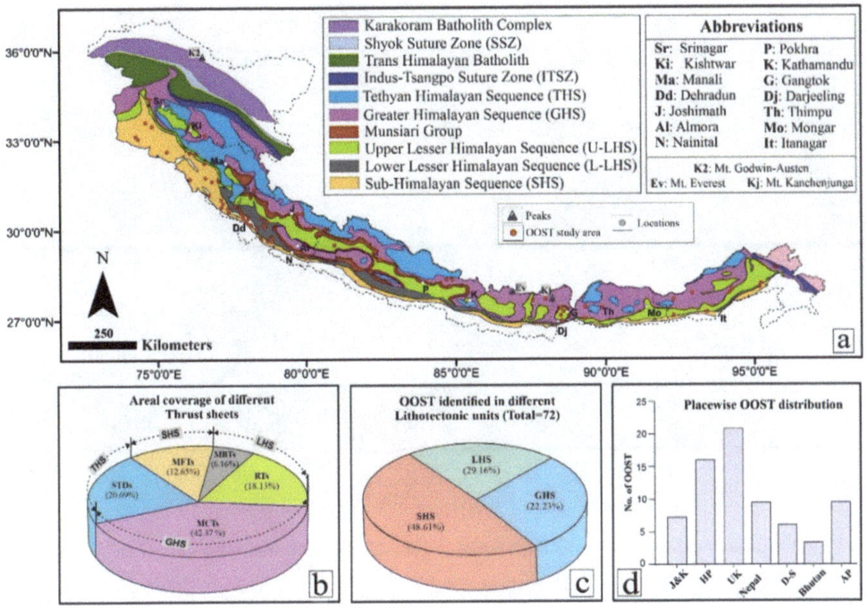

Fig. 1.2 **a**) Himalayan lithotectonic units (after Valdiya, 2002; Jain & Banerjee, 2020) in parts of India, Nepal and Bhutan; **b**) a pie chart showing the areal distribution of different thrust sheets along and across the Himalayan range. Here, the 's' after each thrust name indicates the corresponding thrust's sheet (after Jain & Banerjee, 2020); **c**) a pie chart showing the percentage of Out of Sequence thrust (OOST) fault zone identified within Himalayan lithotectonic units (GHS, LHS, SHS) (after Mukherjee, 2015); **d**) a bar diagram shows the number of OOST identified in the Himalayan region, within different states in India, Nepal and Bhutan (after Mukherjee, 2015)

1.1.2 Himalayan Deformation Mechanisms

A thrust sheet is the basic structural unit in fold-thrust belts such as the Himalaya. Thrust sheets may be internal, transitional or external (Mitra, 1994; Mukul, 1998, 1999). Internal thrust sheets typically deform in a quasi-plastic (QP) manner, with dislocation creep as the dominant deformation mechanism (Long et al., 2011; McNaught & Mitra, 1996; Mitra, 1994; Mukul & Mitra, 1998; Mukul et al., 2004; Parui et al., 2022; Sibson, 1977). Thrust faults that carry internal thrust sheets also exhibit dislocation creep and form mylonites with dynamic recrystallisation as the typical grain-size reduction mechanism (Bhattacharyya & Mitra, 2011; Chakraborty & Mukul, 2019; Chatterjee et al., 2022; Sibson, 1977). External thrust sheets deform in an elastico-frictional (EF) manner with frictional sliding and cataclastic flow as the dominant deformation mechanisms (Ismat, 2013; Ismat & Mitra, 2005; Srivastava & Mukul, 2020; Srivastava et al., 2016, 2018). Thrust faults carrying external thrust sheets exhibit frictional sliding and cataclastic flow with fracturing as the prominent grain-size reduction mechanism (Basa et al., 2019; Haryono, 2025; Srivastava et al., 2016). EF deformation mechanisms are primarily connected to

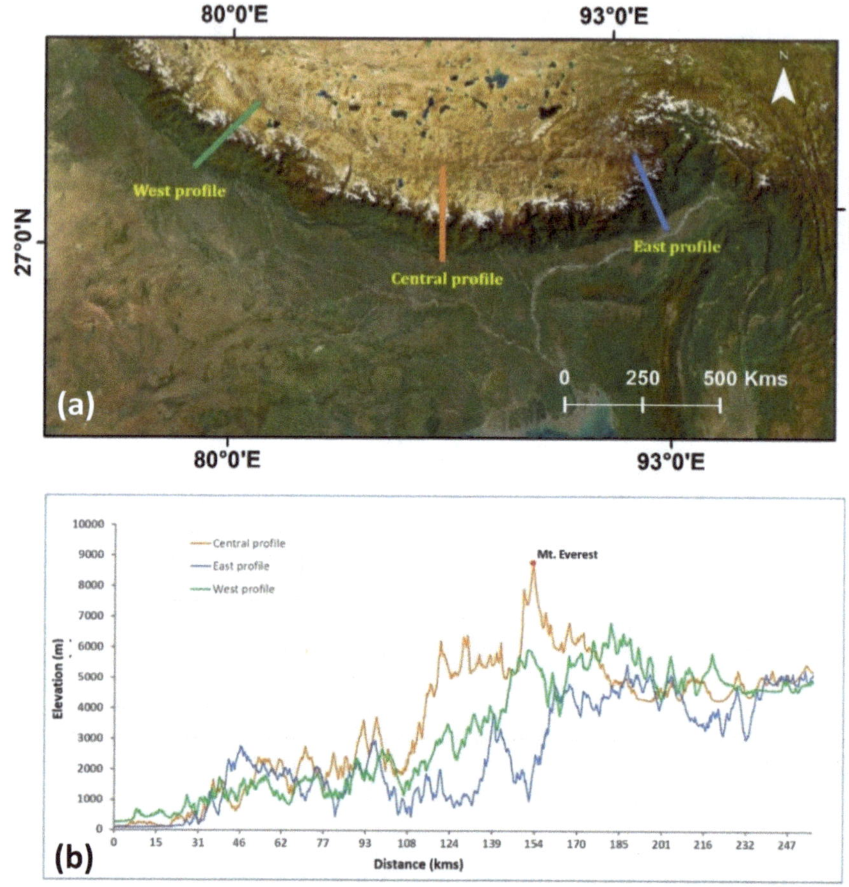

Fig. 1.3 **a**) Map view of the arcuate Himalayan fold-thrust belt with the cross-section lines along which topographic profiles were extracted using ALOS-2 data. **b**) The topographic profiles across the Himalayan arc were extracted from ALOS-2 data, as they exhibit the best vertical uncertainty (Mukul & Mukul, 2021). The highest topography is located in the central Himalayan arc and is characterised by the highest peaks, such as Mt. Everest (8848.86 m). There is also a sharp topographic break ~ 100 km from the foreland. The central part of the Himalayan wedge (100–200 km) is elevated relative to the rest of the mountain belt because of the Lesser Himalayan Duplex (LHD) development and related active uplift

Himalayan natural hazards, such as earthquakes and landslides within fault zones and thrust sheets, particularly during out-of-sequence deformation. The transitional thrust sheets typically exhibit EF-QP deformation mechanisms or characteristics of both external and internal thrust sheets.

The deformation mechanisms within the basal decollement of the active Himalayan wedge also significantly control the seismic hazard. The Himalayan basal decollement, the Main Himalayan thrust (MHT), dips to the north and transitions from an EF deformation regime to QP downdip. This implies that the thrust sheets and thrust deform primarily in the QP regime north of the transition region. This rheological contrast in the Himalayan wedge has important implications for understanding the interactions among deformation, Earth-surface processes, topographic growth and natural hazards in the Himalaya. This contrast is also reflected in the Himalayan fault/shear zones, which are diffuse faults rather than sharp discontinuities (e.g. Bhattacharyya & Mitra, 2011; Chakraborty & Mukul, 2019; Chatterjee et al., 2022; Haque et al., 2026; Mukul et al., 2022; Srivastava et al., 2016, 2018).

1.1.3 Himalayan Fault Zone Rheology

Outcrop-scale faults and shear zones are typically diffuse, simple-shear-dominated, metre-to-kilometre-thick grain-size reduction zones, and rarely exhibit a sharp, distinct planar boundary (e.g. Gudmundsson, 2011; Shipton & Cowie, 2003). The fault zone consists of a highly deformed fault/shear zone core (or slip zone) marked by the maximum reduction in grain size by fracturing (EF) or recrystallisation (QP). This core is surrounded by Fault-Related Damage/Deformation Zones (FRDZ) where deformation intensity decreases and grain size reduction gradually decreases outward, eventually transitioning into the unaffected intact protolith (Caine et al., 1996; Gudmundsson, 2011; Mitchell & Faulkner, 2009; Newman & Mitra, 1993; Shipton et al., 2006; Srivastava et al., 2016).

This implies that fault/shear rheology can be EF, EF-QP or QP (Sibson, 1977; Violay et al., 2015), and that these rheologies vary significantly with depth in the crust. At the upper level of the crust (<10–15 km, T: 0–300 °C, P: 0–4 kbar), rocks primarily deform in an EF manner. They are dominated by frictional sliding and dissolution creep deformation mechanisms (Paterson, 1990). In contrast, at greater depths (>10–30 km, T: > 350 °C, P: > 4 kbar), the fault zone transitions into a QP deformation regime where the dislocation creep deformation mechanisms dominate (Sibson, 1977). This depth-dependent variation in deformation mechanisms is fundamental to understanding fault zone strength and mechanical behaviour, and how they interact and affect Himalayan natural hazards.

1.2 Seismic Hazard in the Himalayan Wedge

The 'symptomatic' critical wedge model at younger (10–100 years; Mukul, 2005; Mukul et al., 2007) time scales provides a framework for linking Himalayan deformation, structural architecture and topographic growth to the Himalayan seismic hazard. The N-S variation in rheology and deformation mechanisms within the basal decollement (MHT) of the active Himalayan wedge also significantly controls the seismic risk in the wedge.

Seismic hazard in the Himalaya is associated with slip transfer along the MHT from the hinterland to the foreland. The continuous slip transfer along the MHT is aseismic at depths below the crustal brittle(EF)-ductile(QP) transition (e.g. Bilham et al., 2001; Lave & Avouac, 2000; Grandin et al., 2012; Jade et al., 2014; Gautam et al., 2017; Mukul et al., 2018). The slip is arrested at the brittle(EF)-ductile(QP) transition ~ 100 km north of the foreland near the surface trace of the MCT and converts to shortening. This leads to maximum uplift in the region and the eventual formation of the Lesser Himalayan Duplex (LHD) over geological time scales (Grandin et al., 2012; Mitra & Boyer, 2020; Mukul et al., 2018). This zone is marked by interseismic seismicity (Fig. 1.4) along the Himalayan arc (e.g. Hubbard et al., 2021; Mukul et al., 2014; Pandey et al., 1999); the MHT south of this zone is locked and typically records minimal slip (Bilham et al., 2001). This is usually the region between 100 and 200 km north of the Himalayan front, where maximum topographic growth is associated with the Lesser Himalayan Duplex (Figs. 1.1 and 1.3).

The above decollement tectonics implies that all earthquakes in thin-skinned style, active wedges like the Himalaya, originate along the decollement at the EF-QP transition, which effectively acts as the 'earthquake nursery'. Therefore, identifying the EF-QP transition in active wedges is the key to forecasting all earthquake hypocentres in space. This zone is marked by interseismic seismicity in the Himalaya (Fig. 1.4). Therefore, all major and great earthquake hypocentres in the Himalaya will likely be located in the region ~ 100–200 km north of the Himalayan front within the active Lesser Himalayan Duplex (e.g. Bilham, 2019; Bilham et al., 2001; Mukul et al., 2018). This is an essential insight for planning human settlements and infrastructure in the Himalaya and other active wedges.

The magnitude of the earthquakes and the consequent seismic hazard depend on the extent of slip transfer along the locked basal decollement south of the EF-QP region. A great earthquake will be required to transfer slip to the mountain front (Bilham, 2019; Bilham et al., 2001, 2011) and will involve the slip of the entire wedge along the decollement without any internal deformation in the wedge. This is possible only in an active critical wedge (Mukul et al., 2007) and will result in an in-sequence earthquake. If the slip transfer stops short of the mountain front, the earthquake will be an out-of-sequence earthquake, resulting in internal deformation within a subcritical wedge (Mukul et al., 2007) (Figs. 1.4 and 1.5). The 25th April, 2015, 7.8 magnitude Gorkha earthquake was a recent example of a large out-of-sequence earthquake, in which slip transfer along the MHT stopped short of the Nepal Himalayan front (e.g. Parameswaran et al., 2015). The 4th April, 1905, 7.8 Kangra earthquake (Hussain et al., 2009; Husson et al., 2004), the 20th October, 1991, 6.8 Uttarkashi (Thakur & Sushil, 1994; Valdiya, 1991) and the 29th March, 1999, 6.6 Chamoli earthquakes (Kayal et al., 2003; Rastogi, 2000) are also examples of significant out-of-sequence Himalayan earthquakes.

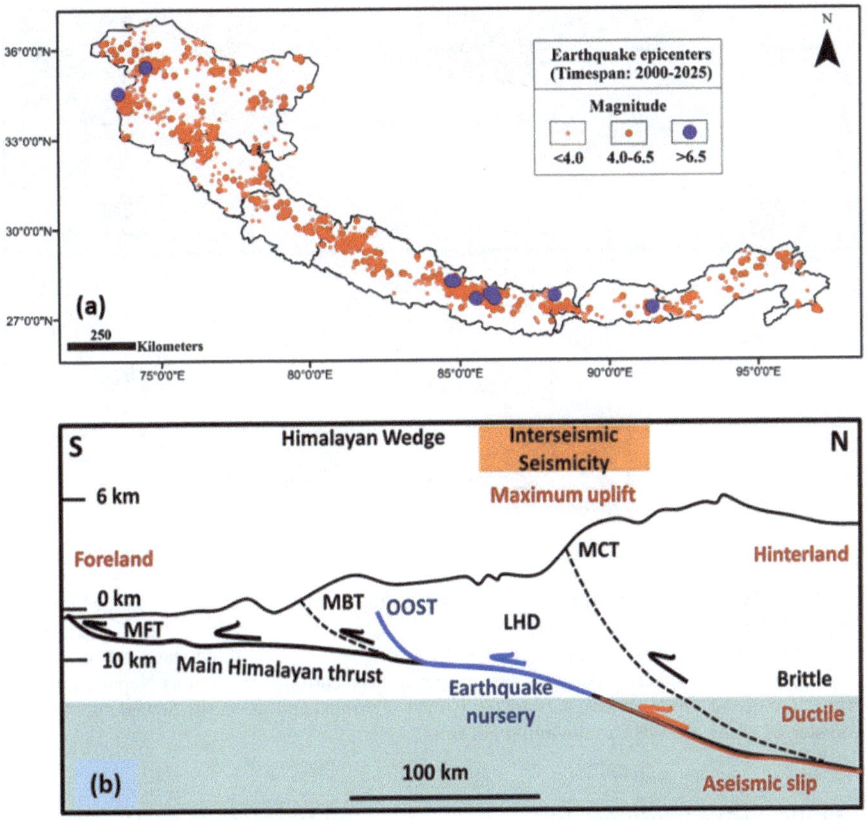

Fig. 1.4 **a**) Earthquake epicentre data (from earthquake.usgs.gov) between 2000 and 2025 showing interseismic Himalayan earthquakes in the zone of EF-QP transition along the Main Himalayan thrust. **b**) Schematic representation of the active Himalayan wedge showing the brittle(EF)-ductile(QP) transition near the Lesser Himalayan Duplex (LHD) where earthquake hypocentres originate in the 'Earthquake nursery'. OOST—a hypothetical out-of-sequence thrust within the Himalayan wedge. The zone of interseismic seismicity shown in the map view **a** is marked in the cross-sectional view **b**

1.3 Landslide Hazard in the Himalayan Wedge

Landslides in the Himalayan wedge constitute the second significant natural hazard and are typically triggered by natural causes such as large earthquakes and rainfall (Crozier, 2010; Guzzetti et al., 1999; Kokutse et al., 2016; Yang et al., 2021). However, anthropogenic activities and infrastructure development greatly aid these natural triggers, and the impact of anthropogenic factors on landslides has intensified. The Himalayan (2007–2022) landslide inventory from NASA (Fig. 1.6a) (Kirschbaum et al., 2010, 2015) indicates a dominant concentration of landslides in the external thrust sheets in the frontal part of the Himalayan wedge south of

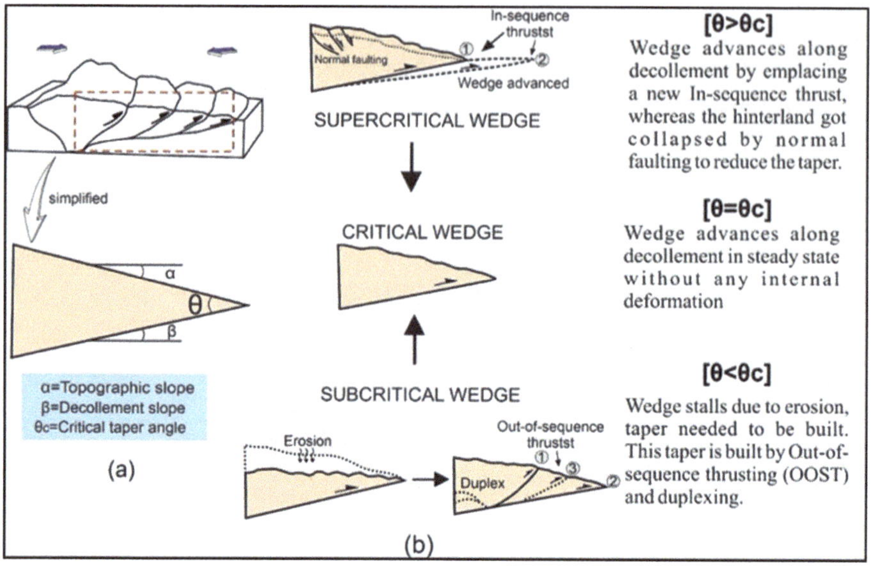

Fig. 1.5 Critical wedge model for active wedges like the Himalaya (after Mitra, 1997; Mitra et al., 2026; Mukul et al., 2007). **a**) A typical simplified wedge where θ is the wedge taper, α is the topographic slope and β is the dip of the basal décollement. **b**) The three stages of an actively deforming wedge. A subcritical wedge can trigger a large earthquake, such as Nepal's 2015 7.8 Gorkha earthquake. Great earthquakes are likely only in critical and supercritical wedges when slip is transferred along the MHT to the mountain front

the brittle-ductile transition, especially in the central Himalaya in Nepal, Darjiling-Sikkim and the NW Himalaya. This trend is further confirmed by the Geological Survey of India (2000–2019) inventory in the Darjiling-Sikkim and the NW Himalaya (Fig. 1.6b). This is intriguing as the MFT is locked in the frontal Himalayan wedge. Also, the NE Himalaya east of Sikkim seem to have fewer landslides in both inventories (Fig. 1.6) than the rest of the Himalaya.

Further statistical analysis of the datasets reveals that the relative frequency (No. of landslides per individual thrust sheet area) of landslide occurrences (timespan: 2000–2022) in the Lesser to Sub-Himalayan MFT, MBT and RT thrust sheets (65–67%) is much higher than that of the MCT and STD sheets (33–35%) (Fig. 1.7). Also, the RT sheet shows the highest landslide frequency (~ 30%), followed by the MCT sheet (~ 27%) (Fig. 1.7b1 and b2). The RT and MCT sheets are affected by the Lesser Himalayan Duplex, which is an active structure causing maximum uplift in the Himalayan wedge. These observations suggest that the landslides are more pronounced in the central Himalayan wedge that has undergone EF deformation. Moreover, they are concentrated in the region experiencing the maximum active uplift. Therefore, slope stability in the Himalaya appears to be more strongly influenced by EF rock rheology and active deformation, which increase slope steepness through active uplift. These insights also suggest that the architecture of diffused

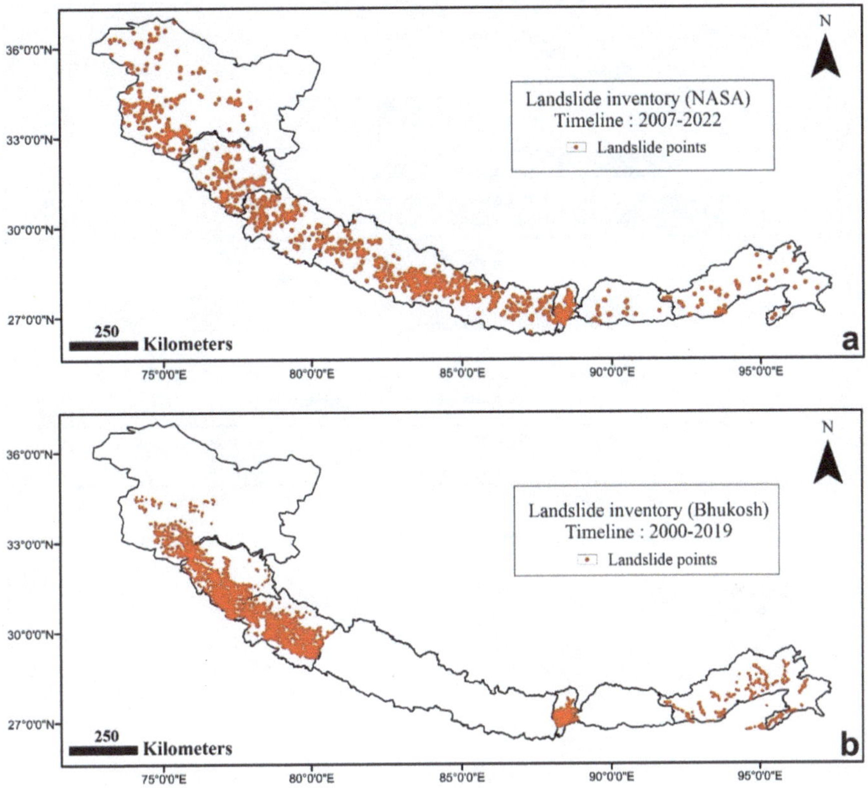

Fig. 1.6 Landslide distribution in the Himalayan wedge; **a**) landslide inventory map downloaded from the NASA website (landslides.nasa.gov), and **b**) landslide inventory map from the Geological Survey of India, Bhukosh portal (bhukosh.gsi.gov.in) in the Indian Himalaya

EF fault zones may play a significant role in determining slope stability, alongside active deformation, rainfall and seismic triggers. This is particularly true in the locked frontal Himalayan wedge, where there is no active deformation and landslide frequency is high. The EF fault core, where grain-size reduction is maximum, and fault gouge dominates, is likely to be the most susceptible to landslides compared to fault damage zones. However, most landslide studies are carried out on fractured rocks that are likely part of damage zones. The QP deformed rocks generally form rheologically stable slopes in the High Himalaya, unless later EF faults deform them or slopes become too steep. Addressing these issues through an integrated, process-based approach appears crucial for developing concept-based landslide susceptibility models. Understanding how fault-related deformation zones, crustal uplift, rainfall and earth-surface processes interact and affect slope stability can improve the predictive accuracy of Himalayan landslide models.

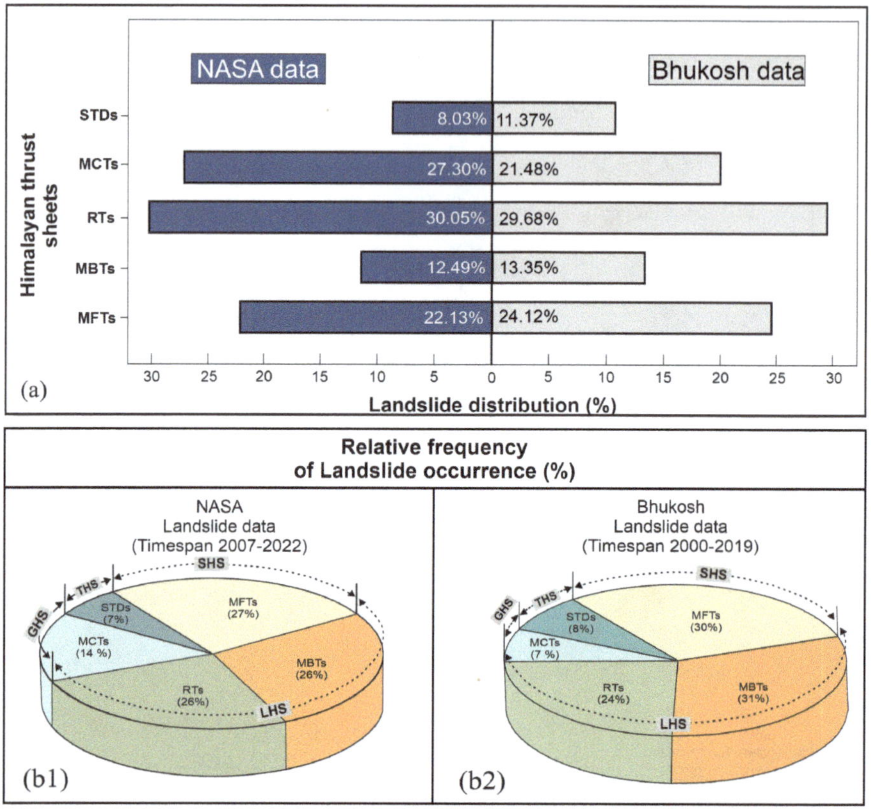

Fig. 1.7 **a**) Landslide distribution of two different landslide inventories and timelines. One common observation across these two inventories is that the Lesser to Sub-Himalaya (RT, MBT, MFT) region records the highest number of landslide events. The relative frequency of landslide occurrence is shown here by two pie diagrams (**b1**) and (**b2**). The figure (**b1**) represents data from NASA, and (**b2**) represents data from the Bhukosh website. The data presented have been generated using ArcGIS Pro

1.4 Flood Hazard in the Himalayan Wedge

Flooding within the Himalayan wedge and its adjoining foreland is a complex hazard influenced by Himalayan tectonics and the resulting rugged, varied topography, active river networks, heavy monsoonal precipitation and glacial dynamics (e.g. Bookhagen & Burbank, 2010; Immerzeel et al., 2010; Mukul et al., 2025). Most of the glaciers are located in the MCT hanging wall at high elevations, resulting from uplift related to the formation and evolution of the underlying Lesser Himalayan Duplex (Figs. 1.1 and 1.8). Consequently, the main Himalayan River systems also originate in this region, although they exhibit significant along-strike variability. Western rivers such as the Indus, Sutlej and Ganga are predominantly snow-fed and sourced from high-altitude glaciated terrains. In contrast, central and eastern rivers such as the Kosi, Teesta and Brahmaputra also originate in high-altitude glaciers but are influenced by monsoon precipitation and glacial melt, resulting in seasonal, highly variable discharge patterns (e.g. Singh & Lal Kansal, 2022; Thayyen & Gergan, 2010). Steep gradients and narrow valleys in the higher Himalaya frequently led to flash floods, particularly during cloudburst events (Dimri et al., 2017). Glacial Lake Outburst Floods (GLOFs), on the other hand, are sudden, powerful floods caused by the breach of glacial lakes that form when glaciers retreat. Exceptionally high peak discharges characterise them, often much larger than those of hydro-meteorologically induced floods, such as cloudbursts (Cenderelli & Wohl, 2001; Cook et al., 2018). GLOFs are low-probability, high-impact events that occur with little to no warning. Consequently, they pose formidable challenges for disaster management, given their destructive potential that can extend far from their source (Jain et al., 2025). The unpredictable, sudden onset of GLOFs requires a paradigm shift away from current reactive disaster responses towards proactive, preventive and early warning solutions.

1.4.1 Himalayan Glacial Lake Outburst Floods (GLOFs)

GLOFs constitute an abrupt and catastrophic outflow of a large volume of water from a glacial lake by any mechanism (Emmer, 2017). They are a significant and increasingly common geohazard in high-mountain regions worldwide, especially in the dynamic Himalayan cryosphere (Negi, 1991; Shroder, 2011). GLOFs typically occur when glacial lakes formed by glacial retreat and shrinkage breach their banks. These lakes form from the meltwater that fills depressions previously occupied by glacier ice. They are unstable glacial lakes contained by natural barriers such as morainic deposits, bedrock or glacial ice (Emmer, 2017). The increasing frequency and intensity of GLOFs in the Himalaya are attributed to global climate change and the resulting accelerated glacier retreat (Jain et al., 2025). In the northern Himalayan range (Fig. 1.8a), rising global temperatures (0.2 °C/decade) (United Nations report) have accelerated glacial melt, leading to the rapid formation and expansion of these lakes. A study conducted in the central Himalaya using historical Landsat data revealed that the glacial lakes expanded rapidly by 17.11% from 1990 to 2010 (Nie et al., 2013). It was observed that the pre-existing glacial lakes located between elevations of ~ 4500–5600 m on both the northern and southern regions of the central Himalaya contributed most to the expansion (Nie et al., 2013). Many studies have documented the expansion of glacial lakes across various areas of the Himalaya. In the Hindu Kush region of the Himalaya, the total area of 4549 glacial lakes was 398.9 km^2 in 1990, increasing to 455.5 km^2 in 2015, with ~ 400 new lakes added (Mal et al., 2021). The South Lhonak lake in the Sikkim Himalaya increased in area by ~ 4 times since 1989, before the 2023 GLOF tragedy (Sattar et al., 2025), and by ~ 17.5 times since 1962 (Jain et al., 2025). Nepal's Imja Lake started as a small pond in 1962 and grew to 0.77 km^2 by 2000 and 1.56 km^2 by 2020 (Bajracharya et al., 2007). In the Chenab River basin (Western Himalaya), the area of glacial lakes increased by ~ 75% between 1910 and 2022 (Das et al.,

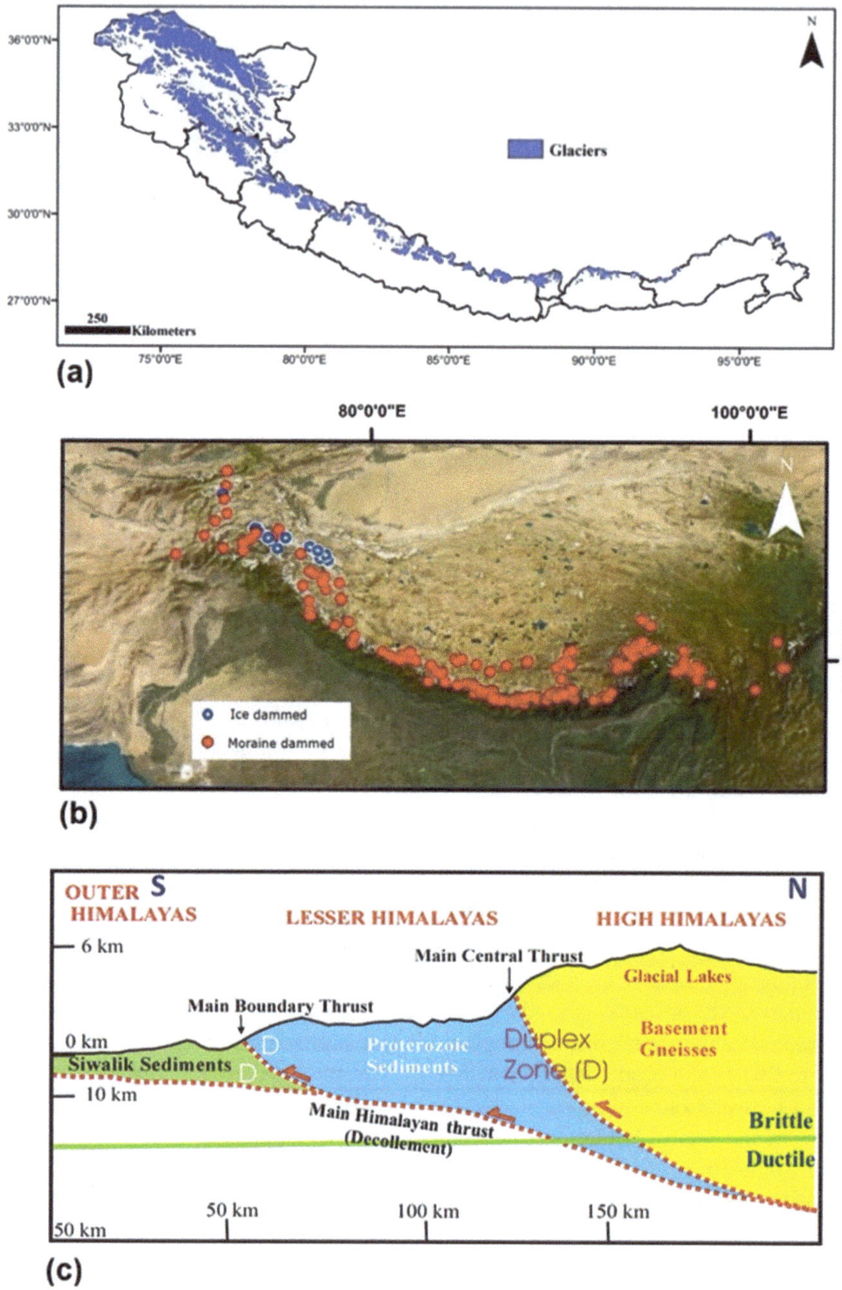

Fig. 1.8 **a**) Himalayan Glacier map released in 2023 from RGI Version 7.0 (2023): (www.glims.org/RGI/); **b**) distribution of moraine and ice-dammed glacial lakes in the Himalaya (ICIMOD, 2022); **c**) a schematic cross-section through the Himalayan wedge showing the Main Himalayan thrust and the rheological contrast in the wedge across the Main Boundary thrust and the Main Central thrust (MCT). The Lesser Himalayan Duplex zone near the MCT surface trace is shown. The Glacial Lakes occur in the High Himalaya at elevations of ~ 6 km

2024). Similarly, in the Rongxer basin (Central Himalaya), the area of the glacial lakes increased by ~ 31% from 1990 to 2020 (Zhang et al., 2024). Over 5000 glacial lakes have sprung up throughout the Himalaya due to prolonged glacier melt, many of which are surrounded by potentially unstable morainic dams (Sati, 2020; Veh et al., 2020). Moraine dams are inherently unstable, and the presence of buried ice cores can drastically increase the instability of the dams and cause them to fail by the melting of the ice (Emmer & Cochachin, 2013; Jain et al., 2025) due to factors like heavy rainfall, avalanches or seismic activity, releasing large volumes of water downstream that may also form temporary landslide lakes that burst to give rise to Landslide-dammed Lake Outburst Floods (LLOFs). With continued global climate change, the Himalaya face increasing risks of GLOFs, threatening mountain communities and critical infrastructure (Mishra et al., 2022; Ritse et al., 2025; Veh et al., 2019).

The glacial lakes are categorised into four types: erosion lakes, ice-dammed lakes, moraine-dammed lakes and supraglacial lakes (Yao et al., 2018). The majority of glacial lakes in the Himalaya are moraine-dammed and are concentrated in the central and eastern Himalaya (Rather et al., 2024; Shrestha et al., 2010; Veh et al., 2020). The ice-dammed lakes are comparatively fewer and are located mainly in the western Himalaya (Fig. 1.8b) (ICIMOD, 2022). The moraine dams comprise loose, permeable and heterogeneous sediments deposited at the glacier's edges. Due to their unconsolidated nature, moraine dams are naturally unstable and prone to failure even under minor triggering factors (Emmer & Cochachin, 2013). Ongoing mountain-building processes increase the risk of GLOFs. Rapid mass wasting is promoted by groundwater discharge, vertical bedrock uplift induced by the Lesser Himalayan Duplex, intense fluvial erosion and steep slope gradients acting on naturally unstable sediment. By supplying unstable material for moraine dams and serving as a direct trigger mechanism, such large-scale mass movements into glacial lakes directly contribute to GLOF processes. Also, glacier retreat has led to glacial areas being converted into periglacial weathering zones, resulting in increased fragmentation of loose materials, rock-mass displacement and permafrost degradation. Due to these changes, the frequency and severity of related hazards, such as landslides and debris flows, have significantly increased. Interactions with unique topographical features produce a synergistic effect that accelerates slope instability and permafrost degradation, directly causing GLOFs. This indicates that GLOF risk arises from the complex, dynamic interaction between topography and climate change.

The morphology of glacial lakes and their natural dams also plays a vital role in determining the occurrence of a GLOF event. The volume of water stored in the lake, its shape and the structural strength of the moraine dams are essential for determining the GLOF event and its magnitude (Jain et al., 2025). Also, water can gradually seep through the dam's structure due to internal erosion, leading to its degradation. Furthermore, the slope of the moraine dams also contributes to their vulnerability to failure. Glacial lakes held by dams with frontal slopes greater than 10° are identified as critical for failure (Zhang et al., 2024). Hence, a GLOF's risk depends on the stability of moraine dams, which is determined by the moraine's internal structure and material composition. The gradual melting of the glacial ice and the seepage of water through the dam over time make it more susceptible to failure. This indicates that the risk of GLOF is not always due to sudden triggers but can result from a slow, ongoing weakening of dam structures, accelerated by climate change. Moreover, the Higher Himalaya, where most glacial lakes are located, are characterised by high relief, steep slopes and evolving topography. Landslides, avalanches, rockfall and icefalls occur regularly, and they are common triggers for GLOFs. Large masses of material on impact can displace substantial volumes of water, leading to moraine-dam breaches or overflows and triggering GLOF events (Jain et al., 2025; Sattar et al., 2025). Therefore, the leading risk factor for GLOF in the Himalaya is the instability of the moraine dams that hold the glacial lakes. Hence, monitoring and exploring engineering solutions to strengthen the moraine dams are crucial.

1.4.2 Insights from the Recent Glacial Lake Outburst Floods

GLOFs have occurred frequently in recent years. The 2013 Kedarnath disaster in Uttarakhand, triggered by intense rainfall and glacial melt (e.g. Allen et al., 2019), stands as a famous example of such a catastrophic event. Past GLOFs, such as the 1985 Dig Tsho event in Nepal (Vuichard & Zimmermann, 1987) and the 2023 South Lhonak Lake outburst in Sikkim and Darjiling (Sattar et al., 2025), provided essential insights into GLOFs and their destructive effects. We discuss critical insights from the most recent 2023 South Lhonak Lake GLOFs in the Sikkim Himalaya, located in the Kanchenjunga glacier system, which has been studied in detail.

The October 2023 Sikkim GLOF was caused by the collapse of the lateral moraine (Fig. 1.9) into the South Lhonak Lake, generating a displacement wave of ~ 20 m. The estimated volume of the displaced lateral moraine was ~ 14.7 million m^3, which triggered the drainage of ~ 50 million m^3 of water, equivalent to the volume of 25,000 Olympic-sized swimming pools, with a peak discharge of 48,500 m^3 per second (Sattar et al., 2025). As South Lhonak Lake is located in the Higher Himalaya, in the hanging wall of the MCT at an elevation of 5200 m, it holds vast potential energy due to its large volume of stored water. GLOFs convert potential energy into kinetic energy over a short interval of time (Cenderelli & Wohl, 2003). This gives them a remarkable capacity for erosion and sediment transport, which can develop into extremely destructive flow-type movements, such as debris or sheet flows. As moraine dams fail, this large volume of water rushes down the valley at high velocity, converting all the stored potential energy into kinetic energy and eroding and transporting large quantities of sediments and debris. The accelerating water volume, along with a large mass of eroded material, generates a tremendous force that can alter the flow path and destroy anthropogenic structures such as dams and bridges. As flood waters reach more expansive valleys and low-slope areas, sediments are deposited in river beds and banks, causing long-term changes in the river channel morphology. The existing river terraces are eroded during the GLOF, destroying infrastructure built on them.

Fig. 1.9 Satellite imagery of a portion of South Lhonak Lake illustrating the collapsed lateral moraine and the resulting expansion of the lake due to the displaced material

A critical topographic factor at South Lhonak Lake was the lake's tapering shape towards the outlet (Fig. 1.10) (Sattar et al., 2021), which amplified the impulse wave generated by the landslide, leading to greater overtopping. Given the topography of South Lhonak Lake and its surrounding moraine, even though the landslide was relatively small, the downstream damage was widespread, indicating that the lake's surface morphology and bathymetry can intensify the impact of a GLOF even from a small-scale mass movement.

The morphology of the Himalayan valleys also affects the propagation of GLOFs and their destructive potential. The narrow valley's steep slopes funnel the channel's flow and constrain the pathway, making the GLOF disastrous (Rather et al., 2024). These rapid, high-magnitude floods disrupt the dynamic equilibrium of fluvial systems, destabilise valley slopes and cause significant geomorphic reworking and downstream infrastructural damage (Cook et al., 2018). Even in areas of lower gradient, GLOFs can carve deep channels due to the enormous kinetic energy they carry (Dahlquist & West, 2021). In wider, in more expansive valleys, sediment accumulation is high, and the risk of GLOF-related damage is reduced because floodwaters can spread out, reducing their erosive power. Hence, past GLOF events may also leave distinct topographical/sedimentological signatures that can be used to identify high-risk areas and delineate the extent of the river basin. Bridges and hydropower infrastructure are usually built across narrow valleys to facilitate crossing, optimise costs and support power generation. These locations are in regions with narrow or

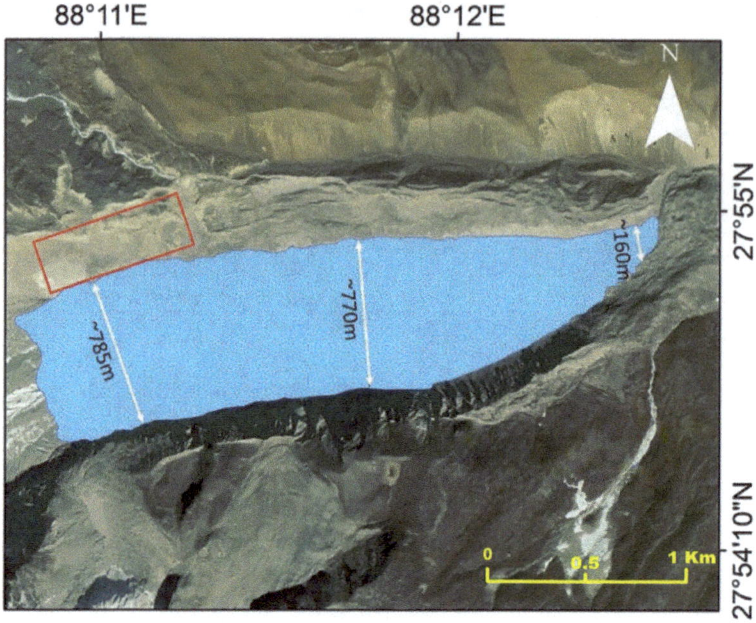

Fig. 1.10 Tapered shape of the South Lhonak Lake, which tapers towards the outlet. The region that collapsed into the lake during the 2023 GLOF (Fig. 1.9) is shown by a rectangle with a red border

restricted channel flow, which amplifies the destructive force of GLOF and makes infrastructure more prone to failure.

Therefore, topographical features that make the sites ideal for infrastructure development also create an environment that amplifies the GLOF's destructive potential. This was evident in the destruction of the 60-m high concrete-faced rock-fill dam of the 1200 MW Teesta Stage III hydropower project in Chungthang, North Sikkim. The dam was breached around midnight on 4 October, 2023, and the power house was submerged, with the bridge connecting it also washed away.

In summary, GLOFs are not a short-term passing event. They are strong geomorphic forces that can reshape channels and valleys over long distances in the Himalayan region. The topography of the Himalayan region also plays an integral role in determining the vulnerability of communities and infrastructure downstream of GLOFs (Somos-Valenzuela et al., 2015). The growth of high-altitude glacial lakes poses high risks to the human settlements, assets and the local ecosystem in low-lying areas downstream (Rather et al., 2024). Hence, building infrastructure in the Himalaya requires a multi-faceted land-use strategy and resilient engineering designs that account for the multi-hazard behaviour of GLOFs. Understanding the hydrological variability and flood-generating mechanisms in the Himalaya is essential for developing effective early-warning systems, resilient infrastructure planning and sustainable river basin management strategies.

1.5 Discussion

The Himalayan region has a dynamic climate coupled with an active tectonic environment. Active tectonics is driven by slip along the Himalayan decollement (MHT) and is directly related to earthquake nucleation and the active topographic build-up in the Lesser Himalayan Duplex. This high-altitude region is home to the planet's tallest peaks, which are associated with glacier systems. Steep slopes in this region, combined with high rainfall, lead to slope instability and landslides. Global warming and climate change are causing glaciers to melt and retreat, forming large, filled proglacial lakes at their termini. Hydrometeorological events such as heavy rainfall and cloud bursts, especially during the monsoon season, can rapidly fill glacial lakes, increasing the risk of overtopping and dam breaches (Emmer & Cochachin, 2013). Earthquakes can generate powerful waves that can trigger mass movements and destabilise moraine dams. Additionally, the time and place of both events are difficult to predict (Emmer & Cochachin, 2013). Hence, the Himalayan deformation has generated high topography in the Himalayan region and created a multi-hazard (seismic, landslide and GLOF) risk scenario.

The persistent threat of multiple, integrated natural hazards such as floods, landslides and earthquakes in the Himalayan region requires a shift from reactive disaster management to proactive mitigation, preparedness and adaptive management (e.g. Mukul, 2024). One effective strategy is to establish region-specific early-warning systems, particularly in high-risk zones. These systems should integrate real-time data on rainfall, river discharge, slope dynamics, seismic activity and glacial lake conditions, using remote sensing technologies, ground-based sensors and participatory community-reporting mechanisms. Equally important is comprehensive hazard zonation mapping that incorporates both climate and tectonics. Infrastructure development must be based strictly on insights from process-oriented approaches to minimise the risk of failure. The worst-case scenario, in which a major earthquake triggers widespread slope failure and breaches multiple glacial lake dams, must be recognised and prepared for. Ultimately, reducing the impact of natural disasters in the Himalaya will take a paradigm shift towards anticipatory, process-informed planning and an integrated, ground-up approach that recognises the region's unique geodynamic, climatic and socio-environmental context.

Acknowledgements We thank the two anonymous referees and the editor, Dr V. M. Tiwari, for their constructive comments, which helped improve the quality of the manuscript. Malay acknowledges funding from the IRCC Research Development Funds at IIT Bombay. V. S. and E. H. acknowledge financial support from IISER Bhopal. We benefitted from discussions with Vimal Singh, Roger Bilham, S. K. Tandon, Peter DeCelles, Gautam Mitra, Jason Barnes and Mary Hubbard.

References

Allen, S. K., Zhang, G., Wang, W., Yao, T., & Bolch, T. (2019). Potentially dangerous glacial lakes across the Tibetan Plateau revealed using a large-scale automated assessment approach. *Science Bulletin, 64*(7), 435–445. https://doi.org/10.1016/j.scib.2019.03.011

Bajracharya, B., Shrestha, A. B., & Rajbhandari, L. (2007). Glacial lake outburst floods in the Sagarmatha region. *Mountain Research and Development, 27*(4), 336–344.

Basa, A., Ahmed, F., Bhattacharyya, K., & Roy, A. (2019). Evolution and characterisation of fracture patterns: Insights from multi-scale analysis of the Buxa dolomite in the Siang Valley, Arunachal Lesser Himalayan fold-thrust belt. *Journal of Structural Geology, 123*, 54–66. https://doi.org/10.1016/j.jsg.2019.03.004

Beaumont, C., Jamieson, R. A., Nguyen, M. H., & Lee, B. (2001). Himalayan tectonics is explained by the extrusion of a low-viscosity crustal channel coupled to focused surface denudation. *Nature, 414*(6865), 738–742.

Bendick, R., & Bilham, R. (2001). How perfect is the Himalayan arc? *Geology, 29*(9), 791–794.

Bhattacharyya, K., & Mitra, G. (2011). Strain softening along the MCT zone from the Sikkim Himalaya: Relative roles of quartz and micas. *Journal of Structural Geology, 33*(6), 1105–1121.

Bilham, R. (2019). Himalayan earthquakes: A review of historical seismicity and early 21st century slip potential. *Geological Society Special Publication, 483*(1), 423–482. https://doi.org/10.1144/sp483.16

Bilham, R., Gaur, V. K., & Molnar, P. (2001). Himalayan seismic hazard. *Science, 293*, 1442–1444.

Bilham, R., Szeliga, W., Bali, B. S., Khan, A., Wahab, A., & Khan, F. (2011). Velocity field in the NW Himalayan Syntaxis: Implications for future seismicity. *Transactions, American Geophysical Union.*

Bookhagen, B., & Burbank, D. W. (2010). Toward a complete Himalayan hydrological budget: Spatiotemporal distribution of snowmelt and rainfall and their impact on river discharge. *Journal of Geophysical Research: Earth Surface, 115*, F03019. https://doi.org/10.1029/2009JF001426

Boyer, S. E., & Elliott, D. (1982). Thrust systems. *American Association of Petroleum Geologists Bulletin, 66*, 1196–1230.

Burbank, D. W., Leland, J., Fielding, E., Anderson, R. S., Brozovic, N., Reid, M. R., & Duncan, C. (1996). Bedrock incision, rock uplift, and threshold hillslopes in the northwestern Himalayas. *Nature, 379*(6565), 505–510. https://doi.org/10.1038/379505a0

Burton, I., & Kates, R. W. (1964). The perception of natural hazards in resource management. *Natural Resources Journal, 3*, 412–441.

Caine, J. S., Evans, J. P., & Forster, C. B. (1996). Fault zone architecture and permeability structure. *Geology, 24*(11), 1025. https://doi.org/10.1130/0091-7613(1996)024%3c1025:fzaaps%3e2.3.co;2

Cenderelli, D. A., & Wohl, E. E. (2001). Peak discharge estimates of glacial-lake outburst floods and "normal" climatic floods in the Mount Everest region, Nepal. *Geomorphology, 40*(1–2), 57–90.

Cenderelli, D. A., & Wohl, E. E. (2003). Flow hydraulics and geomorphic effects of glacial-lake outburst floods in the Mount Everest, Nepal. *Earth Surface Processes and Landforms: the Journal of the British Geomorphological Research Group, 28*(4), 385–407.

Chakraborty, S., & Mukul, M. (2019). New insights into the position and geometry of the Main Central Thrust from Sikkim, eastern Himalaya. *The Journal of Geology, 127*(3), 289–322.

Chatterjee, N., Mukul, M., & Sarkar, A. (2022). A single extensional, diffuse, ductile fault zone in the Goriganga section, western Himalaya: Part of the upper South Tibetan Detachment System? *Tectonophysics, 2022*(841), Article 229561.

Cook, K. L., Andermann, C., Gimbert, F., Adhikari, B. R., Hovius, N. (2018). Glacial lake outburst floods as drivers of fluvial erosion in the Himalaya. *Science (New York, N.Y.), 362*(6410), 53–57. https://doi.org/10.1126/science.aat4981

Crozier, M. J. (2010). Deciphering the effect of climate change on landslide activity: A review. *Geomorphology (Amsterdam, Netherlands), 124*(3–4), 260–267. https://doi.org/10.1016/j.geomorph.2010.04.009

Dahlquist, M. P., & West, A. J. (2021). Central Himalayan rivers record the topographic signature of erosion by glacial lake outburst floods. *Earth Surface Dynamics Discussions, 2021*, 1–26.

Das, S., Das, S., Mandal, S. T., Sharma, M. C., & Ramsankaran, R. (2024). Inventory and GLOF susceptibility of glacial lakes in Chenab basin, Western Himalaya. *Geomatics, Natural Hazards and Risk, 15*(1), 2356216.

DeCelles, P. G., Gehrels, G. E., Quade, J., Kapp, P. A., Ojha, T. P., & Upreti, B. N. (1998). Neogene foreland basin deposits, erosional unroofing, and the kinematic history of the Himalayan fold-thrust belt, western Nepal. *Geological Society of America Bulletin, 110*, 2–21.

DeCelles, P. G., & Mitra, G. (1995). History of the Sevier orogenic wedge in terms of critical taper models, northeast Utah and southwest Wyoming. *Geological Society of America Bulletin, 107*(4), 454. https://doi.org/10.1130/0016-7606(1995)107%3c0454:hotsow%3e2.3.co;2

Dimri, A. P., Chevuturi, A., Niyogi, D., Thayyen, R. J., Ray, K., Tripathi, S. N., Pandey, A. K., & Mohanty, U. C. (2017). Cloudbursts in the Indian Himalayas: A review. *Earth-Science Reviews, 168*, 1–23. https://doi.org/10.1016/j.earscirev.2017.03.006

Duncan, J. M., & Wright, S. G. (2005). *Soil strength and slope stability*. Wiley.

Emmer, A. (2017). Glacier retreat and glacial lake outburst floods (GLOFs). In *Oxford Research Encyclopedia of Natural Hazard Science*.

Emmer, A., & Cochachin, A. (2013). The causes and mechanisms of moraine-dammed lake failures in the Cordillera Blanca, North American Cordillera, and Himalayas. *AUC GEOGRAPHICA, 48*(2), 5–15.

Froude, M. J., & Petley, D. N. (2018). Global fatal landslide occurrence from 2004 to 2016. *Natural Hazards and Earth Systems Sciences, 18*, 2161–2181. https://doi.org/10.5194/nhess-18-2161-2018

Gautam, P. K., Gahalaut, V. K., Prajapati, S. K., Kumar, N., Yadav, R. K., Rana, N., & Dabral, P. C. (2017). Continuous GPS measurements of crustal deformation in Garhwal-Kumaun Himalaya. *Quaternary International, 462*, 124–129.

Grandin, R., Doin, M. P., Bollinger, L., Pinel-Puysségur, B., Ducret, G., Jolivet, R., & Sapkota, S. N. (2012). Long-term growth of the Himalaya inferred from interseismic InSAR measurement. *Geology, 40*(12), 1059–1062.

Grujic, D., Hollister, L. S., & Parrish, R. R. (2002). Himalayan metamorphic sequence as an orogenic channel: Insight from Bhutan, Earth Planet. *Science Letters, 198*, 177–191. https://doi.org/10.1016/S0012-821X(02)00482-X

Gudmundsson, A. (2011). *Rock fractures in geological processes*. Cambridge University Press.

Guzzetti, F., Carrara, A., Cardinali, M., & Reichenbach, P. (1999). Landslide hazard evaluation: A review of current techniques and their application in a multi-scale study, Central Italy. *Geomorphology (Amsterdam, Netherlands), 31*(1–4), 181–216. https://doi.org/10.1016/s0169-555x(99)00078-1

Haque, E., Kote, N., Gayakwad, B., Srivastava V. (2026). Out-of-sequence thrusting (OOST) in the Himalaya: Insights from the Geil-Khola thrust fault zone in the Ramgarh thrust sheet Darjeeling Himalaya India. *Journal of Structural Geology*, 105774. https://doi.org/10.1016/j.jsg.2026.105774

Haryono, I. (2025). Influence of fracture characteristics and the use of network connectivity index on synthetic rock mass strength characterisation of moderately jointed granite in Hong Kong. In *The HKIE Geotechnical Division 45th Annual Seminar 2025*.

Hubbard, M., Mukul, M., Gajurel, A. P., Ghosh, A., Srivastava, V., Giri, B., Seifert, N., & Mendoza, M. (2021). Orogenic segmentation and its role in Himalayan mountain building. *Frontiers in Earth Science, 9*, Article 641666.

Hussain, A., Yeats, R. S., & Lisa, M. (2009). Geological setting of the 8 October 2005 Kashmir earthquake. *Journal of Seismology, 13*, 315–325.

Husson, L., Mugnier, J.-L., Leturmy, P., & Vidal, G. (2004). Kinematics and sedimentary balance of the Sub-Himalayan zone, western Nepal. In K. R. McClay (Ed.), *Thrust tectonics and hydrocarbon systems*. American Association of Petroleum Geologists Memoir (Vol. 82, pp. 115–130).

ICIMOD. (2022). *GLOF database of High Mountain Asia*. ICIMOD. https://doi.org/10.26066/RDS.1973283

Immerzeel, W. W., van Beek, L. P. H., & Bierkens, M. F. P. (2010). Climate change will affect the Asian water towers. *Science (New York, N.Y.), 328*(5984), 1382–1385. https://doi.org/10.1126/science.1183188

Ismat, Z. (2013). Block-supported cataclastic flow within the upper crust. *Journal of Structural Geology, 56*, 118–128. https://doi.org/10.1016/j.jsg.2013.08.010

Ismat, Z., & Mitra, G. (2005). Folding by cataclastic flow at shallow crustal levels in the Canyon Range, Sevier orogenic belt, west-central Utah. *Journal of Structural Geology, 23*, 355–378.

Iverson, R. M. (2000). Landslide triggering by rain infiltration. *Water Resources Research, 36*(7), 1897–1910. https://doi.org/10.1029/2000wr900090

Jade, S., Mukul, M., Gaur, V. K., Kumar, K., Shrungeshwar, T. S., Satyal, G. S., Dumka, R. K., Jagannathan, S., Ananda, M. B., Dileep Kumar, P., & Banerjee, S. (2014). Contemporary deformation in the Kashmir-Himachal, Garhwal and Kumaon Himalaya: Significant insights from 1995–2008 GPS time series. *Journal of Geodesy, 88*(2014), 539–557. https://doi.org/10.1007/s00190-014-0702-3

Jain, A. K., & Banerjee, D. M. (2020). The Indian subcontinent: Its tectonics map. *Proceedings of the Indian National Science Academy, 86*, 77–875.

Jain, S. K., Jain, S. K., & Ahmad, R. (2025). Glacial lake outburst floods (GLOFs): Causes, modeling, and mitigation in the Indian Himalayan region. *GeoJournal, 90*(2), 85.

Kamp, U., & Owen, L. A. (2011). Late Quaternary glaciation of northern Pakistan. In *Developments in quaternary sciences* (pp. 909–927). Elsevier.

Kargel, J. S., Leonard, G. J., Shugar, D. H., Haritashya, U. K., Bevington, A., Fielding, E. J., Fujita, K., Geertsema, M., Miles, E. S., Steiner, J., Anderson, E., Bajracharya, S., Bawden, G. W., Breashears, D. F., Byers, A., Collins, B., Dhital, M. R., Donnellan A, Evans, T. L., …, Young, N. (2016). Geomorphic and geologic controls of geohazards induced by Nepal's 2015 Gorkha earthquake. *Science (New York, N.Y.), 351*(6269), aac8353. https://doi.org/10.1126/science.aac8353

Kayal, J. R., Ram, S., Singh, O. P., Chakraborty, P. K., & Karunakar, G. (2003). Aftershocks of the March 1999 Chamoli Earthquake and seismotectonic structure of the Garhwal Himalaya. *Bulletin of the Seismological Society of America, 93*(1), 109–117.

Kirschbaum, D. B., Adler, R., Hong, Y., Hill, S., & Lerner-Lam, A. (2010). A global landslide catalogue for hazard applications: Method, results, and limitations. *Natural Hazards, 52*, 561–575. https://doi.org/10.1007/s11069-009-9401-4

Kirschbaum, D. B., Stanley, T., & Zhou, Y. (2015). Spatial and temporal analysis of a global landslide catalogue. *Geomorphology, 249*, 4–15. https://doi.org/10.1016/j.geomorph.2015.03.016

Kokutse, N. K., Temgoua, A. G. T., & Kavazović, Z. (2016). Slope stability and vegetation: Conceptual and numerical investigation of mechanical effects. *Ecological Engineering, 86*, 146–153. https://doi.org/10.1016/j.ecoleng.2015.11.005

Lave, J., & Avouac, J. P. (2000). A Fluvial incision and tectonic uplift across the Himalayas of Central Nepal. *Journal of Geophysical Research Solid*. https://doi.org/10.1029/2001jb000359

Long, S., McQuarrie, N., Tobgay, T., & Hawthorne, J. (2011). Quantifying internal strain and deformation temperature in the eastern Himalaya, Bhutan: Implications for the evolution of strain in thrust sheets. *Journal of Structural Geology, 33*(4), 579–608.

Mal, S., Allen, S. K., Frey, H., Huggel, C., & Dimri, A. P. (2021). Sector-wise assessment of glacial lake outburst flood danger in the Indian Himalayan region. *Mountain Research and Development, 41*(1), R1.

McNaught, M. A., & Mitra, G. (1996). The use of finite strain data in constructing a retrodeformable cross-section of the Meade thrust sheet, southeastern Idaho. *Journal of Structural Geology, 18*(1996), 573–583.

McQuarrie, N., Robinson, D., Long, S., Tobgay, T., Grujic, D., Gehrels, G., & Ducea, M. (2008). Preliminary stratigraphic and structural architecture of Bhutan: Implications for the along strike architecture of the Himalayan system. *Earth and Planetary Science Letters, 272*, 105–117. https://doi.org/10.1016/j.epsl.2008.04.030

Mishra, A., Nainwal, H. C., Bolch, T., Shah, S. S., & Shankar, R. (2022). Glacier inventory and glacier changes (1994–2020) in the Upper Alaknanda Basin, Central Himalaya. *Journal of Glaciology*, 1–16. https://doi.org/10.1017/jog.2022.87

Mitchell, T. M., & Faulkner, D. R. (2009). The nature and origin of off-fault damage surrounding strike-slip fault zones with a wide range of displacements: A field study from the Atacama fault system, northern Chile. *Journal of Structural Geology, 31*(8), 802–816. https://doi.org/10.1016/j.jsg.2009.05.002

Mitra, G. (1994). Strain variation in thrust sheets of the Sevier fold-and-thrust belt, Idaho-Utah-Wyoming: Implications for section restoration and wedge tape revolution. *Journal of Structural Geology, 16*, 585–602.

Mitra, G. (1997). Evolution of salients in a fold-and-thrust belt: The effects of sedimentary basin geometry, strain distribution, and critical taper. In S. Sengupta (Ed.), *Evolution of geologic structures from macro- to micro-scales* (pp. 59–90). Chapman and Hall.

Mitra, G., Bhattacharyya, K., & Mukul, M. (2010). The lesser Himalayan duplex in Sikkim: Implications for variations in Himalayan shortening. *Journal of the Geological Society of India, 75*, 276–288.

Mitra, G., & Boyer, S. (2020). Duplexes as slip transfer zones: Application to the Himalayan fold-thrust belt. *Journal of Asian Earth Sciences, 190*, Article 104214.

Mitra, G., Boyer, S. E., Kwon, S., Mukul, M., Ismat, Z., Sussman, A. (2026). Lateral Variations in Evolution of a Fold-thrust Belt Hinterland: Passive versus Active Behavior of Dominant Internal Thrust Sheets. *Habitable Planet 2*(2), 303–324. https://doi.org/10.63335/j.hp.2026.0040

Molnar, P., & Lyon-Caen, H. (1988). Some simple physical aspects of the support, structure, and evolution of mountain belts. *Geological Society of America Special Paper, 218*, 179–207. https://doi.org/10.1130/SPE218-p179

Morley, C. K. (1988). Out-of-sequence thrusts. *Tectonics, 7*(3), 539–561. https://doi.org/10.1029/tc007i003p00539

Mukherjee, S. (2015). A review of out-of-sequence deformation in the Himalaya. *Geological Society, London, Special Publications, 412*(1), 67–109. https://doi.org/10.1144/sp412.13

Mukul, M. (1998). A spatial statistics approach to the quantification of finite strain variation in penetratively deformed thrust sheets: An example from the Sheeprock Thrust Sheet, Sevier Fold-and-Thrust Belt, Utah. *Journal of Structural Geology, 20*(4), 371–384.

Mukul, M. (1999). Strain variation in fold-and-thrust belts: Implications for construction of retrodeformable models. *Proceedings of the Indian Academy of Sciences-Earth and Planetary Sciences.*

Mukul, M. (2000). The geometry and kinematics of the Main Boundary thrust and related neotectonics in the Darjiling Himalayan fold-and-thrust belt, West Bengal. *Journal of Structural Geology, 22*(9), 1261–1283.

Mukul, M. (2005). Continental deformation and global positioning system-based geodesy. *Himalayan Geology, 26*(1), 193–198.

Mukul, M. (2010). First-order kinematics of wedge-scale active Himalayan deformation: Insights from Darjiling-Sikkim-Tibet (DaSiT) wedge. *Journal of Asian Earth Sciences, 39*(2010), 645–657.

Mukul, M. (2024). The challenge of integrated natural hazards for continued sustainable development in the Himalayas. *Current Science, 127*(9), 1005–1006.

Mukul, M., Jade, S., Ansari, K., & Matin, A. (2014). Seismotectonic implications of strike-slip earthquakes in the Darjiling-Sikkim Himalaya. *Current Science, 106*(2), 126–131.

Mukul, M., Jade, S., Ansari, K., Matin, A., & Joshi, V. (2018). Structural insights from geodetic Global Positioning System measurements in the Darjiling-Sikkim Himalaya. *Journal of Structural Geology, 114*, 346–356

Mukul, M., Srivastava, V., & Mukul, M. (2025). Integrated natural hazards in the Himalaya: Insights from the Darjiling-Sikkim Himalaya. In S. Ghosh (Ed.), *The Himalaya Dilemma*. Springer. https://doi.org/10.1007/978-3-031-95083-4_2

Mukul, M., Jaiswal, M., & Singhvi, A. K. (2007). Timing of recent out-of-sequence active deformation in the frontal Himalayan wedge: Insights from the Darjiling sub-Himalaya, India. *The Geological Society of America, 35*(11), 999–1002. https://doi.org/10.1130/G23869A.1

Mukul, M., & Mitra, G. (1998). Finite strain and strain variation analysis in the Sheeprock Thrust Sheet: An internal thrust sheet in the Provo salient of the Sevier Fold-and-Thrust belt, Central Utah. *Journal of Structural Geology, 20*(4), 385–405.

Mukul, M., & Mukul, M. (2021). Uncertainties in Digital Elevation Models from Global Advanced Land Observing Satellite (ALOS) digital topography data in the Indian subcontinent. *Journal of Earth System Science, 130*(1), 51.

Mukul, M., Roy, D., Satpathy, S., & Anil Kumar, V. (2004). Bootstrapped spatial statistics: A more robust approach to the analysis of finite strain data. *Journal of Structural Geology, 26*(3), 595–600.

Mukul, M., Srivastava, V., & Matin, A. (2022). Fault rocks and the architecture of diffused brittle faults in the frontal Himalayan fold-thrust belt. *Journal of Nepal Geological Society, 64*, P102.

Mukul, M., Srivastava, V., & Mukul, M. (2017). Out-of-sequence reactivation of the Munsiari thrust in the Relli River basin, Darjiling Himalaya, India: Insights from Shuttle Radar Topography Mission digital elevation model. *Geomorphology, 284*, 229–237.

Negi, S. S. (1991). *Himalayan rivers, lakes, and glaciers*. Indus Publishing.

Nelson, K. D., Zhao, W., Brown, L. D., Kuo, J., Che, J., Liu, X., Klemperer, S. L., Makovsky, Y., Meissner, R., Mechie, J., Kind, R., Wenzel, F., Ni, J., Nabelek, J., Leshou, C., Tan, H., Wei, W., Jones, A. G., Booker, J., … Edwards, M. (1996). Partially molten middle crust beneath southern Tibet: Synthesis of project INDEPTH results. *Science, 274*(1996), 1684–1688.

Newman, J., & Mitra, G. (1993). Lateral variations in mylonite zone thickness as influenced by fluid-rock interactions, Linville Falls fault, North Carolina. *Journal of Structural Geology, 15*(7), 849–863.

Nie, Y., Liu, Q., & Liu, S. (2013). Glacial lake expansion in the Central Himalayas by Landsat images, 1990–2010. *PLoS ONE, 8*(12), Article e83973.

Pandey, M. R., Tandulkar, R. P., Avouac, J. P., Vergne, J., & Heritier, T. (1999). Seismotectonics of the Nepal Himalaya from a local seismic network. *Journal of Asian Earth Sciences, 17*, 703–712.

Parameswaran, R. M., Natarajan, T., Rajendran, K., Rajendran, C. P., Mallick, R., Wood, M., & Lekhak, H. C. (2015). Seismotectonics of the April–May 2015 Nepal earthquakes: An assessment based on the aftershock patterns, surface effects and deformational characteristics. *Journal of Asian Earth Sciences*. https://doi.org/10.1016/j.jseaes.2015.07.030

Parui, C., Bhattacharyya, K., & Ghosh, P. (2022). Penetrative strain and partitioning of convergence-related shallow crustal shortening, across scales, in the Lesser- and Sub-Himalayan thrusts: Insights from the Eastern Himalaya, Sikkim. *Tectonics, 41*(11), e2022TC007210.

Paterson, M. S. (1990). Rock deformation experimentation. In *Geophysical Monograph Series* (pp. 187–194). American Geophysical Union. https://doi.org/10.1029/gm056p0187

Rastogi, B. (2000). Chamoli earthquake of magnitude 6.6 on 29 March 1999. *Journal of the Geological Society of India, 55*, 505–514.

Rather, A. F., Ahmed, R., Bansal, J. K., Mir, R. A., Ahmed, P., Malik, I. H., & Varade, D. (2024). Glacial lake outburst flood risk assessment of a rapidly expanding glacial lake in the Ladakh region of Western Himalaya, using hydrodynamic modelling. *Geomatics, Natural Hazards and Risk, 15*(1), 2413893.

Ritse, V., Hazarika, N., Kulnu, A. S., & Jamir, L. (2025). Shrinking Himalayan glaciers: Glacier inventory, glacial cover mapping and changes in Eastern Himalayas, India. *Journal of Earth System Science, 134*(1). https://doi.org/10.1007/s12040-024-02490-1

Sati, V. P. (2020). *Himalaya on the threshold of change*. Springer.

Sattar, A., Cook, K. L., Rai, S. K., Berthier, E., Allen, S., Rinzin, S., de Vries, Van Wyk de Vries, M., Haeberli, W., Kushwaha, P., Shugar, D. H., Emmer, A., Haritashya, U. K., Fret, H., Rao, P., Gurudin, K. S. K., Rai, P., Rajak, R., Hossain, F, …, Bhat, S. Y. (2025). The Sikkim flood of October 2023: Drivers, causes, and impacts of a multihazard cascade. *Science, 387*(6740).

Sattar, A., Goswami, A., Kulkarni, Anil. V., Emmer, A., Haritashya, U. K., Allen, S., Frey, H., & Huggel, C. (2021). Future Glacial Lake Outburst Flood (GLOF) hazard of the South Lhonak Lake, Sikkim Himalaya. *Geomorphology, 388*, 107783. https://doi.org/10.1016/j.geomorph.2021.107783.

Schelling, D., & Arita, K. (1991). Thrust tectonics, crustal shortening, and the structure of the far-eastern Nepal Himalaya. *Tectonics, 10*, 851–862.

Schuster, R. L., & Highland, L. M. (2001). Socioeconomic and environmental impacts of landslides in the Western Hemisphere. In *The Open-File Report*. US Geological Survey. https://doi.org/10.3133/ofr01276

Shipton, Z. K., Soden, A. M., Kirkpatrick, J. D., Bright, A. M., Lunn, R. J. (2006). How thick is a fault? Fault displacement-thickness scaling revisited. In R. Abercrombie (Ed.), *Earthquakes: Radiated energy and the physics of faulting* (pp. 193–198). AGU.

Shipton, Z. K., & Cowie, P. A. (2003). A conceptual model for the origin of fault damage zone structures in high-porosity sandstone. *Journal of Structural Geology, 25*, 333–344.

Shrestha, A. B., Eriksson, M., Mool, P., Ghimire, P., Mishra, B., & Khanal, N. R. (2010). Glacial lake outburst flood risk assessment of Sun Koshi basin, Nepal. *Geomatics, Natural Hazards and Risk, 1*(2), 157–169.

Shroder, J. F. (2011). Himalaya. In *Encyclopedia of snow, ice and glaciers* (pp. 510–520). Springer.

Sibson, R. H. (1977). Fault rocks and fault mechanisms. *Journal of the Geological Society, 133*(3), 191–213.

Singh, S., & Lal Kansal, M. (2022). Cloudburst-induced flood assessment in the North-Western Himalayan region—A case study of the upper Beas basin. In *The 5th World Congress on Disaster Management* (pp. 89–99). Routledge.

Somos-Valenzuela, M. A., McKinney, D. C., Byers, A. C., Rounce, D. R., Portocarrero, C., & Lamsal, D. (2015). Assessing downstream flood impacts due to a potential GLOF from Imja Tsho in Nepal. *Hydrology and Earth System Sciences, 19*(3), 1401–1412. https://doi.org/10.5194/hess-19-1401-2015

Srivastava, P., & Mitra, G. (1994). Thrust geometries and deep structure of the outer and lesser Himalaya, Kumaun and Garhwal (India): Implications for evolution of the Himalayan fold-and-thrust belt. *Tectonics, 13*, 89–109.

Srivastava, V., & Mukul, M. (2020). Cataclastic strain from external thrust sheets in fold-thrust belts: Insights from the frontal Indian Himalaya. *Journal of Asian Earth Sciences, 188*, Article 104092.

Srivastava, V., Mukul, M., & Barnes, J. B. (2016). Main Frontal thrust deformation and topographic growth of the Mohand Range, northwest Himalaya. *Journal of Structural Geology, 93*, 131–148.

Srivastava, V., Mukul, M., Barnes, J. B., & Mukul, M. (2018). Geometry and kinematics of Main Frontal thrust-related fault propagation folding in the Mohand Range, northwest Himalaya. *Journal of Structural Geology, 115*, 1–18. https://doi.org/10.1016/j.jsg.2018.06.022

Thakur, V. C. (1993). Geology of Western Himalaya. *Physics and Chemistry of the Earth Series* (Vol. 19, Parts I–VI, 366p.). Pergamon Press.

Thakur, V. C., & Sushil, K. (1994). Seismotectonics of the 20 October 1991 Uttarkashi earthquake in Garhwal, Himalaya, North India. *Terra Nova, 6*(1), 90–94.

Thayyen, R. J., & Gergan, J. T. (2010). Role of glaciers in watershed hydrology: A preliminary study of a "Himalayan catchment." *The Cryosphere, 4*(1), 115–128. https://doi.org/10.5194/tc-4-115-2010

Valdiya, K. S. (1991). The Uttarkashi earthquake of 20 October: Implications and lessons. *Current Science, 61*(12), 801–803.

Valdiya, K. S. (2002). Emergence and evolution of Himalaya: Reconstructing history in the light of recent studies. *Progress in Physical Geography, 26*(3), 360–399.

Veh, G., Korup, O., von Specht, S., Roessner, S., & Walz, A. (2019). Unchanged frequency of moraine-dammed glacial lake outburst floods in the Himalaya. *Nature Climate Change, 9*(5), 379–383. https://doi.org/10.1038/s41558-019-0437-5

Veh, G., Korup, O., & Walz, A. (2020). Hazard from the Himalayan glacier lake outburst floods. *Proceedings of the National Academy of Sciences, 117*(2), 907–912.

Violay, M., Gibert, B., Mainprice, D., & Burg, J.-P. (2015). Brittle versus ductile deformation as the main control of the deep fluid circulation in oceanic crust. *Geophysical Research Letters, 42*(8), 2767–2773. https://doi.org/10.1002/2015gl063437

Vuichard, D., & Zimmermann, M. (1987). The 1985 catastrophic drainage of a moraine-dammed lake, Khumbu Himal, Nepal: Cause and consequences. *Mountain Research and Development, 7*(2), 91. https://doi.org/10.2307/3673305

Wobus, C., Heimsath, A., Whipple, K., & Hodges, K. (2005). Active out-of-sequence thrust faulting in the central Nepalese Himalaya. *Nature, 434*(7036), 1008–1011.

Yang, B., Hou, J., Liu, Y., & Zhou, Z. (2021). Dynamic response and failure characteristics of slope with weak interlayer under action of near-fault ground motion. *Shock and Vibration, 2021*(1). https://doi.org/10.1155/2021/5595278

Yao, X., Liu, S., Han, L., Sun, M., & Zhao, L. (2018). Definition and classification system for glacial lakes for inventory and hazard studies. *Journal of Geographical Sciences, 28*(2), 193–205.

Zhang, S., Nie, Y., & Zhang, H. (2024). Glacial lake changes and risk assessment in Rongxer Watershed of China-Nepal economic corridor. *Remote Sensing, 16*(4), 725.

Chapter 2
Nature and Geometry of the Indian Crust Beneath Himalaya

Arun Singh, Chandrani Singh, M. Ravi Kumar, V. M. Tiwari, Gaurav Kumar, Abhisek Dutta, and Shirish Bose

Abstract The convergence of the Indian and Eurasian tectonic plates along the Himalaya exemplifies a classic case of continent–continent collision. Geodynamically, the Himalaya is one of the most active regions on Earth, having experienced numerous significant earthquakes throughout history. In recent decades, various geophysical and seismological studies have been conducted in the Himalaya-Tibet region to understand the structure of the Indian crust better as it descends and the processes of seismogenesis. This synthesis presents the current knowledge regarding the crustal thickness, seismic velocity, subsurface as reflected in gravity anomalies and attenuation structure of the Indian crust beneath the Himalaya. The numerous studies provide evidence of heterogeneous and complex variations of crustal models along the strike and map of decollement surface, which separates the Himalayan crustal wedge from the underthrusting Indian plate, highlighting mid-crustal ramps. The decollement layer is shown to be north-dipping, sheared and anisotropic, featuring alternating velocity jumps. The geometry of the decollement, combined with seismicity, suggests its role in defining the pattern of seismic ruptures. The northward-dipping Moho boundary exhibits complexity, with offsets, overlaps and sharp or gentle dips across different segments of the Himalaya. Crustal thickness estimates indicate variations ranging from approximately 40 km beneath the Lesser Himalaya to about 60 km beneath the Higher Himalaya. While seismic imaging has provided vital information about the deep crust, more dense and long-term networks are necessary for detailed mapping of the major tectonic features.

Keywords Crustal thickness · Velocity structure · Gravity anomalies · Himalayan collision zone

---

A. Singh (✉) · C. Singh · G. Kumar · S. Bose
Department of Geology and Geophysics, Indian Institute of Technology Kharagpur, Kharagpur, India
e-mail: arun@gg.iitkgp.ernet.in

M. Ravi Kumar · A. Dutta
National Geophysical Research Institute, Hyderabad, India

V. M. Tiwari
North East Institute of Science and Technology, Jorhat, India

V. M. Tiwari (ed.), *Himalayan Natural Hazards*, Indian National Science Academy,
https://doi.org/10.1007/978-981-92-3784-5_2

2.1 Introduction

The spectacular mountain belts of the Himalaya are products of a very long (~ 2400 km) and wide collision zone (~ 300 km) (Fig. 2.1), involving the Indian and Eurasian Plates, over the past 70–50 Ma (Yin & Harrison, 2000). The dynamics of collision are ever changing and are important factors in controlling the regional climate, river flows, livelihoods and natural hazards. In the category of hazards, seismic hazard due to the stresses generated within the Indian plate is very high and has resulted in significant loss of lives and infrastructure (Bilham, 2019). Thus, knowledge of dynamic processes remains a crucial objective to the geologists and geophysicists.

Early geological and geophysical investigations established the Himalaya as a classical continent–continent collision zone formed by northward underthrusting of the Indian lithosphere beneath Eurasia (Dewey & Bird, 1970; Molnar & Tapponnier, 1975). The evolution of Himalaya is marked by early oceanic-continental convergence during which closure of Tethyan Sea takes place, to continental–continental convergence (Molnar & Tapponnier, 1975; Yin & Harrison, 2000). The convergence is accommodated by exhumation and shortening of crust. The eastward decreasing trend of crustal shortening along Himalayan arc reveals the complex geodynamic

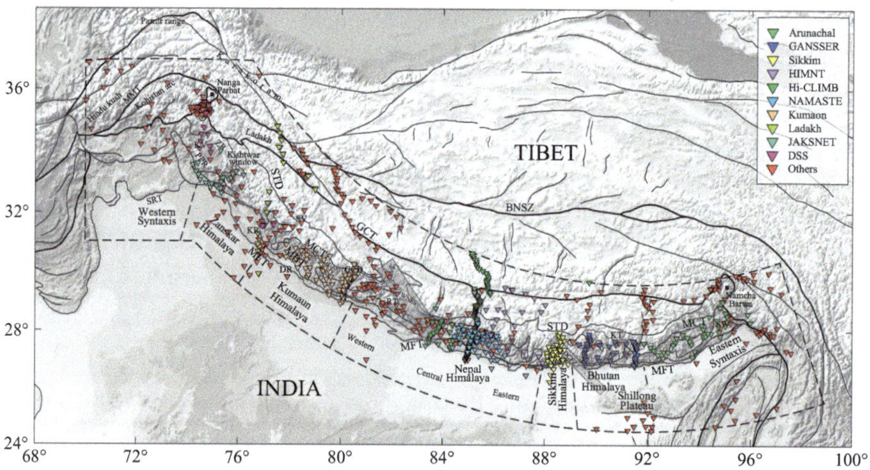

Fig. 2.1 Map showing the tectonic features of the Himalayan arc, extending from the Western to the Eastern Syntaxis. The Himalayan arc and its various segments are marked by dashed lines. Major faults are represented by solid lines. Filled inverted triangles denote seismic station networks, colour coded according to their operating agencies/regions. DCFZ: Dhubri-Chungthang Fault Zone; WNFS: Western Nepal Fault System; MFT: Main Frontal Thrust; MBT: Main Boundary Thrust; MCT: Main Central Thrust; MCT-I: Inner Main Central Thrust; STD: South Tibetan Detachment; GCT: Greater Counter Thrust; BNSZ: Bangong-Nujiang Suture Zone; MMT: Main Mantle Thrust; KV: Kashmir Valley; ZR: Zanskar Range; PPR: Pir-Panjal Range; SRT: Salt Range Thrust; SV: Satluj Valley; KR: Kangra Reentrant; DR: Dehradun Reentrant; CCB: Chiplakot Crystalline Belt; RT: Ramghar Thrust; KT: Kakthang Thrust; SW: Siang Window

scenario (Hinsbergen et al., 2012). In the present, most of the shortening occurs in the frontal part of the collisional zone (Fig. 2.1) (Avouac, 2003; Burgess et al., 2012). Subsequent structural and tectonic studies recognized the Himalayan orogen as a crustal-scale fold–thrust belt characterized by major discontinuities such as the Main Central Thrust (MCT), Main Boundary Thrust (MBT) and Main Frontal Thrust (MFT), reflecting progressive southward propagation of deformation (Gansser, 1974; Hodges, 2000). Moreover, the Main Himalayan Thrust (MHT), which is the low-angle basal decollement plane dipping northward, governs majority of strain accumulation and earthquake generation along the Himalayan arc (Kind et al., 2002; Nábělek et al., 2009; Zhao et al., 1993). Seismological imaging studies have demonstrated that spatial changes in Moho depth and intracrustal velocity structure are closely linked to crustal duplexing, channel flow and lateral variations in lithospheric rheology beneath different Himalayan segments (Nelson et al., 1997; Schulte-Pelkum et al., 2005). The Moho discontinuity represents the crust–mantle boundary and provides first-order constraints on crustal thickness, lithospheric strength, thereby playing a key role in interpreting deformation style and seismic hazard potential in active orogens (Christensen & Mooney, 1995). Furthermore, several other geophysical studies illuminated pronounced along-strike heterogeneity in crustal thickness, indicating a complex geodynamic scenario behind the evolution of Himalaya (Hetényi et al., 2016; Pandey et al., 1995). Such observations emphasize the importance of synthesizing crustal and Moho structure across the entire orogen to better understand deformation partitioning, tectonic segmentation and seismic hazard potential within the Himalayan collision system.

Several notable passive source seismological experiments were carried out across various segments of Himalaya (Table 2.1) to gain knowledge regarding active tectonics and associated features. In the recent decades, a number of seismic experiments were conducted to understand the seismogenic potential of this collision zone (Singh et al., 2015). Results from these experiments provided an opportunity to understand the active tectonics beneath the Himalaya and revealed significant along-strike and across-strike variations in crustal structure and Moho geometry. However, observations from different well-established Himalayan segments remain scattered among numerous independent studies, and a unified synthesis highlighting the nature and geometry of the underthrusting Indian crust across the entire Himalayan arc is still required. In this study, we present a comprehensive picture of the Himalayan geodynamics focusing on the nature and geometry of the underthrusting Indian crust.

Table 2.1 Summary of the notable seismic experiments carried out across the Himalayan arc from east to west

Network code	Operational region	Operational period	Operating agency	References
Arunachal	Arunachal	2010–2016	NGRI	Singh et al. (2021)
GANSSER (XA)	Bhutan	2013–2014	ETH Zurich	Diehl et al. (2017)
Sikkim (SK)	Sikkim	2019–2023	IIT Kharagpur	Singh et al. (2024)
Sikkim	Sikkim	2004–2009	NGRI	Singh et al. (2010)
HIMNT (YL)	Eastern Nepal and Southern Tibet	2001–2002	IRISDMC	Sheehan et al. (2001)
Hi-CLIMB (XF)	Central Nepal and Southern Tibet	2002–2005	IRISDMC	Nábělek (2002)
NAMASTE (XQ)	Central Nepal	2015–2016	IRISDMC	Bai et al. (2019)
HiK-NET (ZO)	Western Nepal	2014–2016	RESIF-SISMOB	Bollinger et al. (2011)
Kumaun	Kumaun and Garhwal Himalaya	2017–present	NGRI and WIHG	Parija et al. (2023) and Mandal et al. (2021)
Y2	Western Tibet	2007–2011	IRISDMC	Steve and Vadim (2007)
LADAKH	Ladakh	2002–2003	NGRI	Rai et al. (2006)
JAKSNET	Jammu and Kashmir	2013–2019	IISER Kolkata, University of Cambridge and SMVDU	Mitra et al. (2023)

2.2 Crustal Structure

2.2.1 Western Syntaxis

Some major tectonic features of the Western syntaxis that mark the western limit of the Himalayan Mountain belt (Fig. 2.1) are the Salt Range Thrust (SRT) or Main Frontal Thrust, Main Mantle Thrust (MMT) or Indus Suture and the Kohistan Arc. The Kohistan Arc is an intra-oceanic island arc of Cretaceous age (Searle et al., 1999), squeezed between the converging India and Eurasian plates with the closure of the Tethys Ocean. East of the Western Syntaxis lies the Kashmir Valley, an active sedimentary basin comprising rocks of all ages. Some of the earliest seismic experiments provided much needed information on this segment of the Himalaya (see Singh et al., 2015).

Results from the DSS seismic profile covering the Higher Himalaya and Kashmir Valley show a very thick crust (Kaila, 1981) (Fig. 2.2). Along with two prominent dipping intracrustal reflectors, the Moho dips NNE from Kashmir to Higher Himalaya, with its depth changing from 45 beneath Kashmir Valley (Kaila, 1981) to 76 km beneath Higher Himalaya, with an offset along Profile P1–P2 (Fig. 2.3). Further north, the Moho depth starts decreasing near Nagaparbat (Kaila, 1981). Receiver function analysis reveals a thick crust, 55–62 km beneath Lesser Himalaya and 65–76 km beneath the Higher Himalaya in Pakistan (Li & Mashele, 2009). A low-velocity layer above the Moho is observed beneath certain seismic stations, reflecting presence of partial melts or volatiles (Li & Mashele, 2009). For the solitary seismic station NIL south of the Main Boundary Thrust (MBT) in Pakistan Himalaya, a thick crust (58 km) and very large (4 km/s) S-wave velocity (Vs) in the intermediate to lower crust was reported based on modelling of receiver functions (Vinnik et al., 2007) (Fig. 2.4). Such large velocities have been attached to the scraping of the Indian lower crust as the Indian lithosphere underthrusts Himalaya (Vinnik et al., 2007). Receiver functions from Kashmir Valley show an upwarp in the Moho ($\sim$ 4 km) with crustal thickness estimates ranging from $\sim$ 40 km south of the Main Central Thrust (MCT) to $\sim$ 50 km on the southwestern flank of Kashmir Valley (Mir et al., 2017). South of the Pir-Panjal Range, similar estimates of crustal thickness (54–60 km) have been observed by Mitra et al. (2023). The Main Himalayan Thrust (MHT) is found to be a sheared low-velocity zone at depths of 12–16 km beneath Kashmir Valley (Mir et al., 2017). A gravity high is observed in regions of thrust earthquakes, including the Kashmir earthquake (8th October, 2005, Mw = 7.6) (Tiwari et al., 2009). This high is related to thrusting of high-density rocks along the syntaxial bend of MBT (Tiwari et al., 2009).

Regions in and around the Western Syntaxis have shown (a) a thick crust and north-dipping Moho, (b) presence of low-velocity zones above Moho, (c) prominent intracrustal reflectors, (d) Moho offsets, (e) upwarp in Moho beneath Kashmir Valley and (f) indications of an MHT.

2.2.2 *Zanskar Himalaya*

The Zanskar Range lies between Ladakh and Higher Himalaya (Fig. 2.1). It is dominated by extensional tectonics, separating the high grade metamorphic sequence of Higher Himalaya and the overlying low-grade sedimentary rocks of the Tethyan Himalaya (Dezes et al., 1999). Another important feature in the region is the Kishtwar Window (Fig. 2.1), a late stage antiformal structure surrounded by the MCT (Stephenson et al., 2000). Seismic stations in this segment of the Himalaya are limited. A network of 24 broadband seismic stations was first installed in 2013 (Sharma et al., 2020), covering the Kishtwar Window and the southern parts of

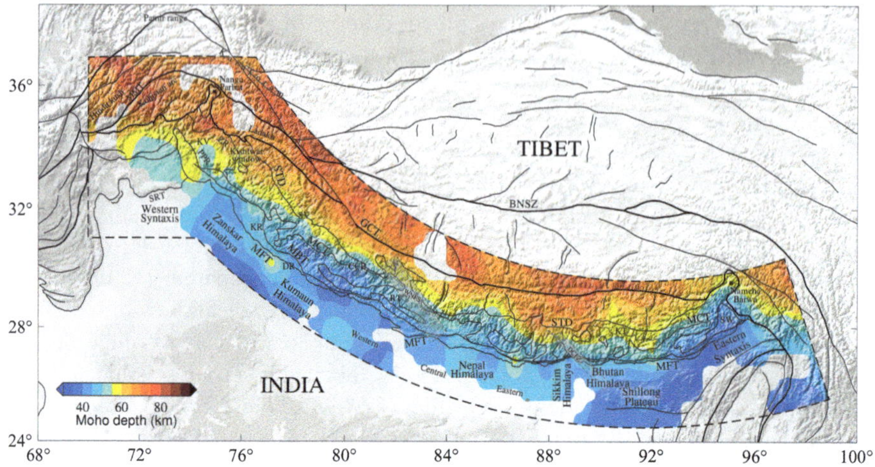

Fig. 2.2 Map showing the lateral variation in Moho depth beneath the Himalaya from the Western to the Eastern Syntaxis. The scale is shown in the legend. The Himalayan arc and its various segments are outlined by dashed lines. Major faults are shown by solid lines. DCFZ: Dhubri-Chungthang Fault Zone; WNFS: Western Nepal Fault System; MFT: Main Frontal Thrust; MBT: Main Boundary Thrust; MCT: Main Central Thrust; MCT-I: Inner Main Central Thrust; STD: South Tibetan Detachment; GCT: Greater Counter Thrust; BNSZ: Bangong-Nujiang Suture Zone; MMT: Main Mantle Thrust; KV: Kashmir Valley; ZR: Zanskar Range; PPR: Pir-Panjal Range; SRT: Salt Range Thrust; SV: Satluj Valley; KR: Kangra Reentrant; DR: Dehradun Reentrant; CCB: Chiplakot Crystalline Belt; RT: Ramghar Thrust; KT: Kakthang Thrust; SW: Siang Window

Himalaya. The Indian crust is found to thicken from $\sim$ 45 km beneath the Sub-Himalaya to $\sim$ 75 km beneath the Kishtwar Window (Mitra et al., 2023). A similar range of crustal thickness (50–60 km) is observed beneath the Higher Himalaya in the recent findings using receiver functions (Das et al., 2025) (Fig. 2.2). A mid-crustal layer with undulations in the Moho is also observed (Das et al., 2025). Joint inversion of Rayleigh wave dispersion and receiver functions delineated the MHT as a flat ramp marked with low velocities (Mitra et al., 2023). A mid-crustal ramp (10–16 km depth) is reported beneath the Kishtwar Higher Himalaya and Zanskar Range (Mitra et al., 2023) (Fig. 2.5), whose position is determined based on a cluster of earthquakes within a low-velocity zone. Higher V_p/V_s values (1.80–1.93) are observed in the Kashmir Valley, and the Kishtwar Window (1.81–1.93) using H–k stacking of receiver functions (Das et al., 2025), and are attributed to intracrustal low-velocity zones and volcanic rocks (Fig. 2.6). A complex Indian crust with a thickness of $\sim$ 40 km at the southern side of the foredeep to $\sim$ 75 km near the Karakorum Fault has been observed using receiver functions (Rai et al., 2006). It was suggested that the Indian plate may reach up to the BNSZ (Rai et al., 2006). Kanna and Gupta (2021) used the same set of stations and reported intracrustal low velocities and high velocities in the partially eclogitized lower crust beneath Ladakh. Results from the Sutlej Valley reveal a $\sim$44 km-thick crust beneath the Himalayan Front (Hazarika et al., 2017), with the thickness reaching $\sim$62 km beneath Tethys Himalaya in the north

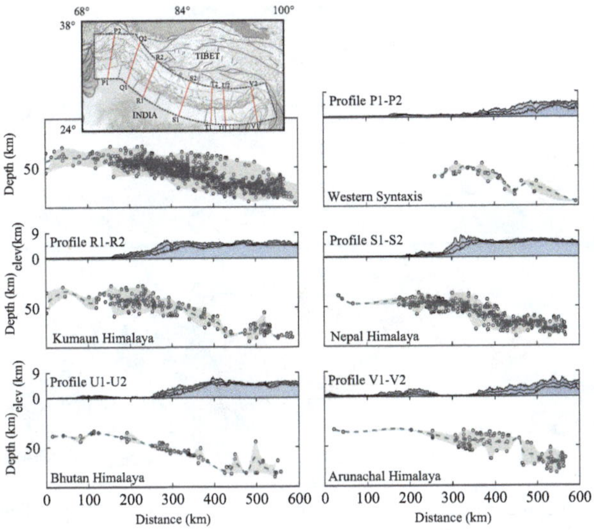

Fig. 2.3 Variations in crustal thickness along profiles representative of different segments of the Himalayan arc. Red lines in the inset map denote the profile transects (P1–P2, Q1–Q2, R1–R2, S1–S2, T1–T2, U1–U2 and V1–V2) along which the Moho depths have been presented. The first panel shows the aggregated Moho depth distribution compiled from all profiles. Individual Moho depth values are shown as black circles. The shaded grey envelope represents the bounds of reported Moho depths. Dashed lines trace the mean Moho depths inferred from this study

(Hazarika et al., 2017). They reported the MHT at depths of ~16–27 km beneath Sub, Lesser and Higher Himalaya (Hazarika et al., 2017). Modelling of gravity anomalies along the Kirtarpur-Manali-Leh-Panamik profile covering the Sub-Himalaya in the south to Karakoram Fault in the north (Chamoli et al., 2011) suggests that the Moho depth increases from ~ 45 to around 60 km beneath the Higher Himalaya (Chamoli et al., 2011).

Attenuation of seismic waves manifests as a decay in the amplitudes of seismic waves with distance, providing information about the crustal rheology that complements the velocity information. Mathematically, it is represented as Q, a dimensionless factor. Seismic attenuation is inversely proportional to Q. A low value of Q indicates high attenuation owing to a highly heterogeneous medium, and a high Q suggests the opposite. Using data from 11 broadband stations operated by NGRI, Rai et al. (2009) reported Q values along an N-S profile in block sampling different tectonic domains across the Himalaya, Ladakh and Karakoram (Fig. 2.7). Efficient propagation of L_g has been reported in the crust of the Himalaya and Tethys with very high Q values. The values are found to be quite higher compared to the others in different parts of the Himalaya (Parija et al., 2023; Singh et al., 2015; Sarkar, 2025; Sarkar et al., 2025). Ladakh is dominated by a low Q value, probably owing to aqueous fluids or partial melts in the crust (Fig. 2.7). However, they have not found any link with the low-viscosity channel present beneath Tibet, as evident in the Garhwal Himalaya (Ashish et al., 2009).

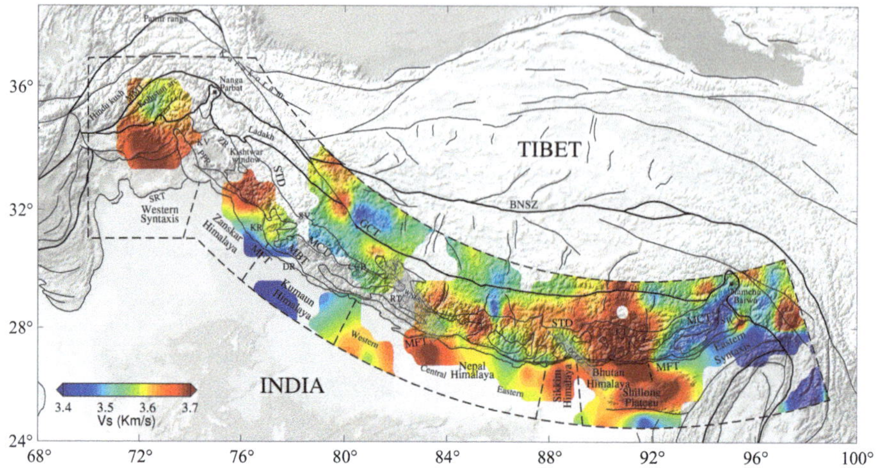

Fig. 2.4 Lateral variation in the average crustal S-wave velocity (V_S) along the Himalaya. The scale of V_S is shown in the legend. The Himalayan arc and its various segments are outlined by dashed lines. Major faults are represented by solid lines. DCFZ: Dhubri-Chungthang Fault Zone; WNFS: Western Nepal Fault System; MFT: Main Frontal Thrust; MBT: Main Boundary Thrust; MCT: Main Central Thrust; MCT-I: Inner Main Central Thrust; STD: South Tibetan Detachment; GCT: Greater Counter Thrust; BNSZ: Bangong-Nujiang Suture Zone; MMT: Main Mantle Thrust; KV: Kashmir Valley; ZR: Zanskar Range; PPR: Pir-Panjal Range; SRT: Salt Range Thrust; SV: Satluj Valley; KR: Kangra Reentrant; DR: Dehradun Reentrant; CCB: Chiplakot Crystalline Belt; RT: Ramghar Thrust; KT: Kakthang Thrust; SW: Siang Window

In this segment of Himalaya, (a) crustal thickness increases towards north (45–75 km) with possible undulations, (b) intracrustal low velocities and high-velocity lower crust are observed beneath Ladakh, (c) the signature of MHT remains speculative since clear arrivals from MHT are not seen in the migrated images of receiver functions and (d) low velocities and cluster of earthquakes show presence of a mid-crustal ramp beneath Kishtwar in the Higher Himalaya.

2.2.3 *Kumaun Himalaya*

Kumaun and Garhwal form central parts of the Himalayan Mountain Range (Fig. 2.1), with MFT, MBT and MCT being the common tectonic features. Multiple strands of MCT are observed (Celerier et al., 2009) and two major thrusts, namely Munsiari and Vaikrita, lie in the central parts of the Kumaun and Garhwal Himalaya (Fig. 2.1). The Vaikrita Thrust is identified as MCT in several studies (Dixit et al., 2024; Valdiya, 1980).

Early seismic experiments yielded the subsurface geometry of Dehradun and Kangra reentrants, with a northward-dipping Sub-Himalayan decollement having dips of 2.5° and 6° beneath Kangara and Dehradun reentrants (Powers et al., 1998).

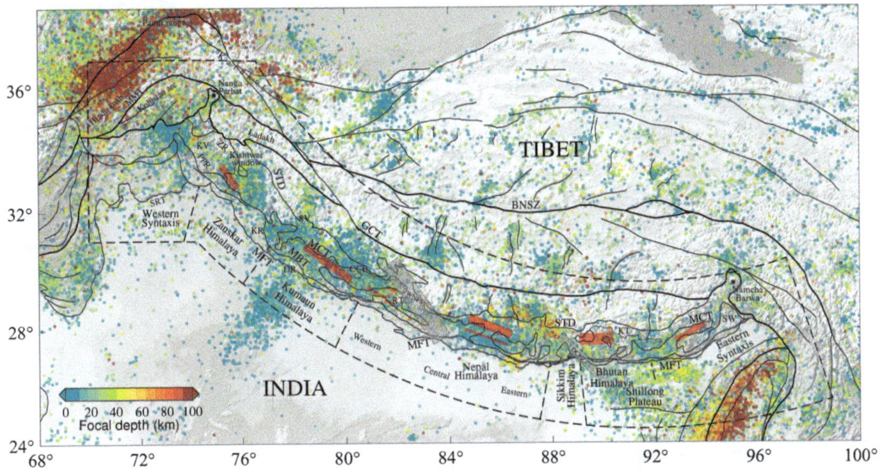

Fig. 2.5 Seismicity along the Himalayan collisional zone. Color-filled circles represent earthquakes. Red rectangles show the possible locations of MHT ramps (reported by Mitra et al. (2023) for Zanskar Himalaya, Caldwell et al. (2013) for Kumaun Himalaya, Hoste-Colomer et al. (2018) for western Nepal Himalaya, Elliott et al. (2016) and Hubbard et al. (2016) for central Nepal Himalaya, Singer (2017) for Bhutan Himalaya and Mangalampalli et al. (2022) for Arunachal Himalaya). The Himalayan arc and its various segments are outlined by dashed lines. DCFZ: Dhubri-Chungthang Fault Zone; WNFS: Western Nepal Fault System; MFT: Main Frontal Thrust; MBT: Main Boundary Thrust; MCT: Main Central Thrust; MCT-I: Inner Main Central Thrust; STD: South Tibetan Detachment; GCT: Greater Counter Thrust; BNSZ: Bangong-Nujiang Suture Zone; MMT: Main Mantle Thrust; KV: Kashmir Valley; ZR: Zanskar Range; PPR: Pir-Panjal Range; SRT: Salt Range Thrust; SV: Satluj Valley; KR: Kangra Reentrant; DR: Dehradun Reentrant; CCB: Chiplakot Crystalline Belt; RT: Ramghar Thrust; KT: Kakthang Thrust; SW: Siang Window

Subsequently, receiver functions have been extensively used to study the geometry of the crust in this segment of the Himalaya (Ashish et al., 2023; Caldwell et al., 2013; Hazarika et al., 2021; Kanna & Gupta, 2020). Delineation of crustal structure along a profile of closely spaced seismic stations in the Garhwal Himalaya revealed a $\sim$ 45 km deep Moho beneath the Sub-Himalaya and a $\sim$ 58 km-thick crust further north beneath the Higher Himalaya (Kanna & Gupta, 2020) (Fig. 2.2). Caldwell et al. (2013) found a relatively flat Moho beneath the Sub-Himalaya and Lesser Himalaya (35–45 km), with the crustal thickness increasing to 50 km beneath the Higher Himalaya (Profile R1–R2; Fig. 2.3). They interpret a flat-ramp-flat geometry of the MHT, which is expressed as a negative impedance contrast. The mid-crustal ramp seems to be 10 km depth and dips at 16° (Caldwell et al., 2013). Similarly, Kanaujia et al. (2025) identified a low-velocity layer associated with the MHT in the Garhwal and Kumaun Himalaya, exhibiting a distinct feature. In the Garhwal Himalaya, MHT displays a flat-ramp-flat geometry with a narrow locking width ($\sim$ 60 km), whereas in the Kumaun Himalaya, MHT shows a wider locking width ($\sim$ 80 km). The top of the MHT is well resolved at $\sim$ 24 km depth (Kanaujia et al.,

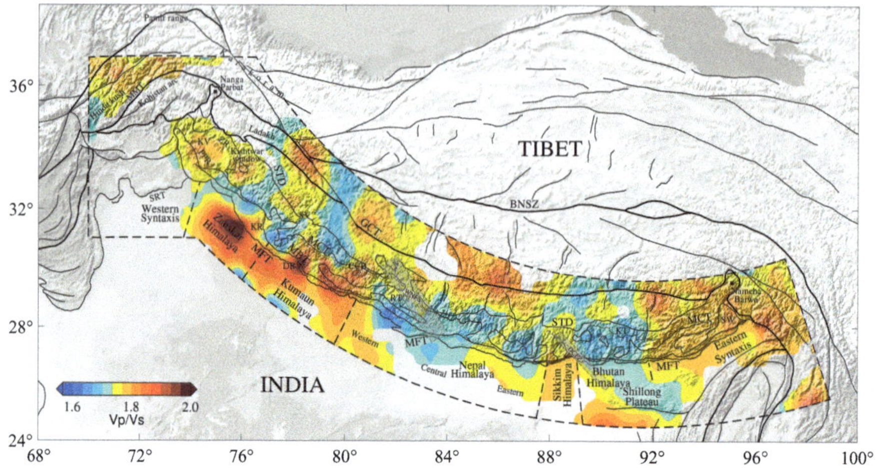

Fig. 2.6 Lateral variation in average crustal P-wave to S-wave velocity ratio (V_p/V_s), in the Himalaya from the Western to the Eastern Syntaxis. The scale of the V_p/V_s shown in the legend. The Himalayan arc and its various segments are outlined by dashed lines. DCFZ: Dhubri-Chungthang Fault Zone; WNFS: Western Nepal Fault System; MFT: Main Frontal Thrust; MBT: Main Boundary Thrust; MCT: Main Central Thrust; MCT-I: Inner Main Central Thrust; STD: South Tibetan Detachment; GCT: Greater Counter Thrust; BNSZ: Bangong-Nujiang Suture Zone; MMT: Main Mantle Thrust; KV: Kashmir Valley; ZR: Zanskar Range; PPR: Pir-Panjal Range; SRT: Salt Range Thrust; SV: Satluj Valley; KR: Kangra Reentrant; DR: Dehradun Reentrant; CCB: Chiplakot Crystalline Belt; RT: Ramghar Thrust; KT: Kakthang Thrust; SW: Siang Window

2024), and seismicity distributed at and above it has been attributed to interseismic stress loading driven by the ongoing India–Eurasia collision (Kanaujia et al., 2023).

Modelling of gravity data along the Kirtarpur-Gaurikund profile lends credence to the flat-ramp-flat geometry of MHT (Chamoli et al., 2023). A relatively high shear-wave velocity was detected for the lower crust (3.9–4.1 km/s) (Kanna & Gupta, 2020). In another study, Mandal et al. (2021) documented a 35–45 km-thick crust beneath the Lesser Himalaya and a 50–55 km one beneath the Higher Himalaya (Fig. 2.2). Two instances of drastic changes in Moho depth and large V_p/V_s values have been observed in the epicentral zones of the 1991 Uttarakhand (Mw 6.6) and 1999 Chamoli (Mw 6.4) earthquakes (Mandal et al., 2021) (Figs. 2.2 and 2.6). Large V_p/V_s values of 1.85 are linked to high fluid pressure associated with the MHT (Fig. 2.6).

Close to the international boundary with Nepal in the Kumaun Himalaya, the crustal thickness estimates are around 40 km (Hazarika et al., 2021) (Fig. 2.2). Inversion of receiver functions also revealed a lower crustal low-velocity zone having a high Poisson's ratio (0.28–0.30) and a ramp structure on the MHT in the vicinity of the Chiplakot Crystalline Belt (Hazarika et al., 2021) (Fig. 2.8). Joint inversion of receiver functions and Rayleigh wave dispersion data also brought out a flat-ramp-flat geometry of MHT, associated with low velocities (3.1–3.4 km/s) (Ashish et al., 2023) (Fig. 2.4). The MHT is seen at depths of 8–24 km, south to north, with a ramp beneath

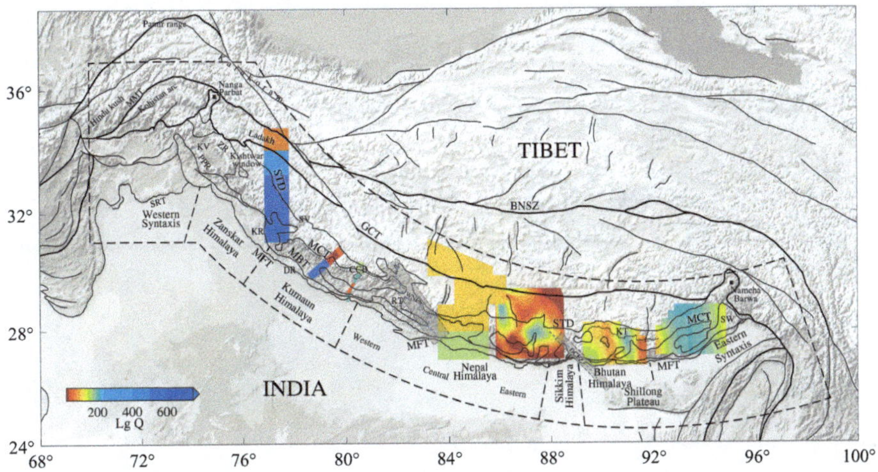

Fig. 2.7 Lateral variation in the quality factor of L_g wave attenuation (L_gQ), highlighting the Himalaya arc region from the Western to the Eastern Syntaxis. The scale of the L_g Q shown in the legend. The Himalayan arc and its various segments are outlined by dashed lines. DCFZ: Dhubri-Chungthang Fault Zone; WNFS: Western Nepal Fault System; MFT: Main Frontal Thrust; MBT: Main Boundary Thrust; MCT: Main Central Thrust; MCT-I: Inner Main Central Thrust; STD: South Tibetan Detachment; GCT: Greater Counter Thrust; BNSZ: Bangong-Nujiang Suture Zone; MMT: Main Mantle Thrust; KV: Kashmir Valley; ZR: Zanskar Range; PPR: Pir-Panjal Range; SRT: Salt Range Thrust; SV: Satluj Valley; KR: Kangra Reentrant; DR: Dehradun Reentrant; CCB: Chiplakot Crystalline Belt; RT: Ramghar Thrust; KT: Kakthang Thrust; SW: Siang Window

the High Himalaya (Ashish et al., 2023). Drastic variations are seen in the nature and geometry of the MHT along the strike of the Himalayan mountain belt. Ambient noise tomography also shows low shear-wave velocities at $\sim$ 10–15 km depth around the MHT (Verma et al., 2023). High velocities near the ramp were attached to duplex structures (Verma et al., 2023). Low V_p and high V_p/V_s in the upper-mid crust were obtained using local earthquake tomography, which also supported a flat-ramp-flat geometry for MHT (Gupta et al., 2021).

The attenuation structure of the Garhwal Himalaya has been reported by Ashish et al. (2009). They found a shield-like low attenuation structure in the Lesser Himalaya and an extremely high attenuation in the Higher Himalaya (Fig. 2.7). Their study is consistent with a low *P-wave* velocity at a shallower depth below High Himalaya. They suggested that a low Q value, along with low velocity, indicates evidence for partial melting associated with Miocene leucogranite plutons that originated during the India-Asian collision. They also discussed the possible connection of mid-crustal level channel flow beneath Tibet. Parija et al. (2023) studied L_g attenuation in the Kumaun Himalaya along an N-S profile extending from the Indo-Gangetic Plain in the south to the Higher Himalaya in the north. They found higher attenuation in the Lesser Himalaya and the lowest attenuation along parts of the Indo-Gangetic Plains and the Sub-Himalaya. The Higher Himalaya and Tethys Himalaya

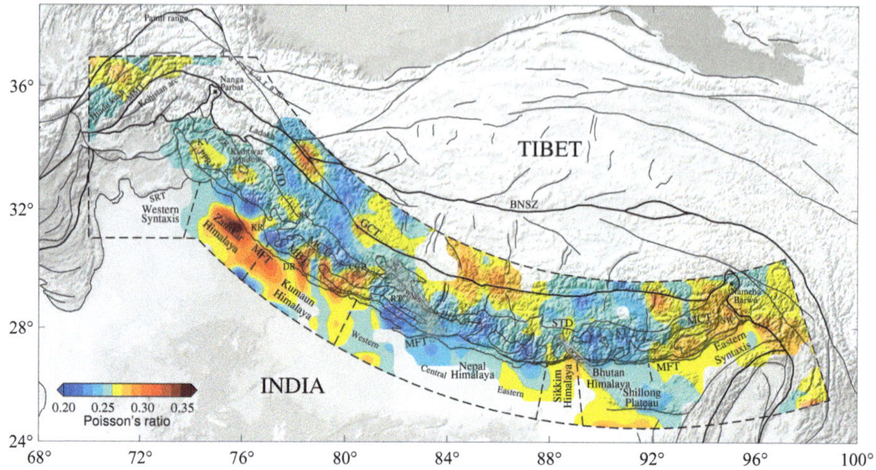

Fig. 2.8 Variation of bulk Poisson's ratio of the crust along the Himalaya arc. The scale of Poisson's ratio is shown in the legend. The Himalayan arc and its various segments are outlined by dashed lines. DCFZ: Dhubri-Chungthang Fault Zone; WNFS: Western Nepal Fault System; MFT: Main Frontal Thrust; MBT: Main Boundary Thrust; MCT: Main Central Thrust; MCT-I: Inner Main Central Thrust; STD: South Tibetan Detachment; GCT: Greater Counter Thrust; BNSZ: Bangong-Nujiang Suture Zone; MMT: Main Mantle Thrust; KV: Kashmir Valley; ZR: Zanskar Range; PPR: Pir-Panjal Range; SRT: Salt Range Thrust; SV: Satluj Valley; KR: Kangra Reentrant; DR: Dehradun Reentrant; CCB: Chiplakot Crystalline Belt; RT: Ramghar Thrust; KT: Kakthang Thrust; SW: Siang Window

show high attenuation. Overall, their study demonstrates a significantly low *Q* structure beneath Kumaun Himalaya owing to both crustal heterogeneities and partial melts because of elevated temperature (Fig. 2.7).

The Kumaun-Garhwal Himalaya shows (a) a very complex nature of crust, with crustal thickness varying between 35 and 58 km, (b) strong evidence of low-velocity zones above the MHT, partially reflecting the presence of fluids/partial melts and (c) a flat-ramp-flat geometry for MHT.

2.2.4 Nepal Himalaya

Like in the rest of the Himalaya, Nepal consists of the same major thrust zones (MFT, MBT and MCT) which run parallel to the strike of the mountain belt (Fig. 2.1). A continuation of the Ramgarh Thrust from Kumaun Himalaya is also seen as a major thrust sheet along the strike of the Nepal Himalaya (Pearson & DeCelles, 2005). The Nepal Himalaya is a well-studied segment in terms of defining the geometry and nature of the Indian plate subduction (He et al., 2018; Nábělek et al., 2009; Schulte-Pelkum et al., 2005; Subedi et al., 2018). The 2015 Mw 7.8 Gorkha-Nepal earthquake has put the spotlight back on this segment (Avouac et al., 2015; Bai et al., 2019).

Lateral and seasonal variations in seismicity were reported in the initial experiments beneath the Nepal Himalaya (Bollinger et al., 2007; Pandey et al., 1999). Seismicity is concentrated close to the mid-crustal ramp, with evidence for bimodal patterns (Dal Zilio et al., 2020). A multistage rupture of the Gorkha-Nepal earthquake of 2015 (Fan & Shearer, 2015), its eastward propagation (Yagi & Okuwaki, 2015) and structural control on the rupture propagation (Bai et al., 2019; Hubbard et al., 2016) were extensively discussed for the Nepal Himalaya. The MHT with a steep ramp is evident in controlling seismicity patterns beneath the Himalaya (Dal Zilio et al., 2020; Elliott et al., 2016) (Fig. 2.5). The geometry of MHT is described as a flat-ramp-flat structure (Elliott et al., 2016) to a gently dipping surface bounded by steep faults in the slip zones (Hubbard et al., 2016). Although the role of MHT is crucial in controlling seismicity patterns, high-resolution seismic imaging to decipher the geometry of this fault zone remains a challenge.

Migrated images of receiver functions computed using data from the HIMNT network were crucial in delineating the crustal geometry and nature beneath Eastern Nepal Himalaya (Schulte-Pelkum et al., 2005). The Moho beneath India is observed at ~ 45 km, reaching depths of ~ 75 km beneath Tibet (Schulte-Pelkum et al., 2005) (Profile S1–S2; Fig. 2.3). Beneath the Himalaya, a gentle down dip is observed (Schulte-Pelkum et al., 2005). An important finding was that the Decollement surface is highly anisotropic in nature (~ 20%) (Schulte-Pelkum et al., 2005). North of the Himalaya, the Indian lower crust may have been probably eclogitized, resulting in fast velocities (Schulte-Pelkum et al., 2005). A high uppermost mantle velocity (8.4 km/s) was observed using local tomography and travel times of regional earthquakes (Monsalve et al., 2008). Metamorphism of lower crustal rocks, resulting in the formation of eclogites, can explain such high velocities (Monsalve et al., 2008; Tiwari et al, 2006). The crustal thickness, estimated from data of the Hi-CLIMB experiment, is 40 km beneath the Ganges Basin and 50 km beneath the Himalaya (Nábělek et al., 2009). The MHT is marked by consistent low velocities atop it, possibly due to underthrusting of low-velocity sediments of the Ganges basin under Lesser Himalayan formations (Nábělek et al., 2009). In western Nepal, the Moho deepens from ~ 40 km to ~ 58 km beneath the Higher Himalayan range (Subedi et al., 2018) (Fig. 2.2). In this region, a strong low-velocity zone is observed beneath the Lesser Himalaya (Subedi et al., 2018) at depths of ~ 12 to ~ 18 km. Estimates from inversion of gravity data reveal a 35 km-thick crust beneath the foothills to 75 km beneath Tibet (Cattin et al., 2001; Hetenyi et al., 2007).

Singh et al. (2015, 2016) have conducted a seismic attenuation study using HIMNT network data operated in eastern Nepal and a local network of 11 seismic stations runs by the CSIR-NGRI in Sikkim Himalaya. Overall, the region belongs to a high attenuation zone. High attenuation is observed along the Lesser Himalaya (Singh et al., 2015). They have found that the low values of Q (Fig. 2.7) appear to coincide with low V_p/V_s Monsalve et al. (2008) (Fig. 2.6) and low Q_P (P-wave Q) and Q_S (S-wave Q) (Sheehan et al., 2014). They discussed that all the observations, along with the reported studies, suggest the presence of trapped fluids in the crust.

The results from Nepal Himalaya show (a) a crustal thickness of 40 km beneath Sub-Himalaya to 60 km beneath Higher Himalaya, (b) a highly anisotropic decollement surface ($\sim$ 20%), (c) a gently northward-dipping Moho, (d) geometry of MHT, a controlling factor in seismicity and (e) a flat-ramp-flat structure for MHT.

2.2.5 *Sikkim Himalaya*

Sikkim Himalaya is at the transition from the central to the eastern Himalaya. The MCT takes a curvilinear turn in here (Fig. 2.1). This region is dominated by strike-slip earthquakes (Hazarika et al., 2010; Uthaman et al., 2024) compared to other segments of the Himalaya where thrust tectonics is prevalent (Manglik et al., 2013). The Dhubri-Chungthang Fault Zone (DCFZ) seems to segment the Indian basement through the Sikkim Himalaya (Diehl et al., 2017). The DCFZ is a zone of moderate seismicity mostly occurring in the mid-crust (Diehl et al., 2017; Uthaman et al., 2024). Results from the INDEPTH experiment were crucial in defining the Indian plate subduction beneath Tibet (Nelson et al., 1997). Strong reflections from MHT at $\sim$ 30 km depth north of STD and a very deep Moho ($\sim$ 70 km) beneath Tethys Himalaya were observed just north of Sikkim (Nelson et al., 1997) (Fig. 2.2). They also advocated a structurally imbricated Indian crust beneath the Higher Himalaya (Nelson et al., 1997).

Modelling of gravity data results has provided the first estimates of crustal thickness beneath the Sikkim Himalaya (Tiwari et al., 2006). It is found that the Moho dips from $\sim$ 36 km near the foothills to around 60 km beneath the Higher Himalaya (Fig. 2.2), with the strongest dips close to the MCT (Tiwari et al., 2006). The first Common Conversion Point (CCP) images were produced using data from 11 broadband seismic stations located along a N-S profile (Singh et al., 2010). Similar to the results from gravity, it is observed that the crustal thickness increases towards the north from $\sim$ 40 km to $\sim$ 61 km beneath the Himalaya, with a dip of 4–10° (Singh et al., 2010). Beneath the Higher Himalaya, a prominent intracrustal layer was observed around $\sim$ 40 km depth and interpreted as a Moho doublet (Singh et al., 2010). A strong anisotropic (17%) low-velocity layer was also detected at depths of 20–30 km (Singh et al., 2010). Subsequently, a complex crust was reported by Paul and Mitra (2017). Recent results using data from a denser network of seismic stations have revealed an imbricated crust with offsets and overlaps in the lower crust (Singh et al., 2024). The MHT is not so well resolved in Sikkim Himalaya (Singh et al., 2024). A significant number of earthquakes are found to be clustered in low-velocity zones in the upper crust along possible locations of MHT (Singh et al., 2024). The crustal thickness is similar to earlier estimates of $\sim$ 40 km beneath the Lesser Himalaya to $\sim$ 60 km beneath the Higher Himalaya (Singh et al., 2024). A high attenuation is observed around MCT in Sikkim Himalaya (Singh et al., 2015). The seismicity reflects complex fault dynamics, with mid-crustal seismicity clusters related to the Dhubri-Chungthang Fault Zone (DCFZ) and a fluid-rich zone above the actively shearing MHT (Uthaman et al., 2024).

The results from Sikkim Himalaya show (a) a ~ 40 km-thick crust beneath the Lesser Himalaya increasing to ~60 beneath the Higher Himalaya, (b) an anisotropic middle crust (17%), (c) intermittent presence of MHT, (d) imbricated crust with offsets and overlaps, (e) dominance of transverse tectonics and (f) possible presence of Dhubri-Chungthang Fault Zone (DCFZ) segmenting the Indian basement through Sikkim.

2.2.6 Bhutan Himalaya

The major thrust zones (MFT, MBT and MCT) are similar to other segments of Himalaya. An out-of-sequence thrust within the Higher Himalayan sequence is known as Kakthang Thrust (KT) in the Bhutan Himalaya (Grujic et al., 2002) (Fig. 2.1). The Bhutan Himalaya is considered to be a quiet zone in terms of seismicity (Gahalaut et al., 2011). Evidence for transcurrent deformation with strike-slip earthquakes, mostly at mid-crustal to lower crust depths, is presented from historical seismicity (Drukpa et al., 2006). There are suggestions of an Mw 7.6 earthquake in Bhutan based on monastery records (Bilham et al., 2001). Seismicity variations along the strike of Himalaya and DCFZ, seismicity concentration at mid-crustal levels summarize the seismotectonics of the Bhutan Himalaya based on the data from GANSSER network of seismic stations (Diehl et al., 2017).

The GANSSER seismic network provided the much needed information on the deep crustal structure beneath the Bhutan Himalaya (Singer et al., 2017a, 2017b) (Fig. 2.1). Migrated images using receiver functions show along-strike differences in crustal structure (Singer et al., 2017a, 2017b). In the western Bhutan Himalaya, a strong dip in the Moho is observed, which reaches to depths of ~ 70 km beneath the Higher Himalaya (Singer et al., 2017a, 2017b) (Profile U1–U2; Fig. 2.3). In eastern Bhutan, the Moho boundary is observed to be flat (~ 50 km) (Singer et al., 2017a, 2017b). Various listric features beneath the MHT were viewed in terms of a strong deformation and crustal imbrication involving the lower crust. The MHT is horizontal beneath the Lesser Himalaya in western Bhutan. A mid-crustal ramp is observed beneath the Higher Himalaya in western Bhutan (Singer et al., 2017a, 2017b). Ambient noise tomography reveals high-velocity anomalies in the upper crust (> 3.6 km/s) (Singer et al., 2017a, 2017b) (Fig. 2.4). Localized duplexes may explain the presence of such high-velocity anomalies linked to the structural complexities of MHT (Singer et al., 2017a, 2017b).

The crustal attenuation models for Arunachal (Sarkar, 2025) and Bhutan Himalaya (Sarkar et al., 2025) are determined using data from a local network of 33 seismic stations operated by NGRI (Singh et al., 2021), and the GANSSER network run by the Swiss Seismological Service, ETH Zurich (Diehl et al., 2017), respectively (Fig. 2.7). Variations in seismic Q structure have been observed from east to west in both regions, which align well with the local tectonic features (Fig. 2.7).

The Q model of eastern Bhutan Himalaya is found to correlate well with the shear-wave velocity structure at the shallow level (Singer, 2017). Relatively higher values of Q are reported around the Paro window in western Bhutan and near the KT, which are consistent with the high-velocity structure obtained by 3-D surface-wave inversion (Singer, 2017). The frequency-dependent parameter indicates a more or less low-deformation rate in the region, though slightly higher deformation is evident in the central part, suggesting scattering-dominated attenuation (Sarkar et al., 2025).

The results from Bhutan Himalaya show (a) variations in Moho geometry in the eastern and western Bhutan Himalaya, (b) a thick crust, with a strong dip in Moho in western Bhutan and (c) MHT with a ramp beneath Higher Himalaya.

2.2.7 *Arunachal Himalaya*

The tectonics of the Arunachal Himalaya is complex, the region being at the junction of major continental plates (Fig. 2.1). The major thrusts run parallel to each other, with the MCT showing curvilinear trends. The easternmost part of the Himalaya collision in Arunachal, known as the Eastern Syntaxis, has much more complex geology with sharp bends in major thrusts (Fig. 2.1).

A major seismic experiment was launched by NGRI in 2010 to capture the structure beneath the Arunachal Himalaya (Singh et al., 2021). A total of 32 broadband seismic stations were installed across Arunachal Pradesh (Singh et al., 2021) (Fig. 2.1). The crust beneath the Arunachal Himalaya is less complex compared to Sikkim and Bhutan Himalaya (Singh et al., 2021). The Moho dips northward, and the average crustal thickness is 40–60 km (Singh et al., 2021) (Profile V1–V2; Fig. 2.3). Signatures of the MHT are absent in the migrated seismic images (Singh et al., 2021) and local earthquake tomography images (Mahesh et al., 2023). The mid-crust is highly anisotropic in nature (Singh et al., 2021). Harmonic analysis of RFs shows strong indications of dipping and anisotropic layers within the crust (Singh et al., 2021). Later, with additional seismic stations providing relatively dense coverage, an intermittent MHT was observed at depths of $\sim$ 10–20 km (Mangalampalli et al., 2022). The Siang window was also covered in other studies (Kundu et al., 2022). A thick crust ($\sim$ 38–54 km) with Poisson's ratio between 0.23 and 0.31 was observed in the Siang window using Receiver functions (Kundu et al., 2022) (Fig. 2.8).

The eastern Arunachal Himalaya demonstrates characteristics of intrinsic attenuation corroborated by high attenuation (Fig. 2.7), a weak frequency dependence, a high Poisson's ratio, as well as elevated temperature (Fig. 2.8). The Tawang and Siang windows exhibit relatively higher attenuation, while low attenuation is observed in the central parts (Fig. 2.7). The western part of the Arunachal Himalaya shows scattering attenuation substantiated by high attenuation with a strong frequency dependence. It appears to be a continuing feature from eastern Bhutan Himalaya, which is highly heterogeneous in nature because of several geological structures like MCT, MBT, KT and STD. The Q model of eastern Bhutan Himalaya is found to correlate well with the shear-wave velocity structure at shallower levels (Singer et al, 2017a, 2017b).

In summary, the Arunachal Himalaya is characterized by (a) a less complex crust with thickness between ~ 40–60 km, (b) no clear signatures of MHT and mid-crustal ramp and (c) strong anisotropic mid crust associated with LVLs.

2.3 Discussion

Various models have been proposed to explain the formation of the Himalaya-Tibet collision zone (Dal Ding et al., 2022; Zilio et al., 2021). These include (a) complete underthrusting of Indian continental plate beneath Tibet (Argand, 1924), (b) distributed homogeneous shortening (England & McKenzie, 1982), (c) extrusion and lower crustal flow (Beaumont et al., 2001), (d) Himalayan segmentation with proto tectonic features controlling the growth of orogeny (Dal Zilio et al., 2021), (e) slab tearing, breakoff and delamination (Ding et al., 2022; Kelly et al., 2020; Molnar, 1988; Pusok & Stegman, 2020; Webb et al., 2017). The collision has resulted in ~ 3600 km of shortening (van Hinsbergen et al., 2012), out of which nearly 1400–2350 km is well explained by geological records (van Hinsbergen et al., 2012; Yin & Harrison, 2000). To explain, the remaining shortening various models have been proposed, including multistage subduction (Pusok & Stegman, 2020; van Hinsbergen et al., 2012). Lack of high-resolution datasets resulted in different opinions and adaptations of models specific to particular regions.

The Himalayan collision zone is witness to many deadly earthquakes (Bilham, 2019). The role of MHT in generation of seismicity is discussed and debated. The MHT has the potential to create great earthquakes of magnitude 8 and above (Bilham, 2019; Michel et al., 2021). The mid-crustal ramp on MHT is a common feature in various segments of the Himalaya (Ashish et al., 2023; Caldwell et al., 2013; Elliott et al., 2016; Singer et al., 2017a, 2017b; Singh et al., 2024). The ramps are locales for clusters of earthquakes and, in some cases, reasons for rupture termination (Sathiakumar & Barbot, 2021). The slip potential controlled through the geometry of MHT may result in the strongest possible release of energy. The 2015 Gorkha-Nepal Earthquake (Mw 7.8) ruptured the MHT, and the rupture length seems to be controlled by its geometry (Bai et al., 2019). Slip inversion models suggest that the geometry of MHT is a controlling factor in rupture length of great earthquakes (Hubbard et al., 2016). In most cases, a flat-ramp-flat geometry of MHT is advocated (Ashish et al., 2023; Caldwell et al., 2013; Elliott et al., 2016), while in certain cases, a double ramp is also introduced (Wang et al., 2017). With the existing data, the geometry of MHT remains poorly resolved, with possible reasons of a very complex nature of this surface or low resolution of seismic data. Understanding the nature of this decollement surface is essential to our understanding of seismogenesis in Himalaya.

The Arc-Parallel Gravity Anomaly (APaGA) (Fig. 2.9) discloses a complex feature of deep structural lateral variability beneath the orogen (Hetényi et al., 2016). The flipping nature of APaGA from foreland to orogenic front describes the flexure geometry of the undergoing Indian plate. A positive to negative transition of APaGA associates well with short flexural wavelength in Bhutan Himalaya segment while opposite trend of APaGA in the Nepal Himalaya segment correlates with long flexural wavelength of Indian plate. The Arunachal Himalaya segment shows similar trend with Nepal and Kumaun Himalaya, while Zanskar Himalaya and Bhutan Himalaya exhibit same trend. These trends suggest that Indian plate in the Nepal and Arunachal Himalaya begins to bend further south of the topographic front. In contrast, in the Zanskar and Bhutan Himalaya segment, Indian plate starts bending closer to the topographic front. Overall, APaGA also indicates lateral variability of subduction pattern of Indian plate beneath Himalaya.

Most of the geological cross-sections in the Himalaya collision zone are focussed on upper-to-middle parts of the crust (Yin, 2006) (Fig. 2.10). The structure of the Indian basement, mainly the geometry of the lower crust, is ignored in the geological models. The downwelling Indian continental crust is highly deformed, marked by ongoing collision tectonics. Various instances of overlapping zones and offsets in the Moho boundary are observed beneath Himalaya (e.g. Kaila, 1981; Mandal et al., 2021; Singh et al., 2024). The lower crustal earthquakes also bring out its heterogeneous nature (Monsalve et al., 2006; Uthaman et al., 2024). Also, the Moho is not always flat and gently dipping boundary as envisioned in many simplified geological cross-sections. We require a thorough understanding and detailed modelling to

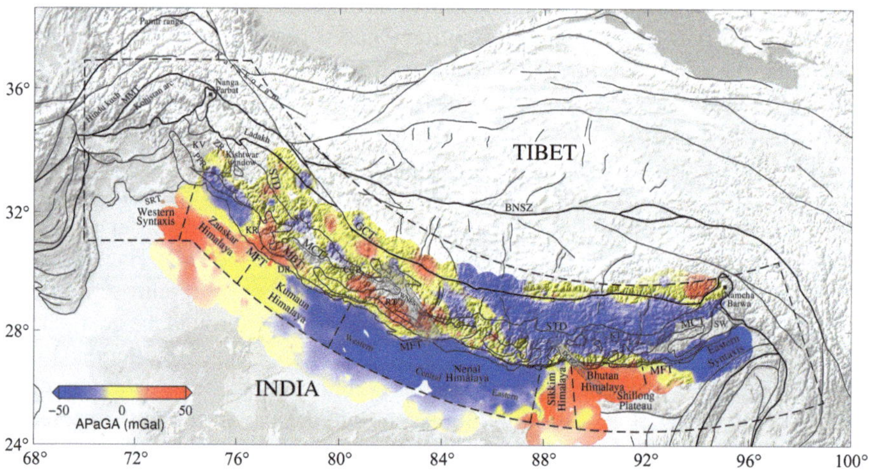

Fig. 2.9 Arc-parallel Gravity anomaly modified after (Hetényi et al., 2016). Red and blue correspond to positive and negative anomalies, respectively, in topography and gravity relative to the mean profile oriented perpendicular to the Himalayan arc. The Himalayan arc and its various segments are outlined by dashed lines. DCFZ: Dhubri-Chungthang Fault Zone; WNFS: Western Nepal Fault System; MFT: Main Frontal Thrust; MBT: Main Boundary Thrust; MCT: Main Central Thrust; MCT- I: Inner Main Central Thrust; STD: South Tibetan Detachment; GCT: Greater Counter Thrust; BNSZ: Bangong-Nujiang Suture Zone; MMT: Main Mantle Thrust; KV: Kashmir Valley; ZR: Zanskar Range; PPR: Pir-Panjal Range; SRT: Salt Range Thrust; SV: Satluj Valley; KR: Kangra Reentrant; DR: Dehradun Reentrant; CCB: Chiplakot Crystalline Belt; RT: Ramghar Thrust; KT: Kakthang Thrust; SW: Siang Window

understand the role of lower crust in mass budget estimates of the Himalayan collision zone.

Future lies in installation of a dense network of seismic stations covering each segment of the Himalaya, from foothills to the Higher Himalaya. Although various issues could be resolved with the existing network of seismic stations, the nature and

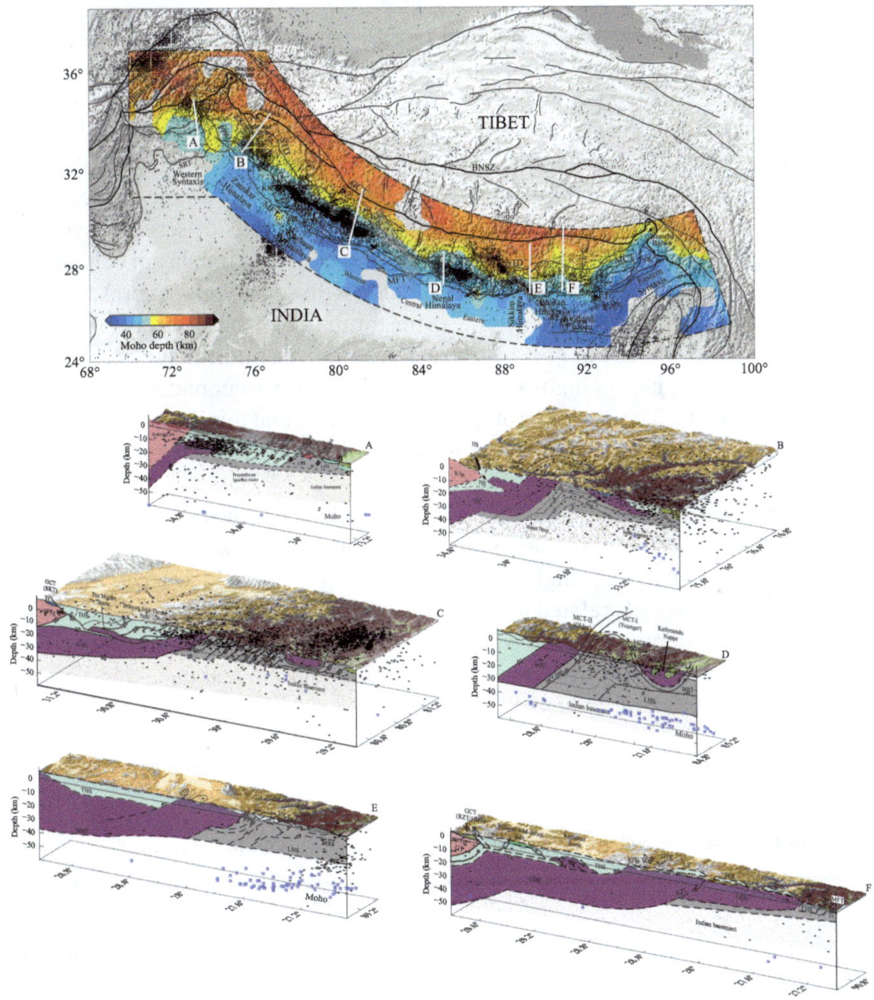

Fig. 2.10 Geological cross-sections, seismicity and Moho depth distribution along six profiles (A, B, C, D, E and F) in different regions. The geological cross-sections have been modified after Yin (2006). The region maps illustrate the profile locations and lateral variations in the Moho depth across the study region. Color-filled inverted triangles and diamonds denote broadband seismic and deep seismic sounding stations, respectively, used to constrain Moho depth. Earthquakes are represented by black circles, while blue square symbols indicate the estimated Moho depths along each profile

geometry of the subducting Indian plate need refinement. Our understanding of MHT and its role in seismogenesis needs an advanced integrated approach using various state-of-the-art techniques and high-resolution mapping using a dense network of seismic stations with inter-station distance of less than 1 km.

2.4 Conclusions

We utilized data collected from various sources to create maps of crustal thickness, velocity and attenuation structures in the Himalayan collision zone. The crust thickens as we move from the foothills to the Higher Himalaya. Our syntheses indicate clear variations in average crustal velocities and Poisson's ratio values along the strike of the mountain range. The Moho boundary exhibits offsets and overlaps, suggesting a complex crust–mantle transition throughout the Himalayan belt. Recent results from receiver function analyses, combined with seismicity patterns, indicate that the lower crust plays a significant role in the formation of the Himalayan orogeny. The decollement serves as a major energy-releasing structure during significant earthquakes; however, the lack of high-resolution imaging hinders our understanding of its nature and geometry. The compiled information suggests that this layer is anisotropic, characterized by low velocities and the presence of ramps where major Himalayan thrusts are connected to the MHT.

The synthesis also underlines that integrating diverse datasets and employing similar methodologies would be necessary for robust crustal models for the Himalaya collision zone. Furthermore, high-resolution imaging is crucial to clarify the ambiguities in existing models related to Himalayan subduction and the formation of the Tibetan Plateau. Adopting advanced techniques such as full-waveform inversion and machine learning for analysis of geophysical and seismological data could greatly enhance the resolution of findings.

Acknowledgements All the figures are made using GMT software (Wessel et al., 2024). VMT acknowledges ANRF JCB Fellowship.

Conflict of Interest The authors declare that they have no known conflict of interest.

Author Contributions All the authors have contributed equally.

Data Availability All the data produced in this paper are available online. Proper citations are given in the main text.

References

Argand, E. (1924). La tectonique de l'Asia. *Int. Geol. Cong. Rep. Sess., 13*, 170–372.

Ashish, Padhi, A., Rai, S. S., & Gupta, S. (2009). Seismological evidence for shallow crustal melt beneath the Garhwal High Himalaya, India: Implications for the Himalayan channel

flow. *Geophysical Journal International, 177*, 1111–1120. https://doi.org/10.1111/j.1365-246X.2009.04112.x
Ashish, Saha, G., & Rai, S. (2023). 3-D crustal structure in Kumaon–Garhwal Himalaya using joint inversion of receiver functions and surface wave group velocity. *Geophysical Journal International, 233*, 2101–2123. https://doi.org/10.1093/gji/ggad044
Avouac, J. P. (2003). *Mountain building, erosion, and the seismic cycle in the Nepal Himalaya* (pp 1–80), Elsevier.
Avouac, J. P., Meng, L., Wei, S., Wang, T., & Ampuero, J. P. (2015). Lower edge of locked Main Himalayan thrust unzipped by the 2015 Gorkha earthquake. *Nature Geoscience, 8*, 708–711. https://doi.org/10.1038/ngeo2518
Bai, L., Klemperer, S. L., Mori, J., Karplus, M. S., Ding, L., Liu, H., Li, G., Song, B., & Dhakal, S. (2019). Lateral variation of the Main Himalayan Thrust controls the rupture length of the 2015 Gorkha earthquake in Nepal. *Science Adventure, 5*, eaav0723.
Beaumont, C., Jamieson, M. H., Nguyen, M. H., & Lee, B. (2001). Himalayan tectonics explained by extrusion of a low-viscosity crustal channel coupled to focused surface denudation. *Nature, 414*, 738–742. https://doi.org/10.1038/414738a
Bilham, R. (2019). Himalayan earthquakes: A review of historical seismicity and early 21st century slip potential. *Geological Society, London, Special Publications, 483*, 423–482. https://doi.org/10.1144/SP483.16
Bilham, R., Gaur, V. K., & Molnar, P. (2001). Himalayan seismic hazard. *Science, 293*, 1442–1444.
Bollinger, L., Lyon-Caen, H., & RESIF. (2011). *HiK-NET temporary experiment (RESIF-SISMOB)*. RESIF - Réseau Sismologique et géodésique Français. https://doi.org/10.15778/resif.zo2014
Bollinger, L., Perrier, F., Avouac, J. P., Sapkota, S., Gautam, U., & Tiwari, D. R. (2007). Seasonal modulation of seismicity in the Himalaya of Nepal. *Geophysical Research Letters, 34*.https://doi.org/10.1029/2006GL029192
Burgess, W. P., Yin, A., Dubey, C. S., Shen, Z. K., & Kelty, T. K. (2012). Holocene shortening across the Main Frontal Thrust zone in the eastern Himalaya. *Earth and Planetary Science Letters, 357*, 152–167.
Caldwell, W. B., Klemperer, S. L., Lawrence, J. F., Rai, S. S., & Ashish. (2013). Characterizing the Main Himalayan Thrust in the Garhwal Himalaya, India with receiver function CCP stacking. *Earth and Planetary Science Letters, 367*, 15–27. https://doi.org/10.1016/j.epsl.2013.02.009.
Cattin, R., Martelet, G., Henry, P., Avouac, J., Diament, M., & Shakya, T. (2001). Gravity anomalies, crustal structure and thermo-mechanical support of the Himalaya of Central Nepal. *Geophysical Journal International, 147*, 381–392.
Celerier, J., Harrison, T. M., Webb, A. A. G., & Yin, A. (2009). The Kumaun and Garwhal Lesser Himalaya, India: Part 1. Structure and stratigraphy. *Geological Society of America Bulletin, 121*, 1262–1280. https://doi.org/10.1130/B26344.1
Chamoli, A., Pandey, A., Dimri, V., & Banerjee, P. (2011). Crustal configuration of the Northwest Himalaya based on modeling of gravity data. *Pure and Applied Geophysics, 168*, 827–844. https://doi.org/10.1007/s00024-010-0149-2
Chamoli, A., Rana, S. K., Dwivedi, D., & Pandey, A. K. (2023). Crustal structure and fault geometries of the Garhwal Himalaya, India: Insight from new high-resolution gravity data modeling and PSO inversion. *Tectonophysics, 859*, Article 229904. https://doi.org/10.1016/j.tecto.2023.229904
Christensen, N. I., & Mooney, W. D. (1995). Seismic velocity structure and composition of the continental crust: A global view. *Journal of Geophysical Research: Solid Earth, 100*(B6), 9761–9788.
Dal Zilio, L., Hetenyi, G., Hubbard, J., & Bollinger, L. (2021). Building the Himalaya from tectonic to earthquake scales. *Nature Reviews Earth & Environment, 2*. https://doi.org/10.1038/s43017-021-00143-1
Dal Zilio, L., Jolivet, R., & van Dinther, Y. (2020). Segmentation of the main Himalayan thrust illuminated by Bayesian inference of interseismic coupling. *Geophysical Research Letters, 47*. https://doi.org/10.1029/2019GL086424

Das, A., Hazarika, D., Neog, P., Kumar, N., & Yadav, D. K. (2025). Delineation of crustal structure and composition in the Northwest Himalaya and adjoining Indo-Gangetic Plain. *Tectonophysics, 904*, Article 230717. https://doi.org/10.1016/j.tecto.2025.230717

Dewey, J. F., & Bird, J. M. (1970). Mountain belts and the new global tectonics. *Journal of Geophysical Research, 75*(14), 2625–2647.

Dezes, P. J., Vannay, J. C., Steck, A., Bussy, F., & Cosca, M. (1999). Synorogenic extension: Quantitative constraints on the age and displacement of the Zanskar shear zone (northwest Himalaya). *Geological Society of America Bulletin, 111*, 364–374. https://doi.org/10.1130/0016-7606(1999)111%3c0364:SEQCOT%3e2.3.CO;2

Diehl, T., Singer, J., Hetenyi, G., Grujic, D., Clinton, J., Giardini, D., & Kissling, E. (2017). Seismotectonics of Bhutan: Evidence for segmentation of the Eastern Himalayas and link to foreland deformation. *Earth and Planetary Science Letters, 471*, 54–64. https://doi.org/10.1016/j.epsl.2017.04.038

Ding, L., Kapp, P., Cai, F., Garzione, C., Xiong, Z., Wang, H., & Wang, C. (2022). Timing and mechanisms of Tibetan Plateau uplift. *Nature Reviews Earth & Environment, 3*, 1–16. https://doi.org/10.1038/s43017-022-00318-4

Dixit, R., Srivastava, D. C., Deshmukh, G. G., & Jain, A. K. (2024). Structural anatomy of the Munsiari and Vaikrita thrust zones, Garhwal Himalayas, India. *Tectonophysics, 880*, 230–334. https://doi.org/10.1016/j.tecto.2024.230334

Drukpa, D., Velasco, A. A., & Doser, D. I. (2006). Seismicity in the Kingdom of Bhutan (1937–2003): Evidence for crustal transcurrent deformation. *Journal of Geophysical Research: Solid Earth, 111*. https://doi.org/10.1029/2004JB003087

Elliott, J. R., Jolivet, R., Gonzalez, P. J., Avouac, J. P., Hollingsworth, J., Searle, M. P., & Stevens, V. (2016). Himalayan megathrust geometry and relation to topography revealed by the Gorkha earthquake. *Nature Geoscience, 9*, 174–180. https://doi.org/10.1038/ngeo2623

England, P., & McKenzie, D. (1982). A thin viscous sheet model for continental deformation. *Geophysical Journal International, 70*, 295–321. https://doi.org/10.1111/j.1365-246X.1982.tb04969.x

Fan, W., & Shearer, P. M. (2015). Detailed rupture imaging of the 25 April 2015 Nepal earthquake using teleseismic P waves. *Geophysical Research Letters, 42*, 5744–5752. https://doi.org/10.1002/2015GL064587

Gahalaut, V., Rajput, S., & Kundu, B. (2011). Low seismicity in the Bhutan Himalaya and the stress shadow of the 1897 Shillong Plateau earthquake. *Physics of the Earth and Planetary Interiors, 186*, 97–102. https://doi.org/10.1016/j.pepi.2011.04.009

Gansser, A. (1974). Himalaya. *Geological Society, London, Special Publications, 4*(1), 267–278.

Grujic, D., Hollister, L. S., & Parrish, R. R. (2002). Himalayan metamorphic sequence as an orogenic channel: Insight from Bhutan. *Earth and Planetary Science Letters, 198*, 177–191. https://doi.org/10.1016/S0012-821X(02)00482-X

Gupta, S., Perugu, M., Kanna, N., Krishnavajhhala, S., & Paul, A. (2021). 3-D seismic velocity structure of the Kumaun-Garhwal (Central) Himalaya: Insight into the Main Himalayan Thrust and earthquake occurrence. *Geophysical Journal International, 229*. https://doi.org/10.1093/gji/ggab449

Hazarika, D., Hajra, S., Kundu, A., Bankhwal, M., Kumar, N., & Pant, C. C. (2021). Imaging the Moho and Main Himalayan Thrust beneath the Kumaon Himalaya: Constraints from receiver function analysis. *Geophysical Journal International, 224*, 858–870. https://doi.org/10.1093/gji/ggaa478

Hazarika, D., Wadhawan, M., Paul, A., Kumar, N., & Borah, K. (2017). Geometry of the Main Himalayan Thrust and Moho beneath Satluj valley, northwest Himalaya: Constraints from receiver function analysis. *Journal of Geophysical Research: Solid Earth, 122*, 2929–2945. https://doi.org/10.1002/2016JB013783

Hazarika, P., Kumar, M. R., Srijayanthi, G., Raju, P. S., Rao, N. P., & Srinagesh, D. (2010). Transverse tectonics in the Sikkim Himalaya: Evidence from seismicity and focal-mechanism data.

Bulletin of the Seismological Society of America, 100, 1816–1822. https://doi.org/10.1785/0120090339

He, P., Lei, J., Yuan, X., Xu, X., Xu, Q., Liu, Z., Mi, Q., & Zhou, L. (2018). Lateral Moho variations and the geometry of the Main Himalayan Thrust beneath the Nepal Himalayan orogen revealed by teleseismic receiver functions. *Geophysical Journal International, 214*, 1004–1017. https://doi.org/10.1093/gji/ggy192

Hetényi, G., Cattin, R., Berthet, T., Le Moigne, N., Chophel, J., Lechmann, S., Hammer, P., Drukpa, D., Sapkota, S.N., Gautier, S., & Thinley, K. (2016). Segmentation of the Himalayas as revealed by arc-parallel gravity anomalies. *Scientific Reports, 6*(1), 33866.

Hetenyi, G., Cattin, R., Brunet, F., Bollinger, L., Vergne, J., Nabelek, J. L., & Diament, M. (2007). Density distribution of the India plate beneath the Tibetan plateau: Geophysical and petrological constraints on the kinetics of lower-crustal eclogitization. *Earth and Planetary Science Letters, 264*, 226–244. https://doi.org/10.1016/j.epsl.2007.09.036

Hodges, K. V. (2000). Tectonics of the Himalaya and southern Tibet from two perspectives. *Geological Society of America Bulletin, 112*(3), 324–350.

Hoste-Colomer, R., Bollinger, L., Lyon-Caen, H., Adhikari, L. B., Baillard, C., Benoit, A., Bhattarai, M., Gupta, R. M., Jacques, E., Kandel, T., & Timsina, C. (2018). Lateral variations of the midcrustal seismicity in western Nepal: Seismotectonic implications. *Earth and Planetary Science Letters, 504*, 115–125.

Hubbard, J., Almeida, R., Foster, A., Sapkota, S. N., Burgi, P., & Tapponnier, P. (2016). Structural segmentation controlled the 2015 Mw 7.8 Gorkha earthquake rupture in Nepal. *Geology, 44*, 639–642. https://doi.org/10.1130/G38077.1

Kaila, K. L. (1981). Structure and seismotectonics of the Himalaya-Pamir Hindukush Region and the Indian Plate Boundary. *American Geophysical Union.* https://doi.org/10.1029/GD003p0272

Kanaujia, J., Kumar, M. R., Bandari, N., & Vijayaraghavan, R. (2025). Tomographic evidence for segmentation of the Main Himalayan Thrust in the Uttarakhand Himalaya. *Tectonophysics, 900*, Article 230655.

Kanaujia, J., Ravi Kumar, M., Sunilkumar, T., Naresh, B., Devi, A., Vijayaraghavan, R., Suresh, G., Saha, S., Srinivas, D., & Solomon Raju, P. (2024). Nature of the Himalayan Seismicity Belt in the Garhwal-Kumaun Segment and its tectonic implications. *Seismological Research Letters, 95*, 2386–2398.

Kanaujia, J., Ravi Kumar, M., Vijayaraghavan, R., & Solomon Raju, P. (2023). An optimum 1D velocity model for the Garhwal-Kumaun Himalaya using Monte Carlo style inversion. *Seismological Research Letters, 94*, 2244–2256.

Kanna, N., & Gupta, S. (2020). Crustal seismic structure beneath the Garhwal Himalaya using regional and teleseismic waveform modelling. *Geophysical Journal International, 222*, 2040–2052. https://doi.org/10.1093/gji/ggaa282

Kanna, N., & Gupta, S. (2021). Seismic crustal shear velocity structure across NW Himalaya and Ladakh-Karakoram using receiver function modelling: Evidence of the Main Himalayan Thrust. *Physics of the Earth and Planetary Interiors, 311*, Article 106642. https://doi.org/10.1016/j.pepi.2020.106642

Kelly, S., Beaumont, C., & Butler, J. P. (2020). Inherited terrane properties explain enigmatic post-collisional Himalayan-Tibetan evolution. *Geology, 48*, 8–14.

Kind, R., Yuan, X., Saul, J., Nelson, D., Sobolev, S. V., Mechie, J., Zhao, W., Kosarev, G., Ni, J., Achauer, U., & Jiang, M. (2002). Seismic images of crust and upper mantle beneath Tibet: Evidence for Eurasian plate subduction. *Science, 298*(5596), 1219–1221.

Kundu, A., Hazarika, D., Yadav, D. K., & Ghosh, P. (2022). Crustal thickness and Poisson's ratio variations in the Siang Window and adjoining areas of the Eastern Himalayan Syntaxis. *Journal of Asian Earth Sciences, 231*, Article 105225. https://doi.org/10.1016/j.jseaes.2022.105225

Li, A., & Mashele, B. (2009). Crustal structure in the Pakistan Himalaya from teleseismic receiver functions. *Geochemistry, Geophysics, Geosystems*, 10. https://doi.org/10.1029/2009GC002700

Mahesh, P., Srijayanthi, G., & Kumar, M. R. (2023). Seismotectonics and 3-D seismic velocity structure of the Arunachal Himalaya. *Tectonophysics, 869*, Article 230098. https://doi.org/10.1016/j.tecto.2023.230098

Mandal, P., Dakuri, S., Suresh, G., & Srinagesh, D. (2021). Modelling of crustal composition and Moho depths and their Implications toward seismogenesis in the Kumaon–Garhwal Himalaya. *Science and Reports, 11*. https://doi.org/10.1038/s41598-02193469-1

Mangalampalli, R. K., Bommoju, P. R., Perugu, M., & Venkatesh, V. (2022). Scattered wave imaging of the Main Himalayan Thrust and mid-crustal ramp beneath the Arunachal Himalaya and its relation to seismicity. *Journal of Asian Earth Sciences, 236*, Article 105335. https://doi.org/10.1016/j.jseaes.2022.105335

Manglik, A., Pavan Kumar, G., & Thiagarajan, S. (2013). Transverse tectonics in the Sikkim Himalaya: A magnetotelluric study. *Tectonophysics, 589*, 142–150. https://doi.org/10.1016/j.tecto.2012.12.040

Michel, S., Jolivet, R., Rollins, C., Jara, J., & Dal Zilio, L. (2021). Seismogenic potential of the main Himalayan thrust constrained by coupling segmentation and earthquake scaling. *Geophysical Research Letters, 48*. https://doi.org/10.1029/2021GL093106

Mir, R. R., Parvez, I. A., Gaur, V. K., Ashish, Chandra, R., & Romshoo, S. A. (2017). Crustal structure beneath the Kashmir Basin adjoining the Western Himalayan Syntaxis. *Bulletin of the Seismological Society of America, 107*, 2443–2458. https://doi.org/10.1785/0120150334

Mitra, S., Sharma, S., Wanchoo, S. K., & Priestley, K. (2023). 3D crustal structure of the Jammu and Kashmir Himalaya: Signatures of mid-crustal ramp and lesser Himalayan duplex. *Journal of Geophysical Research: Solid Earth, 128*. https://doi.org/10.1029/2023JB027350

Molnar, P. (1988). A review of geophysical constraints on the deep structure of the Tibetan Plateau, the Himalaya and the Karakoram, and their tectonic implications. *Philosophical Transactions of the Royal Society a: Mathematical, Physical and Engineering Sciences, 326*, 33–88. https://doi.org/10.1098/rsta.1988.0080

Molnar, P., Tapponnier, P. (1975). Cenozoic tectonics of Asia: Effects of a continental collision: Features of recent continental tectonics in Asia can be interpreted as results of the India-Eurasia collision. *Science, 189*(4201), 419–426. https://doi.org/10.1126/science.189.4201.419

Monsalve, G., Sheehan, A., Rowe, C., & Rajaure, S. (2008). Seismic structure of the crust and the upper mantle beneath the Himalayas: Evidence for eclogitization of lower crustal rocks in the Indian plate. *Journal of Geophysical Research, 113*.https://doi.org/10.1029/2007JB005424

Monsalve, G., Sheehan, A., Schulte-Pelkum, V., Rajaure, S., Pandey, M. R., & Wu, F. (2006). Seismicity and one-dimensional velocity structure of the Himalayan collision zone: Earthquakes in the crust and upper mantle. *Journal of Geophysical Research, 111*. https://doi.org/10.1029/2005JB004062

Nábělek, J. (2002). Collaborative research: Lithospheric scale dynamics of active mountain building along the Himalayan-Tibetan collision zone. https://doi.org/10.7914/SN/XF_2002

Nábělek, J., Hetényi, G., Vergne, J., Sapkota, S., Kafle, B., Jiang, M., Su, H., Chen, J., Huang, B. S., & Team, T. H. C. (2009). Underplating in the Himalaya-Tibet collision zone revealed by the Hi-CLIMB experiment. *Science, 325*(5946), 1371–1374. https://doi.org/10.1126/science.1167719

Nelson, K., Zhao, W., Brown, L., Kuo, J., Che, J., Liu, X., Klemperer, S., Makovsky, Y., Meissner, R., Mechie, J., Kind, R., Wenzel, F., Ni, J., Nabelek, J., Leshou, C., Tan, H., Wei, W., Jones, A., Booker, J., & Edwards, M. (1997). Partially molten middle crust beneath Southern Tibet: Synthesis of Project INDEPTH results. *Science, 274*, 1684–1688. https://doi.org/10.1126/science.274.5293.1684

Pandey, M. R., Tandukar, R. P., Avouac, J. P., Lave, J., & Massot, J. P. (1995). Interseismic strain accumulation on the Himalayan crustal ramp (Nepal). *Geophysical Research Letters, 22*(7), 751–754.

Pandey, M., Tandukar, R., Avouac, J., Vergne, J., & Heritier, T. (1999). Seismotectonics of the Nepal Himalaya from a local seismic network. *Journal of Asian Earth Sciences, 17*, 703–712. https://doi.org/10.1016/S1367-9120(99)00034-6

Parija, M., Kumar, S., & Arjun, V. H. (2023). Delineation of partial melts and crustal heterogeneities within the crust beneath Kumaon Himalaya, India from Lg wave attenuation. *Scientific Reports, 13*. https://doi.org/10.1038/s41598-023-36269-z

Paul, H., & Mitra, S. (2017). Three-dimensional crustal architecture beneath the Sikkim Himalaya and its relationship to active deformation. *Journal of Geophysical Research: Solid Earth, 122*, 7860–7878. https://doi.org/10.1002/2017JB014506

Pearson, O. N., & DeCelles, P. G. (2005). Structural geology and regional tectonic significance of the Ramgarh thrust, Himalayan fold-thrust belt of Nepal. *Tectonics, 24*. https://doi.org/10.1029/2003TC001617

Powers, P. M., Lillie, R. J., & Yeats, R. S. (1998). Structure and shortening of the Kangra and Dehra Dun reentrants, Sub-Himalaya, India. *Geological Society of America Bulletin, 110*, 1010–1027. https://doi.org/10.1130/0016-7606(1998)110%3c1010:SASOTK%3e2.3.CO;2

Pusok, A. E., & Stegman, D. R. (2020). The convergence history of India-Eurasia records multiple subduction dynamics processes. *Science Advances, 6*.https://doi.org/10.1126/sciadv.aaz8681

Rai, S. S., Ashish, Padhi, A., & Sarma, P. R. (2009). High crustal seismic attenuation in Ladakh–Karakoram. *Bulletin of the Seismological Society of America, 99*, 407–415. https://doi.org/10.1785/0120070261

Rai, S. S., Priestley, K., Gaur, V. K., Mitra, S., Singh, M. P., & Searle, M. (2006). Configuration of the Indian Moho beneath the NW Himalaya and Ladakh. *Geophysical Research Letters, 33*. https://doi.org/10.1029/2006GL026076

Sarkar, S. (2025). *Crustal and upper mantle structure beneath Eastern Himalaya (Arunachal and Bhutan) using regional seismic wave attenuation tomography*. Ph.D. thesis, IIT Kharagpur.

Sarkar, S., Singh, C., Dutta, A., Bose, S., Kumar, M. R., & Singh, A. (2025). Regional Lg-wave attenuation tomography of Bhutan Himalaya: Implications for crustal deformation. *Soil Dynamics and Earthquake Engineering, 197*, Article 109495. https://doi.org/10.1016/j.soildyn.2025.109495

Sathiakumar, S., & Barbot, S. (2021). The stop-start control of seismicity by fault bends along the Main Himalayan Thrust. *Communications Earth & Environment, 2*, 87. https://doi.org/10.1038/s43247-021-00153-3

Schulte-Pelkum, V., Monsalve, G., Sheehan, A., Pandey, M. R., Sapkota, S., Bilham, R., & Wu, F. (2005). Imaging the Indian subcontinent beneath the Himalaya. *Nature, 435*, 1222–1225. https://doi.org/10.1038/nature03678

Searle, M. P., Khan, M. A., Fraser, J. E., Gough, S. J., & Jan, M. Q. (1999). The tectonic evolution of the Kohistan-Karakoram collision belt along the Karakoram Highway transect, north Pakistan. *Tectonics, 18*, 929–949. https://doi.org/10.1029/1999TC900042

Sharma, S., Mitra, S., Sharma, S., Priestley, K., Wanchoo, S. K., Powali, D., & Ali, L. (2020). A Report on Broadband Seismological Experiment in the Jammu and Kashmir Himalaya (JAKSNET). *Seismological Research Letters, 91*, 1915–1926. https://doi.org/10.1785/0220190389

Sheehan, A., Francis, W., & Roger, B. (2001). Himalayan seismotectonics, Nepal and Tibet. https://doi.org/10.7914/SN/YL_2001

Sheehan, A. F., de la Torre, T. L., Monsalve, G., Abers, G. A., & Hacker, B. R. (2014). Physical state of Himalayan crust and uppermost mantle: Constraints from seismic attenuation and velocity tomography. *Journal of Geophysical Research, 119*, 567–580. https://doi.org/10.1002/2013JB010601

Singer, J. (2017). *Tectonics of the Bhutan Himalaya: New insights from seismic tomographic images of the lithosphere*. Ph.D. thesis, ETH Zurich.

Singer, J., Kissling, E., Diehl, T., & Hetenyi, G. (2017b). The underthrusting Indian crust and its role in collision dynamics of the Eastern Himalaya in Bhutan: Insights from receiver function imaging. *Journal of Geophysical Research, 122*, 1152–1178. https://doi.org/10.1002/2016JB013337

Singer, J., Obermann, A., Kissling, E., Fang, H., Hetenyi, G., & Grujic, D. (2017a). Along-strike variations in the Himalayan orogenic wedge structure in Bhutan from ambient seismic noise tomography. *Geochemistry, Geophysics, Geosystems, 18*, 1483–1498. https://doi.org/10.1002/2016GC006742

Singh, A., Kumar, G., Singh, C., Kumar, M. R., Uthaman, M., Vadlamani, R., Dutta, A., Saikia, D., & Dubey, A. K. (2024). Evidence for an imbricated lower crust and its role in accommodating crustal shortening in Sikkim Himalaya. *Tectonics, 43*. https://doi.org/10.1029/2023TC008044

Singh, A., Kumar, M. R., & Raju, P. S. (2010). Seismic structure of the underthrusting Indian crust in Sikkim Himalaya. *Tectonics, 29*. https://doi.org/10.1029/2010TC002722

Singh, A., Saikia, D., & Kumar, M. R. (2021). Seismic imaging of the crust beneath Arunachal Himalaya. *Journal of Geophysical Research: Solid Earth, 126*. https://doi.org/10.1029/2020JB020616

Singh, C., Mondal, P., Singh, S., Mohanty, D. D., Jaiswal, N., & Kumar, M. R. (2015). Lg attenuation tomographic models of Himalaya and southern Tibet. *Tectonophysics, 664*, 176–181. https://doi.org/10.1016/j.tecto.2015.09.009

Singh, S., Singh, C., Biswas, R., Mukhopadhyay, S., & Sahu, H. (2016). Attenuation characteristics in eastern Himalaya and southern Tibetan Plateau: An understanding of the physical state of the medium. *Physics of the Earth and Planetary Interiors, 257*, 48–56.

Stephenson, B., Waters, D., & Searle, M. (2000). Inverted metamorphism and the Main Central Thrust: Field relations and thermobarometric constraints from the Kishtwar Window, NW Indian Himalaya. *Journal of Metamorphic Geology, 18*, 571–590. https://doi.org/10.1046/j.1525-1314.2000.00277.x

Steve, R., & Vadim, L. (2007). Collaborative Research: Imaging the Upper Mantle Beneath the Western Tibetan Plateau [Data set]. International Federation of Digital Seismograph Networks. https://doi.org/10.7914/SN/Y2_2007

Subedi, S., Hetenyi, G., Vergne, J., Bollinger, L., Lyon-Caen, H., Farra, V., Adhikari, L. B., & Gupta, R. M. (2018). Imaging the Moho and the main Himalayan thrust in Western Nepal with receiver functions. *Geophysical Research Letters, 45*, 13222–13230. https://doi.org/10.1029/2018GL080911

Tiwari, V. M., Rajasekhar, R. P., & Mishra, D. C. (2009). Gravity anomaly, lithospheric structure and seismicity of Western Himalayan Syntaxis. *Journal of Seismology, 13*, 363–370. https://doi.org/10.1007/s10950-008-9102-6

Tiwari, V., Vyghreswara Rao, M., Mishra, D., & Singh, B. (2006). Crustal structure across Sikkim, NE Himalaya from new gravity and magnetic data. *Earth and Planetary Science Letters, 247*, 61–69. https://doi.org/10.1016/j.epsl.2006.03.037

Uthaman, M., Singh, C., Singh, A., Hetenyi, G., Dutta, A., Kumar, G., & Dubey, A. K. (2024). Complex multi-fault dynamics in Sikkim Himalaya: New insights from local earthquake analysis. *Geochemistry, Geophysics, Geosystems, 25*. https://doi.org/10.1029/2023GC011363

Valdiya, K. (1980). The two intracrustal boundary thrusts of the Himalaya. *Tectonophysics, 66*, 323–348. https://doi.org/10.1016/0040-1951(80)90248-6

van Hinsbergen, D. J. J., Lippert, P. C., Dupont-Nivet, G., McQuarrie, N., Doubrovine, P. V., Spakman, W., & Torsvik, T. H. (2012). Greater India basin hypothesis and a two-stage cenozoic collision between India and Asia. *Proceedings of the National Academy of Sciences, 109*, 7659–7664. https://doi.org/10.1073/pnas.1117262109

Verma, S. K., Kumar, N., Hazarika, D., Paul, A., Yadav, D. K., & Pal, S. K. (2023). Shear wave crustal velocity structure in the Garhwal-Kumaon Himalaya based on noise cross-correlation of Rayleigh wave. *Tectonophysics, 866*, Article 230047. https://doi.org/10.1016/j.tecto.2023.230047

Vinnik, L., Singh, A., Kiselev, S., & Kumar, M. R. (2007). Upper mantle beneath foothills of the western Himalaya: Subducted lithospheric slab or a keel of the Indian shield? *Geophysical Journal International, 171*, 1162–1171. https://doi.org/10.1111/j.1365246X.2007.03577.x

Wang, X., Wei, S., & Wu, W. (2017). Double-ramp on the Main Himalayan Thrust revealed by broadband waveform modeling of the 2015 Gorkha earthquake sequence. *Earth and Planetary Science Letters, 473*, 83–93. https://doi.org/10.1016/j.epsl.2017.05.032

Webb, A. A. G., Guo, H., Clift, P. D., Husson, L., Muller, T., Costantino, D., Yin, A., Xu, Z., Cao, H., & Wang, Q. (2017). The Himalaya in 3D: Slab dynamics controlled mountain building and monsoon intensification. *Lithosphere, 9*, 637–651.

Wessel, P., Esteban, F., & Delaviel-Anger, G. (2024). The generic mapping tools and animations for the masses. *Geochemistry, Geophysics, Geosystems, 25.* https://doi.org/10.1029/2024GC011545

Yagi, Y., & Okuwaki, R. (2015). Integrated seismic source model of the 2015 Gorkha, Nepal, earthquake. *Geophysical Research Letters, 42*, 6229–6235. https://doi.org/10.1002/2015GL064995

Yin, A. (2006). Cenozoic tectonic evolution of the Himalayan orogen as constrained by along-strike variation of structural geometry, exhumation history, and foreland sedimentation. *Earth-Science Reviews, 76*, 1–131. https://doi.org/10.1016/j.earscirev.2005.05.004

Yin, A., & Harrison, T. M. (2000). Geologic evolution of the Himalayan-Tibetan orogen. *Annual Review of Earth and Planetary Sciences, 28*, 211–280. https://doi.org/10.1146/annurev.earth.28.1.211

Zhao, W., Nelson, K. D., Che, J., Quo, J., Lu, D., Wu, C., & Liu, X. (1993). Deep seismic reflection evidence for continental underthrusting beneath southern Tibet. *Nature, 366*(6455), 557–559.

Chapter 3
A Synthesis of the Along-Arc Variation in the Crustal Geoelectric Structure of the Himalayan Collision Belt

A. Manglik and G. Pavankumar

Abstract This chapter presents a brief account of the Magnetotelluric (MT) studies carried out along the Himalayan arc and its surrounding Tibet and Ganga Basin regions, with the focus on the crustal resistivity structure. A synthesis of the published results affirms the variations in the crustal geoelectric structure along the arc. In the trans-Himalayan region, the Karakoram and the Ladakh batholiths appear as high resistivity blocks underlain by conductive channels linked to the partial melt zones in south-western Tibet. In the Himachal Pradesh and Uttarakhand sectors, the Main Himalayan Thrust (MHT) ramp is associated with high conductivity, possibly due to fluid-filled weak zones, and is seismically active. The same pattern of the MHT ramp conductor continues till central Nepal. This flat-ramp-flat model, delineated by the MT results, is consistent with the seismological models. However, the scenario changes significantly in the Sikkim sector, where instead of a clear ramp structure, the lithosphere appears to be consisting of two compositionally different blocks separated by the Main Central Thrust Zone. The crustal structure of the southern Tibet region along the Himalayan arc is dominated by a widespread conductive zone at > 20 km depth linked to the presence of partial melts. In the foothill region of the western Nepal Himalaya, the lithospheric structure of the Indian plate is distinct, marked by the presence of a super-thick sedimentary basin. There are many unexplored sectors along the Himalayan arc which need to be covered by MT exploration through concerted efforts, including international cooperation.

Keywords Himalaya · Crustal structure · Magnetotellurics · Electrical resistivity

A. Manglik (✉) · G. Pavankumar
CSIR-National Geophysical Research Institute, Hyderabad, Telangana, India
e-mail: ajay.ngri@csir.res.in

V. M. Tiwari (ed.), *Himalayan Natural Hazards*, Indian National Science Academy,
https://doi.org/10.1007/978-981-92-3784-5_3

3.1 Introduction

The Himalayan Collision Belt (HCB), a product of the ongoing convergence between the Indian and Eurasian plates, is a part of the much larger Himalaya-Tibet system that has evolved during the past 50 Ma leading to a prominent tectonically and seismically active continental belt. Various major thrust fault systems of the HCB, e.g., the Main Frontal Thrust (MFT), the Main Boundary Thrust (MBT), the Main Central Thrust (MCT), and the Main Himalayan Thrust (MHT), collectively accommodate the ongoing convergence and play a crucial role in shaping the tectonic and seismic character of the region. There are several other small-scale and local faults as well as transverse structures that modulate the stress state and seismicity along the arc. It is widely regarded that the MHT is the master décollement surface, slip on which leads to major-to-strong earthquakes, and it has potential for great earthquakes. In this widely accepted model for the genesis of the Himalayan seismicity, a ramp in the MHT at 10–25 km depth coinciding with the sharp elevation rise is the locale for nucleation of these devastating earthquakes from which rupture propagates southward along the MHT (Ni & Barazangi, 1984). This model explains the shallow focal depths of the Himalayan earthquakes confined to the wedge and its base. However, seismicity in the eastern Nepal and Sikkim sector of the arc is characteristically different as this region experiences strike-slip dominated earthquakes and deep crustal seismicity well below the Himalayan wedge. Such observations suggest along-arc variations in the wedge and crustal structure and their coupling, composition, thermomechanical state, and the nature of stress accumulation with resulting impact on the seismic hazard and tectonic deformation processes. Therefore, it is imperative, as a first step, to map the lithospheric structure of the arc at a reasonable spatial resolution by integrated multi-parametric geophysical approach.

Over the past several decades, many seismological and Magnetotelluric (MT) experiments have been carried out in the Himalaya-Tibet region to address the above issues. In this chapter, we summarize the MT experiments carried out within the Himalayan arc and its surrounding regions, encompassing southern Tibet and the northern Ganga Basin bordering the Himalaya, and highlight the gap areas devoid of such studies. MT method is a powerful tool based on natural source Electromagnetic (EM) induction for investigating the crustal and upper mantle structure in terms of subsurface electrical conductivity distribution. In this method, the Maxwell's equations of EM induction are solved in the frequency domain under the assumption that inside the earth permittivity of the medium (ε) is very small compared to its electrical conductivity (σ). This assumption reduces the system of equations to a diffusion process in which the depth of investigation, the skin depth (z_s), follows an inverse relationship with the frequency (f) of the signal and electrical conductivity (σ) of the medium ($z_s = \chi/\sqrt{f\sigma}$, where χ is a constant). MT data are acquired in the form of time series of two horizontal electric (e_x, e_y) and three magnetic (horizontal h_x, h_y and vertical h_z) field components, which are processed in frequency domain to estimate MT transfer functions (impedance tensors ($\boldsymbol{Z}$) from horizontal electric and magnetic field components and magnetic transfer functions ($\boldsymbol{T}$) from vertical and horizontal magnetic fields). These are then inverted using geophysical inversion techniques to construct geoelectric models of the subsurface structure.

Electrical conductivity being highly sensitive to the presence of fluids and partial melts in interconnected pore spaces, its subsurface distribution provides critical insights into the lithospheric composition and fault zone properties, including the distribution of fluids and partial melts on a broad scale, which significantly influences tectonic behavior. In recent years, MT studies have significantly advanced our understanding of Himalayan tectonics by revealing deep conductive zones aligned with thrust faults, highlighting fluid-driven deformation, and elucidating the role of crustal segmentation in seismicity. We present a brief synthesis of the MT results for different sectors of the Himalayan arc and its neighboring regions. We have not included the MT work for mapping of shallow subsurface for geothermal energy as it is beyond the scope of the present topic. In addition, there have been some other MT experiments in the NW Himalaya and NE India, which are yet to be published, and hence are excluded from this chapter. The region of the central and northern Tibet has also been explored in some detail by MT (e.g., see Unsworth, 2010), but we restrict here only to the southern Tibet bordering the Himalaya.

It may be noted that the geoelectric sections presented in this chapter (Fig. 3.1; Table 3.1) vary in terms of time series processing and inversion techniques used as well as their depth extent. The depth extent largely depends on the frequency band of the impedance tensors used in the analysis and the electrical conductivity of the subsurface and is governed by the parameter "*skin depth*" mentioned above. The depth ranges of the models chosen by the authors of these studies are mainly to highlight the most significant variations in the resistivity governed by the frequency band of measurements used by them.

3.2 North-Western Himalaya

3.2.1 Kohistan–Ladakh Sector

MT investigations in the Kohistan–Ladakh sector remain limited but have yielded important insights into the region's crustal structure. In the westernmost region of this sector, Shah et al. (2018) obtained a crustal structure of the Kohistan arc, the Main Karakoram Thrust (MKT) and the Karakoram batholith by 2-D inversion of MT data of 14 sites (profile 1, Fig. 3.1). The results reveal the Karakoram batholith as an about 10 km thick resistive block (R2) surrounded by conductors (C1, C2) beneath and north of it (Fig. 3.2a). There is another widespread conductive zone at 5 km depth in the Kohistan arc region. These upper crustal conductors have been inferred as the signatures of the ongoing magmatic activities or hydrothermal fluid. In the Ladakh region, Gokarn et al. (2002a) carried out MT investigations along the Bara-La-Cha-La–Panamik transect (profile 2, Fig. 3.1), but their results showed a zone of no penetration beneath the Tso Morari dome. Subsequently, Arora et al.

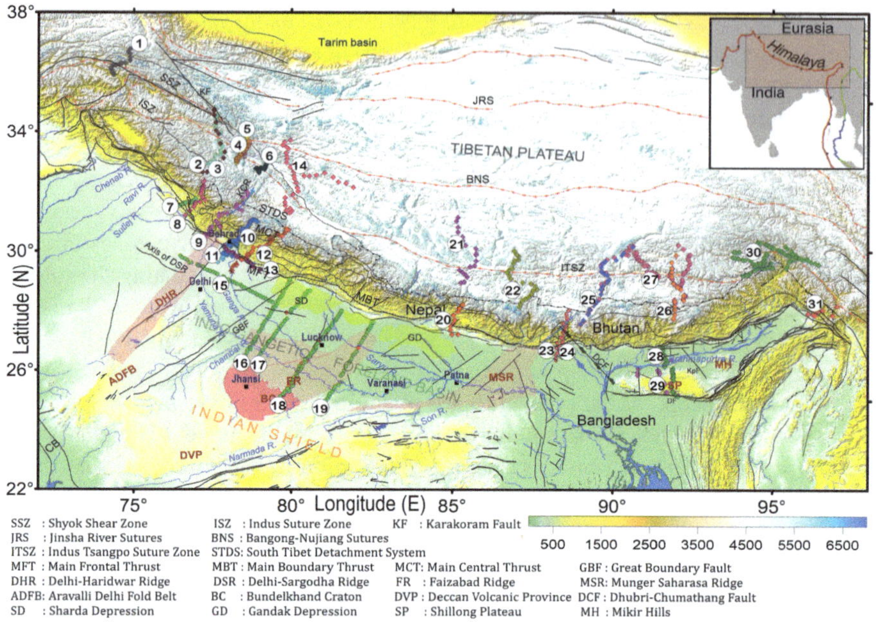

Fig. 3.1 MT profiles along the Himalayan collision belt and surrounding regions superimposed on the elevation map. Subsurface ridges and basin depressions concealed beneath the alluvial cover of the Ganga Basin are also shown (modified after Manglik et al., 2025). The geometry of the concealed extended Bundelkhand craton is taken from Manglik et al. (2024). References for the profiles are listed in Table 3.1

(2007) repeated this transect by Long-Period MT (LMT) sites (profile 3, Fig. 3.1). Their model obtained by 2-D inversion of the LMT data (Fig. 3.2b) revealed the Tso Morari Dome and the Ladakh batholith as highly resistive blocks of about 20 km thickness and the presence of a conductive channel beneath the Ladakh batholith. One branch of this channel rises to the surface at the ISZ while the other branch continues southward beneath the resistive blocks of the Tethyan Sedimentary Zone up to the HHC.

In recent years, two MT studies in Ladakh have provided further insights into the crustal structure of this region. Kumar and Rao (2024) mapped the crustal structure of the region by 2-D inversion of broadband MT data at 15 sites along the Rupshu–Chushul profile (profile 4, Fig. 3.1), which revealed a conductive mid-crust beneath the Ladakh batholith, a conductive channel south of the ISZ, and a crustal conductor beneath the Chushul area. A recent 3-D inversion study by us for this region (profile 5, Fig. 3.1) has brought out some new results on the crustal structure, including off-profile structures, of this region (Pavankumar et al., 2026). Pavankumar et al. (2024) delineated the crustal structure of eastern Ladakh by 2-D and 3-D inversion of MT data of 18 sites acquired along the Ukdungle–Fukche transect (profile 6 in Fig. 3.1). They identified a prominent upper crustal conductor beneath the TMC at ~ 4 km depth (Fig. 3.2c). This anomaly extends northward as a conductive layer beneath

Table 3.1 Related references for the profiles marked in Fig. 3.1

Nr.	References	Nr.	References	Nr.	References
1	Shah et al. (2018)	12	Rawat et al. (2014)	23	Patro and Harinarayana (2009)
2	Gokarn et al. (2002a)	13	Gupta et al. (1994)	24	Manglik et al. (2013); Pavankumar et al. (2014)
3	Arora et al. (2007)	14	Xie et al. (2017)	25	Chen et al. (1996) [100-line, INDEPTH]
4	Kumar and Rao (2024)	15	Manglik et al. (2022)	26	Spratt et al. (2005) [700-line, INDEPTH]
5	Pavankumar et al. (2026)	16	Manglik et al. (2025)	27	Xie et al. (2017)
6	Pavankumar et al. (2024)	17	Suresh et al. (2023)	28	Gokarn et al. (2008)
7	Gokarn et al. (2002b) [digitized from Dhamodharan et al. (2024)]	18	Manglik et al. (2015), Adilakshmi et al. (2021), and Pavankumar et al. (2023)	29	Saha et al. (2007)
8	Veeraswamy et al. (2010)	19	Demudu Babu et al. (2024)	30	Lin et al. (2017)
9	Dhamodharan et al. (2023)	20	Lemonnier et al. (1999)	31	Rawat and Luirei (2021)
10	Israil et al. (2008)	21	Unsworth et al. (2005) [800-line, INDEPTH]		
11	Devi et al. (2019)	22	Ye et al. (2007) [digitized from Unsworth (2010)]		

the Ladakh batholith of 8–12 km thickness. These features are interpreted as zones of partial melt, analogous to those proposed beneath southern Tibet. An interesting observation from this region is that almost all sites show anomalous phase exceeding 90° at periods larger than 10 s indicating the presence of a highly complex structure facilitating current channeling, similar to that reported by Pavankumar and Manglik (2012) from the Main Central Thrust Zone of Sikkim Himalaya. Xie et al. (2017) have provided a crustal model of the western Tibet region east of Ladakh based on 3-D inversion of MT data (profile 14, Fig. 3.1). This model (Fig. 3.2d) reveals the presence of a widespread conductive zone associated with partial melts in the crust north of the Yarlung–Zangbo Sutures (YSZ)/Indus Suture Zone (ISZ). This conductive zone rises to shallow depths south of the ISZ.

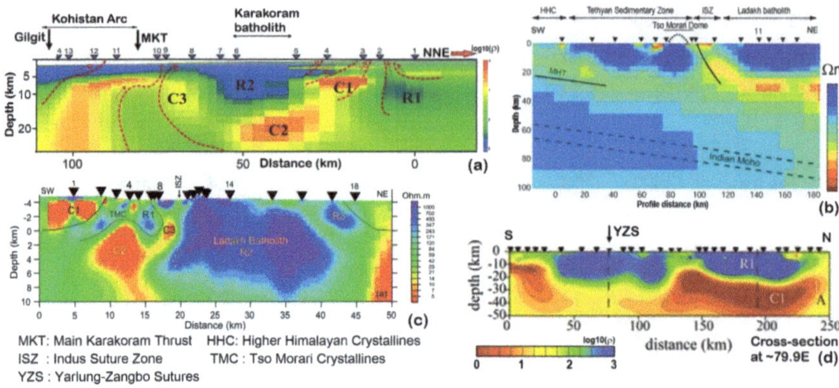

Fig. 3.2 Geoelectric crustal structure along some profiles in the Kohistan–Ladakh–western Tibet region. **a** Kohistan arc (profile 1) (Shah et al., 2018), **b** western Ladakh (profile 3) (Arora et al., 2007), **c** eastern Ladakh (profile 6) (Pavankumar et al., 2024), and **d** western Tibet (profile 14) (Xie et al., 2017)

3.2.2 Himachal Pradesh–Uttarakhand Sector

This sector, spanning about 580 km length of the Himalayan arc, has been covered by many MT profiles (profiles 7–13, Fig. 3.1). In addition to these published profiles, data acquisition has also been done along some other profiles in Himachal Pradesh and Uttarakhand, the results of which are yet to be published.

Gokarn et al. (2002b) covered the Una–Mandi profile (profile 7, Fig. 3.1) in the Kangra recess by 17 MT sites and delineated a conductive layer in the deep crust of the underthrusting Indian plate. Veeraswamy et al. (2010) covered a profile from Bilaspur to Rohtang pass (profile 8, Fig. 3.1) by 16 MT sites and performed 2-D inversion of the data. This profile is basically a southward extension of the Manali–Leh profiles (profiles 2 & 3, Fig. 3.1). They delineated a high conductivity zone in the Dharamshala region, a northward-dipping conductive zone between the MBT and the MC(V)T, and a high resistivity nearly vertical Proterozoic block in the Rampur Window of the Lesser Himalaya (Fig. 3.3a). Further east along the arc, Dhamodharan et al. (2023) present a 2-D geoelectric model (Fig. 3.3b) for the Nahan–Kourick Chango transect (profile 9, Fig. 3.1) along the Satluj River valley revealing the crustal structure. The study delineates the geometry of the MHT, highlighting a flat-ramp-flat structure and an intra-crustal low-resistivity layer in 5–15 km depth range, interpreted as fluid-saturated sediments. This conductive layer is suggested to play a significant role in weakening of the crust and facilitating thrust-type seismicity. Interestingly, this model also shows the presence of a high resistivity block beneath the MHT in the central sector of the profile. However, this region is SE of the Rampur window, for which Veeraswamy et al. (2010) delineated a resistive block.

For the Garhwal Himalaya sector, Israil et al. (2008) conducted a MT survey along the Roorkee–Gangotri profile (profile 10, Fig. 3.1) and acquired MT data at 44 sites.

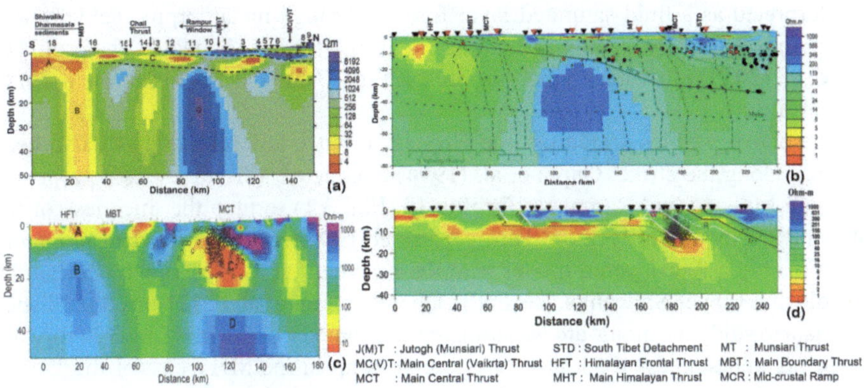

Fig. 3.3 Geoelectric crustal structure along some profiles in the Himachal Pradesh and Uttarakhand sectors of the NW Himalaya. **a** Kangra (profile 8) (Veeraswamy et al., 2010) and **b** Satluj valley (profile 9) (Dhamodharan et al., 2023) regions of Himachal Pradesh. **c** Uttarakhand (profile 10) (Israil et al., 2008) and **d** Chamoli (profile 12) (Rawat et al., 2014) regions of Uttarakhand

They performed 2-D inversion of a subset of this dataset consisting of 27 sites meeting the assumption of two-dimensionality. The results yielded a crustal model showing an ~ 6 km-thick conductive sedimentary layer, representing alluvium and the Shiwaliks overlying the resistive Indian plate, and a conductive (< 10 Ωm) mid-crustal anomaly beneath the MCT, interpreted as fluid-rich or partially molten material (Fig. 3.3c). This mid-crustal conductor correlates with the zone of micro-seismic activity at the MHT ramp. Subsequently, Miglani et al. (2014) remodeled this profile by adding another 10 sites to the dataset (profile 11, Fig. 3.1). Sokolova et al. (2016) performed combined analysis of MT and Magneto-variational (MV) datasets and validated the presence of the mid-crustal conductor, first proposed by Arora et al. (1982), by a large-scale MV array study in the Ganga Basin. Such transverse features have significant role in arc-parallel segmentation of the seismicity along the Himalaya (Arora & Mahashabde, 1987; Arora et al., 2012). Devi et al. (2019) used impedance tensor and tipper data and inverted these by a 3-D inversion algorithm. They interpreted a NNE trending resistive feature beneath the profile as the Delhi–Haridwar Ridge buried beneath the conductive sediments of the Indo-Gangetic Plain.

Further east within the Garhwal Himalaya sector, Rawat et al. (2014) covered another MT profile between Bijnor and Malari (profile 12, Fig. 3.1) by many sites and used 17 MT sites and 17 tipper sites for 2-D inversion. Their model (Fig. 3.3d) is contrastingly different from the other three models (Fig. 3.3a–c) in the sense that the entire crust, sparing some near-surface resistive blocks, is highly conductive in nature. Their study delineated an almost flat ~ 5 km thick Intra-Crustal High Conducting Layer (IC-HCL) at the base of the accretionary wedge till the ramp beneath the inner Lesser Himalaya. However, here also the seismicity is largely concentrated at the ramp, similar to the results of Israil et al. (2008), and the hypocenter of the 1999 Chamoli earthquake lies at this conductive ramp (red star in Fig. 3.3d). The IC-HCL is interpreted as a fluid-saturated layer formed through metamorphic dehydration, facilitating crustal weakening and promoting thrust-type earthquakes. The profile of Xie et al. (2017) in western Tibet (profile 14, Figs. 3.1 and 3.2d) lies to the north of this profile.

The above profiles are across the Himalaya, but a few profiles have also been covered along the strike. Gupta et al. (1994) conducted MT survey along a profile between Mohand and Ramnagar (profile 13, Fig. 3.1) to map the thickness of the Shiwaliks in the foothill region. They performed 1-D inversion of the data and delineated mainly a four-layered model with the basement at 5–8 km depth overlain by Middle and Lower Shiwaliks. Recently, Dhamodharan et al. (2024) used published MT data to study the along-arc variation in the geometry of the MHT by 3-D forward modeling and suggested an along-arc lateral ramp in the MHT east of the Delhi–Haridwar Ridge. Further south, Manglik et al. (2022) covered a profile between Kaithal and Bareilly (profile 15, Fig. 3.1) by 43 sites and mapped the subsurface disposition of the Aravalli–Delhi Fold Belt (ADFB) beneath the alluvial sediments of the Ganga Basin.

3.3 Central Himalaya

Even though the central sector of the Himalayan arc covering Nepal is of ~ 850 km length, only one MT study has been published from this sector. However, there are some published results from southern Tibet and northern Ganga Basin bordering this sector of the arc. The MT profile in Nepal (profile 20, Fig. 3.1) was covered by 12 sites, and 2-D inversion of this dataset yielded a prominent mid-crustal conductor (Fig. 3.4b) coinciding with the MHT ramp and zone of enhanced seismicity (Lemonnier et al., 1999). In southern Tibet, two MT profiles are available, one line 800 of the INDEPTH (INternational DEep Profiling of Tibet and Himalaya) experiment (profile 21, Fig. 3.1) north of the Nepal profile (Unsworth, 2010) and the other to its east (profile 22, Fig. 3.1) (Ye et al., 2007). Xie et al. (2017) performed 3-D inversion of a large dataset over southern Tibet. In Fig. 3.4c, we show the model section of Xie et al. (2017) approximately along profile 21, which reveals a widespread conductive zone in Tibet, attributed to partial melts, at > 20 km depth. This conductive zone thins and rises to shallow depths at the southern end of the profile. In recent years, a series of MT profiles (profiles 16–19, Fig. 3.1) have been carried out in the northern Ganga Basin south of western Nepal Himalaya (Adilakshmi et al., 2021; Demudu Babu et al., 2020, 2024; Manglik et al., 2015, 2025; Suresh et al., 2023). A salient feature of these studies is identification of a super-thick multi-stage sedimentary basin in the Sharda Depression and the presence of sub-crustal conductors beneath the northern Ganga Basin (Fig. 3.4a) (Manglik et al., 2025).

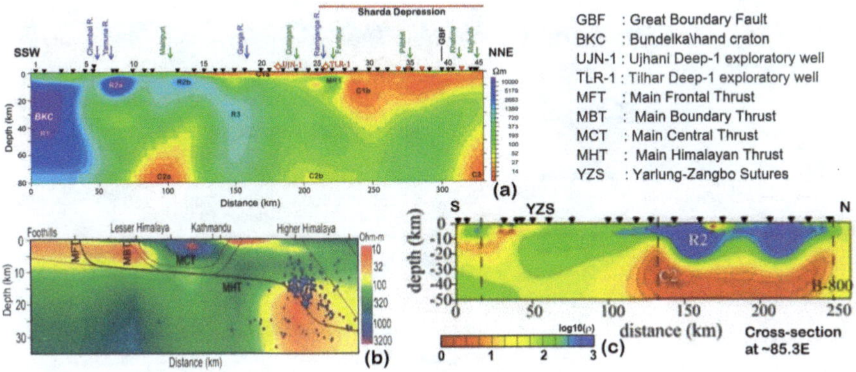

Fig. 3.4 Geoelectric crustal structure along some profiles in the Nepal sector of the Himalaya and its bordering south Tibet and Ganga Basin regions. **a** Ganga Basin south of western Nepal (profile 16) (Manglik et al., 2025), **b** central Nepal profile (profile 20) passing through Kathmandu (Lemonnier et al., 1999), and **c** southern Tibet north of the profile (profile 21) (Xie et al., 2017). In "**c**," YSZ is the name for the ITSZ in southern Tibet

3.4 Eastern Himalaya

3.4.1 Sikkim Sector

The Sikkim sector covers an ~ 90 km length of the Himalayan arc. The seismotectonics of this region changes from thrust-type earthquakes at the ramp in the western and central Himalaya to strike-slip deep crustal earthquake type. Patro and Harinarayana (2009) acquired MT data along a N-S profile between Siliguri and Yumthang valley (profile 23, Fig. 3.1) and inverted the data of 13 sites in the period band of 0.002–500 s to obtain a 2-D model of the geoelectric crustal structure of this region. Their model revealed a collage of resistive and conductive blocks in the crust, including a widespread zone of high electrical conductivity in the depth range of 5–25 km along the profile. Pavankumar et al. (2014) largely followed the same profile (profile 24, Fig. 3.1) by 31 sites, but extended it till 50 km south of Siliguri and covered the Lachen–Guigong sector in the north. They obtained two-dominant regional strike directions, one consistent with the regional trend of the Himalayan arc for the southern segment and the other N30° W consistent with the transverse seismicity trend at the Main Central Thrust Zone (MCTZ). Their 2-D model obtained by inversion of composite MT data in the period range of 0.01–1000 s revealed a complex crustal architecture with distinct segmentation, likely influenced by structural inheritance and ongoing tectonic reorganization. Pavankumar and Manglik (2021) further refined this model by including long-period data (up to 10,000 s) at six sites and also analyzed it along with the published seismic velocity structures. The results (Fig. 3.5a) suggest that the geometry of the MHT and the underthrusting Indian

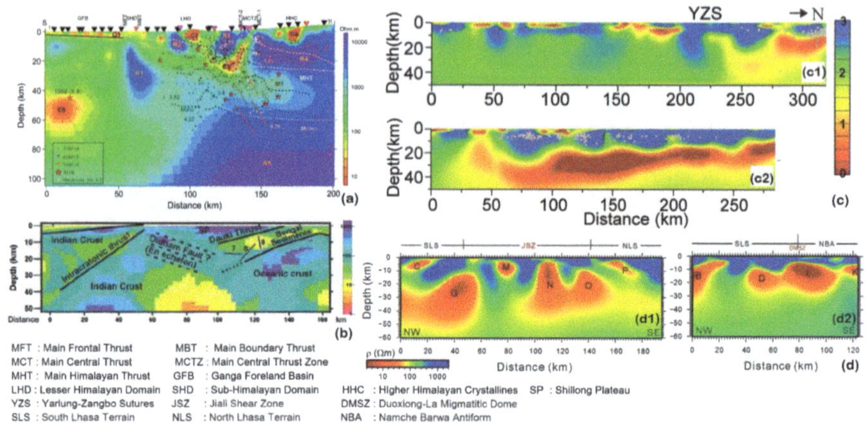

Fig. 3.5 Geoelectric crustal structure of NE Himalaya and SE Tibet regions. **a** Sikkim (profile 24), **b** Shillong plateau and Brahmaputra valley (profile 28) (Gokarn et al., 2008), **c** southern Tibet north of Sikkim and Bhutan (profiles 25–27) (Xie et al., 2021), and **d** Namche–Barwa region of the East Himalayan Syntaxis (profiles group 30) (Lin et al., 2017) with d1 and d2 are the eastern and western profiles

lithosphere is strongly modulated by active deformation associated with the Dhubri–Chungthang Fault (DCF) and other transverse crustal structures. Spatial correlations between these conductive features and clusters of seismicity, especially along the ramp segment of the MHT—underscore the role of deep-seated fluids in influencing fault mechanics and earthquake generation in the eastern Himalaya. Recently, Konda et al. (2023) performed 3-D inversion of the MT data for mapping of off-profile structures. Pavankumar and Manglik (2012) and Manglik et al. (2013) also analyzed the anomalous phase observed at sites within the MCTZ and interpreted it in terms of crustal anisotropy in the region. Wang et al. (2024) performed 3-D modeling of the MT data between the Dingjie-Shenzha and Yadong-Gulu rifts further north of Sikkim.

3.4.2 Assam–Arunachal Pradesh Sector

This sector of the Himalayan arc of ~ 850 km length spanning from Bhutan to the Eastern Himalaya Syntax (EHS) is devoid of any published MT result, though Unsworth (2010) has marked one MT profile between the Mikir Hills in Assam to Tawang in Arunachal Pradesh covered by Indian Institute of Geomagnetism (IIG). Line-700 of the INDEPTH experiment lies to the north of this profile. Gokarn et al. (2008) carried out MT survey along a profile (profile 28, Fig. 3.1) across the Shillong Plateau and the Brahmaputra valley. In the geoelectric section (Fig. 3.5b), they mapped the signature of the Dauki and the Brahmaputra faults as northward-dipping structures and underthrusting of the Bengal basin conductive sediments along the

Dauki fault. In another study, Saha et al. (2007) acquired MT data at 21 sites in the Shillong Plateau and Assam (profile 29, Fig. 3.1) and analyzed these in terms of 1-D and 2-D models. For the EHS, Rawat and Luirei (2021) modeled MT data of 9 sites and obtained a broad regional geoelectric model of the crust highlighting the presence of an intra-crustal low-resistivity layer along the entire profile.

Detailed MT studies have been carried out in the Tibet and Namche–Barwa regions north of the Himalayan arc of this region (profiles 26, 27, 30, Fig. 3.1). Xie et al. (2017) performed 3-D inversion of the MT data of 163 sites acquired along various profiles in south Tibet (profiles 14, 21, 22, 25, 26, 27 in Fig. 3.1), subdividing these into three sub-domains and inverting each sub-domain separately. Xie et al. (2021) inverted MT data of 78 sites covered under different programs including INDEPTH-II (covered by profiles 25–27 in Fig. 3.1) to map mid-crustal partial melting. Their model along profile 25 (Fig. 3.5c1) yields crustal conductive zone north of the YZS (ITSZ), whereas moderate conductivity south of it. However, in the region of profiles 26 and 27, the entire crust below 10 km depth is highly conductive (Fig. 3.5c2). They estimated effective conductivity and viscosity and attributed these to the presence of partial melts. Lin et al. (2017) presented a high-resolution 3-D MT model for the Namche–Barwa region by using 83 MT sites data (profile 30, Fig. 3.1). The models along the syntaxis-parallel profiles (Fig. 3.5d) reveal a distinct high conductivity zone in the mid-to-lower crust beneath the syntaxis, interpreted as zones of partial melt or fluid accumulation associated with crustal weakening. These conductive features align with major thrust structures and zones of high seismicity, suggesting fluid-assisted deformation.

3.5 Synthesis of Some Contiguous Profiles Across the HCB

Previous sections described individual profiles in different sectors of the Himalayan arc. Many of these profiles cover only a part of the entire HCB due to logistic and other constraints. However, some profiles align well and, when put together, form long geoelectric sections of the HCB covering contiguous tectonic blocks from the MFT to the trans-Himalaya and southern Tibet. Figure 3.6 shows the geoelectric sections along such contiguous profiles in a compact view. Profile sections are arranged from the west to the east along the Himalayan arc and have been scaled to have uniform horizontal profile length scale and no vertical exaggeration for ease of visual comparison. These sections reveal both across- and along-arc variations in the geoelectric crustal structure. It can be seen that the flat-ramp-flat nature of the décollement surface (the MHT) in the northwestern and central Himalaya (Fig. 3.6a–d) changes to a more complex crustal structure in the Sikkim sector (Fig. 3.6e). Similarly, the crustal conductor below 20 km depth in southern Tibet north of the NW and central Himalaya (Fig. 3.6c, d) is widespread, but the crustal conductivity of southern Tibet becomes moderate north of Sikkim and Bhutan (Fig. 3.6e). The Rampur Window in Himachal Pradesh is marked by the presence of a highly resistive block within the Indian plate crust (Fig. 3.6a, b), but the crustal structure of the underthrusting Indian plate yields a highly conductive layer in Uttarakhand (Fig. 3.6c). These geoelectric models, though widely varying in terms of data acquisition, pre-processing, frequency bands used, estimation of impedance tensors, distortion analyses, and applied inversion techniques, provide a broad view of the crustal complexities along the Himalayan arc. However, development of a unified model of the along-arc variations in the crustal structure of the Himalaya would require a benchmark approach for re-processing and modeling of the MT data available with various research groups.

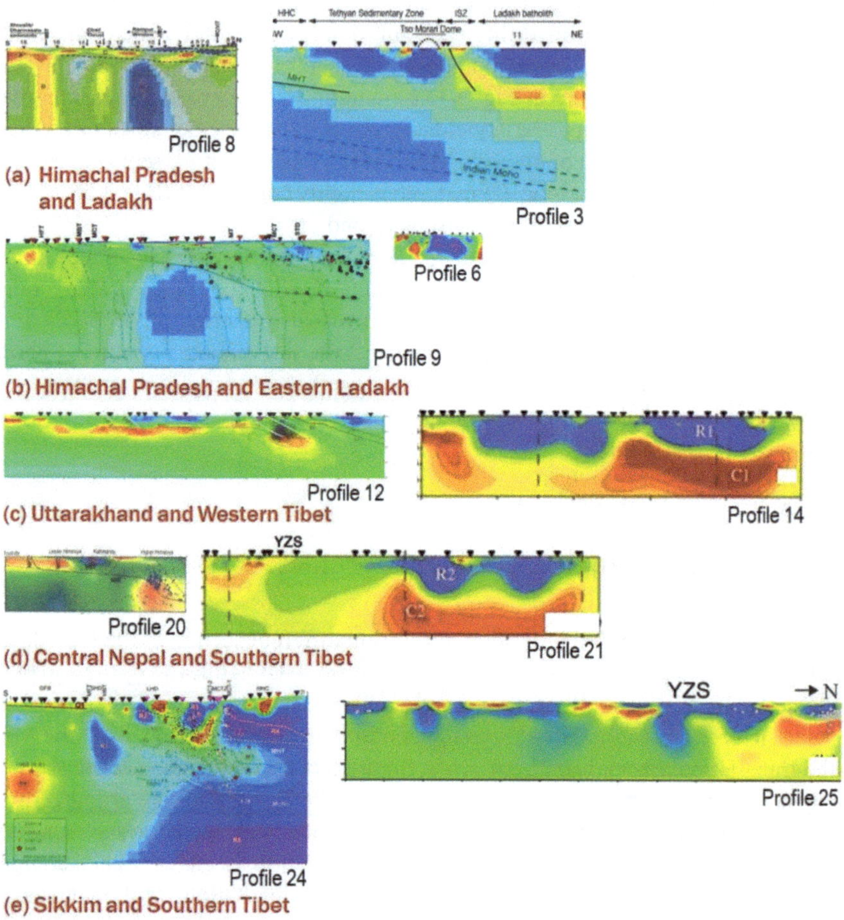

Fig. 3.6 Geoelectric crustal structure for some well-aligned profiles to show variations across the Himalayan collision belt in different sectors of the Himalayan arc. Profile numbers refer to those marked in Fig. 3.1. Profile sections are arranged from the west **a** to the east **e** along the Himalayan arc

3.6 Summary and Future Direction

This chapter presents a brief account of the MT studies carried out along the Himalayan arc and its surrounding Tibet and Ganga Basin regions in the north and south, respectively, with a focus on crustal resistivity structure. A synthesis of the published results discussed in this chapter reveals variations in the crustal geoelectric structure along the arc. In the trans-Himalayan region of Karakoram and Ladakh, both these batholiths appear as high resistivity blocks with their thicknesses ranging between 12 and 20 km, and both are underlain by conductive channels linked to the partial melt zones in south-western Tibet. In the Himachal and Uttarakhand sector, the Indian plate seems to be resistive, except for profile 12, and is overlain by conductive sedimentary successions. The MHT ramp in this region is associated with high conductivity, possibly due to fluid-filled weak zones, and is seismically active. The same pattern of the Indian plate resistivity and MHT ramp conductor continues till central Nepal. This flat-ramp-flat model delineated by the MT results is consistent with the seismological model, proposed to explain the genesis of significant earthquakes in the Himalaya (Ni & Barazangi, 1984). However, the scenario changes significantly in the Sikkim sector where instead of a clear flat-ramp-flat structure, the lithosphere appears to be consisting of two compositionally different blocks separated by the MCTZ. The occurrence of deep crustal seismicity of strike-slip nature can be explained in terms of this lithospheric resistivity contrast. Thus, the MT models also highlight the change in the tectonic style along the arc.

The crustal structure of the southern Tibet region along the Himalayan arc is dominated by a widespread conductive zone at > 20 km depth overlain by resistive blocks till the ITSZ/YZS, linked to the presence of partial melts. This conductive zone appears to be thin and it rises upward south of the ITSZ, though the conductivity becomes from very high in eastern Ladakh to moderate west of Sikkim. This widespread crustal conductive zone is once again seen in the south-eastern Tibet region (east of 90°E longitude) and in the Eastern Himalayan Syntaxis. The Ganga Basin also presents some new results, especially for the western Nepal sector of the Himalaya where a super-thick sedimentary basin has been mapped at the foothills region. Such lateral variations in the lithospheric structure of the Indian plate underthrusting the Himalaya are significant for any studies focused on the Himalayan hazard.

It can be seen in Fig. 3.1 that many regions of the Himalayan arc are still devoid of any MT study. The Himachal Pradesh–Uttarakhand sector has relatively good coverage, but the Jammu and Kashmir, Nepal, Bhutan, and Arunachal Pradesh sectors lack such studies. Despite the Nepal sector being ~ 850 km long with some devastating earthquakes in the past century (e.g., 1934 Bihar–Nepal and 2015 Gorkha earthquakes), there is only one MT profile till now in this sector. Therefore, concerted efforts (including international cooperation) are required to cover these unexplored sectors in terms of electromagnetic exploration through deployment of MT arrays along the arc. Integration of multidisciplinary datasets is also important to advance our knowledge about the structurally and geologically complex segments of the

Himalayan arc and associated tectonic processes, which are crucial inputs for any earthquake hazard assessment and mitigation efforts.

Acknowledgements We thank Dr. V.M. Tiwari, Director, CSIR-NEIST, for inviting AM to contribute this article and two anonymous Reviewers for their insightful comments and suggestions on the manuscript. This study has been carried out under the in-house project MLP7004-28(AM) of CSIR-National Geophysical Research Institute (CSIR-NGRI) and has been approved by CSIR-NGRI for publication as contribution number NGRI/Lib/2025/Pub-91.

Statements and Declarations

Competing Interests The authors have no competing interests to declare that are relevant to the content of this article.

Funding No funding was received to assist with the preparation of this manuscript.

References

Adilakshmi, L., Manglik, A., Thiagarajan, S., & Suresh, M. (2021). Crustal structure of the Indian plate underneath the alluvial plains of the central Ganga basin by broadband magnetotellurics. *Tectonophysics, 802*, Article 228746. https://doi.org/10.1016/j.tecto.2021.228746

Arora, B. R., Gahalaut, V. K., & Kumar, N. (2012). Structural control on along-strike variation in the seismicity of the northwest Himalaya. *Journal of Asian Earth Sciences, 57*, 15–24.

Arora, B. R., Lilley, F. E. M., Sloane, M. N., Singh, B. P., Srivastava, B. J., et al. (1982). Geomagnetic induction and conductive structures in northwest India. *Geophysical Journal of the Royal Astronomical Society, 69*, 459–475.

Arora, B. R., & Mahashabde, M. V. (1987). A transverse conductive structure in the northwest Himalaya. *Physics of the Earth and Planetary Interiors, 45*, 119–127.

Arora, B. R., Unsworth, M. J., & Rawat, G. (2007). Deep resistivity structure of the northwest Indian Himalaya and its tectonic implications. *Geophysical Research Letters*, 34, L04307. https://doi.org/10.1029/2006GL029165

Chen, L., Booker, J. R., Jones, A. G., Wu, N., Unsworth, M. J., et al. (1996). Electrically conductive crust in southern Tibet from INDEPTH magnetotelluric surveying. *Science, 274*, 1694–1696.

Demudu Babu, M., Manglik, A., Pavankumar, G., Suresh, M., & Thiagarajan, S. (2024). Transverse Faizabad ridge in the foreland Ganga basin: A mosaic of fragmented crustal blocks as revealed by Magnetotellurics. *Tectonophysics, 874*, Article 230238. https://doi.org/10.1016/j.tecto.2024.230238

Demudu Babu, M., Manglik, A., Thiagarajan, S., & Suresh, M. (2020). Electrical resistivity image of a basement ridge in the foreland central Ganga basin. *Journal of Applied Geophysics, 179*, Article 104097. https://doi.org/10.1016/j.jappgeo.2020.104097

Devi, A., Israil, M., Gupta, P. K., Varshney, S. K., & Meqbel, N. (2019). Transverse tectonics structures in the Garhwal Himalaya corridor inferred from 3D inversion of magnetotelluric profile data. *Pure and Applied Geophysics, 176*, 4921–4940.

Dhamodharan, S., Raju, K., & Patro, P. K. (2024). Lateral electrical conductivity variations along the Main Himalayan thrust in the northwestern Himalayas: Insights from 3D Magnetotelluric forward modeling. *Physics of the Earth and Planetary Interiors, 353*, Article 107213. https://doi.org/10.1016/j.pepi.2024.107213

Dhamodharan, S., Rawat, G., & Bagri, D. S. (2023). Deep electrical resistivity imaging of crustal structures along the Satluj River valley, northwest Himalaya, India. *Tectonophysics, 869*, Article 230096. https://doi.org/10.1016/j.tecto.2023.230096

Gokarn, S. G., Gupta, G., Rao, C. K., & Selvaraj, C. (2002a). Electrical structure across the Indus Tsangpo suture and Shyok suture zones in NW Himalaya using magnetotelluric studies. *Geophysical Research Letters, 29*(8), 1251. https://doi.org/10.1029/2001GL014325

Gokarn, S. G., Gupta, G., Walia, D., Sanabam, S. S., & Hazarika, N. (2008). Deep geoelectric structure over the Lower Brahmaputra valley and Shillong Plateau, NE India using magnetotellurics. *Geophysical Journal International, 173*(1), 92–104.

Gokarn, S. G., Rao, C. K., & Gupta, G. (2002b). Crustal structure in the Siwalik Himalayas using magnetotelluric studies. *Earth, Planets and Space, 54*, 19–30.

Gupta, G., Gokarn, S. G., & Singh, B. P. (1994). Thickness of the Siwalik sediments in the Mohand–Ramnagar region using magnetotelluric studies. *Physics of the Earth and Planetary Interiors, 83*, 217–224.

Israil, M., Tyagi, D. K., Gupta, P. K., & Niwas, S. (2008). Magnetotelluric investigations for imaging electrical structure of Garhwal Himalayan corridor, Uttarakhand, India. *Journal of Earth System Science, 117*, 189–200.

Konda, S., Patro, P. K., Reddy, K. C., & Babu, N. (2023). Three-dimensional magnetotelluric signatures and rheology of subducting continental crust: Insights from Sikkim Himalaya, India. *Journal of Geodynamics, 155*, Article 101961. https://doi.org/10.1016/j.jog.2023.101961

Kumar, A., & Rao, C. K. (2024). Geothermal resources and deep tectonic in Leh Ladakh (NW Himalaya), India: Inference from magnetotelluric studies. *Geothermics, 122*, Article 103083. https://doi.org/10.1016/j.geothermics.2024.103083

Lemonnier, C., Marquis, G., Perrier, F., Avouac, J. P., Chitraka, G., Kafle, B., Sapkota, S., Gautam, U., Tiwari, D., & Bano, M. (1999). Electrical structure of the Himalaya of Central Nepal: High conductivity around the midcrustal ramp along the MHT. *Geophysical Research Letters, 26*, 3261–3264.

Lin, C., Peng, M., Tan, H., Xu, Z., Li, Z.-H., et al. (2017). Crustal structure beneath Namche Barwa, eastern Himalayan syntaxis: New insights from three-dimensional magnetotelluric imaging. *Journal of Geophysical Research: Solid Earth, 122*(7), 5082–5100.

Manglik, A., Adilakshmi, L., Suresh, M., & Thiagarajan, S. (2015). Thick sedimentary sequence around Bahraich in the northern part of the central Ganga foreland basin. *Tectonophysics, 653*, 33–40.

Manglik, A., Pavankumar, G., Suresh, M., Chakravarthi, N. N., & Demudu Babu, M. (2024). Northern limit of the concealed Bundelkhand Craton: A synthesis of new magnetotelluric results from the Ganga Basin. *Journal of the Geological Society of India, 100*(2), 181–188.

Manglik, A., Pavan Kumar, G., & Thiagarajan, S. (2013). Transverse tectonics in the Sikkim Himalaya: A magnetotelluric study. *Tectonophysics, 589*, 142–150.

Manglik, A., Suresh, M., Chakravarthi, N. N., & Pavankumar, G. (2025). Crustal geoelectric structure of the Indian plate at the western extremity of the Bundelkhand craton and Sharda Depression in the Ganga Basin. *Tectonophysics, 899*, Article 230653. https://doi.org/10.1016/j.tecto.2025.230653

Manglik, A., Suresh, M., Nagarjuna, D., Pavankumar, G., Babu, M. D., et al. (2022). Subsurface expressions of the Aravalli-Delhi Fold Belt in the Western Ganga Basin by magnetotellurics. *Journal of the Geological Society of India, 98*, 1721–1727.

Miglani, R., Shahrukh, M., Israil, M., Gupta, P. K., Varshney, S. K., et al. (2014). Geoelectric structure estimated from magnetotelluric data from the Uttarakhand Himalaya, India. *Journal of Earth System Science, 123*, 1907–1918.

Ni, J., & Barazangi, M. (1984). Seismotectonics of the Himalayan collision zone: Geometry of the underthrusting Indian plate beneath the Himalaya. *Journal of Geophysical Research: Solid Earth, 89*, 1147–1163.

Patro, P. K., & Harinarayana, T. (2009). Deep geoelectric structure of the Sikkim Himalayas (NE India) using magnetotelluric studies. *Physics of the Earth and Planetary Interiors, 173*, 171–176.

Pavankumar, G., & Manglik, A. (2012). Electrical anisotropy in the main central thrust zone of the Sikkim Himalaya: Inference from anomalous MT phase. *Journal of Asian Earth Sciences, 57*, 120–127.

Pavankumar, G., & Manglik, A. (2021). Complex tectonic setting and deep crustal seismicity of the Sikkim Himalaya: An electrical resistivity perspective. *Physics and Chemistry of the Earth, 124*, Article 103077. https://doi.org/10.1016/j.pce.2021.103077

Pavankumar, G., Manglik, A., Demudu Babu, M., Kandregula, R. S., & Barman, A. (2024). A potential deep geothermal reservoir in Eastern Ladakh as inferred from the upper crustal geoelectric structure of the region. *Physics of the Earth and Planetary Interiors, 356*, Article 107263. https://doi.org/10.1016/j.pepi.2024.107263

Pavankumar, G., Manglik, A., Suresh, M., & Adilakshmi, L. (2023). Electrical Moho and upper mantle metasomatism in the Archaean Bundelkhand craton of the Indian peninsular shield. *Tectonophysics, 854*, Article 229786. https://doi.org/10.1016/j.tecto.2023.229786

Pavankumar, G., Manglik, A., & Thiagarajan, S. (2014). Crustal geoelectric structure of the Sikkim Himalaya and adjoining Gangetic foreland basin. *Tectonophysics, 637*, 238–250.

Pavankumar, G., Manglik, A., Chakravarthi, N. N., Demudu Babu, M., & Kandregula, R. S. (2026). Electromagnetic signatures of the tectono-magmatic processes in eastern Ladakh and their implications for geothermal energy potential of the region. *Physics of the Earth and Planetary Interior, 371*, 107503. https://doi.org/10.1016/j.pepi.2026.107503

Rawat, G., Arora, B. R., & Gupta, P. K. (2014). Electrical resistivity cross section across the Garhwal Himalaya: Proxy to fluid-seismicity linkage. *Tectonophysics, 637*, 68–79.

Rawat, G., & Luirei, K. (2021). Deep crustal resistivity structure of the Lohit Valley in the Eastern Himalayan Syntaxial region. *Geological Journal, 57*, 4920–4928.

Saha, D. K., Naskar, D. C., Bhattacharya, P. M., & Kayal, J. R. (2007). Geophysical and Seismological Investigations for the Hidden Oldham Fault in the Shillong Plateau and Assam Valley of Northeast India. *Journal of the Geological Society of India, 69*, 359–372.

Shah, S. T. H., Zhao, J., Xiao, Q., Bhatti, Z. I., Khan, N. G., et al. (2018). Electrical resistivity structures and tectonic implications of Main Karakorum Thrust (MKT) in the western Himalayas: NNE Pakistan. *Physics of the Earth and Planetary Interiors, 279*, 57–66.

Sokolova, E. Y., Israil, M., Gupta, P., Koshurnikov, A. V., Smirnov, M. Y., & Cherevatova, M. V. (2016). Crustal electrical conductivity of the Indian continental subduction zone: New data from the profile in the Garhwal Himalaya. *Izvestiya, Physics of the Solid Earth, 52*, 271–290.

Spratt, J. E., Jones, A. G., Nelson, K. D., & Unsworth, M. J. (2005). Crustal structure of the India-Asia collision zone, southern Tibet, from INDEPTH MT investigations. *Physics of the Earth and Planetary Interiors, 150*, 227–237.

Suresh, M., Manglik, A., & Thiagarajan, S. (2023). Basement and lithospheric structure of the central Ganga basin between the Bundelkhand craton and the Sharda Deep by magnetotellurics. *Tectonophysics, 863*, Article 229991. https://doi.org/10.1016/j.tecto.2023.229991

Unsworth, M. (2010). Magnetotelluric studies of active continent–continent collisions. *Surveys in Geophysics, 31*, 137–161.

Unsworth, M. J., Jones, A. G., Wei, W., Marquis, G., Gokarn, S. G., et al. (2005). Crustal rheology of the Himalaya and Southern Tibet inferred from magnetotelluric data. *Nature, 438*, 78–81.

Veeraswamy, K., Abdul Azeez, K. K., Basava, S., Dhanunjay Naidu, G., & Harinarayana, T. (2010). Electrical structure across Lesser and Higher Himalaya, India. *Chinese Journal of Geophysics, 53*(3), 576–584.

Wang, G., Fang, H., Lei, Q., Xu, X., Liang, H., et al. (2024). Magnetotelluric imaging of the central Himalayan crust away from rift zones. *Geochemistry Geophysics Geosystems, 25*, e2023GC011132. https://doi.org/10.1029/2023GC011132

Xie, C., Jin, S., Wei, W., Ye, G., Jing, J., et al. (2021). Middle crustal partial melting triggered since the mid-Miocene in southern Tibet: Insights from magnetotelluric data. *Journal of Geophysical Research: Solid Earth, 126*, e2021JB022435. https://doi.org/10.1029/2021JB022435

Xie, C., Jin, S., Wei, W., Ye, G., Zhang, L., et al. (2017). Varying Indian crustal front in the southern Tibetan Plateau as revealed by magnetotelluric data. *Earth, Planets and Space, 69*, 147. https://doi.org/10.1186/s40623-017-0734-z

Ye, G., Jin, S., Wei, W., & Unsworth, M. J. (2007). Research of the conductive structure of crust and the upper mantle beneath the South-Central Tibetan Plateau. *Journal of China University of Geosciences, 18*(4), 334–343.

Chapter 4
Deep Earth Structure and Seismogenesis in the Uttarakhand Himalaya

Prantik Mandal

Abstract The primary objective of this chapter is to elucidate the fundamental influence exerted by the deep geological and geophysical structure on the genesis and recurrence of earthquakes in the Uttarakhand segment of the Indian Himalaya. The Uttarakhand Himalaya is a landscape of extreme tectonic suspense, where the inexorable northward migration of the Indian plate at 18 mm/yr is met by a structurally partitioned and fluid-weakened Main Himalayan Thrust (MHT). This chapter presents the deep-seated structure of the segment. The segment is characterized by a complex ramp-flat-ramp architecture with older transverse "ribs"—the Delhi–Haridwar Ridge and the Moradabad Fault—that divide the crust into separate mechanical blocks. High-resolution seismic tomography and receiver function imaging reveal extensive low-velocity domains at depths of 8–22 km, with high V_P/V_S ratios and high crack densities indicative of metamorphic fluids. Such fluid-rich reservoirs act as the great lubricators of the mid-crust, lowering the failure threshold and controlling the nucleation of moderate to large events. The remarkable build-up of a 3 m slip deficit since 1803, even with release provided by historical earthquakes, highlights a segment primed for a major $M_w \geq 7.5$ earthquake. Ultimately, closing the gap between the geological complexity of the region and the urgent needs for seismic resilience and hazard mitigation can only be achieved by moving from static observation to dynamic, process-oriented geophysics that combines real-time geodetic monitoring with deep structural insights.

Keywords NW Himalaya · Seismogenesis · Velocity structure

P. Mandal (✉)
CSIR-National Geophysical Research Institute, Hyderabad, India

V. M. Tiwari (ed.), *Himalayan Natural Hazards*, Indian National Science Academy,
https://doi.org/10.1007/978-981-92-3784-5_4

4.1 Regional Geological Framework

4.1.1 *Tectonic Architecture and Seismic Hazard in the Uttarakhand Himalaya*

The Uttarakhand Himalaya forms a tectonically active corridor within the Himalayan arc, shaped by the persistent convergence of the Indian and Eurasian plates since the Eocene (~ 55 Ma). Present-day geodetic measurements indicate convergence rates of 15–18 mm/yr, predominantly accommodated along the Main Himalayan Thrust (MHT), a gently north-dipping décollement underlying the orogenic front and acting as the principal seismogenic interface (Valdiya, 1980; Seeber & Armbruster, 1971; Bilham, 2019).

Recent advances in local earthquake tomography, P-receiver function (PRF) imaging, and joint inversion have revealed that the MHT beneath the Kumaon–Garhwal Himalaya is characterized by a 10–20% reduction in Vp and Vs, and a 10–15% increase in Vp/Vs at depths of 8–22 km (Mandal et al., 2022). These anomalies, coinciding with high crack density ($\varepsilon \approx 0.011$–$0.012$) and saturation rate ($\xi \approx 0.38$–$0.39$), suggest the presence of aqueous and metamorphic fluids (Mandal, 2024). Elevated pore-fluid pressures reduce effective friction, rendering the MHT a critical locus for earthquake nucleation.

Historically, the region has experienced three moderate to strong earthquakes: the 1803 Mw 7.8 Garhwal, 1991 Mw 6.8 Uttarkashi, and 1999 Mw 6.5 Chamoli events, all nucleating within the mid-crustal MHT zone (Mandal et al., 2021). However, geodetic and paleoseismic evidence indicate that the accumulated strain—estimated at ~ 3 m since 1803 (Bilham, 2019)—remains only partially released, highlighting a major seismic gap with potential for future Mw $\geq$ 7.5 events.

Crustal structure studies reveal variable Moho depths (28–55 km) and lithospheric thicknesses up to 180 km, with evidence for double Moho structures indicative of crust–mantle imbrication during India–Asia collision (Mandal, 2023). These features have been resolved through common conversion point (CCP) imaging of broadband seismic data from a dense network deployed by CSIR-NGRI (Cook et al., 2021; Mandal, 2023).

Furthermore, the MHT is laterally segmented by three prominent NNE-trending transverse features: the Delhi–Haridwar Ridge, Moradabad Fault, and Great Boundary Fault. These structures disrupt the continuity of the décollement, acting as barriers to rupture propagation (Mandal et al., 2023). This segmentation likely limits earthquake magnitudes to $\leq$ Mw 7.8, in contrast to Mw $\geq$ 8.0 events in less segmented regions such as Nepal, Kashmir, and Assam (Li & Song, 2018; Dal Zilio et al., 2020).

In summary, the Uttarakhand Himalaya is underlain by a fluid-saturated, structurally segmented, and mechanically weak MHT, rendering the region highly vulnerable to future large earthquakes. The integration of multi-parameter seismic imaging with geodetic strain accumulation provides a robust framework for seismic hazard assessment in this densely populated corridor of the central Himalayan arc.

4.1.2 Scientific Context and Framework

The Uttarakhand Himalaya is situated between the rupture zones of the 1905 Kangra and 1934 Bihar–Nepal earthquakes, within the so-called Central Himalayan Seismic Gap (Valdiya, 1980; Seeber & Armbruster, 1971). Despite hosting the 1803 Mw 7.8 Garhwal, 1991 Mw 6.8 Uttarkashi, and 1999 Mw 6.5 Chamoli earthquakes, the region has not experienced a full release of accumulated tectonic strain, with GPS-derived slip deficits estimated at ~ 3 m since 1803 (Bilham, 2019; Mandal et al., 2022).

Seismic imaging studies (Rai et al., 2006; Hazarika et al., 2017; Mandal et al., 2021, 2022; Monika et al., 2024; Mandal, 2024) have demonstrated that the MHT beneath Kumaon–Garhwal is a shallow, north-dipping low-velocity zone (8–22 km depth) with a 10–20% drop in P- and S-wave velocities, high Vp/Vs ratios (1.74–1.92), elevated crack density ($\varepsilon \approx 0.011$–$0.012$), and saturation rates ($\xi \approx 0.38$–0.39), consistent with the presence of metamorphic and aqueous fluids. These fluids reduce effective normal stress, weakening the fault and facilitating earthquake rupture (Caldwell et al., 2013; Hazarika et al., 2017; Mandal, 2024).

Receiver function analyses across 55 broadband stations have mapped a complex, undulating Moho (35–55 km) and a lithosphere–asthenosphere boundary (130–180 km) beneath this region (Fig. 4.1; Mandal, 2023). The presence of a double Moho structure—interpreted as interlocked crust-mantle blocks from the Indian plate—suggests ongoing crustal interpenetration due to continent–continent collision (Mandal, 2023, 2024).

Significant segmentation of the MHT is attributed to inherited transverse lithospheric features, including the Delhi–Haridwar Ridge, Moradabad Fault, and Great Boundary Fault, as revealed by basement topography and receiver function images (Mandal et al., 2023; Dal Zilio et al., 2020; Manglik et al., 2022). This segmentation, not uniformly present across the Himalaya, may explain the absence of Mw $\geq$ 8.0 earthquakes in Uttarakhand, in contrast to Nepal, Kashmir, or Assam.

Deep mantle dynamics also influence regional seismicity. Receiver function modeling of the Mantle Transition Zone (MTZ) beneath Uttarakhand reveals low-velocity anomalies near the 410- and 660-km discontinuities, suggestive of hydrated or partially molten layers—possibly remnants of the subducted Neotethyan slab (Mandal et al., 2024; Rai et al., 2009b). These anomalies support the hypothesis that deep mantle processes contribute to the regional geodynamic regime influencing crustal seismicity.

Collectively, these observations underscore the importance of a deeply rooted tectonic framework involving fluid-filled weak zones, mid-crustal segmentation, and inherited lithospheric fabric in governing seismicity. Understanding these controls is critical for anticipating future earthquake behavior and developing physics-based hazard models for the Himalayan arc.

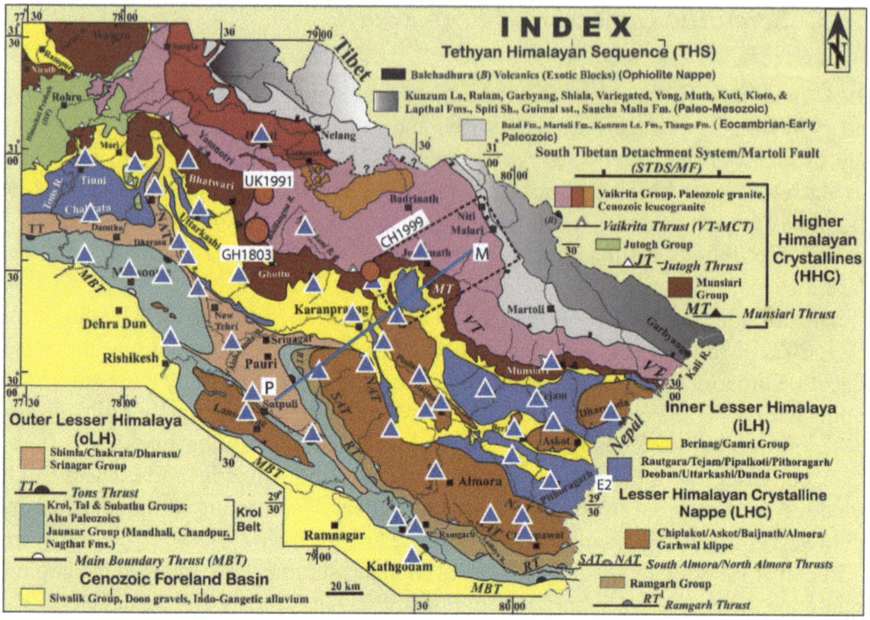

Fig. 4.1 Geological map of the Uttarakhand Himalaya between Tons and Kali Rivers, showing distribution of main tectonic units. Modified after Valdiya (1980), Célérier et al. (2009), Webb et al. (2011), Jain et al. (2013), Thakur et al. (2018) and Mukherjee et al. (2019). Solid thick blue line marks the location of the NE–SW trending PM profile, whose location is same as the Profile of Monika et al. (2024) and Mandal (2024). Red filled circles mark the epicenters of 1803 Garhwal, 1991 Uttarkashi, and 1999 Chamoli Events of $M_W \geq 6.5$. Open red triangles mark the broadband stations

4.1.3 *Tectonic and Geological Framework*

Regional Tectonic Configuration

The Himalayan orogen is the world's most prominent and seismically active continental collision zone, resulting from the ongoing northeastward convergence between the Indian and Eurasian plates since ~ 55 Ma. This collision has led to crustal shortening, thickening, and the evolution of a thrust-dominated structural system comprising the Main Frontal Thrust (MFT), Main Boundary Thrust (MBT), Main Central Thrust (MCT), and South Tibetan Detachment (STD)—all rooting into the north-dipping MHT (Valdiya, 1980; Seeber & Armbruster, 1971). Manglik et al. (2022) have shown the strike-parallel spatial distribution of the MHT, transverse faults (viz., Delhi–Haridwar Ridge (DHR), Moradabad fault (MF), Great Boundary Fault (GBF)), and historic rupture zones across the whole Himalaya.

The Indian plate continues to underthrust southern Eurasia at 15–18 mm/yr, with motion largely accommodated along the MHT (Bilham, 2019). Structurally, the MHT dips shallowly northward (~ 4°–6°) beneath the Lesser and Greater Himalaya, steepening beneath the Higher Himalaya and southern Tibet (Chen et al., 2004; Larson et al., 1999).

Receiver function and tomographic imaging beneath Uttarakhand indicate that the MHT is segmented by inherited transverse lithospheric structures—Delhi–Haridwar Ridge, Moradabad Fault, and Great Boundary Fault (Manglik et al., 2022; Mandal et al., 2021, 2023). These features, rooted in the Indian basement, segment the MHT into discrete blocks, influencing rupture length and earthquake magnitude potential.

Structural Thrust Zones

- **Main Frontal Thrust (MFT)**: The youngest and southernmost thrust, accommodating recent surface deformation and marking the boundary between the Indo-Gangetic Plain and Siwalik Hills (Valdiya, 1980).
- **Main Boundary Thrust (MBT)**: Separates Siwalik molasse sediments from Lesser Himalayan crystalline rocks; considered reactivated in the current stress regime (Valdiya, 1980).
- **Main Central Thrust (MCT)**: A ductile shear zone separating the Lesser Himalayan sequence from the Greater Himalayan Crystalline Complex; merges into the MHT at mid-crustal depths (Mandal et al., 2021; Yin, 2006).
- **South Tibetan Detachment (STD)**: A normal-sense detachment fault forming the upper boundary of the Greater Himalayan Sequence, accommodating crustal extrusion during uplift.

All these thrusts root into the MHT, imaged as a low-velocity zone with high Vp/Vs (1.74–1.92), crack density (~ 0.011–0.012), and saturation rate (~ 0.38–0.39) at 8–22 km depth in Uttarakhand (Mandal, 2024; Mandal et al., 2022).

The distribution of thrust zones and tectonic units is shown in Fig. 4.1.

Schematic Representation

The deep structural architecture of the Garhwal–Kumaon Himalaya, from the Munsiari to the Vaikrita thrust zones, is best described as a complex multi-layered lithospheric system. Integrated P-wave receiver function (PRF) and tomographic modeling (Mandal, 2023, 2024) reveal a Main Himalayan Thrust (MHT) with a complex northeast-dipping ramp-flat-ramp geometry. This primary interface is supported by large Moho undulations at depths of 28–55 km, indicative of intense crustal thickening caused by convergence (Mandal et al., 2021). Even deeper, the Lithosphere-Asthenosphere Boundary (LAB) dips down to 130–180 km and shows clear evidence for segmentation by inherited transverse structures. In addition, the MTZ shows well-developed low-velocity zones at the upper (350–385 km) and lower (590–640 km) boundaries, which are probably remnants of subducted slab material, indicating that the present seismogenesis of the region is intrinsically linked to deep geodynamic processes (Mandal et al., 2024).

4.1.4 Conclusion

The Uttarakhand Himalaya is a master class in tectonic suspense! The persistent 18 mm/yr northward march of the Indian plate runs into a ramp in the mid-crust that is broken and weakened by fluid. This regional framework is not a mere inert stage but a dynamic, segmented engine of deformation, dominated by a Main Himalayan Thrust (MHT) looking more and more diverse. High-conductivity, fluid-saturated zones near the brittle-ductile transition imply a lubricated plate interface that may facilitate nucleation, but the structural "ribs" of inherited basement ridges present significant persistent obstacles to along-strike rupture. These transverse features like the Delhi–Haridwar Ridge may have been responsible for limiting Himalayan ruptures to moderate sizes in the past, but the startling slip deficit of almost 200 years since the 1803 catastrophe looms big. What remains is a landscape where the crust is primed by fluid pressure and partitioned by ancient geometry. A large chunk of the Kumaon–Garhwal stretch is left in a state of high elastic overdue waiting for the ultimate failure of its locked patches.

4.2 Seismogenic Architecture

The seismotectonic characterization of the Uttarakhand Himalaya hinges on the application of diverse geophysical imaging techniques. These methods have been instrumental in unraveling the geometry and properties of the Main Himalayan Thrust (MHT), crustal thickness, lithospheric segmentation, and fluid distribution at seismogenic depths. This chapter provides a concise overview of the primary tools employed—seismic tomography, receiver function analysis, magnetotelluric imaging, and GPS geodesy—highlighting their strengths, limitations, and key contributions from Indian research institutions.

4.2.1 Earthquake Data and Seismic Network

In 2017, the CSIR-National Geophysical Research Institute (CSIR-NGRI) established a comprehensive seismic monitoring network to study this dynamic region. The network consisted of 76 three-component broadband seismographs, strategically placed with an average inter-station spacing of 19.8 km (Cook et al., 2021; Fig. 4.2a and Table 4.1). Each station was equipped with a Reftek 3-component broadband sensor and a 24-bit recorder. Between 2017 and 2020, this network captured valuable seismic data, recording 713 local earthquakes of local magnitude (ML) ranging from 0.8 to 5.6 (Fig. 4.2a–d). Additionally, 200 aftershocks from the 1999 Mw 6.5 Chamoli earthquake were incorporated (Fig. 4.2b; Mandal et al., 2001). The total dataset comprised 913 events, including 14,074 P-wave and 13,013 S-wave high-quality arrival times (Mandal et al., 2022; Fig. 4.2a; Table 4.1). The P- and S-velocity models of Mandal et al. (2022), which have been determined through the VELEST (Kissling, 1995) simultaneous inversion of arrival time data of 528 earthquakes from 10 to 35 three-component stations, are used in the tomographic analysis (Fig. 4.2c). Technical analysis revealed an average crustal Vp/Vs ratio of 1.72, calculated using the Wadati plot of P-arrivals and (S-P) times (Fig. 4.2d).

The above network has also provided high-quality 3-component digital broadband waveforms for around 1000 teleseismic events with moment magnitudes ranging from 5.5 to 8.0 (Cook et al., 2021; Fig. 4.2b). These events occurred between 2017 and 2023 and had epicentral distances between 30° and 90°. The waveforms were used to construct radial and transverse P-receiver functions (PRFs). This study utilises the Common Conversion Point (CCP) stacking method and depth migration of stacked radial PRFs to image the subsurface crustal structure down to 46 km depth beneath the Uttarakhand Himalaya (Eagar, 2012; Duekar and Sheehan, 1997). An effort has been made to interpret the modeling results in relation to active geodynamic processes occurring under the region.

4.2.2 Seismic Tomography

Seismic tomography utilizes the travel-time anomalies of body waves from local and regional earthquakes to construct 2D and 3D images of seismic velocity structure in the crust and upper mantle. The research employs Benz et al. (1996)'s seismic velocity tomography algorithm to analyze the complex geological structure of the Uttarakhand Himalaya. Building on improvements by Pujol and Shen (1999), the method calculates earthquake focal depths with accuracies within 1 km for depths and 0.5 km for epicenter positions. The inversion process included a damped least-squares approach, and the model was iteratively refined to minimize travel time residuals (Mandal & Pujol, 2006; Mandal et al., 2022; Mandal, 2024). The grid parameterisation for our tomography investigation employs a spatial domain spanning 330 km in the east–west direction and extending to a depth of 40 km. This configuration comprises 11,979 grid nodes with uniform 5 km spacing both horizontally and vertically, ensuring adequate resolution for the tomographic inversion process.

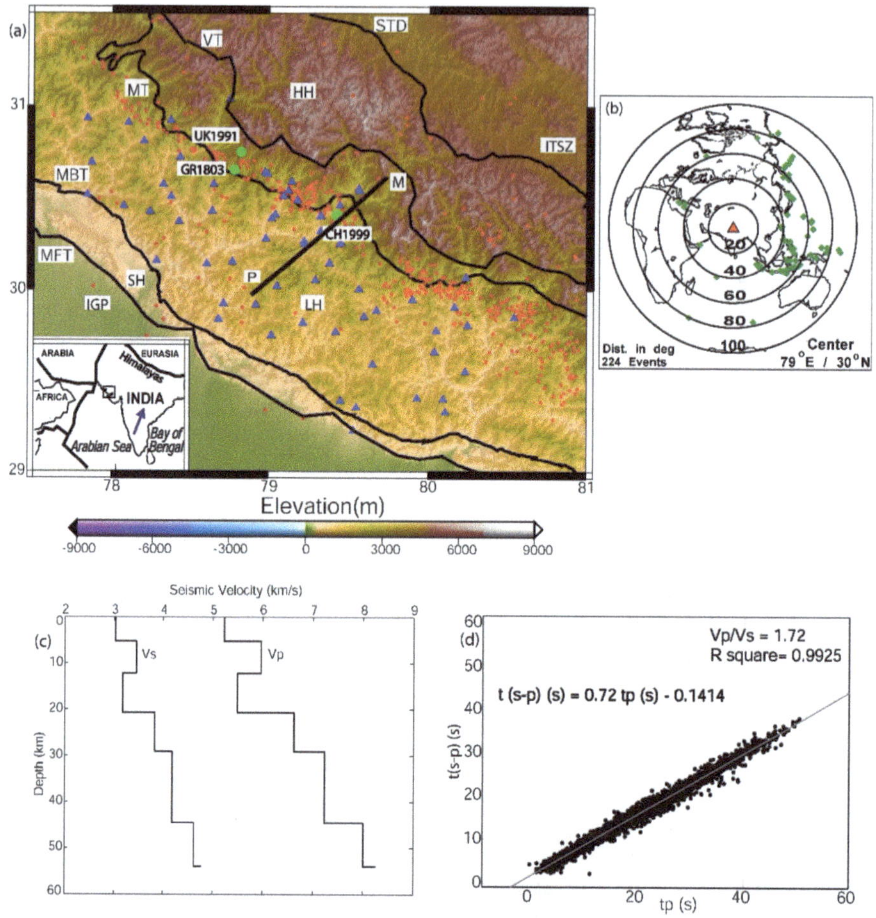

Fig. 4.2 **a** Station location map of the Uttarakhand region. Filled blue triangles: broadband stations; small filled red circles: earthquake relocations from simultaneous inversion. Thick black line: NE–SW trending PM profile. Major faults: MT (Munsiari Thrust), VT (Vaikrita Thrust), MZ (MCT Zone), RT (Ramgarh Thrust), TT (Ton Thrust), MBT (Main Boundary Thrust), MFT (Main Frontal Thrust). Himalayan subdivisions: HH (High Himalaya), LH (Lesser Himalaya), SH (Siwalik Himalaya), IGP (Indo-Gangetic Plain), AsK (Askot Klippe), CCB (Chiplakot crystalline belt), LK (Lansdown Klippe), AK (Almora Klippe). Green filled circles: epicentral locations of the 1803 M_w7.8 Garhwal, 1991 M_w6.8 Uttarkashi, and 1999 M_w6.5 Chamoli earthquakes. Inset: Indian plate and major plate boundaries. **b** Epicentral plot of 200 teleseismic events used for P-receiver function study. Red triangle and green diamond: network center and epicenters of selected teleseismic events. **c** Final 1-D Vp and Vs models (solid black lines). **d** Wadati plot of P-arrivals and (S–P) times, suggesting a Vp/Vs of 1.72

Table 4.1 Summary of seismic network and data coverage (2017–2020)

Parameter	Value
Number of stations	76 broadband (3-component)
Inter-station spacing	19.8 km (average)
Local earthquakes	713 (ML 0.8–5.6)
Aftershocks (Chamoli)	200
Total events analyzed	913
P-wave arrivals	14,074
S-wave arrivals	13,013
Teleseismic events	~ 1,000 (Mw 5.5–8.0, 2017–2023)
Vp/Vs (average crustal)	1.72

Crack Density and Saturation Modeling

Following O'Connell and Budiansky (1974), the study defines key parameters through fundamental equations:

Crack Density (ε):

$$\varepsilon = N\langle a^3 \rangle \tag{4.1}$$

where N is the number of microcracks per unit volume, and $\langle a \rangle$ represents the mean crack radius.

Saturation Rate (v):

$$\nu = \frac{N_2}{N} \tag{4.2}$$

where N_2 is the number of saturated cracks and N is the total number of cracks.

O'Connell and Budiansky (1974) showed that the seismic velocity is not significantly affected by the shape and aspect ratio of spheroidal cracks, as defined by crack density and saturation rate. Based on their theory,

$$\frac{K}{K_0} = 1 - \alpha(1 - \xi)\varepsilon \tag{4.3}$$

$$\frac{E}{E_0} = 1 - \beta(1 - 2\xi)[3(1 - \xi) + \gamma]\varepsilon \tag{4.4}$$

$$\frac{G}{G_0} = 1 - \delta(1 - \xi)[1 - \lambda + \mu]\varepsilon \tag{4.5}$$

$$\varepsilon = f\,(\text{seismic velocities, Poisson's ratio}) \tag{4.6}$$

where K, E, G are the bulk, Young's, and shear moduli of the cracked volume, and K_0, E_0, G_0 are those of the uncracked volume; $\alpha, \beta, \gamma, \delta, \lambda, \mu$ are constants depending on crack geometry and elastic properties.

The relationship between seismic velocity and elastic moduli is:

$$V_p = \left[\frac{K + \frac{4}{3}G}{\rho}\right]^{1/2} \quad (4.7)$$

$$V_s = \left[\frac{G}{\rho}\right]^{1/2} \quad (4.8)$$

$$V_p/V_s = \left[\frac{K + \frac{4}{3}G}{G}\right]^{1/2} \quad (4.9)$$

where V_p and V_s are the average P and S wave velocities, respectively, for a volume that contains cracks, and ρ is density.

Hence, Eqs. (4.3)–(4.9) can be utilized to calculate crack density (ε) and saturation rate (ξ) based on seismic velocity and Poisson's ratio (Mishra & Zhao, 2003; Tadokoro et al., 1999; Zhao & Mizuno, 1999).

Porosity Estimation

Utilizing Lees et al. (2000)'s method, derived from Pickett's (1963) porosity theory, the study estimates porosity through velocity relationships:

$$\Psi = V_p \times V_s \quad (4.10)$$

$$\ln\Psi = \ln V_p \times \ln V_s \quad (4.11)$$

$$\frac{d\Psi}{\Psi} = \frac{dV_p}{V_p} + \frac{dV_s}{V_s} \quad (4.12)$$

Thus, Eqs. (4.3)–(4.12) can be utilized to calculate porosity.

The comprehensive dataset includes 65 three-component broadband seismographs monitoring 913 JHD relocated events of Mw 1.5–5.6 from 2017 to 2022, capturing 14,074 P-wave and 13,013 S-wave arrivals. The research utilized a 3D grid model spanning 330 km east–west and 40 km in depth, comprising 11,979 grid nodes. A grid spacing of 5 km was used for both horizontal and vertical directions (Mandal, 2024; Mandal et al., 2022).

4.2.3 Results from the Tomographic Study

In the Uttarakhand region, this technique has been used effectively to map low-velocity zones (LVZs) that coincide with the MHT at 8–22 km depth, where earthquakes such as the 1991 Mw 6.8 Uttarkashi and 1999 Mw 6.5 Chamoli occurred (Rai et al., 2006; Hazarika et al., 2017; Mandal et al., 2022; Mandal, 2024; Monika et al., 2024). These LVZs exhibit 10–20% reductions in Vp and Vs and a 10–15% increase in Vp/Vs, attributed to aqueous or metamorphic fluid presence (Fig. 4.3a–f). These low-velocity and high Vp/Vs anomalies have direct implications for earthquake generation mechanisms discussed in this Chapter (Sect. 4.2 and 4.3), particularly in terms of fluid-assisted rupture initiation.

The study validated findings by comparing modeled porosity, crack density, and saturation rate with magnetotelluric and velocity tomography studies along the NE–SW trending PM profile (Fig. 4.3a–g), referencing works by Rawat et al. (2014), Caldwell et al. (2013), Monika et al. (2024), Rai et al. (2006, 2009), and Mandal (2024).

Receiver Function Analysis

Receiver function (RF) analysis, particularly P-receiver functions (PRFs), is employed to estimate crustal and lithospheric discontinuities based on converted seismic phases (P-to-S) at impedance contrasts. Stacking and analysis of ~ 1,500 high-quality radial PRFs from 52 broadband stations (Fig. 4.3a-b) in the Uttarakhand Himalaya (out of a total of 76 broadband stations) revealed key crustal structures through Common Conversion Point (CCP) imaging. Three prominent converted phases corresponding to the Main Himalayan Thrust (MHT), Moho, and lithosphere-asthenosphere boundary (LAB) were identified (Figs. 4.4a–p, 4.5a–i and 4.6a–b). The observation of a Moho doublet in the Uttarkhand Himalaya through radial P-receiver functions (PRFs) provides robust evidence for complex crustal and upper mantle architecture in this sector of the orogen, echoing analogous findings from the Tibetan Plateau (Kind et al., 2002; Li et al., 2011). Distinct positive arrivals at 4–7 s and 7–9 s, corresponding to 35–55 km and 55–75 km depths (Figs. 4.4, 4.5, 4.6 and 4.7), respectively delineate a double Moho structure. This feature reflects the imbrication and underthrusting of Indian crust and upper mantle resulting from India-Eurasia collision processes, as previously interpreted by Yuan et al. (1997) and Zhao et al. (2001). Joint inversion of PRFs and Rayleigh wave group velocity dispersion (Mandal et al., 2023) further substantiates these interfaces by resolving sharp shear-velocity contrasts at the inferred Moho depths (Fig. 4.4). The reproducibility of these phases across multiple stations, verified through synthetic modeling using propagator matrix and time-domain deconvolution (Kennett, 1983; Ammon & Ligorria, 1999), emphasizes their tectonic significance (Figs. 4.5 and 4.6). It is noteworthy, however, that studies such as Rai et al. (2006), Oreshin et al. (2011), and Hazarika (2017) report regional variation or ambiguity regarding the double Moho, highlighting the value

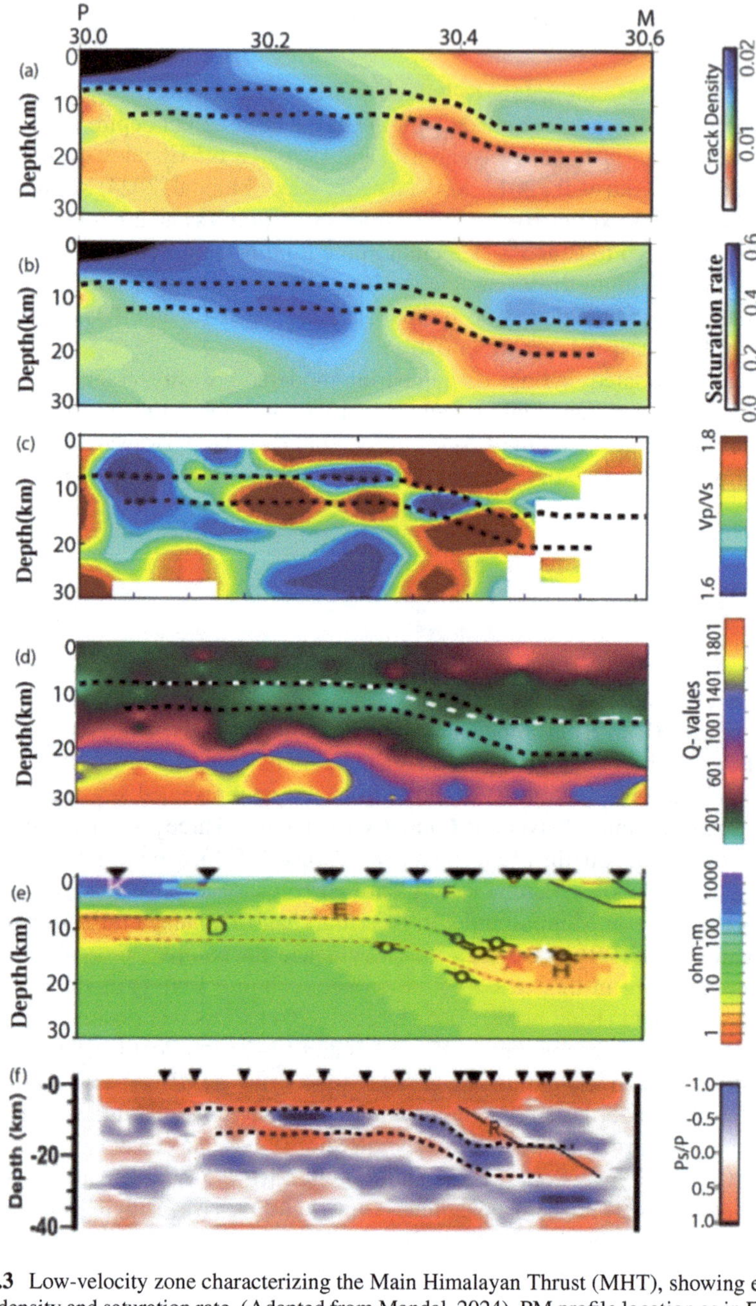

Fig. 4.3 Low-velocity zone characterizing the Main Himalayan Thrust (MHT), showing elevated crack density and saturation rate. (Adapted from Mandal, 2024). PM profile location as in Fig. 4.2a

of dense arrays and multi-parameter approaches for resolving Himalayan crustal structure.

Further, the CCP imaging identifies a ramp-flat-ramp geometry of the MHT as a negative Ps conversion at 8–22 km depth, suggesting a fluid-rich, low-velocity zone

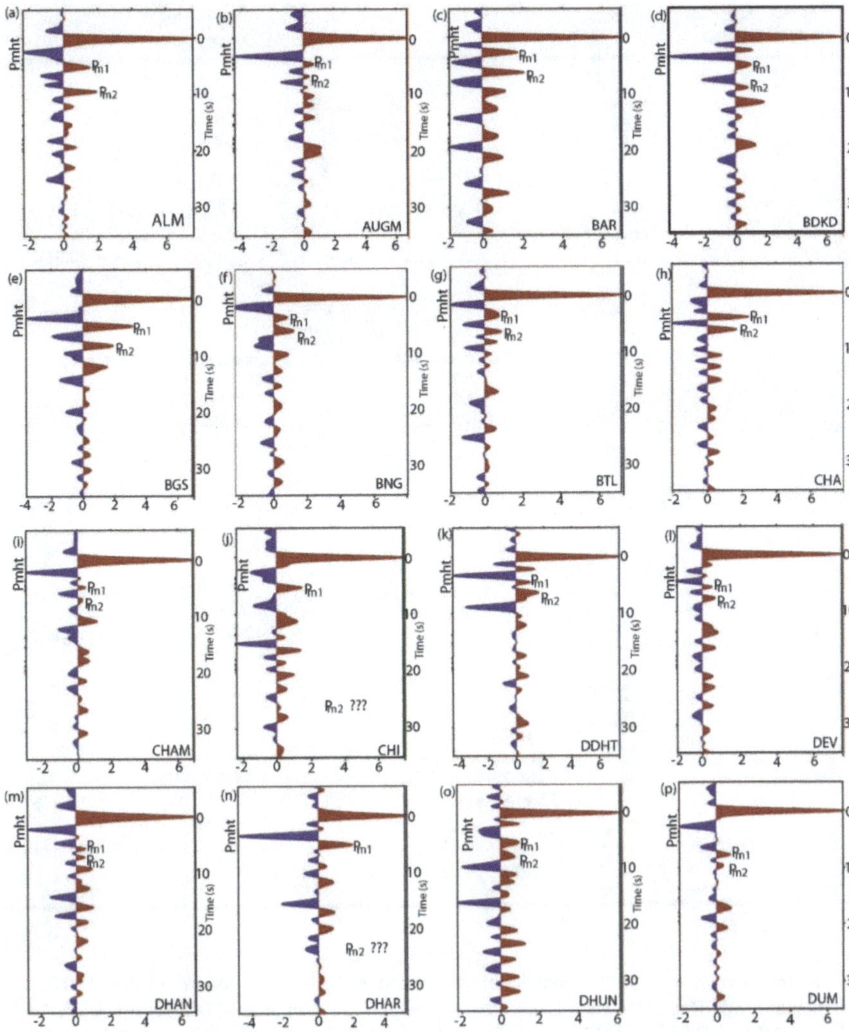

Fig. 4.4 Plot showing the radial PRF gather at **a** ALM, **b** AUGM, **c** BAR, **d** BDKD, **e** BGS, **f** BNG, **g** BTL, **h** CHA, **i** CHAM, **j** CHI, **k** DDHT, **l** DEV, **m** DHAN, **n** DHAR, **o** DHUN, and **p** DUM. The gather radial PRFs at 14 out of 16 stations show a clear and strong negative phase from MHT (P_{mht}) and two positive arrivals (P_{m1} and P_{m2}) associated with the Moho and a deeper Moho, respectively, which supports the idea of the presence of double Moho beneath the region. While at CHI and DHAR stations, the gather PRF shows one clear conversion from the Moho and a negative conversion from the MHT (P_{mht})

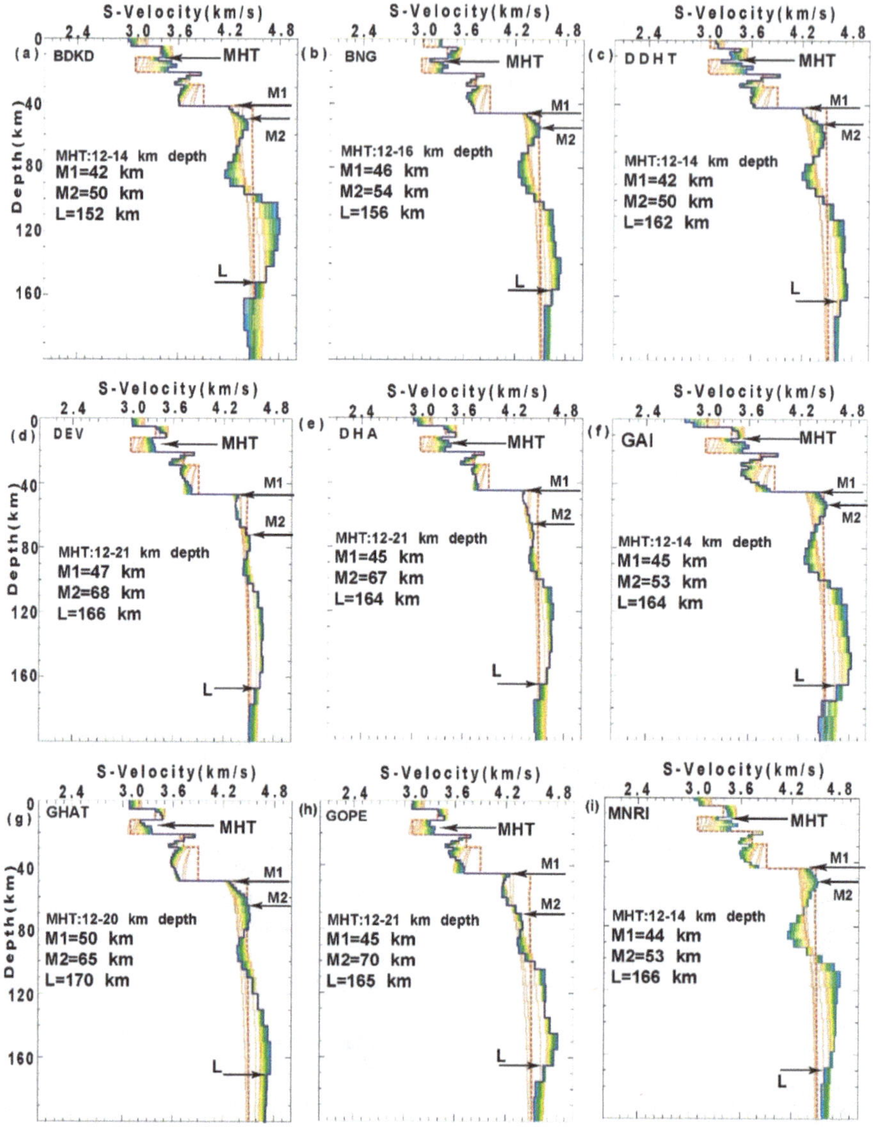

Fig. 4.5 Inverted shear velocity models obtained through the joint inversion of radial P-RFs and fundamental mode Rayleigh wave group velocity dispersion data at **a** BDKD, **b** BNG, **c** DDHT, **d** DEV, **e** DHA, **f** GAI, **g** GHAT, **h** GOPE and **i** MNRI stations (Mandal et al., 2023). Different colors represent different Vs models used for the joint inversion. The initial shear velocity model is shown by a thick red dotted line, while the final shear velocity model is shown by a thick blue line. Furthermore, MHT, M, and L mark the thickness of the Main Himalayan Thrust in km, Moho depths in km, and lithospheric thickness in km, respectively

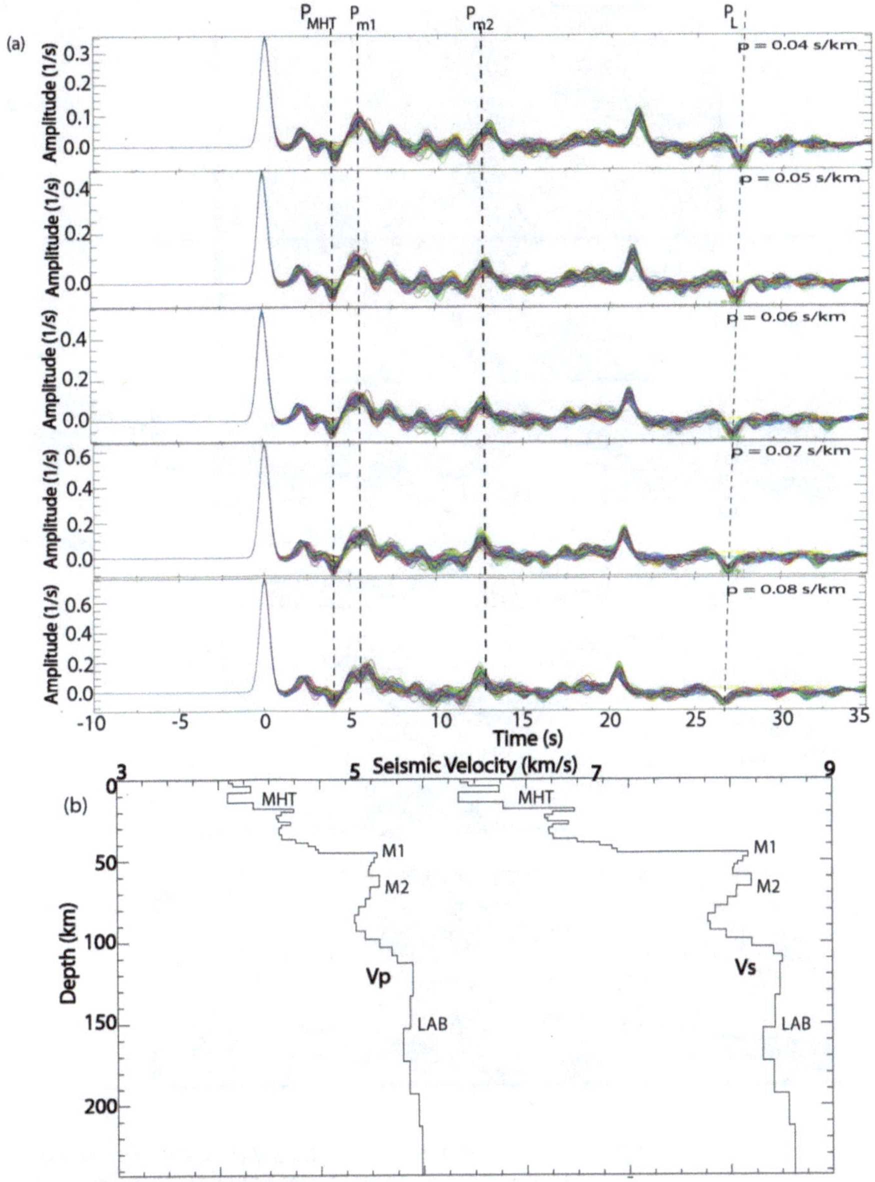

Fig. 4.6 **a** Synthetic PRFs for 100 velocity models for each horizontal slowness value from 0.04 to 0.08, and **b** the average velocity model modeled by the joint inversion of radial PRFs and Rayleigh wave group velocity dispersion data

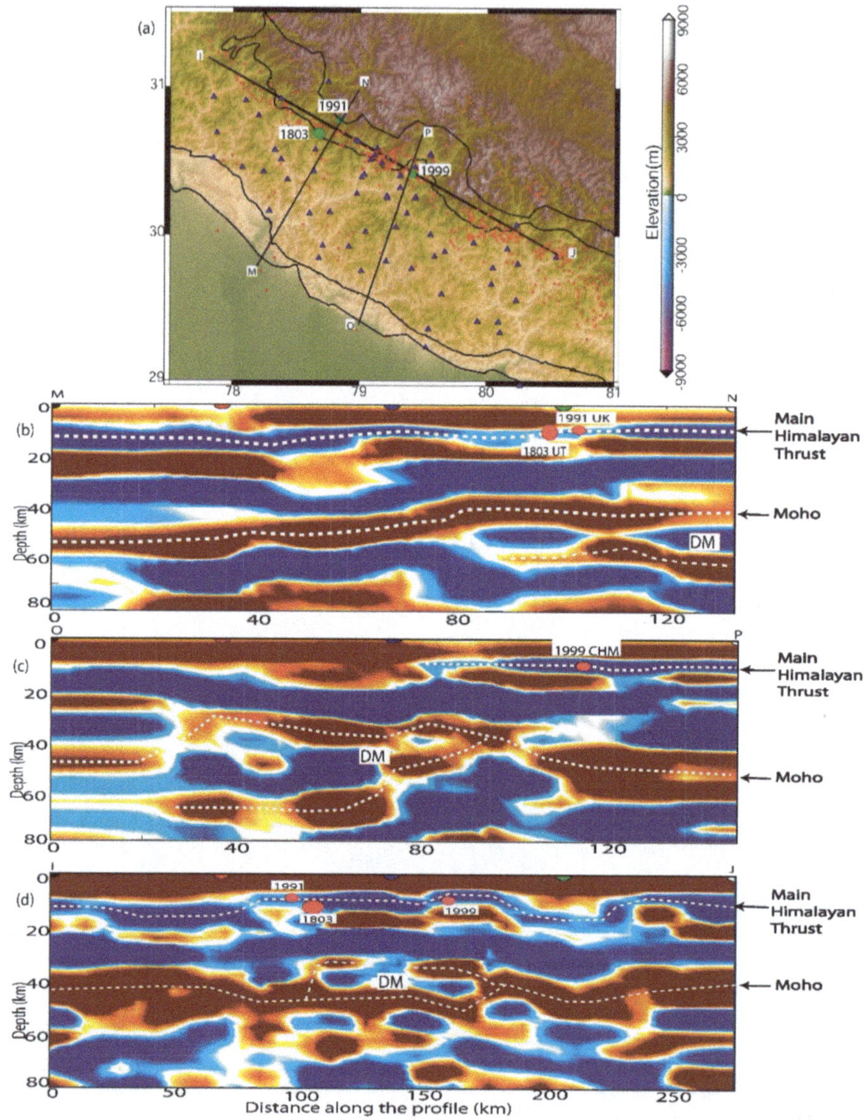

Fig. 4.7 2-D images from CCP stacking of radial PRFs along NE–SW (MN, OP) and NNW–SSE (IJ) profiles (locations in Fig. 4.2a). Radial PRF images along **b** MN, **c** OP, **d** IJ. White dotted lines: MHT and Moho depths. DM: double Moho structure. Pink circles: epicentral locations of 1991 Mw6.8 Uttarkashi, 1999 Mw6.5 Chamoli, and 1803 Mw7.8 Garhwal events. For correlation with seismicity characteristics across depth levels, see Table 4.1

likely responsible for recent microseismicity (Figs. 4.4, 4.5, 4.6 and 4.7). The NE-dipping LAB is imaged as a negative Ps phase at 130–180 km depth (Fig. 4.7b–d). These results, consistent with studies by Mandal (2023 and 2024), highlight crustal complexity and potential fluid involvement in Himalayan seismogenesis.

The H–K and CCP stacking of radial PRFs (Zhu & Kanamori, 2000) at 52 stations (out of a total of 76 broadband stations) performed by CSIR-NGRI further reveal transverse structural segmentation in the crust and upper mantle associated with inherited basement features like the Delhi–Haridwar Ridge and Moradabad Fault (Fig. 4.8a–b; Mandal et al., 2021, 2023). These structures correlate with distinct variations in seismic velocity and crack density, delineating seismogenic asperities and barriers (Mandal, 2024; Mandal et al., 2023).

Further, depth migration of PRFs reveals low-velocity layers in the Mantle Transition Zone (MTZ) between 350–385 km (LVL-410) and 590–640 km (LVL-660), interpreted as remnant subducted lithosphere and partial melting zones (Mandal et al., 2024; Rai et al., 2009b). These imaging results provide critical constraints on the spatial distribution of crustal features discussed in the Sect. 4.2, particularly in relation to the depth localization of seismicity along the MHT.

Magnetotellurics (MT)

Magnetotelluric methods infer the electrical conductivity structure of the crust and upper mantle by measuring natural variations of electric and magnetic fields at the Earth's surface. MT studies in the Himalayan arc have detected high-conductivity zones coinciding with the MHT and deeper shear zones, implying the presence of fluids and partial melts (Fig. 4.9a–c; Israil et al., 2008; Rawat et al., 2014; Caldwell et al., 2013; Monika et al., 2024). Such observations complement seismic low-velocity anomalies and strengthen the fluid-induced weakening hypothesis for earthquake nucleation.

The NE–SW geoelectric models of Rawat et al. (2014) and Israil et al. (2008) reveal mid-crustal fluid zones through the identification of prominence of low-resistivity anomalies at mid-crustal depths beneath the Main Central Thrust (MCT) in the Garhwal Himalaya. Both employed broadband magnetotelluric (MT) profiling, with robust impedance tensor decomposition (notably the Groom-Bailey method in Israil et al., 2008) to correct for galvanic distortion and determine regional geoelectric strike. Their 2D inversion consistently models show a resistive upper crust (interpreted as Indian basement) overlain by conductive sediments, but crucially, a distinct conductive zone at depths of 8–22 km beneath the MCT (Fig. 4.9a–c). This anomalous conductivity, spatially correlated with microseismicity, is interpreted as evidence for intertwined mid-crustal fluids or partial melts (Fig. 4.9b–c). These fluids are derived from metamorphic dehydration reactions or crustal melting, and their presence is considered to weaken the mid-crust, facilitating ductile flow and accommodating Himalayan deformation. Thus, the MT-derived geoelectric models (Fig. 4.9b–c) provide compelling geophysical evidence for the existence and tectonic significance of mid-crustal fluid zones in the Himalayan orogen.

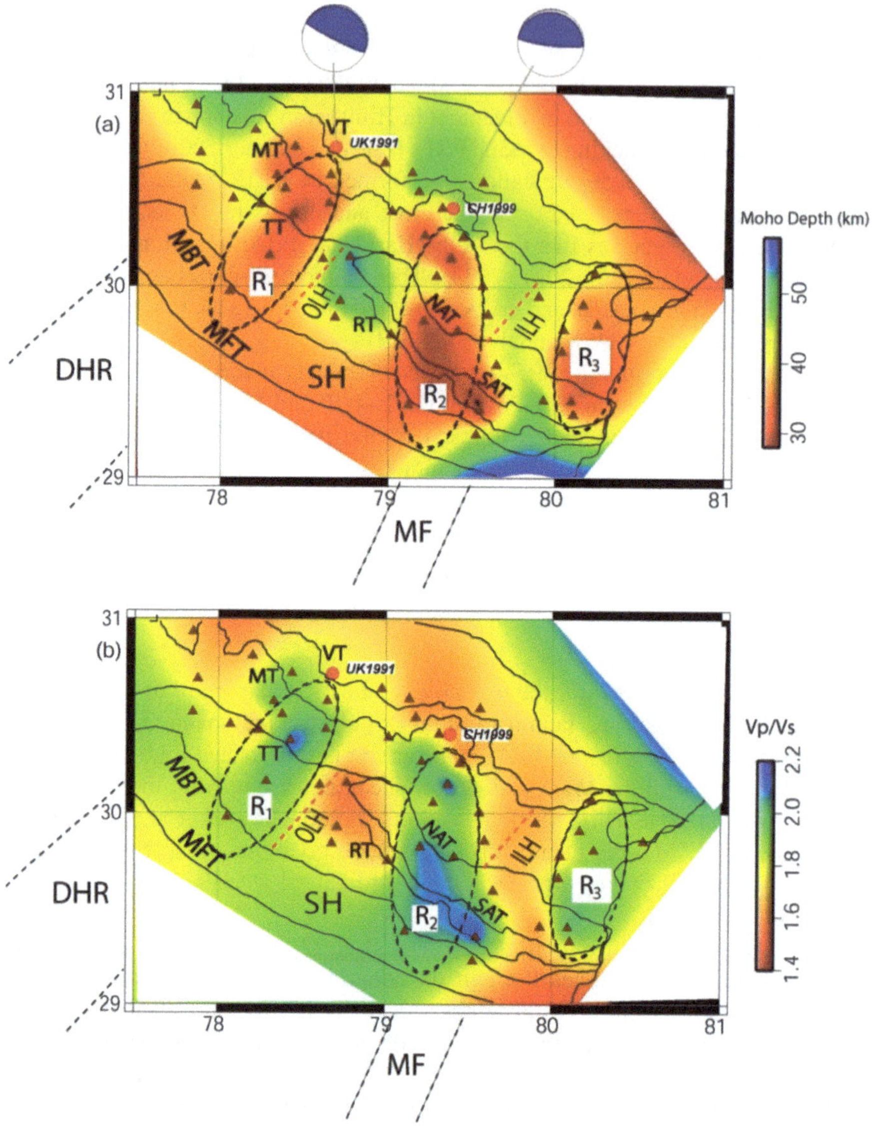

Fig. 4.8 Moho depth variations and crustal Vp/Vs distribution map across Uttarakhand Himalaya (adapted from Mandal et al., 2021). For correlation with seismicity characteristics across depth levels, see Table 4.1

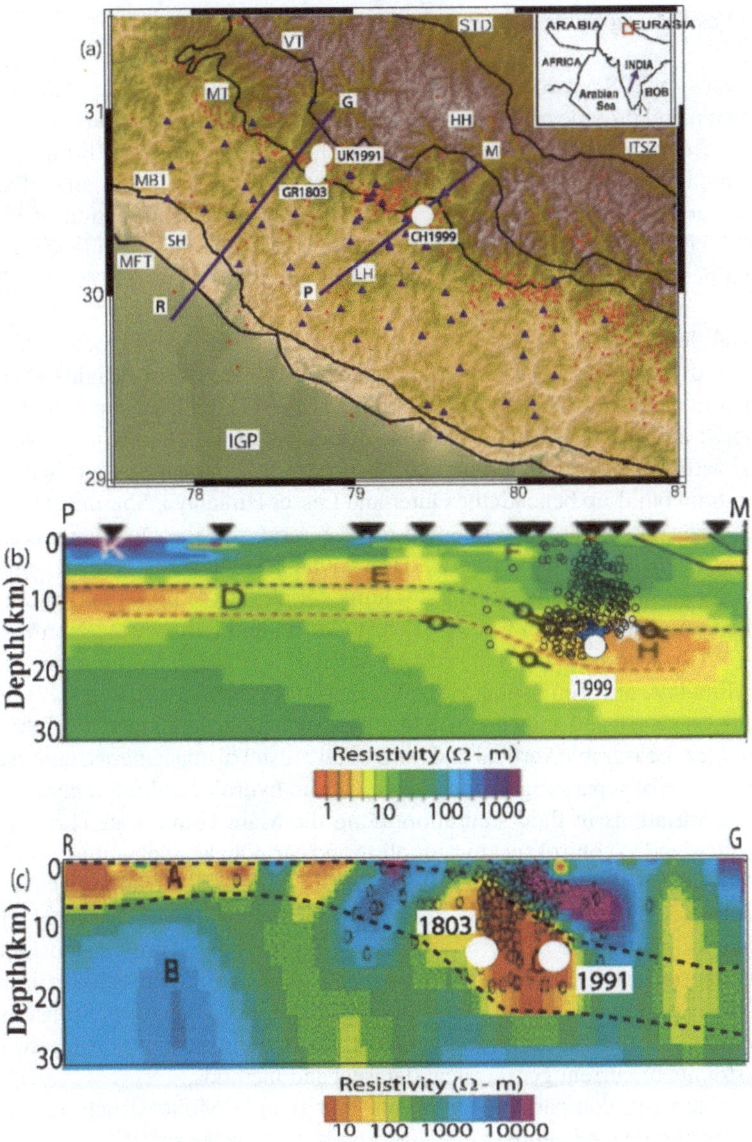

Fig. 4.9 **a** Station location map of Uttarakhand: blue triangles—broadband stations, white circles—earthquake relocations, black line—major thrust faults. NW–SE trending profiles RG (Roorkee–Gangotri) and PM (Pauri–Munsyari) across rupture zones of 1991 Uttarkashi (M_W6.8) and 1999 Chamoli (M_W6.5) earthquakes (thick blue lines). Epicenters of 1991, 1999, and 1803 (Mw7.8) earthquakes in white circles. **b** Geo-electric model along PM profile (XX' profile of Monika et al., 2024). **c** Geo-electric model along RG profile (after Israil et al., 2008)

Global Positioning System (GPS) Geodesy

High-precision GPS networks deployed across the Himalayan belt quantify surface deformation and provide critical estimates of interseismic strain accumulation. Data from the Uttarakhand region indicate a convergence rate of ~ 15–18 mm/yr and a strong coupling along the MHT, with an estimated 3 m slip deficit since the 1803 Garhwal earthquake (Bilham, 2019). This reinforces the identification of Uttarakhand as a potential for a future $M_W \geq 7.5$ event. This geodetic evidence supports the identification of seismic gaps and rupture potential discussed in this Chapter (Sects. 4.2 and 4.3).

Crustal deformation studies using continuous and campaign-mode GPS observations across the Kumaon–Garhwal Himalaya provide critical insights into strain accumulation and tectonic loading. Gautam et al. (2017) reported slip deficit rates of ~ 18 mm/yr across a ~ 100-km-wide locked segment of the Main Himalayan Thrust (MHT), with crustal velocities increasing northward, indicating significant interseismic strain build-up beneath the Outer and Lesser Himalaya. Sharma et al. (2020) estimated crustal convergence rates up to 13.8 mm/yr in the Dehradun–Badrinath sector, with reduced motion across the Himalayan Frontal Thrust (HFT), suggesting its locked status and seismic potential. Dumka et al. (2014) found maximum strain accumulation near the Main Central Thrust (MCT) and north of the North Almora Thrust (NAT), with horizontal shortening rates of ~ 6.7 ± 2.5 mm/yr between the Lesser Himalaya and Peninsular India. Furthermore, Gahalaut et al. (2017) demonstrated that seasonal and anthropogenic factors, such as the loading of the Tehri reservoir, produce measurable vertical and horizontal crustal displacements, underscoring the importance of separating tectonic signals from hydrological influences.

Spatial variations in fluid saturation along the Main Himalayan Thrust (MHT) are hypothesized to control seismic coupling and earthquake nucleation, as indicated by geophysical imaging, MT, and GPS studies (Table 4.2). Seismic velocity data, resistivity, and MT imaging reveal fluid-rich zones, which lower frictional strength and influence earthquake behavior. Integrating these methods provides insights into fluid distribution and its impact on fault coupling, although interpretation is limited by the depth resolution and uncertainties of each technique. Overall, multidisciplinary evidence supports the fluid control hypothesis, but results must be considered within the constraints of current geophysical datasets and methods.

Our systematic comparison highlights that double Moho structures, while a striking seismic image feature in collision zones, show weak spatial correlation with seismicity patterns and segmentation (Table 4.3). Instead, variations in MHT geometry and segmentation appear to play more direct roles in controlling maximum earthquake magnitudes and the partitioning of slip along the megathrust. High fluid signatures may further influence rupture processes, though their effect is context-dependent. This suggests that the geodynamic relevance of double Moho signatures lies more in their record of long-term crustal evolution, rather than as controls on seismogenic segmentation. These trends are consistent with observations elsewhere in the Himalaya and other orogenic belts (e.g., Alps), where deep structure is necessary but not sufficient for controlling seismic hazard.

Table 4.2 Resolution, uncertainty, and limitations of different Geophysical Methods

Data type	Resolution (km)	Uncertainty	Key limitations
Seismic tomography	5–10	± 0.2 km/s	Sparse station spacing in NE Uttarakhand
Receiver function imaging	~ 10	± 2 km	Lateral heterogeneity, ambiguous Moho phases
Magnetotellurics	Variable	± 10% in resistivity	Conductive noise, non-uniqueness
GPS geodesy	Station-based	± 1 mm/yr	Anthropogenic/hydrologic signal contamination

Table 4.3 Comparison of results from Uttarkhand and other regions

Region	MHT geometry	Fluid signature	Max. observed Mw	Segmentation
Uttarakhand	Ramp-flat-ramp	High	7.8	Strong
Nepal	Ramp-flat	Moderate	8.2	Weak
Sikkim	Flat-ramp	Low	6.9	Moderate
Tibet	Multiple ramps	Variable	8.2	Variable

4.2.4 Conclusion

The seismic character of the Uttarakhand Himalaya is increasingly being understood as a complicated interaction between inherited crustal architecture and fluid-driven mechanics. Recent high-resolution multi-parametric imaging, tomography, receiver functions, and magnetotellurics show the Main Himalayan Thrust (MHT) is not a uniform sliding surface but a segmented engine, partitioned by transverse basement faults and lubricated by fluid-rich domains. This compartmentalization appears to serve as a structural barrier that promotes frequent release of strain of moderate magnitude but may inhibit the along-strike propagation of great, contiguous ruptures. But the crucial question is, are those segmentation barriers permanent barriers or ephemeral obstacles that could be overcome in a rare cascading super-event? To do so, we need to move to dense, persistent monitoring using high-resolution seismic arrays and time-lapse geodesy to capture the transient signals of fault loading. Only through such rigorous, multidisciplinary scrutiny can we hope to refine our forecasts and understand the seismic potential of this poised Himalayan segment.

4.3 Earthquake Recurrence, Segmentation and Mantle Dynamics

The Uttarakhand segment of the Himalaya represents a geodynamically complex and seismically hazardous portion of the central Himalayan arc. Its tectonic architecture is fundamentally controlled by the deep structure of the Main Himalayan Thrust (MHT), variations in crustal and lithospheric thickness, and the distribution of fluids and partial melts in the mid to lower crust. A comprehensive understanding of these deep structural features is essential for deciphering the mechanisms of earthquake nucleation, rupture dynamics, and seismic hazard potential in this densely inhabited region. This chapter synthesizes recent advances in seismic imaging, joint inversion, and geophysical modeling to present a detailed account of the deep structure beneath Uttarakhand Himalaya.

4.3.1 Introduction

The central Himalayan arc, particularly the Uttarakhand segment, is characterized by active tectonics and significant seismic hazard. The region's tectonic framework is governed by the Main Himalayan Thrust (MHT), which acts as the principal interface for seismic coupling between the Indian and Eurasian plates. Understanding the geometry, segmentation, and physical properties of the MHT, as well as associated crustal and lithospheric features, is crucial for assessing earthquake potential and hazard in the region.

4.3.2 Geometry and Segmentation of the Main Himalayan Thrust (MHT)

The MHT is a shallow, north-dipping décollement that serves as the primary zone of seismic coupling. High-resolution receiver function analysis and seismic tomography have revealed that the MHT beneath the Kumaon–Garhwal Himalaya exhibits a complex ramp-flat-ramp geometry, with major seismogenic segments occurring at depths of 8–22 km (Hazarika et al., 2017; Mandal et al., 2022; Mandal, 2023 and 2024; Kanaujia et al., 2025). The presence of ramps in the MHT is inferred from seismicity clustering and sharp velocity contrasts, particularly beneath Uttarkashi and Chamoli, the sites of the 1991 (M_W6.8) and 1999 (M_W6.5) earthquakes, respectively (see Figs. 4.3, 4.7, and 4.9). These ramps serve as stress accumulation sites and are considered likely nucleation points for future earthquakes (Mandal et al., 2021). In contrast, the flat segments may represent creeping zones or areas of elevated fluid pressure that decouple the plates.

The geometry of the Main Himalayan Thrust (MHT) in the Kumaon and Garhwal sectors exhibits critical along-strike and along-dip variability, which fundamentally modulates both interseismic locking and coseismic rupture characteristics (Gautam et al., 2017). Notably, the mid-crustal ramp—where the MHT rises from a gently north-dipping basal décollement to a steeper segment—plays a decisive role. The GPS investigation by Gautam et al. (2017) and the seismic imaging study by Caldwell et al. (2013) both show that the ramp in the Garhwal Himalaya is rather wide and has a modest dip of about 12–15°. It goes down-dip for 30–40 km. This shape makes a wide transition between the completely locked foreground and the deeper, aseismically creeping base. This allows for more evenly distributed interseismic strain and encourages ramp-confined, moderate-magnitude seismicity (Gautam et al., 2017). Conversely, in the Kumaon Himalaya, the MHT ramp is considerably steeper, with dips as high as ~ 35° (Caldwell et al., 2013; Gautam et al., 2017) and is markedly narrower, spanning just 7–10 km downdip. This pronounced steepening at the ramp-flat transition localizes stress accumulation and facilitates abrupt unlocking, permitting larger, through-going ruptures that may engage multiple hinterland faults. The back or backward ramp further influences locked-zone width and the depth limit of seismicity, with its geometry modulating the coupling transition and seismic hazard potential (Gautam et al., 2017). Such segmentation, governed by ramp geometry, explains the observed variability in earthquake size, style, and recurrence, highlighting the importance of high-resolution structural constraints in regional seismic hazard assessment.

A joint inversion methodology integrating radial P-receiver functions (PRFs) and Rayleigh wave group velocity dispersion data has been employed to resolve 1-D shear velocity models at 45 seismic stations (out of a total of 76 seismic stations) across the Uttarakhand Himalaya (Mandal et al., 2023). Utilizing Herrmann's (2004) joint inversion code, the study models the MHT, Moho, and lithosphere-asthenosphere boundary (LAB) depths down to 200 km. The inversion reveals substantial lateral variations: MHT thicknesses of 2–9 km, Moho depths of 39–50 km, and lithospheric thicknesses of 136–176 km (see Fig. 4.10). These results, corroborated by CCP stacking, expose three NNE-trending transverse structures (C1–C3), suggesting lithospheric segmentation due to inherited basement ridges such as the Delhi–Haridwar Ridge (DHR), Moradabad Fault (MF), and Great Boundary Fault (GBF). This segmentation likely limits rupture propagation, thus reducing seismic hazard in the region (Fig. 4.8; Mandal et al., 2021, 2023; Manglik et al., 2022; Caldwell et al., 2013).

Recent seismic imaging further reveals lateral segmentation of the MHT caused by inherited basement structures, disrupting the continuity of the thrust and possibly limiting rupture length and earthquake magnitude. This is evidenced by the absence of $M_w \geq 8.0$ events in Uttarakhand as compared to less segmented regions like Nepal and Assam (Mandal et al., 2023).

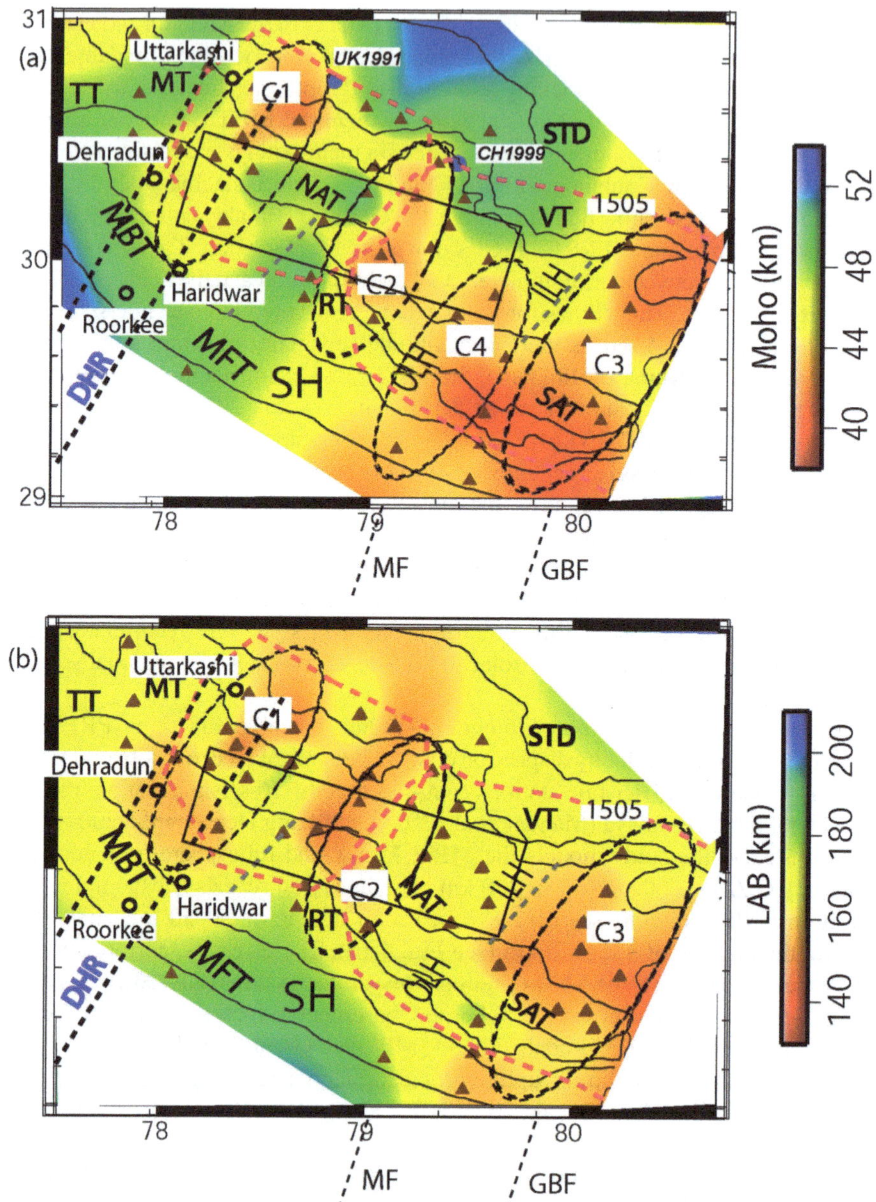

Fig. 4.10 **a** Crustal and **b** lithospheric cross-sections showing major tectonic thrusts and segmentation. Adapted from Mandal et al. (2023)

4.3.3 Crustal Thickness, Moho Depth, and Lithospheric Structure

Crustal thickness and Moho depth beneath the Uttarakhand Himalaya exhibit substantial spatial variability. H–K stacking of radial P-receiver functions (Zhu & Kanamori, 2000) across 42 broadband seismic stations (out of a total of 76 broadband stations) has resolved Moho depths ranging from 28.3 to 52.9 km, with corresponding Vp/Vs values from 1.59 to 2.13, indicating significant heterogeneity in crustal composition (Mandal et al., 2021). The Lower Himalaya generally shows thinner crust (~ 35–45 km), while the Higher Himalaya exhibits a thicker crust (~ 50–55 km), consistent with crustal underplating and shortening. The lithospheric structure, as imaged through joint inversion of P-receiver functions and group velocity dispersion of Rayleigh waves, indicates the presence of a northeast-dipping lithosphere–asthenosphere boundary (LAB) at depths of 130–180 km (Mandal, 2023). The spatial heterogeneity in crustal thickness described here correlates with variations in rupture dynamics and seismic hazard as discussed in this Chapter (Sect. 4.3). The modeled deep structural features of the Uttarakhand Himalaya are listed in Table 4.4.

Mandal et al. (2024) applied depth migration of stacked radial PRFs using CCP (Common Conversion Point) imaging to investigate mantle discontinuities beneath the Uttarakhand Himalaya. Analysis of teleseismic data from 52 broadband stations (out of a total of 76 broadband stations) revealed variations in arrival times for the 410-km (d410) and 660-km (d660) discontinuities, averaging at 406 ± 8 km and 659 ± 10 km, respectively, suggesting a thickened Mantle Transition Zone (MTZ) (255 ± 7 km) (see Fig. 4.11). Notably, low-velocity layers (LVLs) were detected atop both discontinuities—LVL-410 (350–385 km) and LVL-660 (590–640 km)—implying hydration or partial melt due to past subduction. These results indicate preserved remnants of paleo-subducted slabs and suggest lithospheric thickening toward the NE beneath the Lesser Himalaya (Bercovici & Karato, 2023; Mandal et al., 2024; Oreshin et al., 2011; Rai et al., 2009b; Toffelmier & Tyburczy, 2007). These MTZ features are further linked to crustal stress modulation and seismicity patterns detailed in this

Table 4.4 Key deep structural features of the Uttarakhand Himalaya

Feature	Uttarakhand Himalaya characteristics
MHT geometry	Ramp-flat-ramp; segmented by transverse structures (DHR, MF, GBF)
Crustal thickness (Moho)	28–55 km (thinner in south, thicker in north)
Lithospheric thickness (LAB)	130–180 km, dipping northeast
Mantle transition zone (MTZ)	Thickened (~ 255 km), deep d660, low-velocity layers above d410
Fluids/partial melts	High Vp/Vs, low velocities, high crack density and saturation; weakens MHT
Seismicity	Concentrated in upper crust, especially near ramps and transverse structures

Chapter (Sect. 4.2). Thus, the MTZ beneath the region shows evidence of structural thickening (255 ± 7 km), with a deepened d660 discontinuity (~ 659 km) and a low-velocity zone above the d410 (~ 350–385 km), possibly representing hydrated remnants of the subducted Neo-Tethyan slab (Mandal et al., 2024). These features suggest that mantle dynamics play a role in modulating the overlying crustal stress field and thermal structure.

4.3.4 Role of Fluids, Partial Melts, and Fault Weakening

Seismic velocity models across the MHT consistently show low P- and S-wave velocities and elevated Vp/Vs ratios (1.74–1.92) within the 8–22 km depth range (Hazarika et al., 2017; Monica et al., 2024; Mandal et al., 2022). These anomalies, coupled with high crack density (ε ~ 0.011–0.012) and saturation rate (ξ ~ 0.38–0.39), indicate the pervasive presence of metamorphic and aqueous fluids within the MHT zone (Mandal, 2024).

Such fluids drastically reduce the effective normal stress and frictional resistance on fault planes, facilitating earthquake nucleation through fluid overpressure or hydrofracturing mechanisms (Hazarika et al., 2017; Israil et al., 2008; Rawat et al., 2014). In deeper sections, partial melting—inferred from electrical conductivity and low-velocity anomalies—may further weaken the lithosphere and localize deformation along the MHT. The combined presence of fluids and melts explains the frequent occurrence of moderate-magnitude earthquakes along this segment and the absence of seismicity in the lower crust, which may be ductile due to elevated temperatures and melt presence.

4.3.5 Conclusions

Recent multidisciplinary investigations reveal that the Uttarakhand segment of the Himalaya is underlain by a segmented, fluid-enriched Main Himalayan Thrust (MHT), whose geometry and physical state are critical for understanding the region's seismogenesis. Seismic tomography, magnetotelluric imaging, and precise geodetic measurements indicate pronounced lateral variations in crustal and lithospheric thickness, zones of partial melting, and significant pore-fluid overpressure at depth. These factors collectively modulate the mechanical behavior of the fault, influencing both the partitioning and accumulation of tectonic stress. Of particular interest is the evidence that hydration and partial melting within the Mantle Transition Zone (MTZ) may facilitate upward migration of fluids, locally reduce effective stress, and promote segmentation within the MHT. To resolve the spatiotemporal evolution of these heterogeneities, it is essential to deploy denser, multimodal geophysical

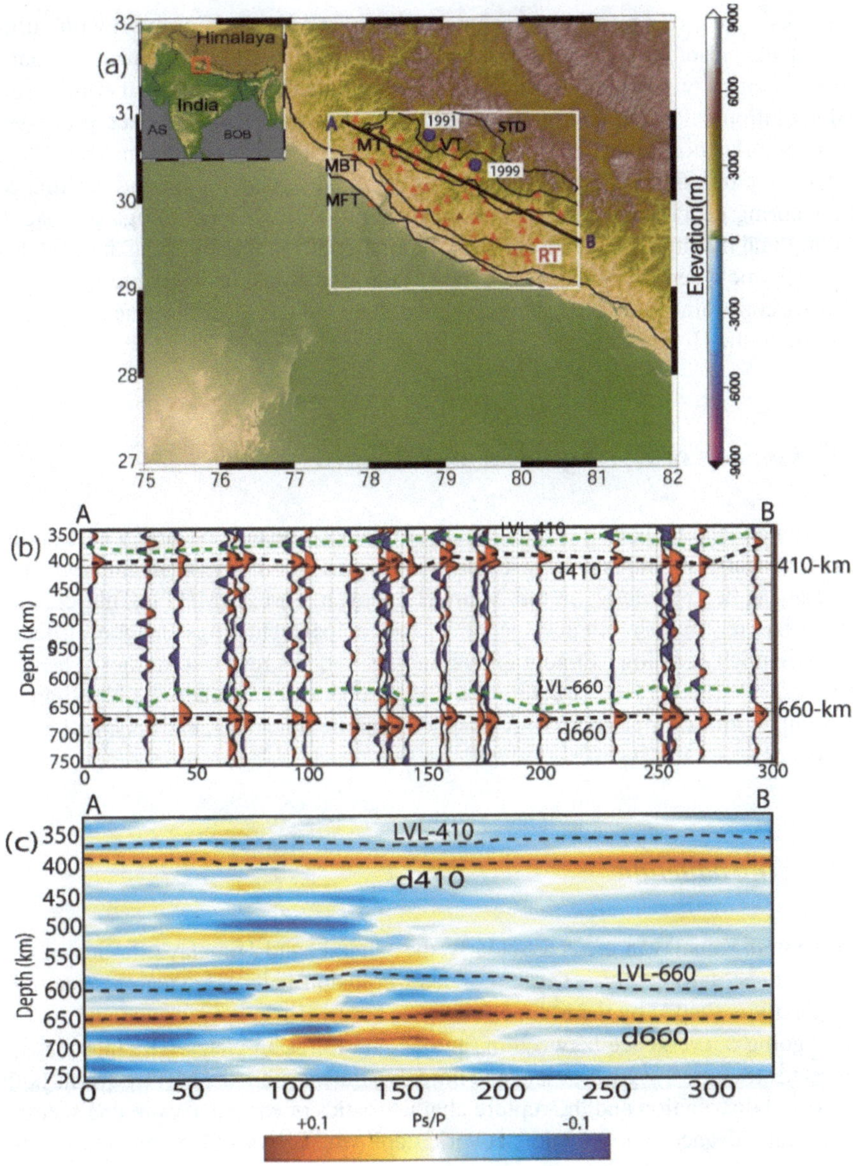

Fig. 4.11 Stacked radial PRFs and CCP images showing the nature of Mantle Transition Zone below the Uttarakhand Himalaya

networks—comprising seismic, magnetotelluric, and geodetic arrays—with sufficient spatial resolution to track evolving fluid pathways and mechanical discontinuities. Critically, integrative models that couple mantle dynamics, thermochemical evolution, and fault-zone mechanics are needed to assess how deep processes exert control over crustal earthquake nucleation and rupture propagation. The role of direct sampling, through ambitious drilling initiatives, must not be underestimated in anchoring geophysical inference to material properties. Only through persistent, longitudinal monitoring and coordinated international collaboration can the complex feedbacks between lithospheric structure, fluid dynamics, and seismic hazard be robustly constrained, thereby advancing predictive models of earthquake recurrence and size in the Himalaya.

4.4 Genesis of Earthquakes and Hazard

The Uttarakhand Himalaya is a region of pronounced seismic hazard, where earthquake genesis is fundamentally controlled by the geometry, segmentation, and mechanical heterogeneity of the Main Himalayan Thrust (MHT). This chapter synthesizes recent seismological, geodetic, and geological findings, integrating dense microseismicity catalogs, tomographic imaging, and site response studies. The role of fluids, inherited crustal structures, and lithospheric segmentation is highlighted, with implications for earthquake recurrence, rupture dynamics, and seismic hazard assessment.

4.4.1 Introduction

The central Himalayan arc, particularly the Uttarakhand Himalaya, is one of the world's most seismically active continental collision belts. The region's tectonic complexity is shaped by the Main Himalayan Thrust (MHT), which accommodates the ongoing convergence between the Indian and Eurasian plates. The interplay of inherited structures, segmentation, and fluid dynamics within the MHT governs both the spatial distribution and the rupture characteristics of earthquakes in this sector.

Recent advances in seismic network deployment, waveform modeling, and geophysical imaging have enabled high-resolution mapping of earthquake hypocenters, crustal velocities, and fluid distributions. These insights are critical for understanding the mechanisms of seismogenesis, the locked-to-creeping transition, and the implications for seismic hazard in densely populated Himalayan foothills.

4.4.2 Spatial and Depth Distribution of Seismicity

The Uttarakhand Himalaya has experienced several damaging earthquakes, including the 1803 M_w7.8 Garhwal, 1991 M_w6.8 Uttarkashi, and 1999 M_w6.5 Chamoli events (Fig. 4.12a). These moderate to large earthquakes nucleated at depths of 15–20 km, directly along the MHT, as confirmed by waveform modeling and precise relocations (Joshi et al., 2014; Monika et al., 2024).

Dense seismic monitoring by CSIR-NGRI since 2017, utilizing 56 broadband stations (out of a total of 76 broadband stations), has cataloged thousands of well-located microearthquakes down to 30 km depth. The majority of seismicity is concentrated between 8 and 22 km, coinciding with a fluid-rich, low-velocity zone along the MHT (Mandal, 2024; Mandal et al., 2022; Fig. 4.12a–b). Receiver function imaging and seismic tomography delineate this seismogenic zone, which correlates with ramp segments of the MHT and zones of enhanced pore-fluid pressure.

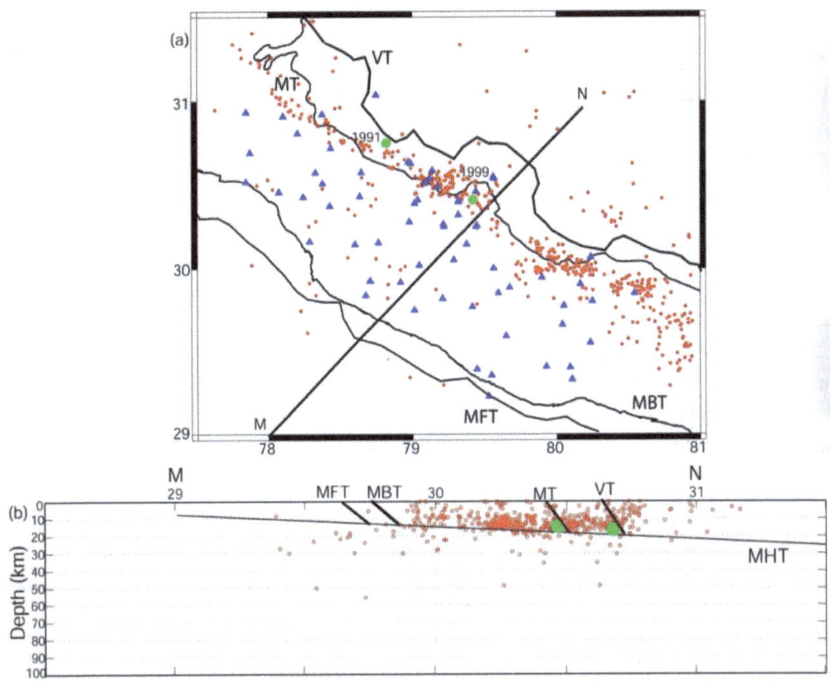

Fig. 4.12 **a** Earthquake relocations (small red circles) determined by simultaneous inversion. Three-component broadband seismographs (blue triangles). Locations of the 1991 M_W6.8 Uttarkashi and 1999 M_W6.5 Chamoli earthquakes (filled green circles). Major thrusts: Vaikrita Thrust (VT), Munsiari Thrust (MT), Main Boundary Thrust (MBT), Main Frontal Thrust (MFT). **b** Depth cross-section of relocated events across AB profile, showing seismicity along and above the MHT. Red open circles: relocated earthquakes. North-dipping major thrusts marked by thick black lines

The lower crust (below ~ 25 km) is largely aseismic, consistent with ductile rheology and elevated temperatures inferred from tomographic models and Mantle Transition Zone anomalies (Fig. 4.12a–b; Table 4.5). This locked-to-creeping transition at the base of the brittle crust defines the down-dip limit of large earthquake rupture, a pattern observed in other segments of the Himalayan arc.

The seismic inventory of the Uttarakhand Himalaya (Table 4.5) confirms that the main seismogenic engine of the region is the 8–22 km depth range of the Main Himalayan Thrust (MHT), a critical interval with low seismic velocities and fluid-saturated patches, where most moderate to large events are born. Below ~ 25 km depth, the brittle crust is replaced by a warmer ductile regime, as evidenced by the rapid disappearance of seismicity into the deeper lithospheric shadows. This tectonic stage is further complicated by a set of inherited "ribs" that divide the crust into distinct mechanical blocks (Delhi–Haridwar Ridge and the Moradabad and Great Boundary Faults), effectively controlling the extent and limit of modern ruptures. The end product of tomography and attenuation studies is a portrait of a primed

Table 4.5 Crustal structure, seismicity, and physical properties of the Uttarakhand Himalaya

Depth range (km)	Structural feature	Seismic velocity anomaly	Fluid/ metamorphic signature	Seismicity characteristics	References
0–8	Upper crust	Near-normal Vp, Vs	Low fluid content	Sparse microseismicity	Caldwell et al. (2013), Mandal (2023), Mandal et al. (2023)
8–22	Main Himalayan Thrust (MHT) zone	10–20% drop in Vp, Vs; 10–15% ↑ Vp/Vs	High fluid pressure, low velocity	Concentrated micro- and moderate earthquake activity (Mw 6–7.8)	-do-
22–25	Lower brittle crust	Transition to higher velocities	Decreasing fluid content	Seismicity decreases sharply	-do-
25–50	Lower crust/Moho	Sharp Vs increase (~ 0.5–0.7 km/s)	Ductile, partial melt possible	Largely aseismic	-do-
50–130	Uppermost mantle	Gradual velocity increase	Ductile, low seismicity	Rare deep events	-do-
130–180	Lithosphere–asthenosphere boundary	1–3% drop in Vs	Possible partial melt, fluids	No significant seismicity	-do-

and partitioned interface where pore-fluid pressures control friction and structural barriers determine whether a rupture is a local crisis or a regional catastrophe.

Recent studies also highlight the role of partial melts and crustal fluids in modulating seismicity, with high Vp/Vs anomalies and low-resistivity zones indicating the presence of fluids and melts that weaken the crust and localize deformation (Israil et al., 2008; Rawat et al., 2014; Hazarika et al., 2017; Mandal, 2022; Mandal, 2024). Fluid overpressure is increasingly recognized as a key factor in earthquake nucleation and rupture propagation in the Himalaya.

4.4.3 Segmentation, Locking, and Rupture Dynamics

The MHT is cut into discrete mechanical blocks by major transverse "ribs"—the Delhi–Haridwar Ridge, the Moradabad Fault and the Great Boundary Fault—which control the spatial distribution and recurrence of earthquakes in Uttarakhand through a first-order lithospheric partitioning. The inherited basement structures create a mosaic of coupling behaviors: locked segments, identified by velocity ramps and low fluid mobility, are high-strain asperities primed for major failures like the 1803 Garhwal event, while fluid-rich creeping domains can act as seismic sponges or sites of silent slip. GPS data reveal a massive slip deficit of ~ 3 m since 1803, making the segment ripe for an $M_w > 7.5$ earthquake. The modeling of the waveforms of local microseismicity further emphasizes this structural quirk, with low stress drops and small source radii indicative of a highly heterogeneous rupture process controlled by fluctuating pore-fluid pressures. The Uttarakhand Himalaya, much like the Nepal and Sikkim arcs in its vicinity, is a prisoner of its deep-seated geometry, where the magnitude of any future rupture is, in the end, controlled by these ancestral structural boundaries.

4.4.4 Site Amplification and Local Hazard Implications

The most destructive expression of the structural heterogeneity of the Uttarakhand Himalaya is at the surface, where local site conditions are powerful amplifiers of seismic energy. In the river valleys and urban basins of Dehradun, Pithoragarh and Almora, the soft sedimentary cover and sharp lateral velocity contrasts produce a lethal recipe for resonance and basin-edge effects, phenomena tragically evident in the 1991 Uttarkashi and 1999 Chamoli events. Instrumental analyses reveal staggering amplification factors, sometimes in excess of 10.0, and predominant frequencies often below 3 Hz. These factors put the region's prevalent low-rise masonry at extreme risk. Furthermore, the typical geophysical shorthand of $V_s \geq 760$ m/s for "bedrock" often breaks down in the Himalaya, where large impedance contrasts and complex subsurface layering require more nuanced, region-specific characterization.

In the end, meaningful microzonation must connect deep-crustal tectonics and shallow geotechnical realities; the integration of V_{s30} models and hazard-consistent ground-motion predictions are not an academic exercise but a fundamental necessity for engineering resilience in these densely populated mountain corridors.

4.4.5 Conclusion

The seismic narrative of the Uttarakhand Himalaya is one of structural caprice and tectonic suspense, where the inevitable northward march of the Indian plate is met by a Main Himalayan Thrust (MHT) that is anything but a uniform sliding surface. High-resolution imaging has replaced our once-simplistic models with a sophisticated landscape of ramp-flat-ramp geometries, where earthquakes are not merely random occurrences but are meticulously choreographed by fluid-saturated, low-velocity domains at depths of 8–22 km. These fluid-rich reservoirs, marked by elevated pore-pressures and high crack density, act as the great lubricators of the mid-crust, lowering the threshold for failure and dictating where the next rupture will nucleate.

This tectonic stage is further partitioned by an ancestral basement architecture—specifically the transverse "ribs" of the Delhi–Haridwar Ridge and the Moradabad Fault—which compartmentalizes the MHT into discrete mechanical blocks. Such segmentation likely serves as a structural brake, potentially explaining why this segment favors moderate-to-large events over the great, contiguous ruptures seen elsewhere in the arc. Yet, the presence of these barriers does not grant immunity; with a staggering slip deficit of approximately 3 m accumulated since the 1803 event, the region is undeniably pregnant with the potential for an $M_w \geq 7.5$ earthquake.

Ultimately, bridging the gap between our structural maps and meaningful hazard mitigation requires a shift toward "process-oriented" geophysics. We must move beyond static snapshots to capture the dynamic interplay of fluid pathways and fault mechanics through dense broadband arrays, geodetic monitoring, and perhaps the ultimate diagnostic tool: deep crustal drilling. In a landscape where the brittle-ductile transition limits the reach of our ruptures but local site conditions amplify their fury, our engineering resilience and early warning systems must be as nuanced as the geology they intend to survive. The Uttarakhand segment is primed and poised; our task is to ensure our predictive capabilities keep pace with its inevitable release.

4.5 Seismic Hazard Assessment

The Uttarakhand Himalaya is among the most tectonically complex and seismically vulnerable regions of the Indian subcontinent. This chapter synthesizes recent advances in structural seismology, probabilistic seismic hazard assessment (PSHA), and early-warning systems, with a focus on the interplay between lithospheric segmentation, earthquake generation, and cascading hazards such as landslides and glacier-related flows. The implications for risk mitigation, infrastructure planning, and disaster resilience are critically evaluated. The need for continual reassessment of hazard models and integrated early-warning frameworks is emphasized.

4.5.1 Introduction

The central Himalayan arc, particularly the Uttarakhand sector, is characterized by a segmented Main Himalayan Thrust (MHT), high strain accumulation, and a history of destructive earthquakes and associated mass-wasting events. This chapter evaluates the current understanding of structural controls on seismogenesis, the efficacy of regional seismic networks for early warning, and the state of PSHA in the region.

4.5.2 Structural Controls on Earthquake Generation

Seismic imaging (Hazarika et al., 2017; Mandal et al., 2021, 2022, 2023; Mandal, 2024; Monika et al., 2024) reveals that the MHT beneath Uttarakhand is a shallow, segmented fault zone comprising alternating ramp and flat structures within a fluid-rich, low-velocity mid-crustal layer (8–22 km depth). Major historical earthquakes (1803 M_w7.8 Garhwal, 1991 M_w6.8 Uttarkashi, 1999 M_w6.5 Chamoli) have nucleated on these ramp segments, where stress is concentrated due to geometric and frictional contrasts.

Transverse lithospheric features, including the Delhi–Haridwar Ridge, Moradabad Fault, and Great Boundary Fault, further segment the MHT into discrete mechanical blocks (Mandal et al., 2023). This segmentation influences rupture propagation and magnitude potential, with some blocks remaining locked and accumulating strain, while others may experience aseismic creep.

A high slip-deficit (~ 3 m since 1803; Bilham, 2019) and persistent microseismicity in the mid-crust confirms that the Uttarakhand segment remains a locked, highly stressed zone, susceptible to future moderate-to-large earthquakes.

4.5.3 Early Warning of Catastrophic Flow Events

The regional seismic network in Uttarakhand, comprising 76 broadband stations, has demonstrated strong capability for early warning of catastrophic landslides and glacier-related mass flows. The 7 February 2021 Ronti Gad event serves as a case study (Fig. 4.13; Cook et al., 2021). The network captured the entire hazard cascade, from initial rockslide and glacier detachment to downstream flooding, with robust signal-to-noise ratios up to 100 km from the source (Fig. 4.13).

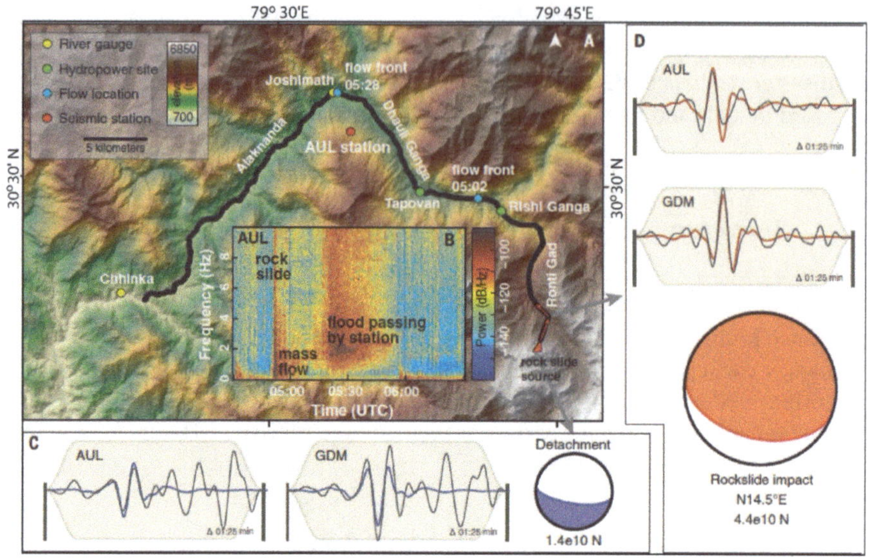

Fig. 4.13 Seismic detection of the 2021 Rishi Ganga flood event

Distinct seismic signatures were observed for each phase: low-frequency signals (<0.15 Hz) for the rockslide, and broadband (0.5–10 Hz) signals for mass flow (Hibert et al., 2019). Real-time tracking using normalized weighted cross-correlation and amplitude-distance analysis enabled rapid flow localization and monitoring within minutes (Cook et al., 2021; Walter et al., 2017). This capability is critical for timely downstream alerts and hazard mitigation.

(Adapted from Cook et al., 2021).

The methodology for rockslide forecasting as described above uses data from a dense regional seismic network to detect and analyze the evolution of complex mass-wasting events in near real time. Broadband seismic stations in Uttarakhand recorded the distinct seismic signatures of three event phases—rockslide, mass flow, and flood (Cook et al., 2021). Event detection and tracking relied on signal-to-noise ratio (SNR) analysis in specific frequency bands, coupled with normalized weighted cross-correlation and amplitude–distance analysis to locate and monitor flow progression. Seismic signals were interpreted alongside remote sensing, river gauge, and eyewitness data, enabling validation of event timing and spatial evolution. These complementary datasets allowed robust mapping of event dynamics, assessment of early warning capabilities, and refinement of detection thresholds tied to network density and noise characteristics (Cook et al., 2021). This integrative, multi-method approach demonstrates the value of regional seismic networks for early warning even in challenging, data-sparse mountain environments.

4.5.4 *Ground-Motion Prediction Equation (GMPE) for the Seismically Hazardous Zone of India Through Supervised XGBoost Algorithm*

In this study, we develop a robust ground motion prediction equation (GMPE) for Peak Ground Acceleration (PGA) in the seismically active Kachchh rift zone, Gujarat, employing a supervised machine learning approach—specifically, the XGBoost algorithm (Chen & Guestrin, 2016). Our methodology leverages a comprehensive dataset compiled from the regional strong motion network, including eight engineering seismoscope (SRR) records from the 2001 M_w7.7 Bhuj mainshock and 237 strong-motion records from 32 significant Bhuj aftershocks of M_w3.3–5.6, recorded between 2002 and 2008. The records, sourced from stations distributed across epicentral distances ranging from 1.0 to 288 km, were curated to diverse site configurations and variability in site conditions. To better capture the physical factors controlling ground motions, the input features used for the XGBoost model comprise moment magnitude (M_w), hypocentral distance, focal depth, and the average shear-wave velocity in the upper meters 30 (V_{s30}) as a proxy for site condition. The instrumentation employed for data collecting was predominantly managed by the National Geophysical Research Institute, alongside accelerographs for strong-motion and engineering seismoscopes strategically positioned by Roorkee University in the Kachchh region. The model training and testing employed the whole dataset and a randomized 30% test to robustly assess predictive ability and generalizability, rated with model performance by correlation between actual and predicted results. This integrated approach, combining advanced statistical learning and region-specific seismic modeling, aims to overcome the limitations of traditional regression-based GMPEs in India to provide regionally optimized hazard estimates.

Mathematical Equation for PGA Prediction

In mathematical notation, the prediction can be formally expressed as:

$$\log_{10}(\text{PGA}) = f_{\text{ML}}\left(M_w, R_{\text{hypo}}, D_f, V_{s30}\right) \tag{4.13}$$

where

- PGA: Peak Ground Acceleration (g)
- f_{ML}: The function learned by the XGBoost machine learning algorithm
- M_w: Moment magnitude of the earthquake
- R_{hypo}: Hypocentral distance (km)
- D_f: Focal depth (km)
- V_{s30}: Average shear-wave velocity in the upper 30 m (m/s)

For clarity, the actual prediction of PGA is then:

$$\text{PGA} = 10^{f_{\text{ML}}(M_w, R_{\text{hypo}}, D_f, V_{s30})} \tag{4.14}$$

where f_{ML} is the computationally-trained, non-parametric regression function created through XGBoost. This function simultaneously accounts for the complex (possibly nonlinear) relationships among these variables.

For context, a traditional regression-based GMPE for the region (e.g., Boore & Atkinson, 2008) typically takes the form:

$$\log_{10}(\text{PGA}) = a + bM_w - c\log_{10}(R_{\text{hypo}} + d) + eV_{s30} + fD_f + \epsilon \tag{4.15}$$

But in our proposed model, the XGBoost algorithm (Chen & Guestrin, 2016) replaces this analytical function (and its coefficients) with an ensemble of regression trees fit directly to the training data.

In summary, the equation representing our XGBoost-based ML GMPE is:

$$\log_{10}(\text{PGA}) = f_{\text{XGB}}(M_w, R_{\text{hypo}}, D_f, V_{s30}) \tag{4.16}$$

where f_{XGB} is the machine-learned predictive function created by the XGBoost model using the seismic record dataset from the Kachchh region.

Results and Discussions

The XGBoost machine learning model (Chen & Guestrin, 2016), developed using an entire dataset of strong-motion records from the Kachchh rift zone, demonstrates robust performance in predicting peak ground (PGA) for mainshocks and aftershocks. The model, trained on moment magnitude (M_w), hypocentral distance, focal depth, and V_{s30}, a proxy site, yields a remarkably high correlation (0.994) between observed and predicted PGA for the full dataset, with a still-strong correlation (0.944) for a randomized test subset—highlighting excellent fit and generalizability. This performance previous GMPEs for the region developed via artificial neural networks and non-linear least squares, underscoring the advantage of advantaged machine learning in complex, non-linear ground-motion behavior. The model's predictive skill for earthquakes of M_w spanning 5.6–7.7, including the 2001 Bhuj mainshock, 1999 Chamoli event, and significant aftershocks, suggesting potential applicability beyond Kachchh, possibly to other seismically active regions of India such as Uttarakhand. Notably, site sorted on V_{s30} (180–1500 m/s) influences predicted attenuation, emphasizing the importance of detailed site characterization for ground-motion prediction. Our supervised machine learning GMPE, utilizing the XGBoost algorithm and a comprehensive PGA dataset with input parameters such as Mw, focal depth, epicentral distance, and V_{s30}, demonstrates excellent predictive accuracy for the Indian subcontinent ($R = 0.994$ for full data, $R = 0.944$ for the test set) (Fig. 4.14). Validation against observed ground motions from major earthquakes in India and Nepal confirms the model's general applicability and robustness. These results highlight the value of machine learning approaches in advancing GMPEs and provide a critical tool for seismic hazard assessment and earthquake-resilient infrastructure across the region (Fig. 4.15). Thus, we suggest the use of sophisticated statistical learning with locally tailored datasets to reduce seismic hazard in areas where classic regression-based GMPE is constrained by data scarcity and oversimplified functional forms.

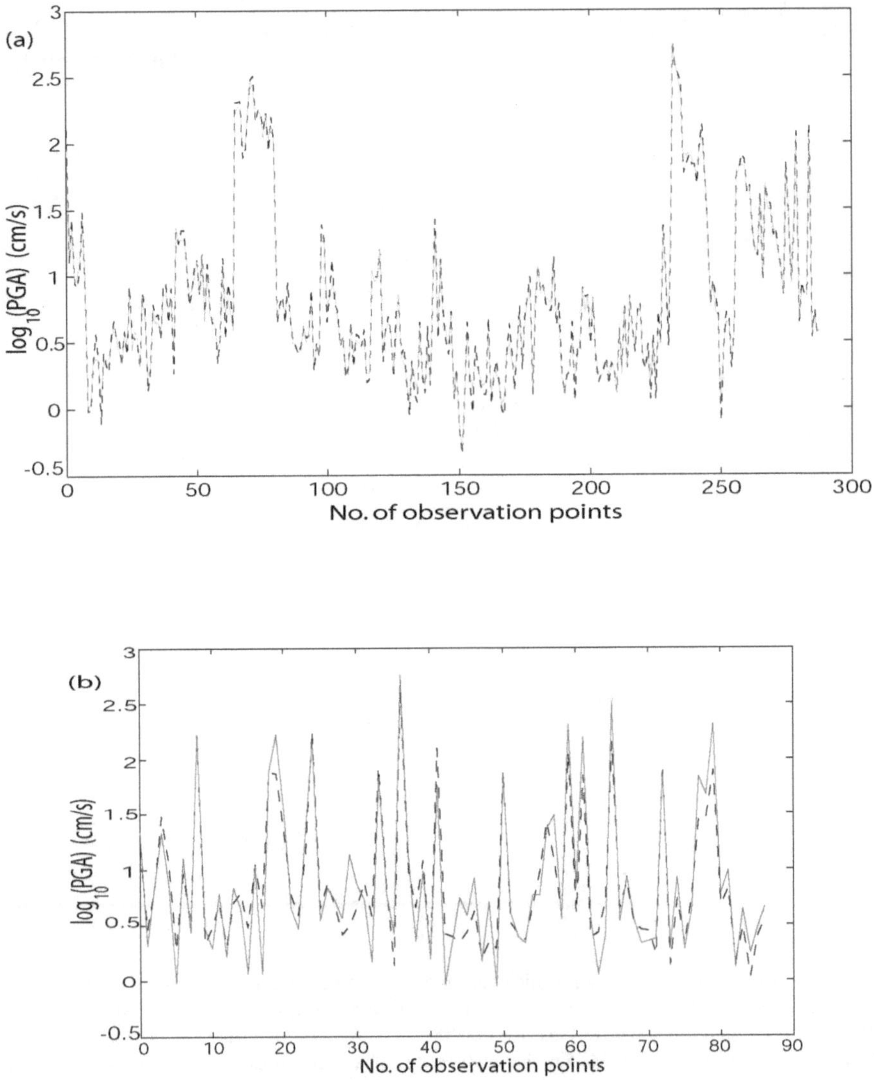

Fig. 4.14 Regression analysis between the output and target of the ML modeling **a** Complete data set (289 dataset points) showing correlation R = 0.994, **b** Testing (30% of all the data points, randomized) showing correlation R = 0.944. Solid gray line connects observed PGA data points while black dotted connects the predicted PGA data points

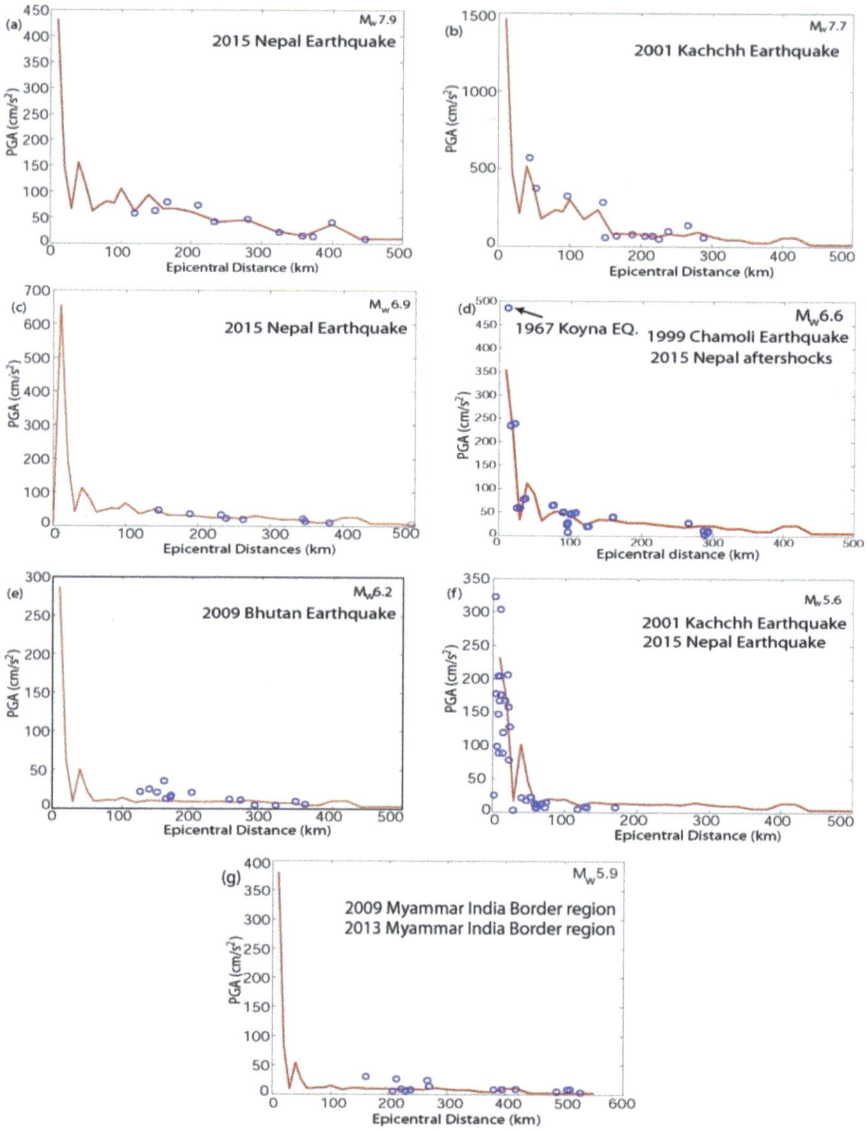

Fig. 4.15 Attenuation curves (solid thick red line) for the predicted PGA using the ML-trained network for **a** $M_W7.9$, **b** $M_W7.7$, **c** $M_W6.9$, **d** $M_W6.6$, **e** $M_W5.2$, **f** $M_W5.6$ and **g** $M_W5.9$ events. Observed PGA data (open blue circles) for the 2001 $M_W7.7$ Bhuj mainshock, 7 March and 6 April 2006 large Bhuj aftershocks of $M_W5.6$, the 1999 $M_W6.5$ Chamoli mainshock, the 2015 $M_W7.9$ Nepal mainshock, the 2015 $M_W6.9$ Nepal aftershock, the 2015 $M_W6.6$ Nepal aftershock, the 7 February 2017 Uttarakhand event of $M_W5.6$ and the $M_W5.9$ Myanmar-India border earthquakes of January 2013 and September 2009, have been plotted to demonstrate the robustness of the prediction

Table 4.6 Summary of PSHA results for Uttarakhand Urban Centres (Joshi et al., 2014)

Location	PGA (g)	Return period (years)
Pithoragarh	0.45	475
Bageshwar	0.40	475
Almora	0.39	475
Rudraprayag	0.42	475
Dehradun	0.25	475

4.5.5 Probabilistic Seismic Hazard Assessment and Risk Implications

PSHA integrates regional seismicity, fault geometry, ground motion prediction equations, and recurrence models to estimate the likelihood of future ground shaking. For Uttarakhand, PSHA indicates peak ground accelerations (PGA) of 0.25–0.45 g for a 475-year return period in urban centers such as Pithoragarh, Bageshwar, Almora, Rudraprayag, and Dehradun (Joshi et al., 2014). Grid-based analyses reveal spatial variability, with PGA values ranging from 0.17 to 0.53 g, highest near major thrusts (MCT, MBT) (Table 4.6; Sharma et al., 2014; Valdiya, 1980; Thakur, 2014). These hazard levels exceed the design standards of much existing infrastructure, highlighting the need for seismic microzonation, site response characterization, and earthquake-resilient construction.

The Probabilistic Seismic Hazard Assessment (PSHA) results for Uttarakhand's urban centers (Table 4.6) underscore a region poised on a high-stakes tectonic edge, where the anticipated Peak Ground Acceleration (PGA) values reflect a formidable threat to both life and infrastructure. Pithoragarh and Rudraprayag, with PGA values exceeding 0.40 g, occupy the front lines of this vulnerability. The risk is not merely a matter of shaking; the critical hydropower lifelines within the Alaknanda and Bhagirathi valleys face the compound threat of seismic failure and the subsequent "knock-on" effects of cascading mass flows (Cook et al., 2021). In urban basins like Dehradun, the underlying Quaternary sediments act as a double-edged sword—not only supporting dense populations but also serving to amplify seismic waves and invite the specter of liquefaction (Rawat et al., 2014). These data demand a shift from static planning to dynamic, scenario-based hazard modeling that accounts for multi-segment ruptures and the specific "fingerprint" of local site responses.

4.5.6 Conclusions

The 2021 Chamoli disaster serves as a reminder that seismic hazard rarely operates in isolation in the Uttarakhand Himalaya. It is the trigger for a cascade of rock failures and glacier-induced floods. Our regional networks have been successful in tracking these catastrophic events in near real time. The real challenge is in translating these signals into life-saving minutes for downstream communities through integration of seismic, remote sensing and river-gauge telemetry.

The development of XGBoost-based Ground-Motion Prediction Equations (GMPEs) has significantly improved the predictive landscape. These machine learning algorithms elegantly go beyond classic linear models and artificial neural networks to model the non-Gaussian complexities of the Himalayan dataset, explicitly weighing the influence of focal depth, epicentral distance and V_{s30}-based site classification. This approach reveals a startling fact: maximum ground accelerations (PGA) in urban zones such as Pithoragarh and Dehradun, for return periods of 475 years, are in the range of 0.25 to 0.45 g, frequently exceeding the tolerances of our current infrastructure designs.

In the end, Uttarakhand's seismic destiny is inscribed in its profound structural segmentation. The highest hazards are not randomly distributed but are tied to proximity to major thrusts and the mechanical blocks defined by inherited transverse "ribs" such as the Delhi–Haridwar Ridge. Surface geology also warps this threat; sedimentary basins can amplify shaking by factors as high as 10.9.

Going forward, the emphasis has to be on "process-oriented" resilience rather than static mapping. This requires denser arrays of strong motion and geodetic data to improve our GMPEs and the systematic inclusion of site-specific amplification in building codes. Only by combining empirical strong-motion datasets with real-time scenario generation can we hope to translate our structural knowledge into a strong protective barrier for the lives and lifelines of this tectonically poised corridor.

4.6 Discussions and Conclusions

The present synthesis of geophysical, geological, and seismological data provides a comprehensive understanding of the seismotectonic framework of the Uttarakhand Himalaya, a region of elevated seismic hazard and structural complexity. Multidisciplinary analyses, including seismic tomography, receiver function imaging, earthquake relocations, and probabilistic seismic hazard assessment (PSHA), reveal the following salient features.

4.6.1 Geometry and Fluidization of the Main Himalayan Thrust (MHT)

Seismic imaging delineates a segmented ramp-flat-ramp geometry of the MHT beneath Uttarakhand. High Vp/Vs ratios (1.74–1.92) and elevated crack density ($\varepsilon \approx 0.011$–$0.012$) at 8–22 km depths suggest pervasive fluid infiltration, creating conditions conducive to earthquake nucleation (Fig. 4.16a–g; Hazarika et al., 2017; Mandal et al., 2022; Mandal, 2024; Monika et al., 2024). The Main Himalayan Thrust

(MHT) is also found to correlate with mid-crustal conductive zones, as revealed by the inversion of magnetotelluric data along profiles across the 1991 Uttarkashi and 1999 Chamoli earthquake regions (Fig. 4.16a–g). This correlation suggests that these earthquakes were likely triggered by the presence of fluids along the MHT.

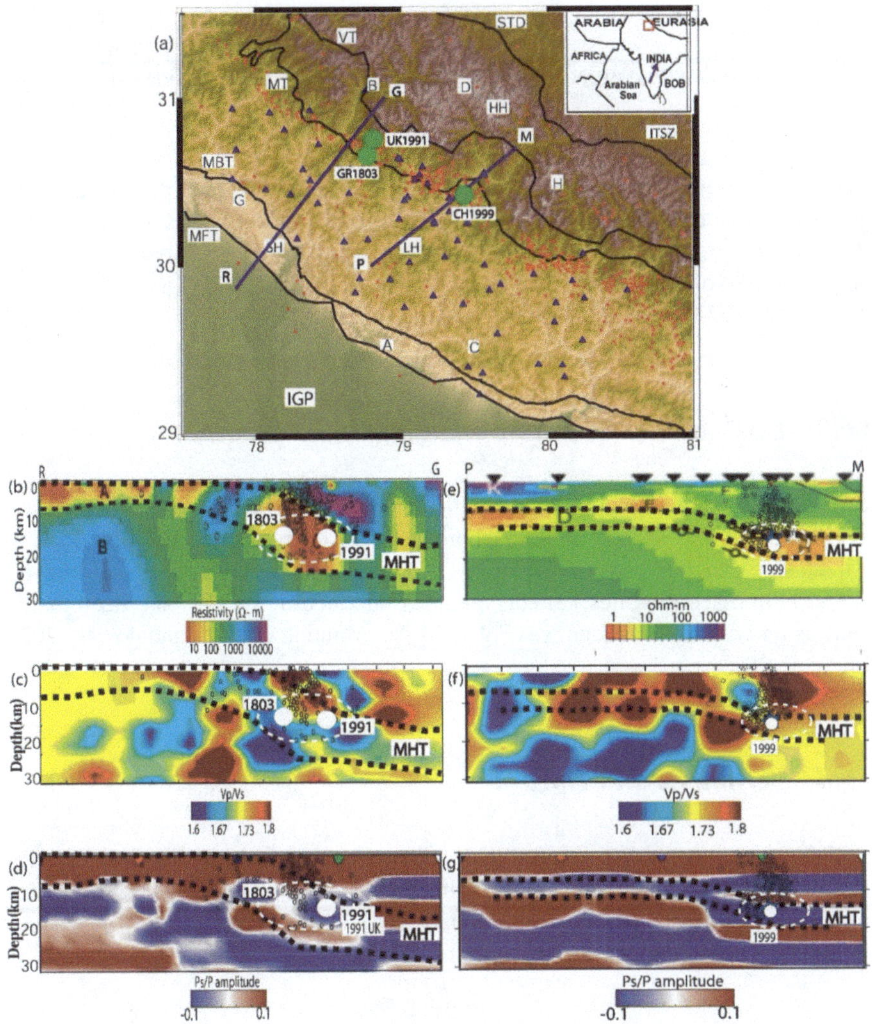

Fig. 4.16 Schematic cross-section of the MHT (marked by thick black dotted lines) beneath Uttarakhand, showing ramp-flat-ramp geometry and zones of high Vp/Vs along **a** NE–SW trending profiles RG and PM (shown by solid green lines), **b** Electrical resistivity (ohm-m) along RG, **c** Vp/Vs along RG, **d** Impedance along RG, **e** Electrical resistivity (ohm-m) along PM, **f** Vp/Vs along PM, and **g** Impedance along PM. The location mainshocks are shown by filled large-size green circles in **a** while they are marked by large-size filled white circles in **b**–**g**

Table 4.7 Major transverse structures in Uttarakhand and their seismotectonic implications

Structure	Location	Seismotectonic role
Delhi–Haridwar Ridge	SW Uttarakhand	Rupture barrier
Moradabad Fault	Central region	Segment boundary
Great Boundary Fault	Eastern region	Asperity, rupture terminus

4.6.2 Structural Segmentation and the Himalayan Seismic Gap

Transverse tectonic features, including the Delhi–Haridwar Ridge, Moradabad Fault, and Great Boundary Fault, disrupt the MHT, producing segmented crustal blocks. These act as barriers or asperities, influencing rupture dynamics and contributing to the persistence of the central Himalayan seismic gap since 1803 (Table 4.7; Mandal et al., 2021, 2023; Bilham, 2019; Manglik et al., 2022).

4.6.3 Crust and Lithosphere Imaging

Receiver function and tomographic studies reveal Moho depths ranging from 28 to 55 km and a northeast-dipping lithosphere–asthenosphere boundary at 130–180 km. A thickened Mantle Transition Zone (MTZ), with velocity anomalies near the 410 and 660 km discontinuities, reflects complex subduction legacies and deep mantle controls on crustal stress regimes (Fig. 4.11a–c; Mandal, 2023; Mandal et al., 2024; Rai et al., 2009b).

4.6.4 Seismicity and Hazard

Earthquake source modeling and PSHA indicate strain accumulation rates of ~ 15–18 mm/yr, with a slip deficit of ~ 3 m since the last major rupture. Hazard models suggest peak ground accelerations of 0.25–0.45 g for 475-year return periods in Dehradun, Pithoragarh, and Almora districts (Table 4.8; Joshi et al., 2014).

Table 4.8 Probabilistic seismic hazard estimates for key districts

District	PGA (g) at 475 yr RP	Slip deficit (m)
Dehradun	0.25	~ 3.0
Pithoragarh	0.45	~ 2.8
Almora	0.39	~ 3.1

4.6.5 Knowledge Gaps and Research Priorities

Our understanding of the Uttarakhand Himalaya has sharpened, but several critical blind spots remain, calling for a more aggressive and integrated research agenda. The current broadband seismic network is too sparse, especially in the east and northeast, so that we are blind to the subtle microseismicity and real-time deformation that presages failure (Mandal et al., 2021). We need to move beyond isolated data sets to a sophisticated "joint inversion" of seismic, magnetotelluric, and geodetic signals to finally resolve the elusive rheology of the Main Himalayan Thrust (MHT) (Fig. 4.16). The gap between detection and emergency action has to be bridged by the implementation of Earthquake Early Warning (EEW) systems, already proven in neighboring Nepal and Himachal Pradesh.

At the same time, urban risk is poorly quantified, as the lethal amplification of soft-sediment basins cannot be considered in our urban planning without comprehensive V_{s30} mapping and microzonation. Finally, the 2021 Chamoli disaster demonstrates the urgent need to extend our monitoring to the high-altitude cryosphere. Identifying glacier-related rock failures necessitates an improved network configuration and improved algorithms capable of filtering the ambient noise to recognize the seismic precursors of mass wasting (Fig. 4.13; Cook et al., 2021). We need to incorporate hydrological and remote sensing data into our seismic arrays to provide meaningful protection against the cascading hazards of this high-energy landscape.

4.6.6 Conclusion

The Uttarakhand Himalaya remains a landscape of profound tectonic suspense, where the relentless convergence of the Indian and Eurasian plates is met by a Main Himalayan Thrust (MHT) that is both structurally partitioned and fluid-weakened. High-resolution imaging has replaced our simplistic models with a sophisticated architecture of ramp-flat-ramp geometries, interrupted by ancient transverse "ribs"—the Delhi-Haridwar Ridge and the Moradabad Fault—that segment the crust into discrete mechanical blocks. At the 8–22 km seismogenic depth, the pervasive signature of metamorphic fluids and high pore-pressures suggests a mid-crustal environment primed for failure. Despite the release offered by historical events, the staggering accumulation of a 3-m slip deficit since 1803 underscores a segment pregnant with the potential for a major $M_w \geq 7.5$ earthquake.

The path toward resilience now demands a transition from static observation to a dynamic, real-time diagnostic approach. We must densify our broadband and geodetic arrays to capture the subtle precursors of both seismic ruptures and high-altitude geomorphic disasters, while integrating geochemical and magnetotelluric data to budget the fluids that lubricate our faults. Ultimately, safeguarding the lives and lifelines of this complex corridor depends on our ability to transform these deep-earth insights into robust early-warning systems and hazard-consistent engineering.

The Uttarakhand arc is poised; our scientific and collaborative efforts must be equally resolute to keep pace with its inevitable release.

Acknowledgements The author expresses sincere gratitude to the Director, Council of Scientific and Industrial Research–National Geophysical Research Institute (CSIR-NGRI), Hyderabad, India, for his invaluable support and for granting permission to publish this work. The research presented here was made possible through funding from several CSIR-sponsored projects, including the Mission-mode and Focused Basic Research (FBR) initiatives under CSIR-NGRI, Hyderabad. Special thanks are extended to Eagar (2002) for providing the FuncLab software, which was instrumental in enabling the receiver function analysis central to this study. All figures presented in the chapter were generated using the Generic Mapping Tools (GMT; Wessel et al., 2019), with open-source code and supporting data available publicly through the official project platform. Elevation data used in the GMT plots were sourced from the publicly available Digital Elevation Model (DEM) provided by NASA.

The author also acknowledges the XGBoost development team for releasing their code to the public domain, facilitating the advanced machine learning analyses conducted in this chapter. Critical seismic data, including peak ground acceleration (PGA) records for the 1999 Chamoli earthquake, were obtained from the Roorkee Strong Motion Network, whose database was essential for robust hazard assessment.

The continued collaboration of these institutions, contributors, and open-data initiatives has been fundamental to advancing the research and hazard modeling efforts summarized in this work.

References

Benz, H. M., Zandt, G., Oppenheimer, D. H., et al. (1996). Three-dimensional P and S wave velocity structure of the Loma Prieta region, California, from inversion of local earthquake arrival times. *Journal of Geophysical Research, 101*, 827–844.

Bercovici, D., & Karato, S. (2023). Whole-mantle convection and the transition-zone water filter. *Nature, 425*, 39–44.

Bilham, R. (2019). Himalayan earthquakes: A review of historical seismicity and current geodetic strain accumulation. *Tectonophysics, 789*, 228549.

Boore, D. M., & Atkinson, G. M. (2008). Ground-motion prediction equations for the average horizontal component of PGA, PGV, and 5%-damped PSA at spectral periods between 0.01 s and 10.0 s. *Earthquake Spectra, 24*(1), 99–138. https://doi.org/10.1193/1.2830434

Caldwell, W. B., Klemperer, S. L., Lawrence, J. F., Rai, S. S., Ashish. (2013). Characterizing the Main Himalayan Thrust in the Garhwal Himalaya, India with receiver function CCP stacking. *Earth and Planetary Science Letters, 367*, 15–27.

Célérier, J., Harrison, T. M., Webb, A. A. G., & Yin, A. (2009). The Kumaun and Garhwal Lesser Himalaya, India: Part 1. Structure and stratigraphy. *Geological Society of America Bulletin, 121*, 1281–1297.

Chen, Q., Freymueller, J. T., Wang, Q., Yang, Z., Xu, C., & Liu, J. (2004). A deforming block model for the present-day tectonics of Tibet. *Journal of Geophysical Research (Solid Earth), 109*(B1). https://doi.org/10.1029/2002JB002151

Chen, T., & Guestrin, C. (2016). XGBoost: a scalable tree boosting system. In *Proceedings of the 22nd ACM SIGKDD International Conference on Knowledge Discovery and Data Mining*, pp. 785–794. https://doi.org/10.1145/2939672.2939785

Cook, K. L., Rekapalli, R., Dietze, M., Pilz, M., Cesca, S., Rao, N. P., Srinagesh, D., Paul, H., Metz, M., Dahm, T., Mandal, P., Suresh, G., Cotton, F., Tiwari, V. M., & Hovius, N. (2021). Detection and potential early warning of catastrophic flow events with regional seismic networks. *Science, 374*, 87–92. https://doi.org/10.1126/science.abj3261

Dal Zilio, L., van Dinther, Y., Gerya, T. V., & Avouac, J. P. (2020). Seismicity and deformation patterns at the Himalayan megathrust: The role of inherited lithospheric structure. *Nature Communications, 11*, 4894.

Devi, A., Israil, M., Singh, A., & Gupta, P. K. (2020). Three-dimensional resistivity structure in MCT zone around Chamoli region, Garhwal Himalaya and its seismogenic implication. *Journal of Applied Geophysics, 178*, 104060. https://doi.org/10.1016/j.jappgeo.2020.104060

Dueker, K. G., & Sheehan, A. F. (1997). Mantle discontinuity structure from midpoint stacks of converted P to S waves across the Yellowstone hotspot track. *Journal of Geophysical Research: Solid Earth, 102*(B4), 8313–8327. https://doi.org/10.1029/96JB03857

Dumka, R. K., Kumar, A., Gahalaut, V. K., et al. (2014). GPS-derived crustal deformation and strain rates across the Kumaun-Garhwal Himalaya, India. *Journal of Asian Earth Sciences, 91*, 243–251.

Eagar, K. C. (2012). FuncLab: A MATLAB interactive toolbox for handling receiver function datasets. *Seismological Research Letters, 83*(3), 596–603. https://doi.org/10.1785/gssrl.83.3.596

Ekström, G., & Stark, C. P. (2013). Simple scaling of catastrophic landslide dynamics. *Science, 339*, 1416–1419.

Gahalaut, V. K., Kalpna, et al. (2017). Influence of Tehri reservoir impoundment on crustal deformation and seismicity in the Garhwal Himalaya, India. *Bulletin of the Seismological Society of America, 107*, 1842–1852.

Gautam, P. K., Gahalaut, V. K., Prajapati, S. K., Kumar, N., Yadav, R. K., Rana, N., & Dabral, C. P. (2017). Continuous GPS measurements of crustal deformation in Garhwal-Kumaun Himalaya. *Quaternary International, 462*, 124–129. https://doi.org/10.1016/j.quaint.2017.05.043

Gupta, A. K., Mandal, P., & Sain, K. (2023). Modelling of earthquake source parameters and scaling relations in the Uttarakhand Himalayan region, India. *Journal of Earth System Science, 132*, 73(1–11).

Hazarika, D., Sen, K., & Kumar, N. (2014). Characterizing the intracrustal low velocity zone beneath northwest India-Asia collision zone. *Geophysical Journal International, 199*, 1338–1353. https://doi.org/10.1093/gji/ggu353

Hazarika, D., Kumar, A., Kumar, N., Kumar, M. R., Kumar, S., Roy, S., Kumar, D., Mandal, P., Saha, D., Saha, R., & Sutar, A. K. (2017). Seismotectonics of the Kumaon-Garhwal Himalaya: Constraints from local earthquake tomography and receiver function imaging. *Journal of Asian Earth Sciences, 138*, 234–247.

Herrmann, R. B. (2004). Joint Inversion of Receiver Function and Surface Wave Dispersion. *Seismological Research Letters, 75*(1), 77–89.

Hibert, C., Michéa, D., Provost, F., Malet, J. P., & Geertsema, M. (2019). Exploration of continuous seismic recordings with a machine learning approach to document 20 yr of landslide activity in Alaska. *Geophs. Jou. Int., 219*, 1138–1147.

Israil, M., et al. (2008). Magnetotelluric imaging of the Garhwal Himalayan collision zone. *Physics of the Earth and Planetary Interiors, 166*, 187–194.

Jain, A. K., Seth, P., Shreshtha, M., Mukherjee, P. K., & Singh, K. (2013). Structurally-controlled melt accumulation: Himalayan migmatites and related deformation—Dhauli Ganga Valley, Garhwal Himalaya. *Journal of Geological Society of India, 82*(4), 313–318. https://www.geosocindia.org/index.php/jgsi/article/view/57429

Joshi, A., Kumar, A., Singh, R. P., et al. (2014). Probabilistic seismic hazard assessment of Uttarakhand Himalaya. *Natural Hazards, 74*, 567–584.

Julia, J., Ammon, C. J., Herrmann, R. B., & Correig, A. M. (2000). Joint inversion of receiver function and surface wave dispersion observations. *Geophysical Journal International, 143*(1), 99–112.

Kanaujia, J., Ravi Kumar, M., Bandari, N., & Vijayaraghavan, R. (2025). Tomographic evidence for segmentation of the Main Himalayan Thrust in the Uttarakhand Himalaya. *Tectonophysics, 900*, 230655. https://doi.org/10.1016/j.tecto.2025.230655

Kayal, J. R., et al. (2003). Seismotectonics of the Himalaya: A review. *Journal of Earth System Science, 112*(4), 579–590. https://doi.org/10.1007/BF02709798

Kennett, B. L. N. (1983). *Seismic wave propagation in stratified media* (pp. 1–1948). New York, Cambridge University Press.

Kennett, B. L. N., & Engdahl, E. R. (1991). Travel times for global earthquake location and phase identification. *Geophysical Journal International, 105*, 429–465. https://doi.org/10.1111/j.1365-246X.1991.tb06724.x

Kind, R., Yuan, X., Saul, J., Nelson, D., Sobolev, S. V., Mechie, J., Zhao, W., Kosarev, G., Ni, J., Achauer, U., & Jiang, M. (2002). Seismic images of crust and upper Mantle Beneath Tibet: Evidence for Eurasian Plate Subduction. *Science, 298*, 1219–1221.

Kissling, E. (1995). Program VELEST user's guide—Short introduction. Inst Geophys, ETH Zurich, Switzerland.

Kumar, R., Singh, A. K., & Prasad, R. (2020). Imaging the upper crustal structure of the Kumaon-Garhwal Himalaya using seismic tomography. *Journal of Earth System Science, 129*(4), 81.

Larson, K. R., Burgmann, R., Bilham, R., & Freymuller, J. (1999). Kinematics of the India-Eurasia collision zone from GPS measurements. *Journal of Geophysical Research, 104*, 1077–1093.

Lees, J. M., & Wu, H. (2000). Poisson's ratio and porosity at Coso-geothermal area, California. *Journal of Volcanology and Geothermal Research, 95*, 157–173.

Li, X., Wei, D., Yuan, X., Kind, R., Kumar, P., & Zhou, H. (2011). Details of the doublet Moho structure beneath Lhasa, Tibet, obtained by comparison of P and S receiver functions. *Bulletin of the Seismological Society of America, 101*(3), 1259–1269.

Li, S., & Song, X. (2018). Segmentation of the Himalayan megathrust and controls on earthquake rupture. *Geophysical Research Letters, 45*, 1321–1329.

Ligorria, J. P., & Ammon, C. J. (1999). Iterative deconvolution and receiver-function estimation. *Bulletin of the Seismological Society of America, 89*(5), 1395–1400.

Mandal, P., Padhy, S., Rastogi, B. K., Kousalya, M., Satyanarayana, H. V. S., Satyamurthy, C., Vijayraghavan, R., & Srinivasan, A. (2001). Aftershock activity and low coda Q_c in the epicentral region of the 1999 Chamoli earthquake of M_w6.4. *Pure and Applied Geophysics, 158*, 1719–1735.

Mandal, P., Srinivas, D., Suresh, G., & Srinagesh, D. (2021). Modelling of crustal composition and Moho depths and their implications toward seismogenesis in the Kumaon-Garhwal Himalaya. *Science and Reports, 11*, 14067. https://doi.org/10.1038/s41598-021-93469-1

Mandal, P., Srinagesh, D., Vijayaraghavan, R., Suresh, G., Naresh, B., Solomon Raju, P., Devi, A., Swathi, K., Singh, D. K., Srinivas, D., Saha, S., Shekar, M., Sarma, A. N. S., & Murthy, Y. V. V. B. S. N. (2022). Seismic velocity imaging of the Kumaon-Garhwal Himalaya, India. *Natural Hazards, 111*, 2241–2260. https://doi.org/10.1007/s11069-021-05135-4

Mandal, P. (2023). The Uttarakhand Himalaya: An image of the main Himalayan thrust, Moho, and lithosphere-asthenosphere boundary. *Journal of Asian Earth-Sciences, 253*, 105724, 1–15. https://doi.org/10.1016/j.jseaes.2023.105724.

Mandal, P., Prathigadapa, R., Srinivas, D., Saha, S., & Saha, G. (2023). Evidence of structural segmentation of the Uttarakhand Himalaya and its implications for earthquake hazard. *Science and Reports, 13*, 2079. https://doi.org/10.1038/s41598-023-29432-z

Mandal, P. (2024). Detection of a cracked and fluid-filled Main Himalayan Thrust (MHT) and its implications toward the genesis of earthquakes in the Uttarakhand Himalaya. *Results Eng., 23*, 102699. https://doi.org/10.1016/j.rineng.2024.102699

Mandal, P., & Pujol, J. (2006). Seismic imaging of the aftershock zone of the 2001 Mw 7.7 Bhuj Earthquake, India. *Geophysical Research Letters, 33*(L05309), 1–4. https://doi.org/10.1029/2005GL025275

Mandal, P., Saha, S., & Prathigadapa, R. (2024). Evidence of low velocity layers at the top and bottom of the Mantle Transition Zone (MTZ) below the Uttarakhand Himalaya, India. *Scientific Reports, 14*, 17239. https://doi.org/10.1038/s41598-024-67941-7

Manglik, A., Kandregula, R. S., & Pavankumar, G. (2022). Foreland Basin Geometry and Disposition of major thrust faults as proxies for identification of segmentation along the Himalayan arc. *Journal of Geological Society of India, 98*, 57–61.

Mishra, O. P., & Zhao, D. (2003). Crack density, saturation rate and porosity at the 2001 Bhuj, India, earthquake hypocenter: A fluid-driven earthquake? *EPSL, 212*, 393–405.

Mittal, H., Wu, Y.-M., Kumar, A., Kumar, A., & Sharma, B. (2016). Evaluating the effects of ground motion parameters on response spectra in Uttarakhand Himalayas, India. *Arabian Journal of Geosciences, 9*(712). https://doi.org/10.1007/s12517-016-2747-5

Monika, K. P., Sandeep, J. A., & Pal, S. K. (2024). 3D attenuation tomography of the Uttarakhand, NW Himalaya: Linkage to fluid or partial melt zones—Seismic hazard. *Soil Dynamics and Earthquake Engineering, 182*, 108699. https://doi.org/10.1016/j.soildyn.2024.108699

Mukherjee, P. K., Jain, A. K., Singhal, S., Singha, N. B., Singh, S., Kumud, K., Seth, P., & Patel, R. C. (2019). U-Pb zircon ages and Sm-Nd isotopic characteristics of the Lesser and Great Himalayan sequences, Uttarakhand Himalaya, and their regional tectonic implications. *Gond. Res., 75*, 282–297.

O'Connell, R. J., & Budiansky, B. (1974). Seismic velocities in dry and saturated cracked solids. *Journal of Geophysical Research, 79*, 5412–5426. https://doi.org/10.1029/JB079i035p05412

Oreshin, S. I., Vinnik, L. P., Makeyeva, L. I., & Kozhevnikov, V. M. (2011). Structure of the upper mantle beneath the Himalayas from P and S receiver functions. *Geophysical Journal International, 185*(2), 775–786.

Pandey, M. R., Tandukar, R. P., Avouac, J. P., Lavé, J., & Massot, J. P. (1995). Evidence for recent interseismic strain accumulation on a mid-crustal ramp in the central Himalaya of Nepal. *Geophysical Research Letters, 22*, 751–758. https://doi.org/10.1029/94GL02971

Pandey, B., & Jakka, R. S. (2022). Selection of an appropriate bedrock for site amplification studies. *Natural Hazards, 111*, 1251–1275. https://doi.org/10.1007/s11069-022-05260-8

Pickett, G. R. (1963). Acoustic character logs and their application in formation evaluation. *Journal of Petroleum Technology, 15*, 659–667.

Pujol, J., & Shen, P. (1999). Three-dimensional velocity structure and earthquake relocations in the New Madrid seismic zone. *Bulletin of the Seismological Society of America, 89*, 1382–1392.

Rai, S. S., Priestley, K., Gaur, V. K., Mitra, S., Singh, M. P., & Searle, M. (2006). Configuration of the Indian Moho beneath the NW Himalaya and Ladakh. *Geophysical Research Letters, 33*, L15308. https://doi.org/10.1029/2006GL026076

Rai, S. S., Priestley, K., Gaur, V. K., & Mitra, S. (2009a). Crustal structure and earthquake focal depths in the Kumaon-Garhwal Himalaya (India) from teleseismic receiver functions. *Geophysical Journal International, 176*(2), 627–638.

Rai, S. S., Suryaprakasam, K., & Gaur, V. K. (2009b). Seismic imaging of the mantle discontinuities beneath India: From Archean Cratons to Himalayan Subduction Zone. In: A. K. Gupta, S. Dasgupta (Eds.), *Physics and chemistry of the Earth's interior*. Springer, New York. https://doi.org/10.1007/978-1-4419-0346-4_9.

Rawat, R., Arora, B. R., & Gupta, P. K. (2014). Electrical resistivity cross-section across the Garhwal Himalaya: Proxy to fluid seismicity linkage. *Tectonophysics, 637*, 68–79.

Saha, G. K., Prakasam, K. S., & Rai, S. S. (2020). Diversity in the peninsular Indian lithosphere revealed from ambient noise and earthquake tomography. *Physics of the Earth and Planetary Interiors, 306*, 106523, 1–17.

Seeber, L., & Armbruster, J. G. (1971). Seismicity of the Himalaya, northern India and Nepal. *Bulletin of the Seismological Society of America, 61*, 775–802.

Sharma, M. L., & Wason, H. R. (1994). Occurrence of low stress drop earthquakes in the Garhwal Himalaya region. *Physics of the Earth and Planetary Interiors, 85*, 265–272. https://doi.org/10.1016/0031-9201(94)90117-1

Sharma, J., Chopra, S., & Roy, K. S. (2014). Estimation of source parameters, Quality Factor (QS), and site characteristics using accelerograms: Uttarakhand Himalaya Region. *Bulletin of the Seismological Society of America, 104*(1), 360–380. https://doi.org/10.1785/0120120304

Srinagesh, et al. (2019). Special section: Seismology in India. *Current Science, 117*(5), 701–762.

Tadokoro, K., Ando, M., & Umeda, Y. (1999). S wave splitting in the aftershock region of the 1995 Hyogo-ken Nanbu earthquake. *Journal of Geophysical Research, 104*, 981–991.

Thakur, V. C. (2014). Active tectonics of the Himalaya. *Geological Society of India Memoir, 66*, 1–40.

Thakur, V. C., Rawat, B. S., & Srivastava, D. C. (2018). Structural framework of the Kumaun Himalaya, India. *Journal of Asian Earth Sciences, 161*, 1–15.

Toffelmier, D. A., & Tyburczy, J. A. (2007). Electromagnetic detection of a 410-km-deep melt layer in the southwestern United States. *Nature, 447*(7147), 991–994. https://doi.org/10.1038/nature05922

Valdiya, K. S. (1980). *Geology of the Kumaun Lesser Himalaya.* Wadia Institute of Himalayan Geology, Dehradun.

Walter, F., Burtin, A., McArdell, B. W., Hovius, N., Weder, B., & Turowski, J. M. (2017). Rapid detection and location of debris flow initiation at Illgraben, Switzerland. *Natural Hazards and Earth System Sciences, 17*(6), 939–955.

Webb, A. A. G., Yin, A., Harrison, T. M., Célérier, J., & Gehrels, G. E. (2011). The Himalaya in 3D: Slab dynamics, orogeny, and subduction retreat. *Earth and Planetary Science Letters, 304*, 493–503.

Wessel, P. et al. (2019). Generic mapping tools version 6. *Geochemistry, Geophysics, Geosystems, 20*, 5556–5564.

Yin, A. (2006). Cenozoic tectonic evolution of the Himalayan orogen as constrained by along-strike variation of structural geometry, exhumation history, and foreland sedimentation. *Earth-Science Reviews, 76*, 1–131.

Yuan, X., Ni, J., Kind, R., Mechie, J., & Sandvol, E. (1997). Lithospheric and upper mantle structure of southern Tibet from a seismological passive source experiment. *Journal of Geophysical Research (Solid-earth), 102*(B12), 27491–27500.

Zhao, W., Mechie, J., Brown, L., Guo, J., Haines, S., Hearn, T., Klemperer, S. L., Ma, Y. S., Meissner, R., Nelson, K. D., Ni, J., Pananont, P., Rapine, R., Ross, A., & Saul, J. (2001). Crustal structure of central Tibet as derived from project INDEPTH wide-angle seismic data. *Geophysical Journal International, 145*, 486–498.

Zhao, D., & Mizuno, T. (1999). Crack density and Saturation rate in the 1995 Kobe earthquake region. *Geophysical Research Letters, 26*(21), 3213–3216.

Zhu, L., & Kanamori, H. (2000). Moho depth variation in Southern California from Teleseismic receiver functions. *Journal of Geophysical Research, 105*, 2969–2980. https://doi.org/10.1029/1999JB900322

Chapter 5
A Glimpse of the Geophysical Research in Northeast India

Bijit Kumar Choudhury, K. V. R. H. Prasad, and V. M. Tiwari

Abstract The Northeastern Region (NER) of India represents a tectonically active and environmentally fragile frontier within the Indian subcontinent. Bounded by the Himalaya to the north, the Indo-Burman ranges to the east and the Shillong Plateau to the south, and it offers a unique setting to study arc–continent collision, lithospheric deformation and surface processes under monsoonal climate regimes. Despite its global geodynamic importance, geophysical investigations of NER remain fragmented. This review synthesises three decades of seismological, magnetotelluric, gravity, GPS and remote sensing studies. It (a) characterises crustal and mantle structures, (b) summarises lithospheric deformation and seismicity, (c) presents the gap areas in geophysical research and (d) proposes the optimum use of integrated geophysical investigation methods.

Keywords Crust · Mantle · Geophysical inversion · Lithosphere · Rheology · Strain

5.1 Introduction

The Northeastern Region (NER) of India is situated in the tri-junction of the Indian Plate, the Eurasian Plate and the Burmese microplate, which makes this geotectonic entity a complex and geodynamically intriguing. This tri-plate interaction has created a tapestry of tectonic domains, viz., the eastern Himalayan orogen to the north, the intraplate tectonic of the Shillong-Mikir Hills Plateau and the subduction tectonics of the Indo-Burman Ranges (IBR) (Acharyya, 1987; Baruah et al., 2021). Together, this mosaic of tectonic domains has developed into an active transition zone between continental collision and oblique subduction, comparable to highly

B. K. Choudhury (✉) · K. V. R. H. Prasad · V. M. Tiwari
CSIR-North East Institute of Science and Technology (CSIR-NEIST), Jorhat, Assam, India
e-mail: bijitchoudhary.neist@csir.res.in

Academy of Scientific and Innovative Research (AcSIR), Ghaziabad, Uttar Pradesh, India

V. M. Tiwari (ed.), *Himalayan Natural Hazards*, Indian National Science Academy,
https://doi.org/10.1007/978-981-92-3784-5_5

tectonically active systems worldwide, namely the Japanese arc and the western North American Cordillera.

NER is reckoned to be one of the most seismically active regions of the world. The region registered two of the largest recorded continental earthquakes in the world, viz., the 12 June 1897 intraplate earthquake of Mw 8.1 (Bilham & England, 2001) and the Assam-Tibet earthquake of Mw 8.4 (Ambraseys & Douglas, 2004). These events accentuate the magnitude of seismotectonic processes and the associated hazard (Bilham & England, 2001; Chen & Molnar, 1990; Oldham, 1899). The Shillong Plateau event reflects the mechanical role of pop-up tectonics, whereas the Syntaxial event of Assam reveals the ongoing process of collisional tectonics and crustal shortening. Thus, quintessentially, the significance of this region stands out in understanding continental deformation, strain accumulation and lithospheric rheology within the broader context of the India–Eurasia collisional system.

Notwithstanding the fact that the region is geodynamically significant, the NER of India remains relatively poorly understood from a geophysical perspective. Previous workers have investigated the region by employing receiver functions, seismic tomography, coda and Lg attenuation, gravity, magnetic, magnetotelluric (MT) and GNSS geodesy data. But these studies are often spatially uneven and method-specific. As such, the crustal and upper mantle structure, strain partitioning between the Himalaya, the Shillong Plateau and the Indo-Burman Ranges, and the role of fluids and weak zones in accommodating crustal deformation are yet to be well understood. This review attempts to present a coherent appraisal of existing geophysical studies focused on NER of India and to identify gaps that offer scope for future studies. This work is an endeavour to synthesise the three decades of geophysical observations to (i) characterise the crustal and upper mantle structure of the region, (ii) provide a comprehensive summary on the pattern of seismicity and deformation and (iii) identify the present constraints in the studies and future opportunities. The novelty of this work lies in its treatment of the problem with an integrated approach, rather than cataloguing results by method or by locality. This review presents a collective evaluation of the implications of geophysical studies for comprehending the India–Eurasia collision and the IBR subduction, with the intraplate tectonics of the Shillong-Mikir Hills Plateau standing as a pivot.

5.2 Geological and Tectonic Framework

The NER of India is defined by a complex tectonic framework, which is marked by a Precambrian basement, active orogenic belts and a foreland basin within a comparatively confined region (Acharyya, 1987; Yin & Harrison, 2000). Tectonically, the region can be classified as the eastern Himalaya collision to the north, the Shillong Plateau in the centre and the Indo-Burman subduction to the eastern margin of the region (Fig. 5.1). Apart from these major tectonic domains, the region is criss-crossed by several transects/faults and lineaments along with basins (Fig. 5.1).

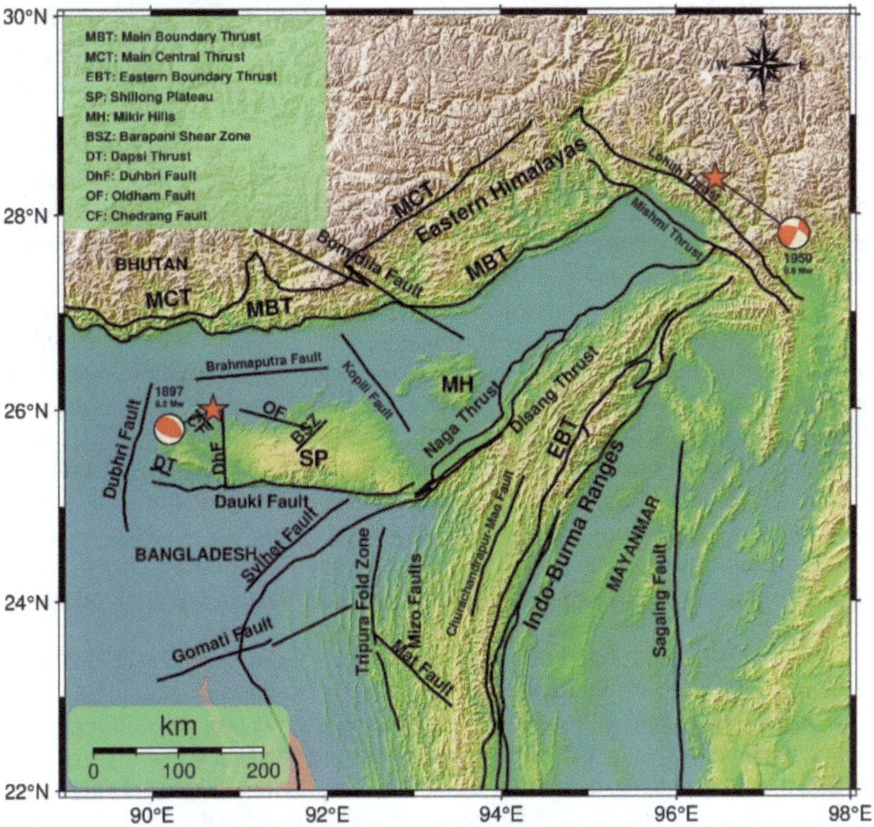

Fig. 5.1 Tectonic setting of Northeast India and surrounding regions. Tectonic features are taken from GSI (2000). Past Large (Mw ≥ 7.0 and great (Mw ≥ 8.0 events are shown (1897–1964)

In northeastern India, the eastern Himalayan collisional zone comprises the eastern Himalaya and the Mishmi Massif in the state of Arunachal Pradesh. The major structural features from north to south may be stated as the Indus Suture Thrust (IST), the Main Central Thrust (MCT), the Main Boundary Thrust (MBT) and the Himalayan Frontal Thrust (HFT) (Fig. 5.1). The Brahmaputra river system (fault) runs along the MBT/HFT in the Assam valley and then takes a sharp turn at 90° E longitude to extend parallel to the Dhubri/Jamuna fault (Kayal, 2008). From the physiographic perspective, the eastern Himalaya may be classified into (i) the Sub-Himalaya comprising the Siwalik Hills, (ii) the Lesser Himalaya and (iii) the Higher Himalaya.

Geologically, the Sub-Himalayan belt is composed of Neogene molasse deposits; the Lesser Himalaya comprises coal-bearing Gondwana rocks, orthoquartzite-carbonate and stable shelf facies along with the Abor volcanics, which are coeval with the Rajmahal traps of Jurassic age (Nandy, 2001). Predominantly, this belt is composed of low- to medium-grade schist, forming the litho unit known as the

Daling Group. The Higher Himalaya comprises dominantly of the 'Central Crystallines' containing high-grade schists, gneisses and patches of granites and the Main Central Thrust (MCT) demarcates the boundary between this Himalaya and the Lesser Himalaya (Nandy, 2001).

The Mishmi Massif connects the eastern Himalaya and the Indo-Burmese mobile belt like a bridge (Nandy, 1976). The lithotectonic units show a structural trend of NW–SE and dip towards NE. As per Nandy (2001), the rocks may be geologically and petrologically classified into three belts, viz., the southwestern belt comprising metamorphosed argillaceous and arenaceous rocks which have been partly migmatised, minor crystalline limestone and garnetiferous amphibolites (Nandy, 2001) and known as the Lohit Group, which is separated from the central belt by NW dipping Tidding thrust. The central belt comprises metamorphic rocks with crystalline limestones. The northeastern belt comprises the diorite-granodiorite complex, with occurrences of metamorphic rocks.

Seismicity in this tectonic domain results from continent–continent collision between the Indian and Eurasian plates. Seismicity is not very intense in the eastern Himalaya and the Mishmi Massif; in fact, a microearthquake study reveals that it is sparse and much lower than that of the Shillong Plateau-Mikir Hills (Kayal, 1996). Mishmi Massif (Eastern Syntaxis Zone) was the seat of the great 1950 Assam earthquake.

The Shillong Plateau is defined by an uplifted Precambrian basement composed of Archean and Proterozoic gneisses and granitoids. The Archaean Gneissic Basement of the plateau is overlain by Proterozoic intracratonic sediments of the Shillong Group along with the intrusive metadolerite Khasi greenstones and subsequently by Upper Proterozoic–early Palaeozoic granite batholiths. The Sylhet Trap occurring in the southern extremity of the Shillong Plateau is bounded between the parallel northern East–West trending Raibah Fault and the southern Dauki fault. The southern and the eastern extension of the Meghalaya Plateau and the southern part of the Mikir Hills show the occurrence of Cretaceous–Eocene shelf sediments. The structural elevation of the plateau relative to the adjoining alluvial plains reveals Cenozoic shortening due to India–Eurasia collision convergence. Rao and Kumar (1997) propounded that the Shillong Plateau was uplifted due to compressive forces from all sides of the plateau. Bilham and England (2001) proposed a 'pop-up' tectonics of the plateau which is the consequence of north–south compression. This happened as the plateau underwent 10 m vertical displacement between the north dipping Dauki fault and the south dipping Oldham fault during the 1897 Shillong plateau earthquake. The Dauki fault demarcates the southern boundary against the Sylhet trough and the Bengal Basin. The Oldham fault forms a part of the northern edge of the plateau, and the NW–SE trending Kopili fault dissects the Shillong Plateau from the Mikir Hills. The Shillong plateau and Mikir hills exhibit higher seismic activity than the Assam valley. There are reports of three large earthquakes of $M > 7.0$ that occurred in the Shillong–Mikir zone, of which the most notable is the 1897 magnitude 8.7 great Shillong Plateau earthquake. As evident from the depth section studies, the seismogenic depth of the plateau is approximately 50 km.

The Indo-Burman arc is a consequence of north-easterly directed movement of the Indian plate, which resulted in its collision with the Burmese plate around early mid-Eocene time. This arc is around 1100 km long and convex towards its western extremity. The Indo-Burman Ranges form a part of this arc, which is defined by oblique subduction of the Indian plate beneath the Burmese plate (Satyabala, 2003). Geologically, this zone comprises flysch and sub-flysch of Eocene to Oligocene age, along with sediments belonging to Palaeozoic-Mesozoic age, with sporadic occurrences of metamorphics and patches of ophiolites. The Indo-Burmese mobile belt is an intercontinental tectonic domain characterised by intermediate-depth earthquakes. It is characterised by a negative Bouger anomaly. Here, this zone has registered as many as 10 large earthquakes of $M > 7.0$ during the last 100 + years since 1897. The conspicuous shallowing of the seismic activity beyond the latitude 26° N is observed, which refers to the fact that the collision process of the eastern Himalaya has been taken over by the subduction process (Mitchell & McKerrow, 1975; Kayal, 1989, 1996). The Indo-Burma Ranges (IBR) shows a rotation of maximum compressive stress from NE–SW to E–W, which is consistent with the deformation partitioning from SSW-directed Sunda-Burma motion to a WSW-directed Burma-India motion (Angelier & Baruah, 2009). The thickness of the Benioff zone varies from 40 to 45 km. Rao and Kalpna (2005) proposed, through focal mechanism studies, a model of slab detachment at the base of the lithosphere. They observed that the stress field in the upper (0–90 km) range of the Indian slab is plate-movement driven, whereas the lower (90–140 km) range is entirely dependent on tensile slab-pull forces. From focal mechanism studies, it is understood that within the Benioff zone (depth ≤ 90 km), and both normal and strike-slip faulting are observed. This is because tensional forces cause the plate to bend or flex. Thrust/reverse faults occur at a deeper level (> 90 km) in the Benioff zone because of the increasing strength of the rocks/gravity loading of the subducted Indian plate at greater depth, which induces a compressional force. The Burmese plate is marked by shallower events and shows thrust faulting as well as strike-slip faulting. The interaction between the overriding and subducting plates generates compressional forces that cause thrust faulting. The strike-slip faulting is caused by regional compression and internal tension within the overriding plate.

The Brahmaputra Foreland Basin is defined by thick alluvial formations, with some Precambrian inselbergs of Basement Rocks extending from Tezpur westwards (Nandy, 2001). The Basement Rocks are overlain by the Tertiary shelf sediments. These Tertiary sediments possess oil reservoirs, which are commonly found in combined stratigraphic-structural traps (Nandy, 2001). The sediment thickness in the Assam valley ranges from 0.5 to 6.5 km, and it increases from southwest to northeast (Lozovsky et al., 2025). The upper Assam foreland basin is experiencing accelerated tectonic subsidence (Kumar et al., 2024). The seismic activity in this basin is sparse. Khattri and Wyss (1978) designated the upper Assam valley as the Assam gap due to its seismic quiescence, which can be the seat of a large impending earthquake. This is supported by the low b-value (Phukan & Mohanty, 2023). But Mukhopadhyay (1984) and Kayal (1996) defined this zone as aseismic.

The Bengal basin is a geosyncline that contains a sediment column of ~ 15 km (Evans, 1964; Talukdar, 1982). The Eocene Hinge Zone is the most important tectonic feature of the Bengal basin, separating the continental shelf to the west from the geosynclinal facies to the east (Evans, 1964). A low level of seismicity is observed in this basin, whereas the Tripura fold belt shows moderate seismicity.

5.3 Crustal and Upper Mantle Structure: Synthesis of Integrated Lithospheric Architecture and Deformation

The NER of India shows strong lateral heterogeneity in crustal and upper mantle structure, as evidenced by receiver function (RF) analyses and seismic tomographic studies. RF studies reveal that the crustal thickness beneath the Shillong Plateau and Mikir Hills is 33–37 km; Assam valley is 35–42 km, which thickens further underneath Eastern Himalaya and Assam syntaxis (40–43 km), and the Indo-Burma Ranges (IBR) stands at 42–50 km (Lozovsky et al., 2025; Mitra et al., 2005; Rai et al., 2016). This is a reflection of the juxtaposition of the Precambrian cratonic lithosphere and Phanerozoic accretionary belts and buttress the interpretation of crustal thinning and possible underplating beneath the Shillong Plateau.

Teleseismic and local earthquake tomography reveals the presence of high-velocity structure and metasomatism or partial melt in the Shillong Plateau and the Indo-Burman wedge (Bhattacharya et al., 2010; Hazarika et al., 2013). Joint inversion of P- and S-wave travel times suggests segmentation of the downgoing Indian lithosphere beneath the Indo-Burma subduction zone, although the continuity and geometry of the slab remain debated. Seismic anisotropy studies using shear-wave splitting indicate significant anisotropy across NER, with the fast polarisation direction varying from NE-SW beneath the Shillong Plateau to E-W beneath the Indo-Burma wedge (e.g., Singh et al., 2016). These patterns are interpreted as evidence for mantle flow influenced jointly by the Himalayan collision and Indo-Burma subduction. Collectively, RF, tomography and anisotropy studies indicate complex crust–mantle coupling, a heterogeneous lithospheric thickness and dynamic upper mantle flow beneath NER.

From the observed gravity, magnetic, magnetotellurics (MT) and seismic data, it is evident that there exists a significant lithospheric-scale contrast throughout NER of India. Singh et al. (2006) noted that the anisotropy in the Himalayan region and its foredeep is due to finite strain induced by collisional tectonics, whereas an asthenopsheric source is said to be the cause of mantle strain in the plate interior. Mohanty and Mondal (2021) have presented evidence of a heterogeneous and unstable anisotropic setting underneath the northeast Indian lithosphere, making it highly complex and reflecting critical mantle dynamics and deformation patterns.

Mohanty et al. (2024) presented evidence from shear-wave splitting observations indicating different deformation patterns in the crust and mantle beneath the Indo-Burmese Wedge (IBW), suggesting active subduction in the IBW. Gravity profiles and seismic imaging across the Brahmaputra foreland reveal flexural subsidence driven by Himalayan and Shillong loads, producing a thick sedimentary basin that modulates seismic wave propagation and crustal deformation (Kumar et al., 2024). Basement faults that influence the distribution of seismicity and deformation pathways are revealed by magnetic studies. The potential-field and MT observations together refer to a heterogeneous lithosphere transitioning from cratonic to accretionary domains, with crust–mantle coupling strongly shaped by subduction, uplift and inherited basement structure. Thus, these data complement seismic and geodetic studies in resolving the lithospheric architecture of the NER of India.

The geophysical studies on NER of India point to a coherent lithospheric system shaped by India–Eurasia collision, India–Sunda oblique subduction and intraplate deformation within the Shillong Plateau. A strongly heterogeneous lithosphere, in which contrasts in crustal thickness, lithospheric strength, mantle hydration and strain accumulation define the style and distribution of deformation, may be inferred from the combined seismological, potential-field, MT, and GNSS datasets. These datasets, when viewed together rather than in isolation, constrain the internal architecture of the lithosphere and the operative deformation mechanisms in this transitional plate boundary domain. High positive Bouger gravity underneath the Shillong Plateau corroborates a crustal root consistent with a mechanically strong cratonic lithosphere (Nandy, 2001), whereas low negative gravity anomalies, sedimentary stacking and thrust geometries in the Indo-Burman ranges reflect an actively shortening accretionary prism (Mitchell, 1993).

These crustal-scale contrasts delineate a first-order boundary between a rigid Precambrian block and a deformable Phanerozoic accretionary system. Strong lateral variations in mantle structure are evident in tomographic images. It is observed that low-velocity zones (LVZs) beneath the Indo–Burma wedge and parts of the lower crust beneath the Shillong Plateau (Hazarika et al., 2013) are consistent with hydrated or partially molten domains, while MT profiles detect conductive bodies beneath Indo–Burma interpreted as slab-derived fluids in the mantle wedge (Selvaraj & Patro, 2016). Together, seismic–MT–gravity results highlight a fundamental rheological contrast between a dry, strong cratonic keel beneath the Shillong Plateau and a fluid-weakened, mechanically compliant subduction wedge beneath the Indo–Burma arc. Seismic anisotropy provides an independent constraint on mantle flow. SKS splitting reveals a transition from NE–SW fast polarisation axes beneath Shillong to E–W beneath Indo–Burma (Singh et al., 2016), indicating a switch from collision-driven to trench-parallel mantle flow. Lower-crustal high-velocity zones beneath Shillong (Rai et al., 2016) are plausibly explained by mafic underplating, which strengthens the lower crust and contributes to plateau uplift and rigidity. GNSS observations provide a kinematic approach to this structural heterogeneity. Approximately ~ 36–50 mm/yr of Indian plate motion is partitioned among: (i) continental underthrusting in the eastern Himalaya, (ii) ~ 3–5 mm/yr of intraplate shortening and strike-slip motion within the Shillong Plateau and (iii) ~ 15–20 mm/yr of oblique India–Sunda convergence in the Indo–Burma wedge (Banerjee et al., 2008; Jade et al., 2011). These velocities define a deformation corridor that transfers stress from the Himalayan thrust system into the Shillong block and onward into the Indo-Burman subduction interface. The localisation of strain energy along the Kopili fault and the megathrust in the IBR region, which aligns with locking depth and independent paleoseismic evidence (Baruah et al., 2021; Rajendran et al., 2004), highlights the mechanical importance of these structures. As a whole, the lithospheric architecture of NER of India may be described as a mosaic of three-pronged system, viz., (i) the subduction wedge (IBR) that accommodates crustal deformation through fluid–assisted processes; (ii) a sturdy cratonic block (Shillong Plateau) that resists deformation and drives stress and mantle flow around its keel; and (iii) a flexural foreland basin (Brahmaputra Basin) that constrains surface deformation and sediment loading. The diversity of deformation patterns in NER of India, ranging from collisional tectonics related to crustal shortening to intraplate deformation of the Shillong Plateau and the oblique subduction of the IBR, is the consequence of interactions with intermediate regions that absorb or redistribute stress. This synthesis reflects that the fundamentally distinct geodynamic features, such as heterogeneous strain fields, conductive mantle anomalies, anisotropy patterns and variable crustal thickness, are a manifestation of an integrated lithospheric system that is constrained by differences in composition, hydration state and rheology. Thus, it is important to understand the principle of coexistence of collisional, subduction and intraplate deformation in a single lithospheric entity. Also, it is intriguing that the partitioning of energy occurs in the same lithospheric framework. Rather, it is pointless to treat seismicity and hazard as discrete endpoints. NER of India may be reckoned as a natural continental geodynamics laboratory and an integrated approach to interpretations points to the fact that previous descriptive investigations are being taken over by process-based quantitative models.

5.4 Seismicity and Seismic Hazard Perspectives

The historical and instrumental seismicity reflects that NER of India shows both shallow crustal earthquakes (< 30 km) and intermediate-depth events (down to ~ 200 km), which are associated with the Himalayan collision, intraplate tectonics and the Indo-Burma subduction zone. Figure 5.2 shows recent seismicity patterns and major tectonic features of the NER of India region. The events of 1897 (Shillong Plateau) and 1950 (Assam-Tibet) demonstrate the capacity of the region to trigger events of $M_W \geq 8$ producing widespread surface deformation and geomorphic reorganisation (e.g. Molnar, 1990; Oldham, 1899). With the advent of modern broadband instrumentation, it has been witnessed that ongoing seismicity is distributed across several structures. Apart from the collisional and subduction zones, the Kopili Fault, Dhubri fault and Dauki Fault in the intraplate zone are also seismically very active. Source mechanism solutions reveal thrust, normal, strike-slip and oblique-slip faulting, reflecting strain partitioning between the collisional Himalaya, the intraplate Shillong Plateau, the Kopili fault in Assam and the Indo–Burma subduction zone. High differential stress and seismic potential are indicated by seismic power-law parameters such as b-values and fractal dimensions (Bhattacharya et al., 2010). On the other hand, GPS studies underscore the coexistence of collisional shortening in the north and oblique subduction with strike-slip shearing in the east (e.g. Kundu & Gahalaut, 2012).

NER of India has been classified among the highest seismic hazard risk regions in the Indian subcontinent due to its proneness to great earthquakes, dense population and vulnerable infrastructure. Seismogenic domains with high seismic potential include: (i) the Shillong Plateau and surrounding faults, (ii) the eastern Himalayan syntaxis, and (iii) the Indo–Burma wedge. Trenching on the Dauki and Kopili faults for paleoseismic studies indicates Quaternary surface-rupturing events, implying ongoing strain accumulation and incomplete seismic release (e.g., Rajendran et al., 2004). Probabilistic seismic hazard assessments (PSHA) for NER of India have been computed to yield peak ground accelerations (PGAs) exceeding 0.4–0.5 g in certain

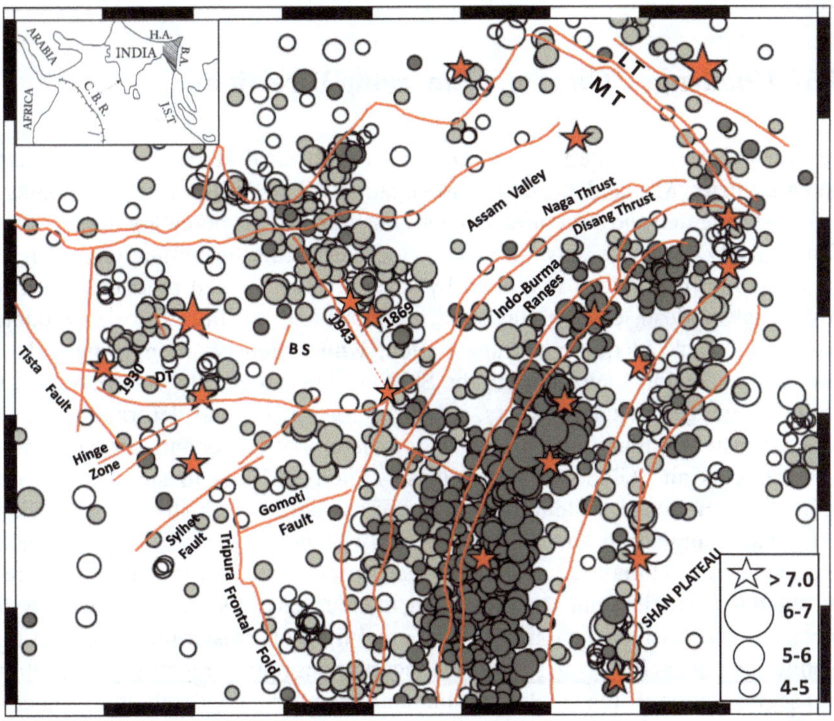

Fig. 5.2 Seismotectonic map of northeastern India (Sharma et al., 2017). The map shows recent seismicity (1990–2016) and major tectonic features of NER of India region. Seismic data are taken from the catalogue of the broadband network of CSIR-NEIST (Jorhat) and CSIR-NGRI-Hyderabad, and the tectonic features are from the GSI (2000). The two great (Mw $\geq$ 8.0), and some 17 large (Mw > 7.0) earthquakes are shown with year of occurrence by stars, respectively. Depth ranges are indicated by open circle: 0–30 km; light grey: 30–60 km; dark grey: > 60 km. Abbreviations used for different tectonic features are: MBT: Main Boundary Thrust; MCT: Main Central Thrust; DT: Dapsi Thrust; BS: Barapni Shear Zone; LT: Lohit Thrust; MT: Mishmi Thrust; SF: Sagaing Fault; OF: Oldham Fault; KF: Kopili Fault; VL: Volcanic Line. Scale: 1° = 111 km

parts of Assam and Meghalaya, consistent with a scenario earthquake of Mw $\geq$ 8 in a thrust and subduction tectonic environment. The enormous thickness of the sediment column in the Brahmaputra foreland basin accentuates the risk of high site amplification. Additionally, the threat of secondary hazards, such as landslides, is very prevalent in the NER of India due to unstable, steep terrain. In totality, seismic hazard in NER of India reflects the collective effects of collisional and subduction processes, the crisscrossing of a plethora of active faults/lineaments and the probability of a large earthquake with uncertain return periods. Therefore, these aspects underscore the significance of real-time monitoring through a dense seismic network. Apart from this, fault characterisation and understanding the physics of the source through the integration of geophysical and geodetic data for a robust seismic hazard assessment of the region are imperative.

5.5 Challenges, Gaps and Emerging Directions

Several efforts are undertaken to conduct geophysical studies, such as seismic, potential-field, MT and geodetic investigations. The NER of India remains the least instrumented and structurally under-constrained plate boundary transition zone. Our capacity to assess seismic hazard, measure strain accumulation and resolve crust–mantle interactions is limited by several scientific and practical challenges. Arunachal Pradesh, eastern Nagaland, Manipur and the Indo-Burman Mountains continue to lack adequate instrumentation, while intensive monitoring is largely confined to corridors surrounding the Shillong Plateau and the Assam Valley. Our ability to image crustal thickness, subduction geometry and deformation patterns near the eastern syntaxis is restricted by the scarcity of broadband seismic networks, MT transects and GNSS arrays in these regions. Efforts to simulate the entire India–Eurasia–Sunda system also require cross-border data sharing with Bhutan, Myanmar, Bangladesh and China, especially along the Indo-Burman megathrust and the Eastern Himalayan Syntaxis. Apart from sparse instrumentation, there are substantial gaps in imaging the lower crust, mantle wedge and slab interface. Receiver functions and seismic tomography have shed light on crustal thickness and upper mantle anomalies. However, significant uncertainties remain regarding the continuity and segmentation of the subducting Indian lithosphere, the depth and geometry of the Main Himalayan Thrust beneath Arunachal, fluid pathways within the mantle wedge and possible lithospheric delamination beneath the Shillong Plateau. High-resolution imaging of these structures is vital for understanding rheology, mantle flow and seismicity in this triple-plate-junction setting. Earlier studies often relied on single-technique approaches, such as receiver functions for Moho-depth estimates, seismic tomography for seismic-velocity structure and gravity data for lithospheric flexure. However, given the unique crust–mantle architecture, an integrated approach combining seismic, MT, gravity and GNSS datasets is now essential to coherently constrain lithology, temperature, fluids and deformation. Comprehensive use of integrated methods, such as coupled seismic–MT inversion, velocity–density modelling and thermo-mechanical simulations, remains limited in the NER of India, yet is critical for distinguishing among competing models of underplating, slab tearing and lithospheric thinning.

Technological opportunities also remain underutilised. For instance, InSAR time series can be used to investigate creeping segments, landslide deformation and seasonal subsidence in the Assam foreland basin; ambient-noise tomography can provide insight in regions with low seismicity; distributed acoustic sensing (DAS), combined with fibre-optic cables, can enable dense, low-cost networks; and machine-learning tools can expedite catalogue homogenisation, tremor detection and hazard forecasting. These emerging techniques can help bridge the spatial and temporal gaps left by conventional networks and, when combined with GNSS, offer new perspectives on aseismic deformation and earthquake-cycle processes. Advancements in geophysical research in the NER of India can also be achieved through stronger regional and international collaboration. Because NER India lies close to Bhutan, Bangladesh, China and Myanmar, cooperation in establishing geophysical networks and sharing data across the Eastern Himalaya, the Indo-Burman Arc and the Tibetan Plateau would significantly improve structural imaging and hazard assessment. Similar efforts in the western and central Himalaya (e.g., India–US, India–France, Nepal–US) have been successful. Replicating such a model, particularly with Myanmar for megathrust segmentation and Bhutan/China for syntaxis structure, could create a win–win scenario for geophysical R&D in the NER of India. To bring about meaningful progress in geophysical research in the NER of India, it is essential to densify broadband seismic networks and MT deployments and to establish GNSS networks in Arunachal Pradesh, Nagaland, Manipur and especially along the Indo-Burman Arc. Approaches such as joint inversion frameworks integrating seismic, MT, gravity and GNSS datasets, as well as imaging slab geometry, mantle hydration and lower-crustal structure beneath the Shillong Plateau and the Indo-Burma Arc, should be adopted. Research on quantifying locking depths, slip rates and strain partitioning using continuous GNSS and InSAR; developing thermo-mechanical models linking mantle flow, crustal rheology and seismicity; and assessing megathrust and intraplate earthquake potential in the Shillong Plateau, Kopili Fault and Indo-Burma domains should be prioritised, along with mechanisms for regional data sharing and cross-border research programmes. Thus, these future directions broadly underscore that the NER of India is not merely a tectonic boundary, but a natural laboratory for studying continental deformation, fluid-assisted subduction and intraplate seismicity. Rapid progress will be possible only through an integrated geophysical approach, improved observational coverage and strong regional cooperation.

5.6 Conclusions

The NER of India is a geodynamically significant domain in which continental collision, subduction and intraplate block deformation interact within a single lithospheric framework. Three decades of geophysical research as a whole transformed understanding of the region's crustal and upper mantle structure, revealing sharp lateral contrasts in crustal thickness, lithospheric strength, mantle hydration and deformation mechanisms and patterns. These contrasts stem from the fact that there is a coexistence of a thick, mechanically robust Shillong Plateau, a hydrated and rheologically weaker Indo–Burman accretionary wedge and a flexural foreland basin system beneath the Brahmaputra valley. This review documents how diverse geophysical methods converge and where they diverge. Geophysical methods such as receiver functions, surface waves and gravity data interpretation could resolve crustal thinning beneath Shillong Plateau; MT and seismic tomography could decipher a hydrated mantle wedge underneath IBR; and geodetic analysis revealed the current strain partitioning between collisional, intraplate and oblique-convergent tectonic domains. On the other hand, the geometry of the subducting Indian lithosphere, the depth extent of lower-crustal delamination and the rheological properties of the mantle wedge remain poorly understood, underscoring the need for deeper lithospheric imaging and joint inversion techniques. These studies are consequential to a refined geodynamic model, while at the same time recognising the gap in geophysical network density and insufficient methodological integration. The approaches like seismic-MT–gravity inversions to reduce non-uniqueness in crust–mantle models; coupled GNSS–InSAR analyses to refine strain accumulation and slip partitioning; and thermo-mechanical modelling to link mantle flow, lithospheric strength, and surface deformation need to be given priority to better comprehend lithospheric-scale processes with observed seismicity, surface deformation, and geomorphology.

We conclude that geophysical research in NER of India requires an understanding of how collision, subduction and intraplate deformation mechanisms coexist and partition seismic energy within a single lithospheric framework. The densification of deployment of geophysical instruments, dataset integration and process-based modelling will play a cardinal role in resolving the unaddressed problems in lithospheric architecture and deformation dynamics, which will improve our knowledge of plate boundary evolution, rheology and the coupling between deep Earth processes and surface observations.

References

Acharyya, S. K. (1987). Cenozoic stratigraphy and structure of the Assam-Arakan basin, NE India. *Journal of the Geological Society of India, 29*(1), 19–27.

Ambraseys, N. N., & Douglas, J. (2004). Magnitude calibration of north Indian earthquakes. *Geophysical Journal International, 159*, 165–206.

Angelier, J., & Baruah, S. (2009). Seismotectonics in Northeast India: A stress analysis of focal mechanism solutions of earthquakes and its kinematic implications. *Geophysical Journal International, 2009*(178), 303–326.

Banerjee, P., Bürgmann, R., Nagarajan, B., & Apel, E. (2008). Intraplate deformation of the Indian subcontinent. *Geophysical Research Letters, 35*(18), L18301. https://doi.org/10.1029/2008GL035468

Baruah, S., Hazarika, D., Gogoi, N., Kayal, J. R., & Baidya, P. R. (2021). Seismotectonics and earthquake hazards in the northeast region of India. *Journal of Earth System Science, 130*(1), 58. https://doi.org/10.1007/s12040-021-01619-5

Bhattacharya, P. M., Kayal, J. R., Baruah, S., & Arefiev, S. S. (2010). Earthquake source zones in Northeast India: Seismic tomography, fractal dimension and b value mapping. *Pure and Applied Geophysics* https://doi.org/10.1007/s00024-010-0084-2.

Bilham, R., & England, P. (2001). Plateau "pop-up" in the great 1897 Assam earthquake. *Nature, 410*(6830), 806–809. https://doi.org/10.1038/35071057

Chen, W. P., & Molnar, P. (1990). Source parameters of earthquakes and intraplate deformation beneath the Shillong Plateau and the northern Indoburman Ranges. *Journal of Geophysical Research: Solid Earth, 95*(B8), 12527–12552. https://doi.org/10.1029/JB095iB08p12527

Evans, P. (1964). Tectonic framework of Assam. *Journal of Geological Society of India, 5*, 80–96.

Hazarika, D., Baruah, S., & Kayal, J. R. (2013). Tomographic imaging of the crust and upper mantle beneath northeast India: Seismotectonic implications. *Journal of Asian Earth Sciences, 73*, 231–246. https://doi.org/10.1016/j.jseaes.2013.04.010

Jade, S., Bhatt, B. C., Dimri, N., Kumar, A., Mukul, M., Singh, P., & Dhar, A. (2011). GPS-derived deformation rates in the Kumaun-Garhwal Himalaya. *Journal of Geophysical Research: Solid Earth, 116*(B6), B06405. https://doi.org/10.1029/2010JB007918

Kayal, J. R. (1989). Subduction structure at the India/Burma plate boundary: Seismic and gravity evidences. In *28th International Geological Congress, Washington, D.C. USA*, vol. 2, pp. 164–165.

Kayal, J. R. (1996). Precursor seismivity, foreshocks and aftershocks of the Uttarkashi earthquake of October 20, 1991 at Garhwal Himalaya. *Tectonophysics, 263*, 339–345.

Kayal, J. R. (2008). Microearthquake activity in northeast India: Seismotectonic implications. *Seismological Research Letters, 79*(1), 1–10. https://doi.org/10.1785/gssrl.79.1.1

Khattri, K. N., & Wyss, M. (1978). Precursory variation in seismicity rate in the Assam area, India. *Geology, 6*, 685–688.

Kumar, P. C., Kumar, J., & Sain, K. (2024). Cenozoic tectonic subsidence in the Upper Assam Basin, NE India. *Geosystems and Geoenvironment, 3*(1). https://doi.org/10.1016/j.geogeo.2023.100223.

Kundu, B., et al. (2012). Earthquake occurrence processes in the Indo-Burmese accretionary wedge. *Tectonophysics, 524*, 135–145. https://doi.org/10.1016/j.tecto.2011.11.022

Lozovsky, I., Varentsov, I., & Walia, D. (2025). Crustal and upper mantle structure of the Assam Valley Region, NE India: A review of geophysical findings. *Geosciences, 15*, 27. https://doi.org/10.3390/geosciences15010027

Mitchell, A. H. G., & Mckerrow, W. S. (1975). Analogous evolution of Burma Orogen and the Scottish Caledonides. *Geological Society of America Bulletin, 86*, 305–315.

Mitchell, A. H. G. (1993). Cretaceous-Cenozoic tectonic events in the western Myanmar (Burma)–Assam region. *Journal of Southeast Asian Earth Sciences, 8*(3–4), 283–295. https://doi.org/10.1016/0743-9547(93)90025-F

Mitra, S., Priestley, K., Bhattacharyya, A. K., & Gaur, V. K. (2005). Crustal structure and earthquake focal depths beneath northeast India and southern Tibet. *Geophysical Journal International, 160*(1), 227–248. https://doi.org/10.1111/j.1365-246X.2005.02465.x

Molnar, P. (1990). A review of the 1950 Assam earthquake and the tectonics of the eastern Himalaya. *Tectonics, 9*(4), 569–585. https://doi.org/10.1029/TC009i004p00569

Mohanty, D. D., & Mondal, P. (2021). Evidence of heterogeneous and unstable anisotropic settings beneath the Northeast Indian lithosphere from characterization of null splitting measurements. *Acta Geophysica, 69*, 2035–2050.

Mohanty, D. D., Biswal, S., & Yoshizawa, K. (2024). Decoupled deformation between crust and mantle beneath Indo-Burmese Wedge: A new seismotectonic model. *Earth and Planetary Science Letters, 648*, 119089. https://doi.org/10.1016/j.epsl.2024.119089

Mukhopadhyay, M. (1984). Seismotectonics of transverse lineament in the eastern Himalaya and its foredeep. *Tectonophysics, 109*, 227–240.

Nandy, D. R. (2001). *Geodynamics of Northeastern India and the adjoining region*. Northeast Geoscience Association.

Nandy, D. R. (1976). The Assam syntaxis of the Himalaya—A re-evaluation. *Geological Survey of India Miscellaneam, 24*(2), 363–367.

Oldham, R. D. (1899). Report on the great earthquake of 12th June 1897. *Memoirs of the Geological Survey of India, 29*, 1–379.

Phukan, A., & Mohanty, D. D. (2023). Seismic hazard estimation and medium-term earthquake precursor analysis of North East India: An assessment on large earthquake scenario. *All Earth, 35*(1), 46–64. https://doi.org/10.1080/27669645.2023.2191917

Rai, S. S., Priestley, K., Gaur, V. K., & Mitra, S. (2016). Crustal shear velocity structure of northeast India. *Tectonophysics, 670*, 62–73. https://doi.org/10.1016/j.tecto.2015.12.004

Rajendran, C. P., Rajendran, K., & Duarah, B. P. (2004). Active faulting and paleoseismicity in the Shillong Plateau, northeast India. *Tectonophysics, 386*(1–2), 1–20. https://doi.org/10.1016/j.tecto.2004.04.017

Rao, N. P., & Kumar, M. R. (1997). Uplift and Tectonics of the Shillong Plateau, North Esat India. *Journal of Physics of the Earth, 45*, 167–176.

Rao, N. P., & Kalpna. (2005). Deformation of the subducted Indian lithospheric slab in the Burmese arc. *Geophysical Research Letters, 32*, L05301. https://doi.org/10.1029/2004GL022034

Satyabala, S. P. (2003). Oblique plate convergence in the Indo-Burma (Myanmar) Subduction Region. *Pure and Applied Geophysics, 160*, 1611–1650.

Selvaraj, C., & Patro, P. K. (2016). Electrical structure of the crust beneath the Indo-Burma subduction zone from magnetotelluric studies: Implications for seismicity and fluids. *Physics of the Earth and Planetary Interiors, 253*, 19–29. https://doi.org/10.1016/j.pepi.2016.01.005

Sharma, A., Baruah, S., Piccinni, D., Saikia, S., Phukan, M. K., Chetia, M., & Kayal, J. R. (2017). Crustal seismic anisotropy beneath Shillong plateau—Assam valley in North East India: Shear-wave splitting analysis using local earthquakes. *Tectonophysics, 717*, 425–432.

Singh, A., Kumar, M. R., Raju, P. S., & Ramesh, D. S. (2006). Shear wave anisotropy of the northeast Indian lithosphere. *Geophysical Research Letters, 33*, L16302. https://doi.org/10.1029/2006GL026106

Singh, S., Kumar, M. R., & Raju, P. S. (2016). Mantle anisotropy beneath northeast India from shear wave splitting: Implications for collisional and subduction tectonics. *Physics of the Earth and Planetary Interiors, 252*, 39–49. https://doi.org/10.1016/j.pepi.2016.01.001

Talukdar, S. N. (1982). Geology and hydrocarbon prospects of east coast basins of India and their relationship to evolution of the Bay of Bengal. In *Offshore SE Asia, Conference, Singapore, Expl. I, Abs.* (pp. 1–8).

Yin, A., & Harrison, T. M. (2000). Geologic evolution of the Himalayan-Tibetan orogen. *Annual Review of Earth and Planetary Sciences, 28*, 211–280. https://doi.org/10.1146/annurev.earth.28.1.211

Chapter 6
Numerical Simulation of Present-Day Tectonic Stress of the Himalayan-Tibet Region—A Review

R. Yadav and S. Singh

Abstract The Himalayan-Tibet (HT) region is the largest mountain chain formed by orogeny, accommodating the north–south crustal shortening resulting from the collision between the Indian plate and the Eurasian plate over the past 50 million years. This area is tectonically and seismically active, characterized by a fault zone that has experienced both shallow and deep megathrust earthquakes. As a result, we are reviewing the tectonic stress in the HT region and investigating its causes through numerical simulations. The analysis of earthquake focal mechanisms and the world stress map indicates that the Himalayan arc experiences north–south (N-S) compressional stress. In contrast, the Tibet region displays west-northwest to east-southeast (WNW-ESE) stress orientations, which are associated with a combination of normal and strike-slip faulting. Additionally, the assessment of gravity anomalies reveals that the Himalayan arc, Altyn Tagh Fault and eastern Himalaya exhibit high stress magnitudes. Simulation results suggest that earthquakes occur in the upper mantle due to flexure and the convergence of the Indian plate. Furthermore, both the eastern and western Himalaya show high strain energy, consistent with observed seismic activity. The deformation modelled using thin-sheet approximations indicates that the stress in the Himalaya arises from a combination of gravitational potential energy and shear traction derived from mantle convection, while the Tibet region is primarily influenced by gravitational potential energy.

Keywords Geodynamics · Himalaya-Tibet · Tectonic stress · Seismicity

---

R. Yadav (✉)
CSIR-National Geophysics Research Institute, Hyderabad, India
e-mail: ryadav.ngri@csir.res.in

S. Singh
Indian Institute of Technology, Jammu, Jammu, India
e-mail: singh.srishti157@gmail.com

V. M. Tiwari (ed.), *Himalayan Natural Hazards*, Indian National Science Academy,
https://doi.org/10.1007/978-981-92-3784-5_6

6.1 Introduction

The HT region is the world's highest mountain range, with an average elevation of 5 km. It formed from the collision of the Indian Plate and the Eurasian Plate ~ 50 Ma (Molnar & Tapponnier, 1975). Ongoing debates focus on the structural configuration and properties of the lithosphere beneath the HT collision zone, stemming from several competing models. The hypotheses regarding the geological interactions in the region include the idea that the Indian lithosphere is being thrust under Tibet. Additionally, there are possibilities of both low-angle and high-angle subduction of the Indian lithosphere under the Tibetan plateau, as well as southward and northward subduction. Other mechanisms at play include crustal shortening and thickening through pure shear, as well as the delamination or convective removal of the tectonically thickened lithospheric mantle. Each of these models offers distinct interpretations of the complex geological interactions occurring in HT area (Chen et al., 2017; Liang et al., 2012).

The crustal and lithospheric structures are delineated using various geophysical methods like seismic, seismological and potential field, in the HT region (Chen et al., 2025; Li & Song, 2018; Zhao et al., 2020). These methods reveal the structure of the Indian mantle lithosphere beneath the Tibetan Plateau, highlighting notable high-velocity anomalies. Further, effective elastic thickness showed that values of less than 40 km in the Tibetan Plateau, whereas it exceeds 50 km in the Indian region. Moreover, based on the global positioning system (GPS) Indian tectonic plate moves northeast at a rate of 3.5 cm/year and rotates slowly in an anticlockwise direction (Rajewar & Gahalaut, 2023), causing the Himalaya deforms through folding and faulting (Fig. 6.3). Moreover, convergence rates vary from 10 to 20 mm/year from Kashmir to Arunachal Himalaya (Stevence & Avouac, 2015). The Himalaya has various transverse tectonic features across the arc, showing different convergence rates. These rates and tectonic features are responsible for frequent earthquakes in the Himalayan region. Over the past century, six major earthquakes have occurred in the region: 1897 Shillong (Mw 8.1), 1905 Kangra (Mw 7.8–8), 1934 Bihar-Nepal (Mw 8.1), 1950 Assam (Mw 8.4–8.6), 2005 Kashmir (Mw 7.6) and 2015 Gorkha (Mw 7.8). The rupture lengths of these devastating earthquakes in the last century have not exceeded 400 km. Therefore, a substantial part of the arc remains unbroken, and several researchers have opined that one or more great earthquakes are overdue in the Himalayan seismic gap (Bilham, 2019; Bilham et al., 2001). The Kumaon Garhwal Himalaya has been identified as a seismic gap between the 1905 Kangra and 1934 Bihar-Nepal rupture zones (Bilham et al., 2001). Despite experiencing micro to moderate magnitude earthquakes, this region has not had any great earthquakes in the last two centuries. The geophysical, geological, geodetic and microseismicity observations indicate that the built-up stress over interseismic time is capable of generating great earthquakes in seismic gap regions (Berger et al., 2004; Jade et al., 2014; Ghavri & Jade, 2021). The focal earthquake (FM) data of this region suggest predominantly active thrusting in the Himalaya. The strike-slip and normal faulting also appear mainly in the sub-Himalaya to Lesser Himalaya and Higher Himalaya, respectively (Heidbach et al., 2018; Mahesh et al., 2015). However, the stress orientations derived from earthquake data do not explain the source of this stress; therefore, numerical simulation studies are necessary.

These studies indicate that complex structures, such as the composition of the Earth's crust and lithosphere and heterogeneity, significantly influence intricate tectonic stress and deformation. Furthermore, this region experiences numerous seismic activities, including thrust, strike-slip and normal fault types. This chapter reviews tectonic stress and identifies sources through numerical simulations.

6.2 Geology and Tectonic

The HT plateau region, situated between the Indo-Gangetic Plain to the south and the Chinese provinces to the north, includes the Himalaya, Lhasa Terrane (LT), Qiangtang Terrane (QT) and Songpan-Ganze (SG) complex. These geological formations are separated by the Indus-Yarlung Zangbu suture (IYZS), Bangong-Nujiang suture (BNS) and Jinsha suture (JS) (Yin & Harrison, 2000). The IYZS zone trends east–west and is a remnant of the Neo-Tethys Ocean, containing Late Jurassic to Early Cretaceous ophiolite rocks (Dewey et al., 1988). The Tethys Himalaya thrust fold belt is part of the passive continental margin of India, which has an estimated 110–149 km of shortening that initiated in the Eocene and ceased motion in ~ 11 to 9 Ma (Murphy & Yin, 2003). The Himalayan arc (HA) comprises the thrust-like Main Frontal Thrust (MFT), Main Boundary Fault (MBT) and Main Central Thrust (MCT). The HA is divided into Sub-Himalaya (Siwalik), composed of Neogene-Quaternary sedimentary rocks; the Lesser Himalaya, having Proterozoic-Palaeozoic metamorphic and sedimentary rocks; and the Greater Himalaya made up of high-grade metamorphic rocks and granite from the Proterozoic-Palaeozoic periods. Central Tibet comprises various tectonic units known as the LT, BNS, QT, JS and SG terranes, bounded by the IYZS to the south and the Ayimaqin-Kunlun-Mutztagh suture to the north (Fig. 6.1). The southern LT comprises the Gangdese Granitoids. The northern Tibet region consists of the Kunlun-Qaidam and Qilian terranes, situated between the Ayimaqin-Kunlun suture and the southern edge of the Hexi Corridor or Tarim Basins (Fig. 6.2). The northern Tibet region is separated by the sinistral Altyn Tagh fault (ALT) into northeastern and northwestern systems. The northwestern secondary thrust system, located west of the ALT, consists of the Pamir and West Kunlun thrust belts, separated by the Kashgar-Yecheng dextral strike-slip fault. The Pamir thrust belts are situated between the Pamir-Hindu Kush region and the southwestern Tarim Basin (Molnar & Tapponnier, 1975). The west Kunlun thrust belt consists of the eastern Kudi-Gaizi thrust belt (KGT), west Kunlun thrust (WKT) belt and Mazatage thrust (MT) belt, which lies between the Tarim Basin and West Kunlun Shan. The Kunlun-Qaidam terrane has the southern Qaidam thrust (QT), Qimen Tagh thrust belt (QTT), Northern QT belt, northern Kunlun thrust belt (NKT) and southern east Kunlun thrust belt (SEKT), often juxtaposing the Proterozoic metamorphic and Paleozoic sedimentary over Tertiary sediments in the Qaidam Basin. Eastern Tibet was

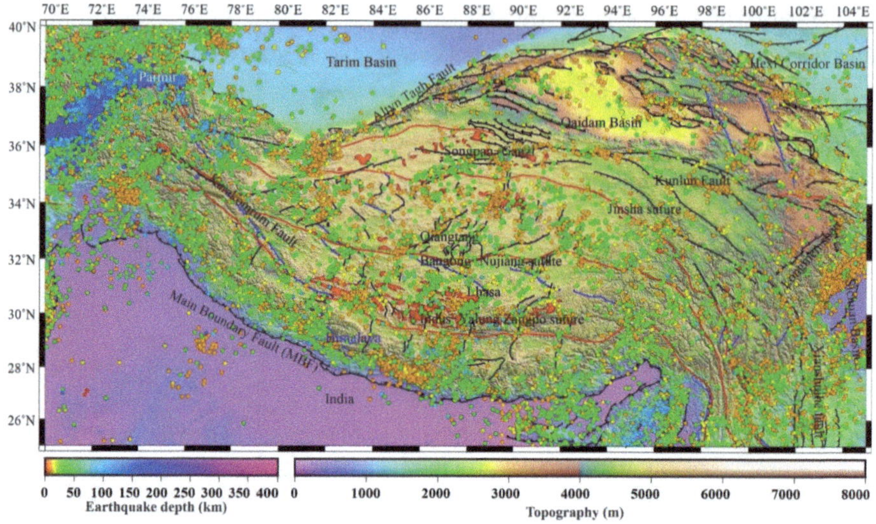

Fig. 6.1 Depth of earthquakes (USGS) is superimposed on the topographic map of the Himalayan-Tibet (HT) region. The eastern and western Himalayan regions experience deeper earthquakes than other regions of the Himalaya

dominated by east to west shortening during the Cenozoic period and lay between western South China and the central plateau. The deformation and geochronology studies show four thrust belts in such as the Ganzi-Litang (GLT), Daxueshan-Gonggashan (DGT), Yupingshan-Yulongshan (YYT) and Longmen Shan (LMST). The northeast striking LMST bounding the eastern margin of the Tibet Plateau is interpreted as a foreland thrust belt that evolved during the Mesozoic.

6.3 Tectonic Stress

(a) Stress from Focal Mechanisms, GPS Velocity and Gravity Anomalies

The SH_{max} orientation of the HT region was computed from the inversion of compiled earthquake FM data (Fig. 6.3) from published articles, reports, the International Seismological Centre (ISC), Global Centroid Moment Tensor (GCMT), National Earthquake Information Centre (NEIC) of the United States Geological Survey (USGS) and Himalaya Nepal Tibet seismic experiment (HIMNMT) (Abhirami et al., 2025; Peitrolungo et al., 2024; Rebetsky & Alekseev, 2022; Wei et al., 2017; Mahesh et al., 2015). The HT was divided into 34 sub-regions to compute the SH_{max} orientation with the inversion of FM (Abhirami et al., 2025) with two approaches named the iterative least squares method and the damped linear inversion method (Fig. 6.4). The results show that the Himalayan arc predominantly compressive stress in the Kashmir, Himachal, Garhwal, Kumaon, Nepal, and Arunachal

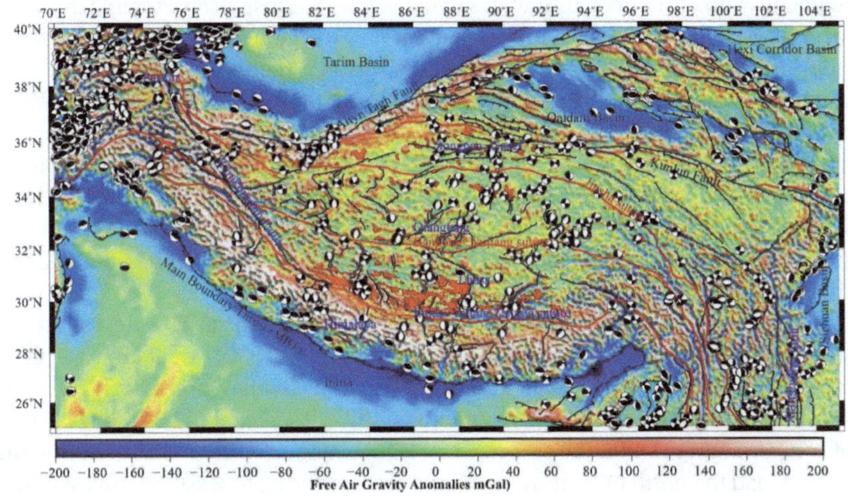

Fig. 6.2 Tectonic and earthquake focal mechanics (USGS) are superimposed on the free-air gravity anomalies map (BGI)

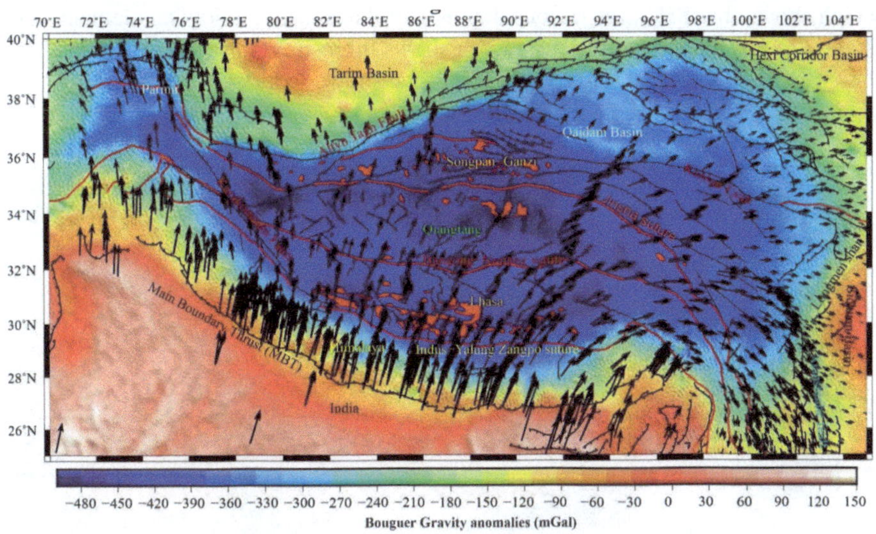

Fig. 6.3 Tectonic and global positions system (GPS) velocities stations (Castro-Perdomo et al., 2025) are superimposed on Bouguer gravity anomalies (BGI)

segments while Sikkim and Bhutan segments have a transpressive regime. Moreover, local studies in the northwestern Himalaya (NWH) region reveal an oblique compression in the Kangra-Chamba-Kishtwar-Zanskar area (between 75°E and 77°E) and a normal compression in the Garhwal-Kumaon-Himachal (Ali et al., 2021; Prasath et al., 2022). Another study (Babu et al., 2024) of NWH is selected eighteen seismically active zones and shows that the Hindukush region is transpression to pure compression, Pamir is pure shear, Nanga Parbat syntaxis is pure extension, Hazara syntaxis is pure compression, Kishtwar-Zanskar transpression and pure compression, Kangra-Chamba and Shimla-Kinnaur are pure compression, Kaurik-Chango rift is radial extension to pure extension, Garhwal and Kumaon are pure compression. However, the tectonic stress in the Karakoram fault and central Tibet regions demonstrate a strike-slip pattern with northeast-southwest (NE-SW) orientations (Abhirami et al., 2025; Prasath et al., 2022). The Gozha-Ashikule fault region reveals pure extension to trans-tension, with SH_{max} azimuth of 150° (Babu et al., 2024). The LT region of southern Tibet exhibits an east–west stress orientation with a normal faulting stress pattern, and the north QT belt zone has the east-northeast to west-southwest trending thrust region (Abhirami et al., 2025). The Tian Shan region and the northern QT belt region are thrust-dominated with east-northeast to west-southwest stress orientations.

The tectonic stress in the HT region is computed using gravity anomalies (Fig. 6.5), which shows that SH_{max} at various depths correlated with geological elements (Xu et al., 2020). The high seismicity region shows high stress, and stable regions have low stress. This study interpreted the SH_{max} and showed a clockwise mass movement

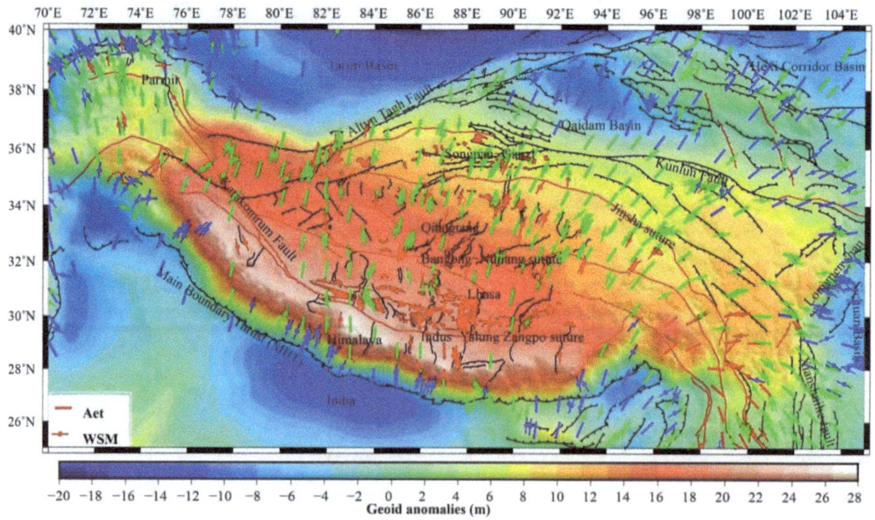

Fig. 6.4 Colour map of geoid anomalies (ICGEM) with superimposed of tectonics and maximum horizontal stress ordinations (SH_{max}) computed from the Focal mechanism and GPS velocities (Abhirami et al, 2025) (Aet) and world stress map (WSM). The colours of symbols are showing that blue is thrust fault, green is strike-slip fault and red is normal fault

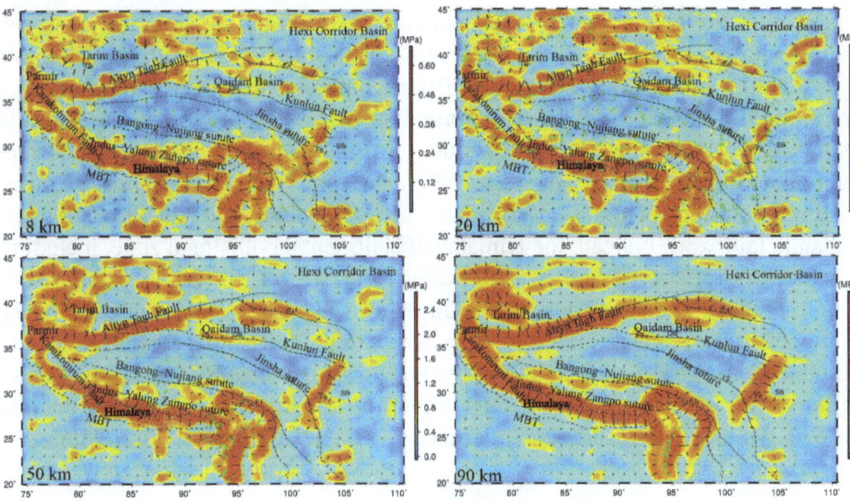

Fig. 6.5 Tectonic stress magnitude and orientations are computed from the gravity anomalies (Xu et al., 2020) at different depths (Here at 8, 20, 50 and 90 km). The high stress at the Himalayan, Hindukush, Pamir, Altyn Tagh Fault, and Longmen Shan regions

pattern in the Tibet region. The eastern Tibet region shows that north–south extrusion between the Indian and Eurasian plates is divided into the stable Sichuan block, one flowing southeast and other moves northwest, resulting in eastward material flow.

(b) **Stress from the Numerical Modelling with Plate Boundary Forces**

The tectonic stresses in the HT region were computed with numerical simulations based on the finite element method, considering plain stress/strain approximations. In plain stress approximation, they have taken the Eurasian plate and implemented the boundary forces like ridge push, slab pull and mantle convection (Liu & Yang, 2003; Sargent 2004; Hieronymu et al., 2008; Yadav & Tiwari, 2018; Duan et al., 2020; Mishra et al., 2024) and implemented the material heterogeneity in terms of Young's modulus and Poisson's ratio. The numerical modelling results from these studies were calibrated against stress orientations from FM inversions, borehole breakouts, hydrofractures and GPS velocities. The tectonic stress of the Eurasian plate, particularly in the HT region, has been modelled by various studies (Sargent 2004; Hieronymu et al. 2008) and shows that the tectonic stress in the HT region is influenced by plate boundary forces, lithospheric thickness and rheology. The numerical simulation study (Duan et al., 2020) conducted five case studies and found that the tectonic stress and orientation are influenced by heterogeneities like the Siberian craton, the North China craton, the Yangtze craton, the TarunKarakum, Lhasa, Qiangtang-Qamdo, Qaidam, Tarim craton, Junggar and Orogenic belt and boundary forces (Fig. 6.6). The strain energy density of this model is high in the western and eastern Himalayan Syntaxes, compared to the middle Himalayan, and consistent with seismicity. The tectonic stress of NWH, including Nepal Himalaya, was modelled along profiles are taken from geophysical studies, showing the significance of the main Himalayan thrust (MHT) and the mid-crustal ramp (MCR) (Chamlagain & Hayashi, 2007; Joshi & Hayashi, 2008, 2010). Compressional and tensional stresses were induced around the MCR, with normal stress developing in the thrust sheets (Chamlagain & Hayashi, 2007). The study (Joshi & Hayashi, 2008) showed that structural and material heterogeneity, mid-crustal ramp and plate boundaries influence the tectonic stress of the Garhwal Himalaya. The numerical stress modelling of the mantle fault zone beneath the Himalayan continental collisional zone, including continent lithospheric flexure, was done by Monsalve et al. (2009) using visco-elastic rheology. The stress around the upper crustal and crustal-mantle boundary was modelled by mountain mass loading and external horizontal forces associated with plate convergence, which deform through brittle processes. The results (stress and deformation) from this model show high differential stress in the upper mantle (Fig. 6.7), which explains stress orientation and seismicity near the Moho (Monsalve et al., 2009).

(c) Stresses Associated with Lithospheric Buoyancy Forces

Continental deformation in the India-Eurasia collision has been a zone of intensive research as it spans thousands of km (Singh and Yadav, 2023; Molnar & Tapponnier, 1975). Hence, the deformation in this area must occur differently from those narrow zones of deformation along the oceanic plate boundaries, as suggested by the theory of plate tectonics. There has been considerable controversy regarding the nature,

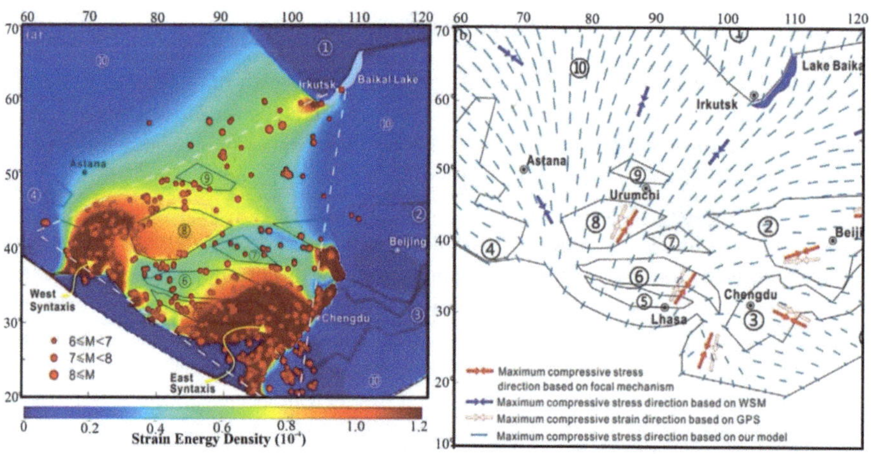

Fig. 6.6 Numerical simulation result of Eurasian plate (Duan et al., 2020) with loading the boundaries conditions where **a** is showing the colour map of strain energy density where eastern and western Himalayan regions have the high strain energy density is zone of high seismicity **b** The computed maximum horizontal stress orientations map is superimposed with in-situ data (WSM). The different bounded polygons are (1) Siberian craton, (2) North China craton, (3) Yangtze craton, (4) Turan Karakum, (5) Lhasa, (6) Qiangtang-Qamdo, (7) Qaidam, (8) Tarim craton, (9) Junggar, and (10) Orogenic belt

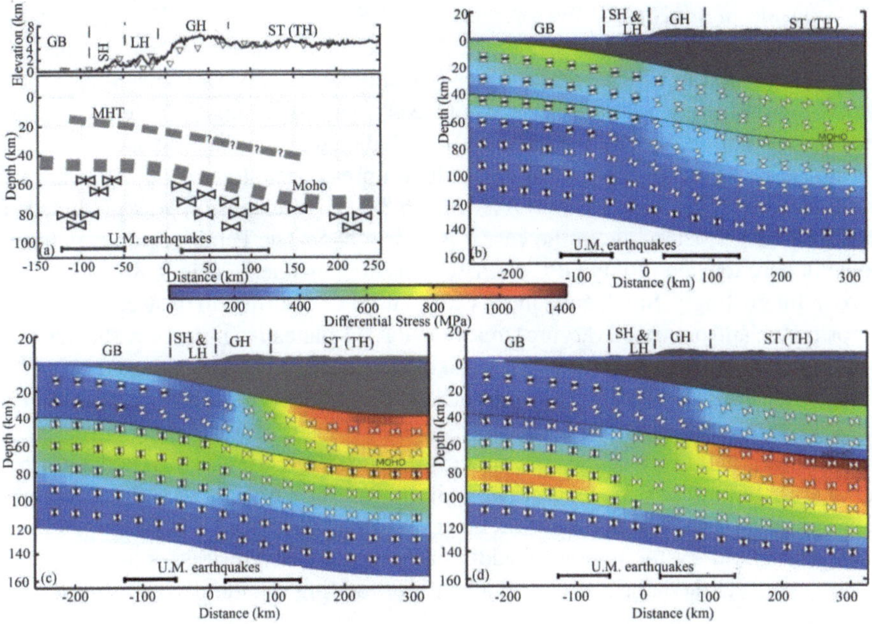

Fig. 6.7 Model of Monsalve et al. (2009) where **a** Geometry of main Himalayan thrust (MHT), Moho depth and topography with superimposed upper mantle earthquakes **b** result of the model is showing the differential stress for elastic crust and topography loading and no horizontal loading **c** result of the model is showing the differential stress for elastic crust and topography loading and horizontal loading **d** result of the model shows the differential stress for visco-elastic crust and topography loading and no horizontal loading. The effective viscosity used in this study was computed using a power law relationship

source and magnitude of the forces that have led to such wide zones of deformation. Primarily, there have been two schools of thought to explain continental kinematics. First school considers lithosphere to be made up of rigid plates/blocks, and the deformation primarily occurs on major faults, while ignoring minor deformations which may occur seismically or on smaller faults (Avouac & Tapponnier, 1993; Matte et al., 1996; Peltzer & Saucier, 1996; Peltzer & Tapponnier, 1988). As opposed to this, the second school suggested deformation to be continuous, which is driven by varying crustal thickness (England & Houseman, 1989; Flesch et al., 2001; Ghosh et al., 2013; Zhang et al., 2004). In the India-Eurasia collision zone, where horizontal dimensions of deformation are much larger than the lithosphere thickness, thin-sheet models have been used to model deformation driven by buoyancy forces. The Thin-sheet approximation assumes that the gradient of shear tractions at the base of the plate is negligibly small compared to the force of gravity arising due to density variations. Flesch et al. (2001) modelled the stresses of the Tibetan plateau by using thin viscous sheet models. They showed that GPE differences are the primary drivers of spatially varying style and magnitude of strain rates in the region. Another study (Ghosh et al., 2006) considered lateral viscosity variations in the lithosphere and found that considering only GPE overpredicts the N-S extension in the Tibetan plateau. Singh and Ghosh (2019) expanded further and modelled the stresses in the HT region, considering stresses arising from two sources: (i) those associated with GPE variations and (ii) another resulting from basal tractions arising from density-driven mantle convection. Singh and Ghosh (2019) predicted extensional stresses in regions of high GPE (Tibetan plateau) along with compressional stresses on MFT and the southern low-lying area of low GPE (Fig. 6.8a). E-W extension was predicted in southwestern Tibet, while in the east it is nearly N-S. The Pamir region was found to be in pure tension if only GPE contributions are considered (Fig. 6.8b) (Jay et al., 2018). Interestingly, Singh and Ghosh (2020) used three different datasets for crustal information and predicted varying stresses in the HT plateau region. They showed that although CRUST1.0 and CRUST2.0 (Laske et al., 2013) predict tension in most of the Tibetan plateau, LITHO1.0 (Pasyanos et al., 2014) model introduces compressional stresses within the region (see Fig. 6.3; Singh & Ghosh, 2020).

Various mantle convection flow models that used densities were calculated from a range of tomography models for two radially varying viscosity structures obtained from studies (Ghosh et al., 2013; Steinberger & Holme, 2008). They computed the final stress field for the region by adding the mantle-derived stresses to those from GPE. The total deviatoric stress obtained after adding mantle-derived stresses led to an increase in magnitudes (Fig. 6.9a). In addition, it introduced N-S compressional stresses in the Tibetan region that cancelled out the N-S extension predicted by the GPE-only model. Hence, the stress regime showed a significant shift towards strike-slip (Fig. 6.9b) from the earlier dominant tensional regime (Fig. 6.8b). Interestingly, Singh and Ghosh (2020) found that the total stress field for GPE and mantle tractions showed a significant change in regime for all three crustal models. Significant compressional stresses from the S40RTS mantle convection model changed the predominant tensional regime to strike-slip for CRUST2 (Fig. 6.10). On the other hand, the introduction of N-S tensional stresses in the Pamir was found to have re-oriented to more E-W directions.

In addition to deviatoric stresses, Singh and Ghosh (2019, 2020) also calculated plate velocities, strain rates, as well as the most compressive horizontal principal stress (SH_{max}) axes. Singh and Ghosh (2020) compared these predicted parameters with observed data of strain rates from Global Strain Rate Model (GSRMv 2.1, Kreemer et al., 2014), SH_{max} from the World Stress Map (WSM, Heidbach et al., 2018) and plate velocities from Kreemer et al. (2014) to evaluate the predicted stresses in the India-Eurasia collision zone (Fig. 6.3). The best fit was obtained for the models predicting strike-slip type of deformation in the India-Eurasia collision zone, such as S40RTScr2 and SAW642ANcr2, which are the joint models of stresses computed using GPE from CRUST2 and mantle tractions derived from S40RTS and SAW642AN models, respectively (Fig. 6.10). Similarly, considering only lithospheric contributions, the LITHO model provided an overall better fit to observed parameters, as it predicted a strike-slip deformation in most of the Tibetan Plateau.

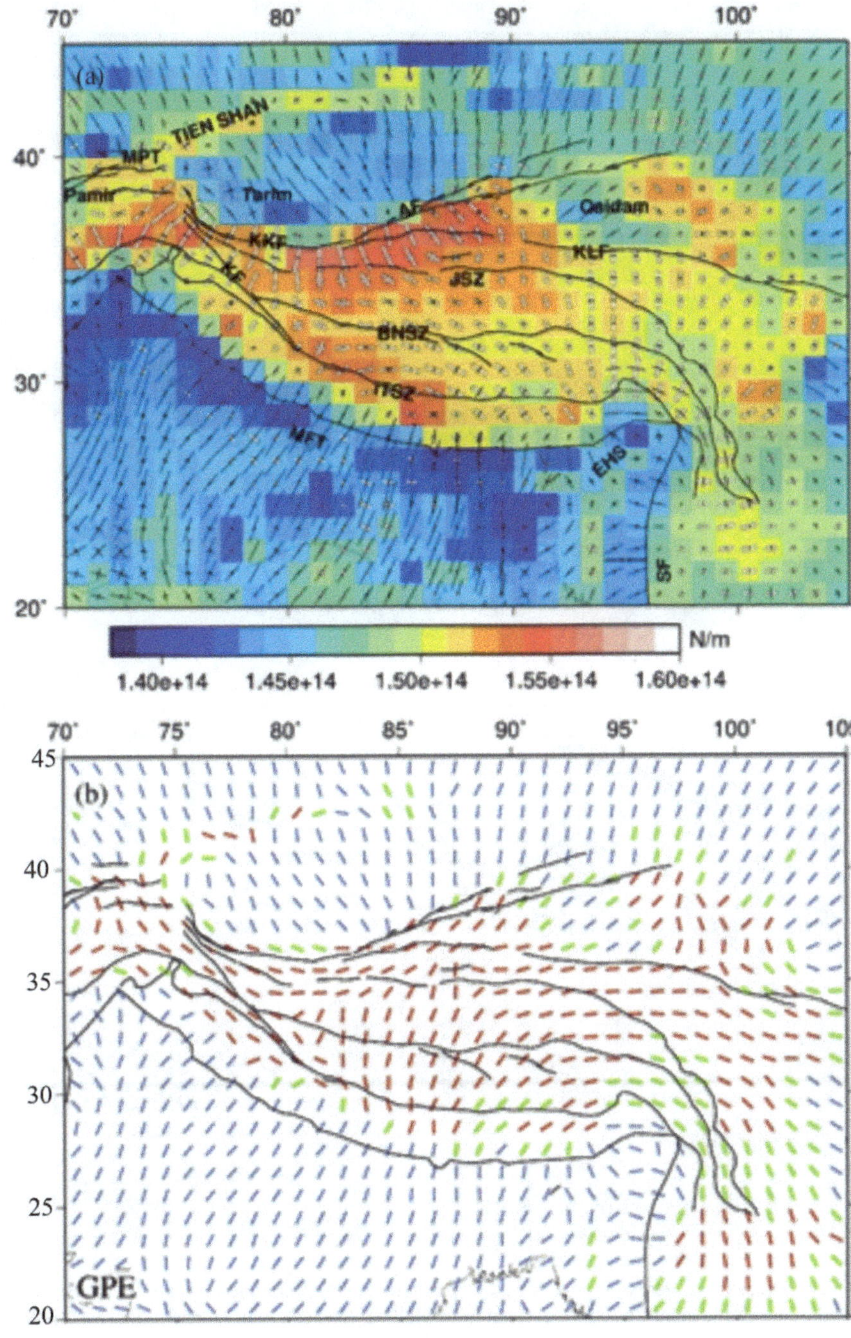

Fig. 6.8 **a** Deviatoric stresses plotted on top of GPE variations. The compressional stresses are denoted by solid black arrows, while white arrows show tensional stresses, **b** predicted S_{Hmax}, red denotes tensional regime, blue is for thrust, and green for strike-slip regime (Singh & Ghosh, 2019)

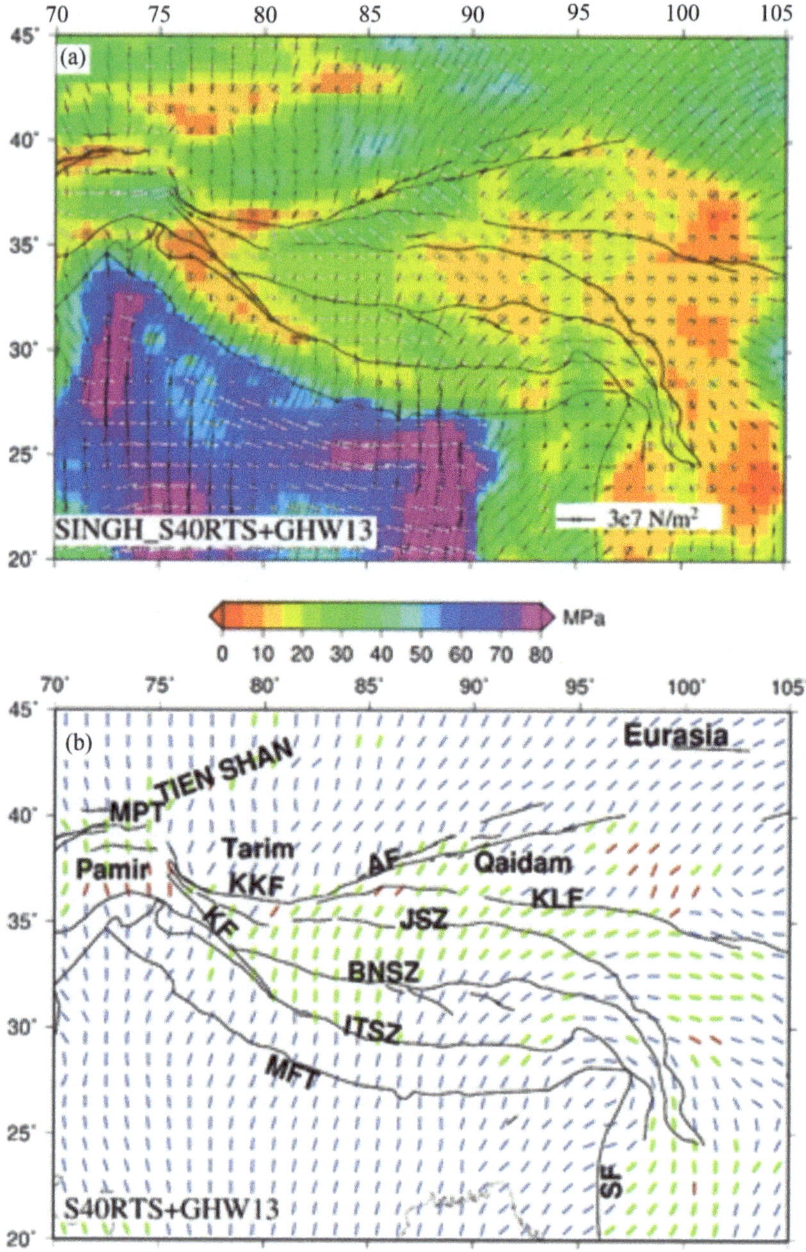

Fig. 6.9 **a** Deviatoric stresses; and **b** S_{Hmax} predicted using combined effects of gravitational potential energy and mantle tractions derived from SINGH_S40RTS tomography model for GHW13 viscosity structure plotted on second invariant of deviatoric stresses. The white arrows denote tensional stresses, and black arrows indicate compressional stresses. Red denotes tensional regime, blue is for thrust, and green is for strike-slip regime (Singh & Ghosh, 2019)

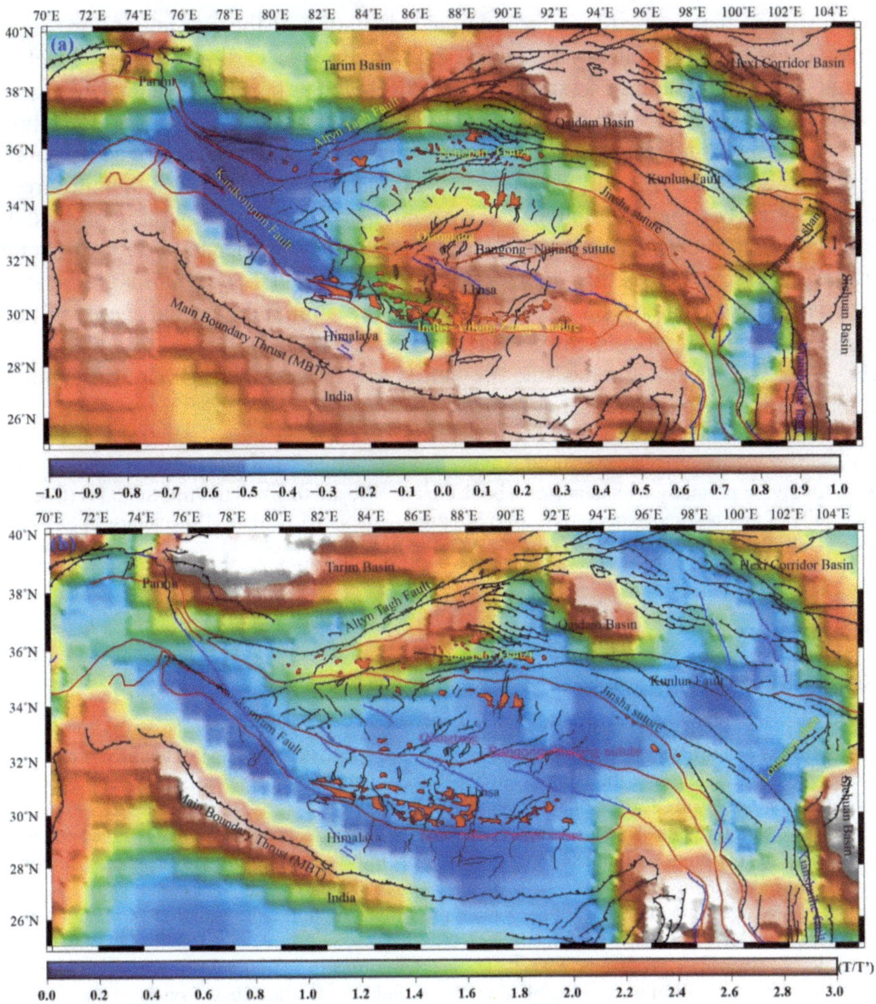

Fig. 6.10 Comparison of the predicted deviatoric stresses from GPE and mantle tractions alone for the SINGH_S40RTS model with GHW13 viscosity structure model. **a** Correlation coefficients between the deviatoric stress tensors predicted from GPE and mantle tractions. **b** The ratio of the second invariant of deviatoric stresses predicted using GPE and mantle tractions (Singh & Ghosh, 2019)

6.4 Summary

The Himalayan-Tibet region has formed as a result of the continental collision between the Indian and Eurasian plates, which began about 50 million years ago. This area is characterised by numerous tectonically and seismically active faults with complex structures, resulting in a range of shallow to deep earthquakes. As a result, many studies have aimed to determine the orientation of the maximum horizontal stress (SH_{max}) and identify its sources. The analysis of earthquake focal mechanisms, gravity anomalies and stress maps indicates that the Himalayan region experiences north–south (N-S) compressive stress. In contrast, the Tibet region exhibits a west-northwest to east-southeast (WNW-ESE) orientation characterized by a mix of normal and strike-slip faults. Numerical simulation studies suggest that the stress in the Himalayan region is influenced by boundary forces, such as ridge push, slab pull and mantle convection, along with lithospheric heterogeneity. However, the primary source of stress that drives deformation in the Tibet region is the heterogeneity of crustal materials. The combination of gravitational potential energy (GPE) and mantle traction models most accurately reproduces the observed orientations of SH_{max} and strain rates, emphasising the significance of considering mantle dynamics in modelling the India-Eurasia collision.

Acknowledgements The authors of this study are thankful to the director of CSIR-NGRI, Hyderabad, for giving permission to submit as a book chapter. We are thankful to Prof. Sunil PS (Cochin University of Science and Technology) who has provided the SH_{max} data. This manuscript has the reference no NGRI/Lib/2026/**Pub-16 in the NGRI library.**

References

Abhirami, R., Sunil, P. S., Sooraj, A., Sreerag, S. S., & Sunil, A. S. (2025). Integrated stress and strain characterisation of the Himalayan-Tibetan collision zone using earthquake and geodetic data. *Tectonophysics, 902*, 230686.

Ali, S. M., Abdelrahman, K., & Al-Otaibi, N. (2021). Tectonic stress regime and stress patterns from the inversion of earthquake focal mechanisms in NW Himalaya and surrounding regions. *The Journal of King Saud University—Science, 33*(2), 101351.

Avouac, J. P., & Tapponnier, P. (1993). Kinematic model of active deformation in central Asia. *Geophysical Research Letters, 20*(10), 895–898.

Babu, V. G., Kumar, N., & Pal, S. K. (2024). Stress regimes in the Himalaya–Karakoram–Tibet, the western part of India-Eurasia collision: Stress field implications based on focal mechanism solution data. *Geophysical Journal International, 239*(3), 1380–1399.

Berger, A., Jouanne, F., Hassani, R., Mugnier, J. L. (2004). Modelling the spatial distribution of present-day deformation in Nepal: How cylindrical is the Main Himalayan Thrust in Nepal? *Geophysical Journal International, 156*(1), 94–114.

Bilham, R., Gaur, V. K., & Molnar, P. (2001). Himalayan seismic hazard. *Science, 293*(5534), 1442–1444.

Bilham, R., & Wallace, K. (2005). Future Mw >8 earthquakes in the Himalaya: Implications from the 26 Dec 2004 Mw = 9.0 earthquake on India's eastern plate margin. *Geological Survey of India Special Publication, 85*, 1–14.

Bilham, R. (2019). Himalayan earthquakes: A review of historical seismicity and early 21st century slip potential. *Geological Society of London, Special Publication, 483*(1), 423–482.

Castro-Perdomo, N., Jónsson, S., Klinger, Y., Masson, F., Becker, T. W., & Johnson, K. (2025). Strain rates along the Alpine-Himalayan Belt from a comprehensive GNSS velocity field. *Journal of Geophysical Research Solid Earth, 130*(12). https://doi.org/10.1029/2025JB031738

Chen, B., Liu, J., Chen, C., Du, J., & Sun, Y. (2025). Elastic thickness of the Himalayan-Tibetan orogen estimated from the fan wavelet coherence method, and its implications for lithospheric structure. *Earth and Planetary Science Letters, 409*, 1–14.

Chen, M., Niu, F., Tromp, J., Lenardic, A., Lee, C. T. A., Cao, W., & Ribeiro, J. (2017). Lithospheric foundering and underthrusting imaged beneath Tibet. *Natural Communication, 8*(1), 15659.

Chamlagain, D., & Hayashi, D. (2007). Neotectonic fault analysis by 2D finite element modeling for studying the Himalayan fold-and-thrust belt in Nepal. *Journal of Asian Earth Science, 29*(2–3), 473–489.

Dewey, J. F., Shackleton, R. M., Chang, C. F., & Sun, Y. Y. (1988). The tectonic evolution of the Tibetan Plateau. *Philosophical Transactions of the Royal Society B London Series A, 327*(1594), 379–413.

Duan, S., Hou, G., & Yang, L. (2020). Geodynamics of the Broad Triangle Area (active seismic zone) in Asia: Stress field modeling. *Journal of Asian Earth Sciences, 188*, 104073.

England, P., & Houseman, G. (1989). Extension during continental convergence, with application to the Tibetan Plateau. *Journal of Geophysical Research, 94*(B12), 17561–17579.

Flesch, L. M., Haines, A. J., & Holt, W. E. (2001). Dynamics of the India-Eurasia collision zone. *Journal of Geophysical Research, 106*(B8), 16435–16460.

Ghavri, S., & Jade, S. (2021). Seismic potential of megathrust in the Kumaun-Garhwal region of NW Himalaya: Implications from geodetic and seismic strain rates. *International Journal of Earth Sciences, 110*(4), 1439–1452.

Ghosh, A., Holt, W. E., Flesch, L. M., & Haines, A. J. (2006). Gravitational potential energy of the Tibetan Plateau and the forces driving the Indian plate. *Geology, 34*(5), 321–324.

Ghosh, A., Holt, W. E., & Wen, L. (2013). Predicting the lithospheric stress field and plate motions by joint modeling of lithosphere and mantle dynamics. *Journal of Geophysical Research, 118*(1), 346–368.

Heidbach, O., Rajabi, M., Cui, X., Fuchs, K., Müller, B., Reinecker, J., Reiter, K., Tingay, M., Wenzel, F., Xie, F., & Ziegler, M. O. (2018). The World Stress Map database release 2016: Crustal stress pattern across scales. *Tectonophysics, 744*, 484–498.

Hieronymus, C. F., Goes, S., Sargent, M., & Morra, G. (2008). A dynamical model for generating Eurasian lithospheric stress and strain rate fields: Effect of rheology and cratons. *Journal of Geophysical Research, 113*(B7).

Jade, S., Mukul, M., Gaur, V. K., Kumar, K., Shrungeshwar, T. S., Satyal, G. S., Dumka R. K., Jagannathan S., Ananda M. B., Kumar P. D., Banerjee, S. (2014). Contemporary deformation in the Kashmir-Himachal, Garhwal and Kumaon Himalaya: Significant insights from 1995–2008 GPS time series. *Journal of Geodesy, 88*(6), 539–557.

Jay, C. N., Flesch, L. M., & Bendick, R. O. (2018). Kinematics and dynamics of the Pamir, central Asia: Quantifying the roles of continental subduction in force balance. *Journal of Geophysical Research: Solid Earth, 123*(9), 8161–8179.

Joshi, G. R., & Hayashi, D. (2008). Numerical modeling of neotectonic movements and state of stresses in the central Seismic Gap region, Garhwal Himalaya. *Journal of Mountain Science, 5*(4), 279–298.

Joshi, G. R., & Hayashi, D. (2010). Development of extensional stresses in the compressional setting of the Himalayan thrust wedge: Inference from numerical modelling. *Natural Science, 2*(07), 667.

Kreemer, C., Blewitt, G., & Klein, E. C. (2014). A geodetic plate motion and Global Strain Rate Model. *Geochemistry, Geophysics, Geosystems, 15*(10), 3849–3889.

Laske, G., Masters, G., Ma, Z., & Pasyanos, M. (2013). Update on CRUST1.0-A 1-degree global model of Earth's crust. *Geophysical Research Abstracts, 15*(15), 2658.

Liang, X., Sandvol, E., Chen, Y. J., Hearn, T., Ni. J., Klemperer, S., Shen, Y., & Tilmann, F. (2012). A complex Tibetan upper mantle: A fragmented Indian slab and no south-verging subduction of Eurasian lithosphere. *Earth and Planetary Science Letters, 333–334*, 101–111.

Liu, M., Yang, Y. (2003). Extensional collapse of the Tibetan Plateau: Results of three-dimensional finite element modeling. *Journal of Geophysical Research, 108*(B8).

Li, J., & Song, X. (2018). Tearing of Indian mantle lithosphere from high-resolution seismic images and its implications for lithosphere coupling in southern Tibet. *Proceedings of the National Academy of Sciences, 115*(33), 8296–8300.

Matte, P., Tapponnier, P., Arnaud, N., Bourjot, L., Avouac, J. P., Vidal, P., Qing, L., Yusheng, P., & Yi, W. (1996). Tectonics of Western Tibet, between the Tarim and the Indus. *Earth and Planetary Science Letters, 142*(3–4), 311–330.

Mahesh, P., Gupta, S., Saikia, U., & Rai, S. S. (2015). Seismotectonics and crustal stress field in the Kumaon-Garhwal Himalaya. *Tectonophysics, 655*, 124–138.

Mishra, A., Yadav, R., & Tiwari, V. M. (2024). Intraplate stress distribution within the Indian Plate: Insights from finite element modelling. *Journal of Asian Earth Sciences, 265*, 106102.

Molnar, P., & Tapponnier, P. (1975). Cenozoic tectonics of Asia: Effects of a continental collision. *Science, 189*(4201), 419–426.

Murphy, M. A., & Yin, A. (2003). Structural evolution and sequence of thrusting in the Tethyan foldthrust belt and Indus-Yalusuture zone, southwest Tibet. *Geological Society of America Bulletin, 115*(1), 21–34.

Monsalve, G., McGovern, P., & Sheehan, A. (2009). Mantle fault zones beneath the Himalayan collision: Flexure of the continental lithosphere. *Tectonophysics, 477*(1–2), 66–76.

Pasyanos, M. E., Masters, T. G., Laske, G., & Ma, Z. (2014). LITHO1. 0: An updated crust and lithospheric model of the Earth. *Journal of Geophysical Research, 119*(3), 2153–2173.

Peltzer, G., & Tapponnier, P. (1988). Formation and evolution of strike-slip faults, rifts, and basins during the India-Asia collision: An experimental approach. *Journal of Geophysical Research, 93*(B12), 15085–15117.

Peltzer, G., & Saucier, F. (1996). Present-day kinematics of Asia derived from geologic fault rates. *Journal of Geophysical Research, 101*(B12), 27943–27956.

Pietrolungo, F., Lavecchia, G., Madarieta-Txurruka, A., Sparacino, F., Srivastava, E., Cirillo, D., de Nardis, R., Andrenacci, C., Bello, S., Parrino, N., & Sulli, A. (2024). Comparison of crustal stress and strain fields in the Himalaya-Tibet Region: Geodynamic implications. *Remote Sensing, 16*, 4765.

Prasath, R. A., Bansal, B. K., & Verma, M. (2022). Stress distribution in the western IndiaEurasia collision zone, its kinematics and seismotectonic implications. *Journal of Asian Earth Sciences, 230*, 105208.

Rajewar, S., & Gahalaut, V. (2023). The rigidity of Indian plate: Constraints from continuous GPS measurements. In: *XXVIII General Assembly of the International Union of Geodesy and Geophysics (IUGG) (Berlin 2023).*

Rebetsky, Y. L., & Alekseev, R. S. (2022). The pattern and possible formation mechanisms of current crustal stress in High Asia. *Russian Geology and Geophysics, 63*(6), 726–742.

Sargent, M. T. (2004). Dynamics of the Eurasian Plate. Doctoral dissertation, Ph. D. thesis, Swiss Federal Institute of Technology Zurich: 81

Singh, S., & Ghosh, A. (2019). Surface motions and continental deformation in the Indian plate and the India-Eurasia collision zone. *Journal of Geophysical Research, 124*(11), 12141–12170.

Singh, S., & Yadav, R. (2023). Numerical modeling of stresses and deformation in the Zagros-Iranian Plateau region. *Solid Earth, 14*(8), 937–959.

Singh, S., & Ghosh, A. (2020). The role of crustal models in the dynamics of the India-Eurasia collision zone. *Geophysical Journal International, 223*(1), 111–131.

Stevens, V. L., & Avouac, J. P. (2015). Interseismic coupling on the main Himalayan thrust. *Geophysical Research Letters, 42*(14), 5828–5837.

Steinberger, B., & Holme, R. (2008). Mantle flow models with core-mantle boundary constraints and chemical heterogeneities in the lowermost mantle. *Journal of Geophysical Research, 113*(B5).

Wei, J., Weifeng, S., & Xiaojing, M. (2017). Tectonic stress pattern in the Chinese mainland from the inversion of focal mechanism data. *Journal of Earth System Science, 126*, 1–22.

Xu, C., Yu, H., Yao, C., Li, J., & Yan, J. (2020). An improved gravity method to horizontal tectonic stresses and its applications in Tibet. *Geodesy and Geodynamics, 11*(6), 468–473.

Xuan, S., & Jin, S. (2022). Moho depth and crustal density structure in the Tibetan Plateau from gravity data modelling. *Journal of Asian Earth Sciences, 233*, 105261.

Yadav, R., & Tiwari, V. M. (2018). Numerical simulation of present day tectonic stress across the Indian subcontinent. *International Journal of Earth Sciences, 107*(7), 2449–2462.

Yin, A., & Harrison, T. M. (2000). Geologic evolution of the Himalayan-Tibetan orogen. *Annual Review of Earth and Planetary Sciences, 28*, 211–280.

Zhang, P.Z., Shen, Z., Wang, M., Gan, W., Burgmann, R., Molnar, P., Wang, Q., Niu, Z., Sun, J., Wu, J., & Hanrong, S. (2004). Continuous deformation of the Tibetan Plateau from global positioning system data. *Geology, 32*(9), 809–812 (2004)

Zhao, G., Liu, J., Chen, B., Kaban, M. K., & Zheng, X. (2020). Moho beneath Tibet based on a joint analysis of gravity and seismic data. *Geochemistry, Geophysics, Geosystems, 21*(2), 02019GC008849.

Chapter 7
Crustal Strain from GNSS Measurements and Earthquakes in the Himalaya

Shubham K. Rajewar, Vineet K. Gahalaut, Neeraj Gurjar, Aditya Mohanty, A. Ajeesha, I. C. Das, T. R. Martha, Priyom Roy, Rajeev Kumar Yadav, and Joshi K. Catherine

Abstract The Himalayan arc is amongst the most seismically active regions of the world, with the majority of earthquakes occurring along the Main Himalayan Thrust (MHT). A few seismic gaps have been identified along the arc, and the central seismic gap has not experienced a major earthquake in the past few hundred years, covering a significant portion of this area. Therefore, it is crucial to study the current strain accumulation rates along the arc. Several studies have employed Global Navigation Satellite System (GNSS) data to examine strain accumulation rates using dense networks; however, some regions still lack sufficient data. In this study, we utilized the most recent compilation of GNSS data along the Himalayan arc to obtain a more robust estimate of strain accumulation rates and their relation with seismicity in the region. Our analysis indicates strain rates ranging from ~ 70 to 100 ηstrain/year, with the principal axes predominantly reflecting compression across the arc due to the collision between the Indian and Eurasian tectonic plates. For seismic hazard assessment, we analysed the correlation between geodetic strain rates and several factors, including seismicity, topography and population density. We found a strong correlation between strain and seismicity, indicating that areas with high seismicity correspond to high strain rates. For hazard assessment, we propose that it is better to consider the locking map rather than the crustal strain map, as the former is derived

S. K. Rajewar · V. K. Gahalaut (✉) · A. Mohanty · A. Ajeesha · R. K. Yadav · J. K. Catherine
CSIR—National Geophysical Research Institute, Hyderabad, India
e-mail: vkgahalaut@yahoo.com

S. K. Rajewar · V. K. Gahalaut · A. Mohanty · A. Ajeesha · R. K. Yadav
Academy of Scientific and Innovative Research (AcSIR), Ghaziabad, India

V. K. Gahalaut
Wadia Institute of Himalayan Geology, Dehradun, India

N. Gurjar
Geodetic and Research Branch, Survey of India, Dehradun, India

I. C. Das · T. R. Martha · P. Roy
National Remote Sensing Centre, Hyderabad, India

V. M. Tiwari (ed.), *Himalayan Natural Hazards*, Indian National Science Academy,
https://doi.org/10.1007/978-981-92-3784-5_7

from earthquake process-based modelling and the latter is only a surface effect of the former.

Keywords Himalayan arc · GNSS · Strain rates · Seismicity · Locking map

7.1 Introduction

The Himalayan arc is amongst the most tectonically active mountain belts in the world, formed due to the collision between the Indian and Eurasian plates (Molnar & Tapponnier, 1975). This process began ~ 50 to 55 million years ago and continues even today, stretching for about 2500 km from northwest to northeast. The current convergence rate between India and Eurasia is about ~ 40 mm/year, with about half of this convergence rate is accommodated in the Himalayan arc through slip on the Main Himalayan Thrust (MHT), as indicated by some studies (e.g., Ader et al., 2012; Banerjee & Bürgmann, 2002; Bettinelli et al., 2006; Bilham et al., 1997; Dal Zilio et al., 2020; Jade et al., 2014; Jouanne et al., 2004; Kumar et al., 2023; Kundu et al., 2014; Larson et al., 1999; Li et al., 2018; Marechal et al., 2016; Mohanty et al., 2025; Panda & Lindsey, 2024; Saji et al., 2025; Sreejith et al., 2018; Stevens & Avouac, 2015; Vernant et al., 2014; Yadav et al., 2019) The MHT exhibits a stick–slip behaviour in the Outer and Lesser Himalaya, while it slips aseismically in the Higher and Tethys Himalaya (Bilham et al., 1997; Gahalaut & Chander, 1997). This leads to the understanding that large earthquakes that are known to recur along the Himalayan front are likely associated with the MHT (Ambraseys & Douglas, 2004). The Himalayan arc have produced several major earthquakes in history are M ~ 8.2 Lo Mustang in 1505 (Bilham, 2004), M ~ 7.6 Kashmir in 1555 (Bilham, 2004), M ~ 8.1 Bhutan in 1714 (Le Roux-Mallouf et al., 2016), M ~ 7.8 Kangra in 1905 (Hough & Roger, 2008; Szeliga & Bilham, 2017), M ~ 7.7 Kathmandu in 1833 (Bilham, 1995), M ~ 8.4 Nepal-Bihar in 1934 (Ambraseys & Douglas, 2004; Bilham, 2019; Molnar, 1990), M ~ 8.7 Assam in 1950 (Bilham, 2004; Coudurier-Curveur et al., 2020), Mw 7.6, Kashmir in 2005 (Avouac et al., 2006) and Mw 7.8, Nepal in 2015 (Galetzka et al., 2015; Lindsey et al., 2015). In the past few decades to a hundred years, the Himalayan arc has not experienced a major earthquake in the three seismic gaps: the Assam seismic gap, the Central seismic gap, and the Kashmir seismic gap (Khattri, 1987). These gaps have significant potential to produce devastating earthquakes in the future, making it crucial to assess the present-day deformation and corresponding interseismic strain rates and seismicity in the Himalayan belt.

In this study, we have compiled over 900 recent GNSS velocities from and around the Himalayan arc to cover the northwest to northeast region of the Himalaya. Our goal is to estimate the strain rate and evaluate the relationship between geodetic strain rates and seismicity for the purpose of seismic hazard assessment.

7.2 A Review of Past GNSS-Derived Strain Rates in the Himalaya

There are several studies that have been carried out in the Pan-Himalayan Arc using GNSS measurements to estimate interseismic strain rate build-up and coupling. A few recent ones of them are discussed here.

Stevens and Avouac (2021) investigated the relationship between geodetic strain rates and seismicity in the India-Asia collision zone. They found a strong correlation between small to moderate independent earthquakes and geodetic strain rates. The authors tested linear and nonlinear models to analyse the relationship between strain rates and seismicity, concluding that the nonlinear model provided a significantly better fit than the linear model. The findings of the nonlinear model suggest that the maximum magnitude of earthquakes could be larger in areas with low strain rates. In contrast, the linear model indicates that the maximum magnitude earthquake is independent of strain rate. This study implies that GNSS-derived strain rates can be utilized for probabilistic seismic hazard assessments, emphasizing that regions with low strain may still experience significant earthquakes.

Dal Zilio et al. (2021) present a comprehensive analysis of how long-term tectonic processes and short-term earthquake cycles interact to shape the Himalaya. It emphasizes the role of the MHT, a major continental megathrust fault, in accommodating crustal shortening and causing large, infrequent earthquakes, such as the 2015 Gorkha and the 1934 Nepal-Bihar earthquakes. The authors use an earthquake cycle model to suggest that partial ruptures primarily occur in the downdip region of the MHT. They highlight that tectonic structures can influence stress accumulation, structural segmentation and the location of earthquake ruptures. The presence of inherited structures seems to affect the segmentation of the Himalaya; however, it remains unclear whether this segmentation is persistent across all timescales. Therefore, the article emphasizes the need for more observations of earthquakes on well-instrumented faults.

Panda and Lindsey (2024) reassess the coupling of the Megathrust MHT using a dense GNSS dataset and a kinematic block model. Their findings indicate a high level of coupling (> 0.8) along the entire length of the MHT, which poses a significant seismic hazard to the region, with a strain energy equivalent to one Mw 8.7 megathrust earthquake every 100 years. Additionally, their results suggest that previously inferred low coupling zones may be artefacts of the block modelling process. This emphasizes the need for accurate modelling of overriding plate kinematics with a dense GNSS network to refine the seismic hazard.

Abhirami et al. (2025) combined GNSS velocity data and earthquake focal mechanisms to study stress and strain patterns across the Himalayan–Tibetan Collision Zone (HTCZ). The focal mechanism data reveal a dominant north–south compressional stress along the Himalayan arc, while the Tibetan Plateau exhibits a combination of strike-slip and normal faulting with a west-northwest to east-southeast stress orientation. They inverted the GNSS velocity data to estimate strain rates, and the resulting dilatation and second invariant strain rate maps highlight significant crustal shortening and the highest deformation concentrated along the Himalayan arc, while the Tibetan Plateau shows signs of extension. The maximum shear strain rates indicate strike-slip activity in the Tibetan Plateau. Their findings suggest that the maximum horizontal stress (SH_{max}) derived from focal mechanisms and the maximum strain rate (εH_{max}) obtained from GNSS velocities correlate strongly within the longitudes of 75°E–95°E. This correlation indicates the high elasticity of the central segment of the HTCZ. Additionally, the clockwise rotation observed in the eastern segment underscores the presence of a viscous lithosphere beneath the Tibetan Plateau.

Saji et al. (2025) focused on the central seismic gap, which includes the Himachal-Kumaun-Garhwal (HKG) regions. Their analysis incorporated new data from 40 GNSS sites along with previously published data. The 2D modelling indicates a convergence rate of ~ 18.5 mm/year, with the HKG region uniformly coupled up to 100 km from the Main Frontal Thrust (MFT). They estimated a geodetic moment rate of ~ 3.8×10^{19} Nm/year for the central seismic gap area. This indicates a significant deficit of seismic slip that has accumulated over the past 500–700 years, making the region capable of producing a magnitude $M_w > 8.7$ earthquake.

Castro-Perdomo et al. (2025) conducted a study of the Alpine-Himalayan belt using 42 GNSS datasets to estimate strain rates. Their results demonstrate a significant agreement between geodetic strain rates and seismic strain rates across the belt. Additionally, they found an azimuthal alignment between the principal axes of geodetic strain rates and the orientations of seismic anisotropy.

7.3 Data and Methods

We utilised over 900 GNSS including our own and published velocities (Ader et al., 2012; Banerjee et al., 2008; Blewitt et al., 2018; Devachandra et al., 2014; Dumka et al., 2018; Gahalaut et al., 2013; Jade et al., 2014, 2017, 2020; Kreemer et al., 2014; Kumar et al., 2023; Mallick et al., 2019; Marechal et al., 2016; Mohanty et al., 2024; Mukul et al., 2018; Steckler et al., 2016; Vigny et al., 2003; Wang & Shen, 2020) from and around the Himalayan arc presented in common reference frame, i.e., ITRF2014 (Altamimi et al., 2016), to assess deformation rate and corresponding strain rates due to India-Eurasia collision; we converted these velocity in a consistent Indian reference frame using the Euler pole estimated by Rajewar et al. (2026) (Fig. 7.1).

Using these velocities, we have calculated Two-dimensional strain rates following the Velocity Interpolation for Strain Rate (VISR) methodology of Shen et al. (2015). This is determined through weighted least square inversion on neighbouring velocities, and the weightage of velocities depends on the distance of the sites using a Gaussian function and their spatial coverage using an azimuthal weighting function. We used an open-source Python-based strain tool available at https://github.com/DSOlab/StrainTool (Anastasiou et al., 2019). We estimated translation, rotation and strain rate components on a $0.5° \times 0.5°$ grid cells. From strain rate tensors, we calculated Maximum shear strain, Dilatational strain and Second Invariant of strain rates,

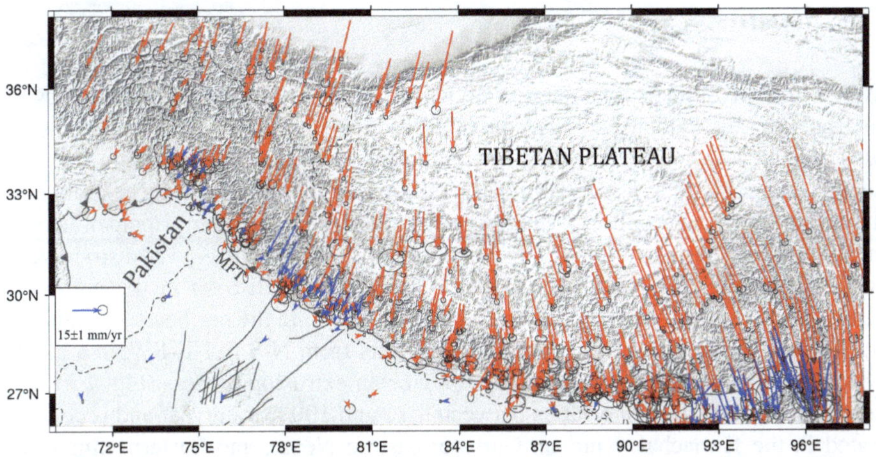

Fig. 7.1 GNSS site velocities (in mm/year) from our present study (blue vectors) and from published literature (red vectors) in the Indian reference frame

as per the following equations.

$$Maximum\ shear = \frac{1}{2}(e_{max} - e_{min}), \tag{7.1}$$

$$Dilatation = \tau_{xx} + \tau_{yy}, \tag{7.2}$$

$$Second\ Invariant = \sqrt{\tau_{xx}^2 + \tau_{yy}^2 + 2\left(\tau_{xy}^2\right)}. \tag{7.3}$$

Here, τ_{xx}, τ_{yy}, τ_{xy} are strain components, e_{max}, e_{min} are principal axes of strain rate.

Furthermore, we have examined the correlation between geodetic strain rate and seismicity in the Himalayan region, following the methodology of (Shen et al., 2007; Zeng et al., 2018). We used the Second Invariant component of strain rate for the purpose. For this study, we used the earthquake catalog of the Global Earthquake Model Global Historical Earthquake Catalog (1000–1903) and the International Seismological Centre-Global Earthquake Model Global Instrumental Earthquake Catalog (1904–2024). We included topography gradient using the Shuttle Radar Topography Mission (SRTM) Digital Elevation Model (DEM) data, using the *grdgradient* module of Generic Mapping Tools (GMT6) (Wessel et al., 2019). Population density data with per grid cell (People/km^2) at 30 arc seconds resolution is available at (*WorldPop:: Population Density*).

7.4 Results

We derived two-dimensional strain rate tensors from over 900 GNSS velocities by interpolating the velocity gradient on a $0.5° \times 0.5°$ grid. From these strain tensors, we calculated various parameters, including the principal axes of strain rate, the second invariant, dilatational (volumetric) strain rate and maximum shear strain rate. The results indicate high deformation rates across the arc. The principal axes of strain rate exhibit high compression ranging from ~ − 80 to − 120 ηstrain/year, with orientations changing from northeast-southwest (NE-SW) in the northwest Himalaya to north–south (N-S) in the Nepal Himalaya. In the northeastern region of the Himalaya, the orientation continuously shifts from N-S to NE-SW, in a clockwise direction, indicating the influence of Tibetan extrusion process (Fig. 7.2). The second invariant strain rate varies between ~ 70 and 100 ηstrain/year and is concentrated in the Himachal, Kumaun-Garhwal eastern Nepal, and Eastern Himalayan Syntaxis regions (Fig. 7.2). Furthermore, we examined dilatational strain rates, which provide insights into the impact of the collision between the Indian and Eurasian plates and the resulting volumetric changes. The negative values indicate predominantly contracting forces acting across the Himalaya, with dilatational strain rates ranging from ~ − 80 to − 120 ηstrain/year and reaching to a maximum of − 150 ηstrain/year, and to the north of it, the Tibetan Plateau shows extensional forces (Fig. 7.3). This strain is concentrated in the same regions as the second invariant strain rate, suggesting significant strain accumulation in these areas. To assess the impact of differential deformation across the Himalaya, we also calculated maximum shear strain rates, which separate volume changes and reflect only shear differential rates. These rates range from ~ 70 to 90 ηstrain/year (Fig. 7.4). Results indicate that compressional forces are dominant compared to shear forces, as evidenced by the higher magnitude of dilatational strain rates relative to maximum shear strain rates caused by the collision between the Indian-Eurasian plate.

7.5 Discussion

7.5.1 Correlations Between Strain and Seismicity, Topography and Population Density

The Himalayan arc is the most active plate boundary around the Indian plate. It extends for about 2500 km and has produced many large-magnitude earthquakes in the past. Several studies claim a good correlation between geodetic strain rate and seismicity rate (e.g., Abhirami et al., 2025; Bird et al., 2015; Carafa et al., 2024; Castro-Perdomo et al., 2025; Chen & Liu, 2024; Masson et al., 2004; Nucci et al., 2025; Shen et al., 2007; Stevens & Avouac, 2021; Zeng et al., 2018). In our analysis of the geodetic strain rate, we found that the Himalayan region exhibits a very high strain rate over a large area along the arc. The sorted strain rate curve analysis indicates

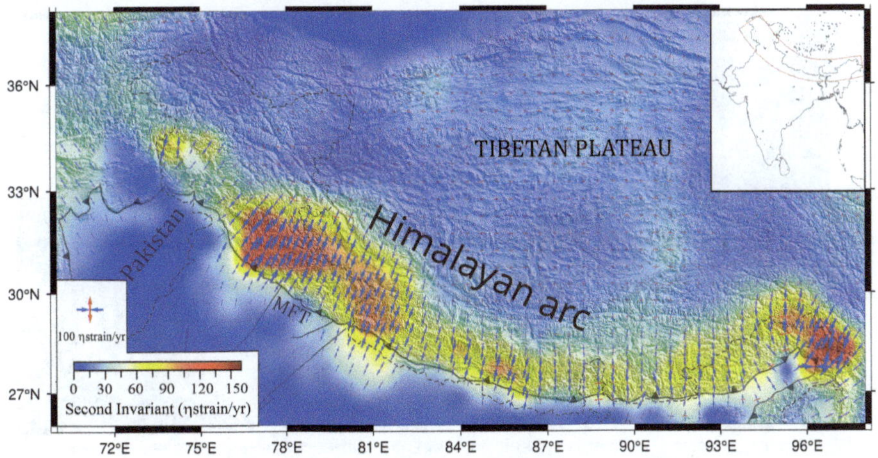

Fig. 7.2 Second invariant of strain rate on a 0.5° × 0.5° grid with principal strain rate axes in blue colour being contractional and red colour being extensional strain rate (ηstrain/year). Paired arrows show the orientation, sign and magnitude of principal strains of the horizontal strain-rate tensor, with red outward-pointing arrows indicating extensional and blue inward arrows indicating contractional strain (see legend for scale)

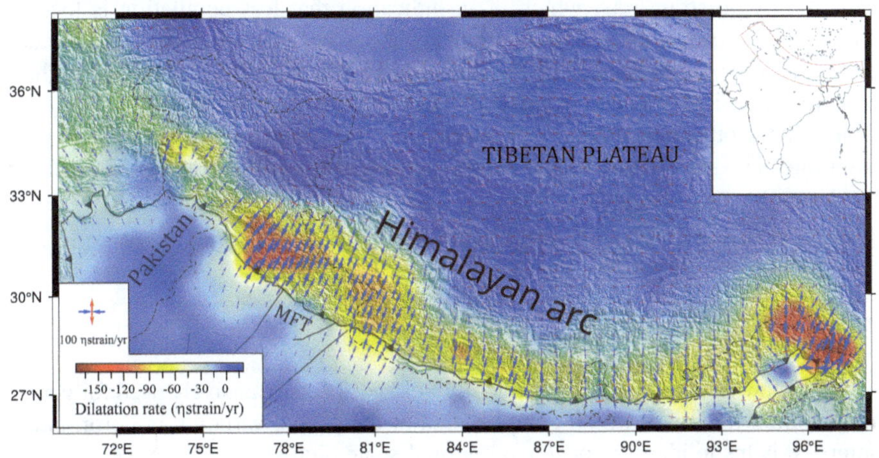

Fig. 7.3 Dilatational strain rate (ηstrain/year) with principal strain rate axes in blue and red for compressional and extensional strain rates, respectively

that 10% of the area shows 25% of the strain, and 50% of the area accounts for about 76% of the total strain. This suggests that a significant portion of the Himalaya is experiencing high strain accumulation. We then compared seismicity with strain rate and observed a strong correlation, indicating that regions with high strain rates also have high seismic activity. Notably, seismic moment release is concentrated in

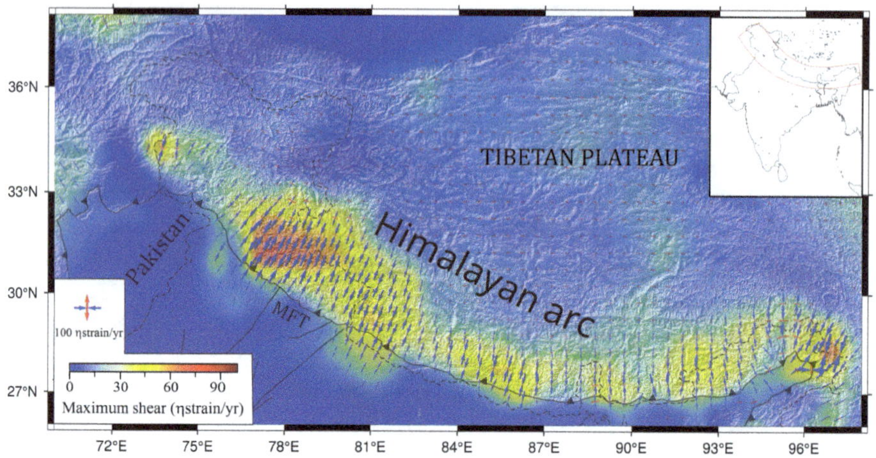

Fig. 7.4 Maximum shear strain rate on a 0.5° × 0.5° grid with principal strain rate axes in blue colour being contractional and red colour being extensional strain rate (ηstrain/year)

areas with high strain, as 50% of the area is responsible for nearly 98% of the total seismic moment release (see Fig. 7.5). However, population density shows a weak correlation with strain rate, which is an indication that less population is located in the high-strain-rate area. This discrepancy arises because the highest strain rates are found in the northern parts of the Lesser to Higher Himalaya, while the highest population density is situated in the Outer Himalaya and the adjoining Indo-Gangetic plains. On the other hand, the topographic gradient exhibits a strong correlation with strain rate. This correlation makes sense, as the highest physiographic transition in the Himalaya coincides with the areas of highest strain.

7.5.2 Strain Rate Versus Locking Maps

In the last few decades, the GNSS velocity field has been extensively used for generating a strain rate map, which gives an idea of the spatial variation in deformation pattern. It helps to identify local and regional-scale high-strain accumulation areas, and it can be used to forecast seismicity rates (e.g., Bird, 2009; Bird & Kreemer, 2015; Bird & Liu, 2007; Bird et al., 2010; Carafa et al., 2024; Rong et al., 2016; Shen et al., 2007; Ward, 2007; Zheng et al., 2018). But the strain rate map does not provide information about locked segments or faults that are critical for assessing seismic hazards. This is because the map is based on surface deformation; locked zones do not generate high strain because of the low velocity gradient. Similarly, creeping zones also lack a velocity gradient. As a result, the strain rate map is unable to differentiate between the highly locked zone and creeping zones (see Fig. 7.7). Relying solely on surface strain may lead to misconceptions about potential future

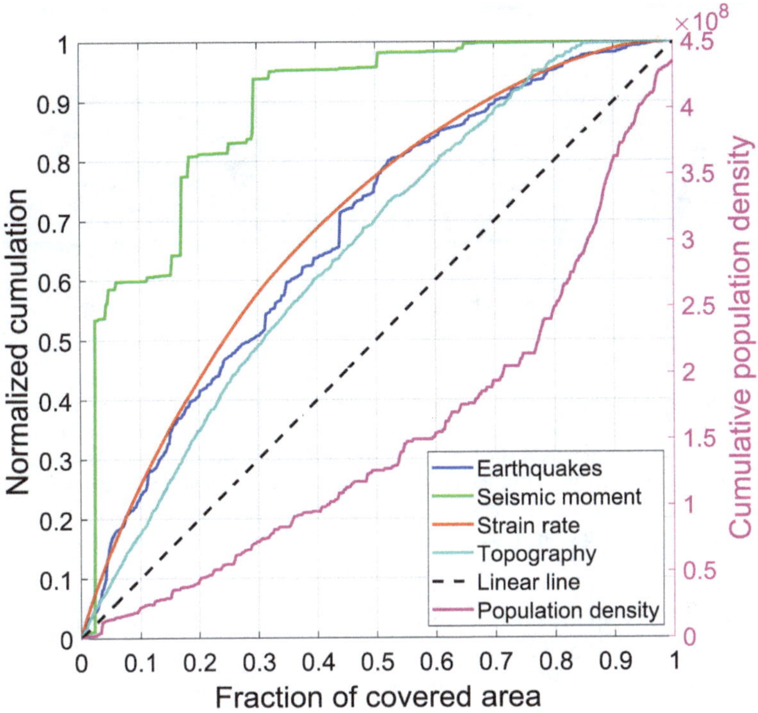

Fig. 7.5 Comparison of correlations between strain rate, seismicity, seismic moment, topography and population density of the Himalaya. The cumulative second invariant strain rate, earthquake count, seismic moment, topography gradient and population density are plotted against the fraction of covered area, sorted by descending strain rates, with the highest strain-rate regions located to the left of the horizontal axis. The diagonal line indicates a random distribution of a given measure in space

earthquake zones. In contrast, a coupling or locking map offers valuable insights into locked fault parameters, including estimations of the convergence rate, slip deficit rate and the width of the locked zone (Figs. 7.6 and 7.7). This can be further elaborated using the elastic dislocation model (Okada, 1992) and the deep/back slip model (Savage, 1983). Additionally, inverting GNSS velocity data allows for the determination of slip distribution along a given subsurface fault geometry by employing weighted least squares inversion with Laplacian regularization (Kositsky & Avouac, 2010). This fault locking is referred to as coupling, and it is defined as the ratio of slip deficit rate during the interseismic period divided by the long-term convergence rate imposed by tectonic shortening. The locking ratio can range from 0 to 1, with 0 indicating no coupling and 1 indicating a highly coupled or locked zone where the slip deficit rate is equivalent to the long-term convergence rate (Fig. 7.6b). In Fig. 7.6, we observe that at the Physiographic Transition zone (PT2), there is a noticeable change in the slip deficit rate. The frontal portion shows a very low slip rate and a corresponding high coupling, suggesting that there is low deformation in the frontal

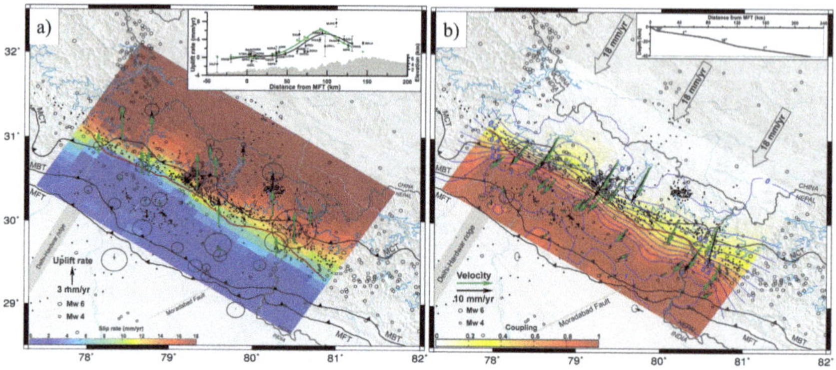

Fig. 7.6 **a** Background colour shows estimated slip rate on the MHT using fault geometry shown in the inset of panel (b), **b** the superimposed colour image and contours represent the coupling derived from the slip deficit rate obtained after inverting the GPS displacement rates. The convergence of 18 mm/year across the Garhwal-Kumaun Himalaya. Black and green vectors are the observed and simulated horizontal velocity estimates. The top right inset shows the configuration of the MHT used in the analysis, with the dip of each segment indicated (Figure source: Yadav et al., 2019)

section of the MHT up to 100 km. However, at PT2, we also note significant seismic activity (as shown in Fig. 7.6) and a high velocity gradient (see Fig. 7.7). In a recent study, Sreejith et al. (2025) utilize InSAR and GNSS data to infer a convergence rate of 20–22 mm/year with 115 km locking width and 5–7 mm/year of uplift in the higher Himalaya, suggesting sufficient strain has accumulated that could lead to hosting two megathrust earthquakes of Mw ~ 8.8. These observations highlight the importance of identifying regions that are fully locked, which may not be evident from seismicity and strain rate observations alone. It is known that most megathrust earthquakes occur in these interseismically locked areas (Loveless & Meade, 2011). So, the locking map can be utilized to estimate the slip deficit based on the convergence rate, which helps in calculating the total seismic moment deficit rate (Ader et al., 2012; Ambraseys & Douglas, 2004; Bilham, 2019; Li et al., 2018; Panda & Lindsey, 2024; Sreejith et al., 2018, 2025; Stevens & Avouac, 2015). This estimation is crucial for predicting both the magnitude and recurrence interval of future earthquakes.

7.5.3 *Strain Rate Magnitude*

Geodetic strain rate is a gradient of the velocity field; when there is a high gradient of the velocity field, it results in a high strain rate, as observed at plate boundary regions, like the Himalaya, where strain rate is typically > 100 ηstrain/year (Abhirami et al., 2025; Kreemer et al., 2014; Mohanty et al., 2025; Panda & Lindsey, 2024; Saji et al., 2025; Sreejith et al., 2025; Stevens & Avouac, 2021). Low gradient in velocity field

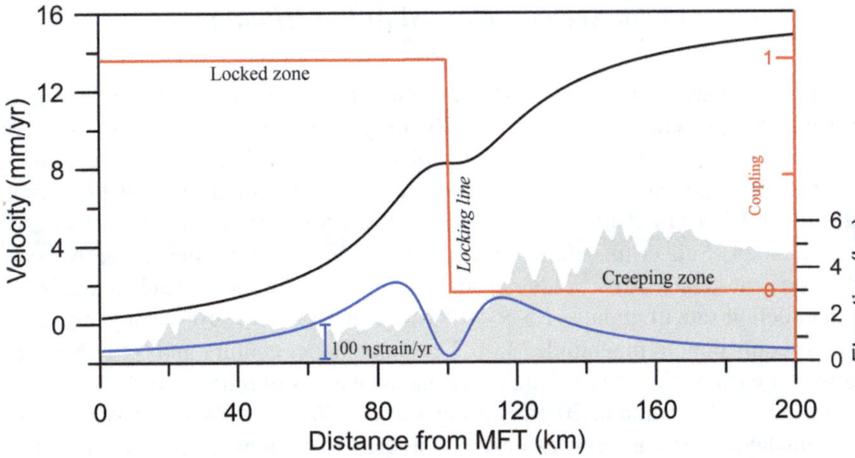

Fig. 7.7 Arc-normal site velocity along a composite profile is presented in the arc-normal direction. The solid black curve represents the impact of full locking of a 100-km-wide MHT beneath the Outer and Lesser Himalaya, with a slip deficit rate of 18 mm/year (Yadav et al., 2019). The solid blue curve, representing strain rate, is a derivative of the velocity (black curve). The red line represents the locking curve, showing 1 (full coupling) and 0 (no coupling)

results in low strain rates as observed in plate interior regions, which is typically < 1–2 ηstrain/year (Calais et al., 2006, 2016; Jade et al., 2017; Kreemer et al., 2014, 2018; Paul et al., 2001). High strain rate is expected at the tectonic plate boundary because of the high rate of deformation as the motion between the two plates is accommodated here, but in the plate interior region, it is not. In plate interior regions, the deformation is extremely slow, transient and may be localized or diffused.

7.5.4 Geodetic Strain Rate Versus Earthquake Strain Rate

There is a good correlation between the geodetic strain derived from short-term GNSS data and earthquake frequency and moment release of the past few decades. The main reason for this good correspondence is that both are focussed close to the base of the seismogenic MHT. Earlier, investigators have compared the seismic moment release during Himalayan earthquakes of the past few centuries and have compared it with the convergence rate accommodated across the Himalayan arc, and they have found a good correspondence in the two quantities (Chen & Molnar, 1977). The consistency in the magnitudes between the two quantities implies that most of the deformation in the Himalayan arc is elastic and recoverable and the possibility of inelastic deformation is about 10% (Avouac, 2003; Stevens & Avouac, 2015).

7.5.5 Non-uniform Seismic Hazard in the Himalaya

The Himalayan arc stretches ~ 2500 km and has a locked width of ~ 100 ± 20 km on MHT with a representative slip rate of ~ 20 mm/year. (Ader et al., 2012; Banerjee & Bürgmann, 2002; Bettinelli et al., 2006; Bilham et al., 1997; Dal Zilio et al., 2020; Jade et al., 2014; Jouanne et al., 2004; Kumar et al., 2023; Kundu et al., 2014; Larson et al., 1999; Li et al., 2018; Marechal et al., 2016; Mohanty et al., 2025; Panda & Lindsey, 2024; Saji et al., 2025; Sreejith et al., 2018, 2025; Stevens & Avouac, 2015; Vernant et al., 2014; Yadav et al., 2019). These numbers lead to a seismic moment deficit rate of about ~ 1.5×10^{20} Nm/year or ~ 1.5×10^{22} Nm per century which is equivalent to magnitude M_w 8.7 earthquake per century and the recurrence interval for earthquakes > 8.0 along the Himalayan arc is of order of 100–1000 years (Ader et al., 2012; Bilham, 2019; Bollinger et al., 2014; Panda & Lindsey, 2024). The Himalayan arc has been extensively studied in regions such as Uttarakhand and the Nepal Himalaya, which have a high density of GNSS data. However, some areas, including Kashmir, Himachal Pradesh, Bhutan and the Northeast region, have limited data available. Research conducted in the Himalayan region indicates varying convergence rates across different areas. In the Kashmir region, the convergence rate is ~ 17 mm/year, with a locking width of about 175 km (Kundu et al., 2014). In the Himachal-Uttarakhand region, the convergence rate is around ~ 18 mm/year, with a locking width of ~ 100 km (Kumar et al., 2023; Mohanty et al., 2025; Yadav et al., 2019). The Nepal Himalaya exhibits a convergence rate of about ~ 17 mm/year in eastern and central Nepal, increasing to around ~ 20 mm/year in western Nepal, with a locking width of 100 km for the MHT. In Bhutan, the convergence rate and locking width vary between the western and eastern regions, with locking widths ranging from 155 to 100 km and a convergence rate between 16–17 mm/year (Marechal et al., 2016; Stevens & Avouac, 2015). At the eastern end of the Himalaya, geomorphic studies indicate convergence rates of ~ 20–23 mm/year (Berthet et al., 2014; Burgess et al., 2012). These estimations indicate a complex deformation process in the Himalaya region that varies from west to east. While a substantial amount of GNSS data has provided valuable insights into this complex process, some areas still lack sufficient GNSS data, which needs to be addressed. In particular, the eastern Himalaya is less studied in terms of GNSS observations. Long-term convergence rates derived from geomorphic studies reveal a higher convergence rate, suggesting that this region requires more detailed investigation with high precision to assess the potential for seismic hazards.

7.6 Conclusions

In this study, we focused on geodetic strain rates and seismicity using the latest GNSS dataset. We have updated the strain map of the Himalayan arc. From these results, we concluded the following points:

- There is a strong correlation between geodetic strain rate, seismicity and topography. High strain rate areas exhibit increased seismicity, and the effects of significant deformation can be seen in the topographic gradient; these factors complement one another.
- We find that geodetic strain rate cannot differentiate the locked and creeping zones of the fault; this limitation arises because it is based on the effect of surface deformation due to the earthquake process.
- A locking map can help to identify regions with fully locked faults that have the potential to accommodate megathrust, because it is based on earthquake process modelling. It can be more useful for a probabilistic seismic hazard model.

Acknowledgements Most of the GNSS sites were installed with the financial support of the Ministry of Science and Technology, the Ministry of Earth Sciences and the CSIR, Government of India. NRSC provided financial support for this work. Thanks to Dr VM Tiwari for inviting us to contribute this paper to the volume and to an anonymous reviewer for very constructive comments. The CSIR-NGRI reference number of the article is NGRI/Lib/2025/Pub-116.

Declarations The authors declare that they have no known competing financial interests or personal relationships that could have appeared to influence the work reported in this article.

References

Abhirami, R., Sunil, P. S., Sooraj, A., Sreerag, S. S., & Sunil, A. S. (2025). Integrated stress and strain characterisation of the Himalayan-Tibetan collision zone using earthquake and geodetic data. *Tectonophysics, 902*, 230686. https://doi.org/10.1016/j.tecto.2025.230686

Ader, T., Avouac, J., Liu-Zeng, J., Lyon-Caen, H., Bollinger, L., Galetzka, J., Genrich, J., Thomas, M., Chanard, K., Sapkota, S. N., Rajaure, S., Shrestha, P., Ding, L., & Flouzat, M. (2012). Convergence rate across the Nepal Himalaya and interseismic coupling on the Main Himalayan Thrust: Implications for seismic hazard. *Journal of Geophysical Research: Solid Earth, 117*(B4), 2011JB009071. https://doi.org/10.1029/2011JB009071

Altamimi, Z., Rebischung, P., Métivier, L., & Collilieux, X. (2016). ITRF2014: A new release of the international terrestrial reference frame modeling nonlinear station motions. *Journal of Geophysical Research: Solid Earth, 121*(8), 6109–6131. https://doi.org/10.1002/2016JB013098

Ambraseys, N. N., & Douglas, J. (2004). Magnitude calibration of north Indian earthquakes. *Geophysical Journal International, 159*(1), 165–206. https://doi.org/10.1111/j.1365-246X.2004.02323.x

Anastasiou, D., Ganas, A., Legrand, J., Bruynix, C., Papanikolao, X., Tsironi, V., & Kapetanidis, V. (2019). Tectonic strain distribution over Europe from EPN Data. *EGU General Assembly 2019, Geophysical Research Abstracts, 21*(Mis 5002697), EGU2019-17744-1.

Avouac, J.-P. (2003). Mountain building, erosion, and the seismic cycle in the Nepal Himalaya. In *Advances in geophysics* (Vol. 46, pp. 1–80). Elsevier. https://doi.org/10.1016/S0065-2687(03)46001-9

Avouac, J.-P., Ayoub, F., Leprince, S., Konca, O., & Helmberger, D. V. (2006). The 2005, *M*w 7.6 Kashmir earthquake: Sub-pixel correlation of ASTER images and seismic waveforms analysis. *Earth and Planetary Science Letters, 249*(3), 514–528. https://doi.org/10.1016/j.epsl.2006.06.025

Banerjee, P., & Bürgmann, R. (2002). Convergence across the northwest Himalaya from GPS measurements. *Geophysical Research Letters, 29*(13), 30-1–30-34. https://doi.org/10.1029/2002GL015184
Banerjee, P., Bürgmann, R., Nagarajan, B., & Apel, E. (2008). Intraplate deformation of the Indian subcontinent. *Geophysical Research Letters, 35*(18), 2008GL035468. https://doi.org/10.1029/2008GL035468
Berthet, T., Ritz, J.-F., Ferry, M., Pelgay, P., Cattin, R., Drukpa, D., Braucher, R., & Hetényi, G. (2014). Active tectonics of the eastern Himalaya: New constraints from the first tectonic geomorphology study in southern Bhutan. *Geology, 42*(5), 427–430. https://doi.org/10.1130/G35162.1
Bettinelli, P., Avouac, J.-P., Flouzat, M., Jouanne, F., Bollinger, L., Willis, P., & Chitrakar, G. R. (2006). Plate motion of India and interseismic strain in the Nepal Himalaya from GPS and DORIS measurements. *Journal of Geodesy, 80*(8–11), 567–589. https://doi.org/10.1007/s00190-006-0030-3
Bilham, R. (1995). Location and magnitude of the 1833 Nepal earthquake and its relation to the rupture zones of contiguous great Himalayan earthquakes. *Current Science, 69*(2), 101–128.
Bilham, R. (2004). Earthquakes in India and the Himalaya: Tectonics, geodesy and history. *Annals of Geophysics*. https://www.earth-prints.org/handle/2122/798
Bilham, R. (2019). Himalayan earthquakes: A review of historical seismicity and early 21st century slip potential. In P. J. Treloar & M. P. Searle (Eds.), *Himalayan tectonics: A modern synthesis* (Vol. 483, p. 0). The Geological Society of London. https://doi.org/10.1144/SP483.16
Bilham, R., Larson, K., & Freymueller, J. (1997). GPS measurements of present-day convergence across the Nepal Himalaya. *Nature, 386*(6620), 61–64. https://doi.org/10.1038/386061a0
Bird, P. (2009). Long-term fault slip rates, distributed deformation rates, and forecast of seismicity in the western United States from joint fitting of community geologic, geodetic, and stress direction data sets. *Journal of Geophysical Research: Solid Earth, 114*(B11). https://doi.org/10.1029/2009JB006317
Bird, P., & Kreemer, C. (2015). Revised tectonic forecast of global shallow seismicity based on version 2.1 of the global strain rate map. *Bulletin of the Seismological Society of America, 105*(1), 152–166. https://doi.org/10.1785/0120140129
Bird, P., Jackson, D. D., Kagan, Y. Y., Kreemer, C., & Stein, R. S. (2015). GEAR1: A global earthquake activity rate model constructed from geodetic strain rates and smoothed seismicity. *Bulletin of the Seismological Society of America, 105*(5), 2538–2554. https://doi.org/10.1785/0120150058
Bird, P., Kreemer, C., & Holt, W. E. (2010). A long-term forecast of shallow seismicity based on the global strain rate map. *Seismological Research Letters, 81*(2), 184–194. https://doi.org/10.1785/gssrl.81.2.184
Bird, P., & Liu, Z. (2007). Seismic hazard inferred from tectonics: California. *Seismological Research Letters, 78*(1), 37–48. https://doi.org/10.1785/gssrl.78.1.37
Blewitt, G., Hammond, W. C., & C. Kreemer. (2018). Harnessing the GPS data explosion for interdisciplinary science. *EOS, 99*. https://doi.org/10.1029/2018EO104623
Bollinger, L., Sapkota, S. N., Tapponnier, P., Klinger, Y., Rizza, M., Van der Woerd, J., Tiwari, D. R., Pandey, R., Bitri, A., & Bes de Berc, S. (2014). Estimating the return times of great Himalayan earthquakes in eastern Nepal: Evidence from the Patu and Bardibas strands of the Main Frontal Thrust. *Journal of Geophysical Research: Solid Earth, 119*(9), 7123–7163. https://doi.org/10.1002/2014JB010970
Calais, E., Han, J. Y., DeMets, C., & Nocquet, J. M. (2006). Deformation of the North American plate interior from a decade of continuous GPS measurements. *Journal of Geophysical Research: Solid Earth, 111*(B6). https://doi.org/10.1029/2005JB004253
Calais, E., Camelbeeck, T., Stein, S., Liu, M., & Craig, T. J. (2016). A new paradigm for large earthquakes in stable continental plate interiors. *Geophysical Research Letters, 43*(20), 10621–10637. https://doi.org/10.1002/2016GL070815

Carafa, M. M. C., Bird, P., Verdecchia, A., Taroni, M., & Doglioni, C. (2024). Empirical evidence for multi-decadal transients affecting geodetic velocity fields and derived seismicity forecasts in Italy. *Scientific Reports, 14*(1), 19941. https://doi.org/10.1038/s41598-024-70816-6

Castro-Perdomo, N., Jónsson, S., Klinger, Y., Masson, F., Becker, T. W., & Johnson, K. (2025). Strain rates along the Alpine-Himalayan belt from a comprehensive GNSS velocity field. *Journal of Geophysical Research: Solid Earth, 130*(12), e2025JB031738. https://doi.org/10.1029/2025JB031738

Chen, W.-P., & Molnar, P. (1977). Seismic moments of major earthquakes and the average rate of slip in central Asia. *Journal of Geophysical Research (1896–1977), 82*(20), 2945–2969. https://doi.org/10.1029/JB082i020p02945

Chen, Y., & Liu, M. (2024). Correlation between strain rate and seismicity in different tectonic settings. *Seismological Research Letters, 95*(4), 2375–2385. https://doi.org/10.1785/0220230306

Coudurier-Curveur, A., Tapponnier, P., Okal, E., Van der Woerd, J., Kali, E., Choudhury, S., Baruah, S., Etchebes, M., & Karakaş, Ç. (2020). A composite rupture model for the great 1950 Assam earthquake across the cusp of the East Himalayan Syntaxis. *Earth and Planetary Science Letters, 531*, 115928. https://doi.org/10.1016/j.epsl.2019.115928

Dal Zilio, L., Hetényi, G., Hubbard, J., & Bollinger, L. (2021). Building the Himalaya from tectonic to earthquake scales. *Nature Reviews Earth & Environment, 2*(4), 251–268. https://doi.org/10.1038/s43017-021-00143-1

Dal Zilio, L., Jolivet, R., & van Dinther, Y. (2020). Segmentation of the Main Himalayan thrust illuminated by Bayesian inference of interseismic coupling. *Geophysical Research Letters, 47*(4), e2019GL086424. https://doi.org/10.1029/2019GL086424

Devachandra, M., Kundu, B., Catherine, J., Kumar, A., & Gahalaut, V. K. (2014). Global positioning system (GPS) measurements of crustal deformation across the frontal Eastern Himalayan syntaxis and seismic-hazard assessment. *Bulletin of the Seismological Society of America, 104*(3), 1518–1524. https://doi.org/10.1785/0120130290

Dumka, R., Choudhury, P., Gahalaut, V. K., Gahalaut, K., & Yadav, R. K. (2018). GPS measurements of deformation caused by seasonal filling and emptying cycles of four hydroelectric reservoirs in India. *Bulletin of the Seismological Society of America, 108*(5B), 2955–2966. https://doi.org/10.1785/0120170355

Gahalaut, V. K., & Chander, R. (1997). On interseismic elevation changes and strain accumulation for great thrust earthquakes in the Nepal Himalaya. *Geophysical Research Letters, 24*(9), 1011–1014. https://doi.org/10.1029/97GL00810

Gahalaut, V. K., Kundu, B., Laishram, S. S., Catherine, J., Kumar, A., Singh, M. D., Tiwari, R. P., Chadha, R. K., Samanta, S. K., Ambikapathy, A., Mahesh, P., Bansal, A., & Narsaiah, M. (2013). Aseismic plate boundary in the Indo-Burmese wedge, northwest Sunda Arc. *Geology, 41*(2), 235–238. https://doi.org/10.1130/G33771.1

Galetzka, J., Melgar, D., Genrich, J. F., Geng, J., Owen, S., Lindsey, E. O., Xu, X., Bock, Y., Avouac, J.-P., Adhikari, L. B., Upreti, B. N., Pratt-Sitaula, B., Bhattarai, T. N., Sitaula, B. P., Moore, A., Hudnut, K. W., Szeliga, W., Normandeau, J., Fend, M., et al. (2015). Slip pulse and resonance of the Kathmandu basin during the 2015 Gorkha earthquake, Nepal. *Science, 349*(6252), 1091–1095. https://doi.org/10.1126/science.aac6383

Hough, S. E., & Roger, B. (2008). Site response of the Ganges basin inferred from re-evaluated macroseismic observations from the 1897 Shillong, 1905 Kangra, and 1934 Nepal earthquakes. *Journal of Earth System Science, 117*(2), 773–782. https://doi.org/10.1007/s12040-008-0068-0

Jade, S., Mir, R. R., Vivek, C. G., Shrungeshwara, T. S., Parvez, I. A., Chandra, R., Babu, D. S., Gupta, S. V., Ankit, Rajana, S. S. K., & Gaur, V. K. (2020). Crustal deformation rates in Kashmir valley and adjoining regions from continuous GPS measurements from 2008 to 2019. *Scientific Reports, 10*(1), 17927. https://doi.org/10.1038/s41598-020-74776-5

Jade, S., Mukul, M., Gaur, V. K., Kumar, K., Shrungeshwar, T. S., Satyal, G. S., Dumka, R. K., Jagannathan, S., Ananda, M. B., Kumar, P. D., & Banerjee, S. (2014). Contemporary deformation in the Kashmir-Himachal, Garhwal and Kumaon Himalaya: Significant insights from 1995–2008 GPS time series. *Journal of Geodesy, 88*(6), 539–557. https://doi.org/10.1007/s00190-014-0702-3

Jade, S., Shrungeshwara, T. S., Kumar, K., Choudhury, P., Dumka, R. K., & Bhu, H. (2017). India plate angular velocity and contemporary deformation rates from continuous GPS measurements from 1996 to 2015. *Scientific Reports, 7*(1), 11439. https://doi.org/10.1038/s41598-017-11697-w

Jouanne, F., Mugnier, J. L., Gamond, J. F., Le Fort, P., Pandey, M. R., Bollinger, L., Flouzat, M., & Avouac, J. P. (2004). Current shortening across the Himalayas of Nepal. *Geophysical Journal International, 157*(1), 1–14. https://doi.org/10.1111/j.1365-246X.2004.02180.x

Khattri, K. N. (1987). Great earthquakes, seismicity gaps and potential for earthquake disaster along the Himalaya plate boundary. *Tectonophysics, 138*(1), 79–92. https://doi.org/10.1016/0040-1951(87)90067-9

Kositsky, A. P., & Avouac, J.-P. (2010). Inverting geodetic time series with a principal component analysis-based inversion method. *Journal of Geophysical Research: Solid Earth, 115*(B3). https://doi.org/10.1029/2009JB006535

Kreemer, C., Blewitt, G., & Klein, E. C. (2014). A geodetic plate motion and global strain rate model. *Geochemistry, Geophysics, Geosystems, 15*(10), 3849–3889. https://doi.org/10.1002/2014GC005407

Kreemer, C., Hammond, W. C., & Blewitt, G. (2018). A robust estimation of the 3-D intraplate deformation of the North American plate from GPS. *Journal of Geophysical Research: Solid Earth, 123*(5), 4388–4412. https://doi.org/10.1029/2017JB015257

Kumar, P., Malik, J. N., Gahalaut, V. K., Yadav, R. K., & Singh, G. (2023). Evidence of strain accumulation and coupling variation in the Himachal Region of NW Himalaya from short term geodetic measurements. *Tectonics, 42*(8), e2022TC007690. https://doi.org/10.1029/2022TC007690

Kundu, B., Yadav, R. K., Bali, B. S., Chowdhury, S., & Gahalaut, V. K. (2014). Oblique convergence and slip partitioning in the NW Himalaya: Implications from GPS measurements. *Tectonics, 33*(10), 2013–2024. https://doi.org/10.1002/2014TC003633

Larson, K. M., Bürgmann, R., Bilham, R., & Freymueller, J. T. (1999). Kinematics of the India-Eurasia collision zone from GPS measurements. *Journal of Geophysical Research: Solid Earth, 104*(B1), 1077–1093. https://doi.org/10.1029/1998JB900043

Le Roux-Mallouf, R., Ferry, M., Ritz, J.-F., Berthet, T., Cattin, R., & Drukpa, D. (2016). First paleoseismic evidence for great surface-rupturing earthquakes in the Bhutan Himalayas. *Journal of Geophysical Research: Solid Earth, 121*(10), 7271–7283. https://doi.org/10.1002/2015JB012733

Li, S., Wang, Q., Yang, S., Qiao, X., Nie, Z., Zou, R., Ding, K., He, P., & Chen, G. (2018). Geodetic imaging mega-thrust coupling beneath the Himalaya. *Tectonophysics, 747–748*, 225–238. https://doi.org/10.1016/j.tecto.2018.08.014

Lindsey, E. O., Natsuaki, R., Xu, X., Shimada, M., Hashimoto, M., Melgar, D., & Sandwell, D. T. (2015). Line-of-sight displacement from ALOS-2 interferometry: Mw 7.8 Gorkha Earthquake and Mw 7.3 aftershock. *Geophysical Research Letters, 42*(16), 6655–6661. https://doi.org/10.1002/2015GL065385

Loveless, J. P., & Meade, B. J. (2011). Spatial correlation of interseismic coupling and coseismic rupture extent of the 2011 MW = 9.0 Tohoku-oki earthquake. *Geophysical Research Letters, 38*(17). https://doi.org/10.1029/2011GL048561

Mallick, R., Lindsey, E. O., Feng, L., Hubbard, J., Banerjee, P., & Hill, E. M. (2019). Active convergence of the India-Burma-Sunda plates revealed by a new continuous GPS network. *Journal of Geophysical Research: Solid Earth, 124*(3), 3155–3171. https://doi.org/10.1029/2018JB016480

Marechal, A., Mazzotti, S., Cattin, R., Cazes, G., Vernant, P., Drukpa, D., Thinley, K., Tarayoun, A., Le Roux-Mallouf, R., Thapa, B. B., Pelgay, P., Gyeltshen, J., Doerflinger, E., & Gautier, S.

(2016). Evidence of interseismic coupling variations along the Bhutan Himalayan arc from new GPS data. *Geophysical Research Letters, 43*(24). https://doi.org/10.1002/2016GL071163

Masson, F., Chéry, J., Hatzfeld, D., Martinod, J., Vernant, P., Tavakoli, F., & Ghafory-Ashtiani, M. (2004). Seismic versus aseismic deformation in Iran inferred from earthquakes and geodetic data: Seismic versus aseismic deformation in Iran. *Geophysical Journal International, 160*(1), 217–226. https://doi.org/10.1111/j.1365-246X.2004.02465.x

Mohanty, A., Yadav, R. K., Rajewar, S. K., Catherine, J. K., Bali, B. S., Mondal, S. K., Singh, Y., Pandey, D. N., & Gahalaut, V. K. (2025). Tectonic deformation in the NW Himalaya from continuous GPS measurements. *Journal of the Geological Society of India, 101*(6), 832–837. https://doi.org/10.17491/jgsi/2025/174169

Mohanty, A., Gahalaut, V. K., Chowdhury, S., Bansal, A. K., Gautam, P., & Catherine, J. (2024). Geodetic constraints on slip rate on the Karakoram fault and its role in the Himalayan arc deformation. *Earth and Planetary Science Letters, 626*, Article 118512. https://doi.org/10.1016/j.epsl.2023.118512

Molnar, P. (1990). A review of seismicity and the rates of active underthrusting and deformation at the Himalaya. *Journal of Himalayan Geology, 1*, 131–154.

Molnar, P., & Tapponnier, P. (1975). Cenozoic tectonics of Asia: Effects of a continental collision. *Science, 189*(4201), 419–426. https://doi.org/10.1126/science.189.4201.419

Mukul, M., Jade, S., Ansari, K., Matin, A., & Joshi, V. (2018). Structural insights from geodetic global positioning system measurements in the Darjiling-Sikkim Himalaya. *Journal of Structural Geology, 114*, 346–356. https://doi.org/10.1016/j.jsg.2018.03.007

Nucci, R., Serpelloni, E., Faenza, L., Garcia, A., & Belardinelli, M. E. (2025). Geodetic strain rates and seismicity rates along the Apennines (Italy). *Journal of Geophysical Research: Solid Earth, 130*(2), e2024JB029848. https://doi.org/10.1029/2024JB029848

Okada, Y. (1992). Internal deformation due to shear and tensile faults in a half-space. *Bulletin of the Seismological Society of America, 82*(2), 1018–1040. https://doi.org/10.1785/BSSA0820021018

Panda, D., & Lindsey, E. O. (2024). Overriding plate deformation controls inferences of interseismic coupling along the Himalayan Megathrust. *Journal of Geophysical Research: Solid Earth, 129*(9), e2024JB029819. https://doi.org/10.1029/2024JB029819

Paul Burgess, W., Yin, A., Dubey, C. S., Shen, Z.-K., & Kelty, T. K. (2012). Holocene shortening across the Main Frontal Thrust zone in the eastern Himalaya. *Earth and Planetary Science Letters, 357–358*, 152–167. https://doi.org/10.1016/j.epsl.2012.09.040

Paul, J., Bürgmann, R., Gaur, V. K., Bilham, R., Larson, K. M., Ananda, M. B., Jade, S., Mukal, M., Anupama, T. S., Satyal, G., & Kumar, D. (2001). The motion and active deformation of India. *Geophysical Research Letters, 28*(4), 647–650. https://doi.org/10.1029/2000GL011832

Rajewar, S. K., Gahalaut, V. K., Bürgmann, R., Puviarasan, N., Sai Krishnan, K. C., Bhu, H., et al. (2026). Present-day tectonic and non-tectonic crustal deformation of and around the Indian plate from GNSS and GRACE measurements. *Journal of Geophysical Research: Solid Earth, 131*, e2025JB032730. https://doi.org/10.1029/2025JB032730

Rong, Y., Bird, P., & Jackson, D. D. (2016). Earthquake potential and magnitude limits inferred from a geodetic strain-rate model for southern Europe. *Geophysical Journal International, 205*(1), 509–522. https://doi.org/10.1093/gji/ggw018

Saji, A. P., Sunil, P. S., Sreejith, K. M., Kumar, K. V., Ponraj, M., & Amirtharaj, S. (2025). Interseismic strain rate and fault coupling along the central Himalayan seismic gap from new GPS data. *Geophysical Journal International*, *242*(2), ggaf186. https://doi.org/10.1093/gji/ggaf186

Savage, J. C. (1983). A dislocation model of strain accumulation and release at a subduction zone. *Journal of Geophysical Research: Solid Earth, 88*(B6), 4984–4996. https://doi.org/10.1029/JB088iB06p04984

Shen, Z.-K., Jackson, D. D., & Kagan, Y. Y. (2007). Implications of geodetic strain rate for future earthquakes, with a five-year forecast of M5 earthquakes in Southern California. *Seismological Research Letters, 78*(1), 116–120. https://doi.org/10.1785/gssrl.78.1.116

Shen, Z., Wang, M., Zeng, Y., & Wang, F. (2015). Optimal Interpolation of Spatially Discretized Geodetic Data. *Bulletin of the Seismological Society of America, 105*(4), 2117–2127. https://doi.org/10.1785/0120140247

Sreejith, K. M., Mallick, R., Fielding, E. J., M. Jasir, M. C., Simons, M., & Agrawal, R. (2025). Geodetic insights to the Himalayan Megathrust kinematics unravel increased earthquake hazard. *Geophysical Research Letters, 52*(20), e2025GL118555. https://doi.org/10.1029/2025GL118555

Sreejith, K. M., Sunil, P. S., Agrawal, R., Saji, A. P., Rajawat, A. S., & Ramesh, D. S. (2018). Audit of stored strain energy and extent of future earthquake rupture in central Himalaya. *Scientific Reports, 8*(1), 16697. https://doi.org/10.1038/s41598-018-35025-y

Steckler, M. S., Mondal, D. R., Akhter, S. H., Seeber, L., Feng, L., Gale, J., Hill, E. M., & Howe, M. (2016). Locked and loading megathrust linked to active subduction beneath the Indo-Burman ranges. *Nature Geoscience, 9*(8), 615–618. https://doi.org/10.1038/ngeo2760

Stevens, V. L., & Avouac, J. P. (2015). Interseismic coupling on the main Himalayan thrust. *Geophysical Research Letters, 42*(14), 5828–5837. https://doi.org/10.1002/2015GL064845

Stevens, V. L., & Avouac, J.-P. (2021). On the relationship between strain rate and seismicity in the India-Asia collision zone: Implications for probabilistic seismic hazard. *Geophysical Journal International, 226*(1), 220–245. https://doi.org/10.1093/gji/ggab098

Szeliga, W., & Bilham, R. (2017). New constraints on the mechanism and rupture area for the 1905 Mw 7.8 Kangra Earthquake, Northwest Himalaya. *Bulletin of the Seismological Society of America, 107*(5), 2467–2479. https://doi.org/10.1785/0120160267

Vernant, P., Bilham, R., Szeliga, W., Drupka, D., Kalita, S., Bhattacharyya, A. K., Gaur, V. K., Pelgay, P., Cattin, R., & Berthet, T. (2014). Clockwise rotation of the Brahmaputra Valley relative to India: Tectonic convergence in the eastern Himalaya, Naga Hills, and Shillong Plateau. *Journal of Geophysical Research: Solid Earth, 119*(8), 6558–6571. https://doi.org/10.1002/2014JB011196

Vigny, C., Socquet, A., Rangin, C., Chamot-Rooke, N., Pubellier, M., Bouin, M.-N., Bertrand, G., & Becker, M. (2003). Present-day crustal deformation around Sagaing fault, Myanmar. *Journal of Geophysical Research: Solid Earth, 108*(B11). https://doi.org/10.1029/2002JB001999

Wang, M., & Shen, Z. K. (2020). Present-day crustal deformation of continental china derived from GPS and its tectonic implications. *Journal of Geophysical Research: Solid Earth, 125*(2). https://doi.org/10.1029/2019JB018774

Ward, S. N. (2007). Methods for evaluating earthquake potential and likelihood in and around California. *Seismological Research Letters, 78*(1), 121–133. https://doi.org/10.1785/gssrl.78.1.121

Wessel, P., Luis, J. F., Uieda, L., Scharroo, R., Wobbe, F., Smith, W. H. F., & Tian, D. (2019). The generic mapping tools version 6 [Software]. *Geochemistry, Geophysics, Geosystems, 20*(11), 5556–5564. https://doi.org/10.1029/2019GC008515

Yadav, R. K., Gahalaut, V. K., Bansal, A. K., Sati, S. P., Catherine, J., Gautam, P., Kumar, K., & Rana, N. (2019). Strong seismic coupling underneath Garhwal-Kumaun region, NW Himalaya, India. *Earth and Planetary Science Letters, 506*, 8–14. https://doi.org/10.1016/j.epsl.2018.10.023

Zeng, Y., Petersen, M. D., & Shen, Z.-K. (2018). Earthquake potential in California-Nevada implied by correlation of strain rate and seismicity. *Geophysical Research Letters, 45*(4), 1778–1785. https://doi.org/10.1002/2017GL075967

Zheng, G., Lou, Y., Wang, H., Geng, J., & Shi, C. (2018). Shallow seismicity forecast for the India-Eurasia collision zone based on geodetic strain rates. *Geophysical Research Letters, 45*(17), 8905–8912. https://doi.org/10.1029/2018GL078814

Chapter 8
Journeying Through the Major Himalayan Indian Earthquakes—The Knowns and Unknowns

Kusala Rajendran and C. P. Rajendran

Abstract The Himalaya has produced several large and great earthquakes during the last two centuries—the 1905 Kangra (M_w 7.8), the 1934 Bihar-Nepal (M_w 8.4), the 1897 Shillong (M_w 8.2, although not a Himalayan earthquake, but one that occurred in the peripheral region), and the 1950 Assam (M_w 8.6), among them. These earthquakes occurred before the advent of modern seismological tools, but they are being revisited, and many insightful accounts of these earthquakes are now available. The 1897 event is discussed for its faulting mechanics, the 1905 Kangra earthquake for its dual-intensity patches, the 1934 Bihar-Nepal earthquake for its surface rupture, and the 1950 earthquake for its source mechanism. Here we present an overview of these earthquakes and highlight some outstanding questions.

Keywords Earthquakes · Himalaya · Shillong plateau · Plate boundary earthquakes · Paleoseismology · Earthquake recurrence

8.1 Introduction

This journey through the great historical Indian earthquakes follows the same route as the Himalaya—from its eastern boundary with Pakistan, to its western boundary with China. The result of the India-Eurasia collision during the Cenozoic, the ~ 2500-km-long Himalayan arc and the contiguous emergent structures have generated some of the most damaging earthquakes in global history—the 1905 Kangra (M_w 7.8), 1934 Bihar-Nepal (M_w 8.4) and 1950 Upper Assam (M_w 8.6), among them. The 1897 Shillong earthquake M_w (8.2) is also included, although it is not a true

K. Rajendran (✉)
Centre of Excellence in Advanced Mechanics in Materials, Indian Institute of Science, Bangalore, India
e-mail: kusalaraj@gmail.com

C. P. Rajendran
School of Natural Sciences and Engineering, National Institute of Advanced Studies, Bangalore, India

V. M. Tiwari (ed.), *Himalayan Natural Hazards*, Indian National Science Academy,
https://doi.org/10.1007/978-981-92-3784-5_8

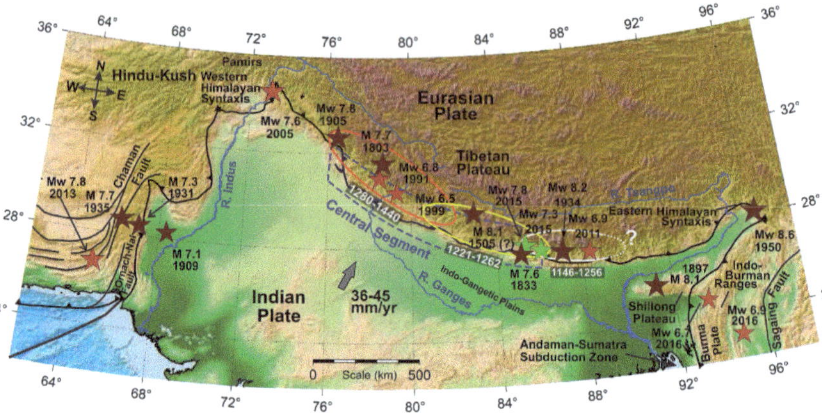

Fig. 8.1 Himalaya plate boundary (black line) and the large/great earthquakes. Red and brown stars are historical earthquakes of magnitude 6.9–8.6. Green stars: more recent events in the same magnitude range. Grey arrow: India-Eurasia plate rate (Stevens & Avouac, 2015), Rupture zones of the 1100 (white), 1255 (yellow) and 1344 CE (red) events are shown (figure after Rajendran and Rajendran, 2022)

plate boundary event, but under the influence of convergent tectonics (Fig. 8.1). Some of the Himalayan earthquakes have occurred during the pre-instrumentation era and details about them are primarily based on archival records (e.g., Iyengar et al., 1999; Ambraseys & Jackson, 2003). Archaeological and geological evidence provide information about prehistoric earthquakes (e.g.,, a, b; Jayangondaperumal et al., 2017; Kumar et al., 2006; Lave et al., 2005; Malik et al., 2016; Rajendran et al., 2013, 2015, 2016; Wesnousky et al., 2017a, b).

Although the Himalaya has been seismically highly active during the last century, it has been relatively quiet, with no great earthquakes since 1950, and it is generally reported that a great earthquake is imminent. GPS models estimate that ten years of stress accumulation is enough to generate the moment to drive a magnitude M_w 8 earthquake, and a century of stress build-up could produce an M_w 8.6 earthquake. However, since 1600, only two $M_w > 8$ earthquakes (1934 and 1950) and only eight earthquakes with $7 < M_w < 8$ have been documented in the Himalaya (e.g., Biham 2019). In brief, most investigations conclude that some segments of the Himalaya plate boundary, especially its central segment, is overdue for large earthquakes. This paper revisits the 1905 Kangra, 1934 Bihar-Nepal, 1950 Assam and the 1897 Shillong earthquakes and reviews some debated issues, which are discussed in detail by Rajendran and Rajendran (2022)

8.2 Brief Overview of the Seismotectonic Setting of the Himalaya

The approximately 2500-km-long arcuate collision boundary that developed from the Cenozoic India-Eurasia collision is dominated by arc-normal compression and thrust faulting along the arc, but the syntaxes deform predominantly by strike-slip faulting. As an orogenic wedge formed by a stack of thrust sheets scraped off the Indian crust, the tectonics of the Himalaya is dominated by major thrust belts that get progressively younger southward (Le Fort, 1975). The Main Central Thrust (MCT) and the younger Main Boundary Thrust (MBT) are presently being passively displaced above the basal Main Himalayan Thrust (MHT). The Main Frontal Thrust (MFT), the youngest, also known as the Himalayan Frontal Thrust (HFT), emerges along the Siwalik Hills (Mugnier et al., 2013). The imbricated thrust stack, comprising the MCT, MBT and MFT, along with the MHT, defines the overall structural architecture of the Himalayan fold-and-thrust belt. All these thrust systems merge with the main basal décollement, the MHT, a mid-crustal reflector at a depth of about 40 km beneath the southern Tethyan Himalaya near Tibet (e.g., Zhao et al., 1993). In this general structural geometry, large earthquakes originate on the MHT, at the depth of the brittle-ductile transition (Fig. 8.2). Large/great earthquakes also occur on the syntaxis, the M_w 8.6, 1950 Assam, close to the western syntaxis and the M_w 7.7, 1935 Chaman Fault earthquake, close to the eastern syntaxis. While MHT generates mostly thrust-faulting events, syntaxial structures give rise to strike-slip or oblique-thrust events. Next, we discuss the individual earthquakes.

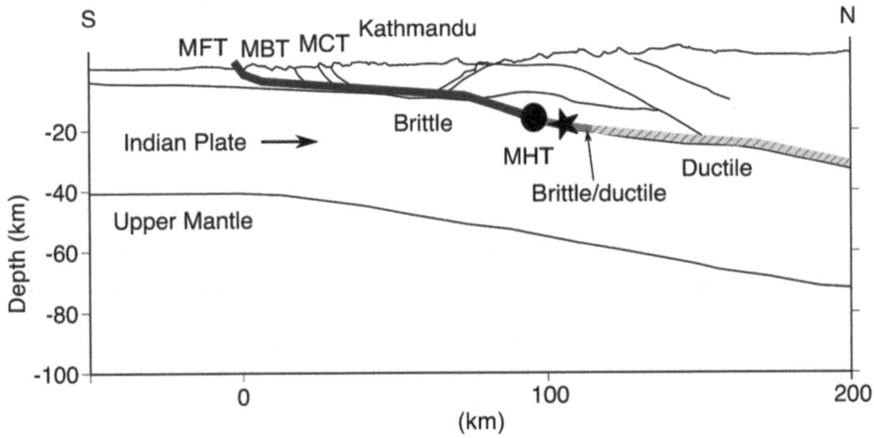

Fig. 8.2 Schematic cross-section through central Nepal. Dark band shows the inferred rupture zone, and the grey band is the ductile regime. The black dot denotes the tip of the dislocation, and the star shows the source of great earthquakes (after Mugnier et al., 2013)

8.3 The Kangra Earthquake of 1905

With more than 20,000 people killed and nearly 100,000 dwellings destroyed in the Kangra Valley, the earthquake on 4 April 1905, occurred without foreshocks, and it surpassed all other twentieth-century Indian earthquakes in scale and extent of destruction (Middlemiss, 1905, 1910) (Fig. 8.3a). Using observations from 18 stations, Ambraseys and Bilham (2000) estimated its surface-wave magnitude as 7.83 ± 0.18. Like most other Himalayan earthquakes, the Kangra event also raised questions that were not adequately addressed, owing to data limitations and the knowledge available at the time. The absence of surface rupture and the second region of high intensity near Dehra Dun, 200 km away, were two issues demanding explanation. We revisit the Kangra earthquake and comment on these two issues investigated by later researchers.

8.3.1 Tectonic Framework

The Kangra earthquake took place in the northwestern part of the foreland-fold and thrust belt within the Sub-Himalaya. The Jwalamukhi (JMT) and Bilaspur Thrusts (BT), the two contractional structures that run somewhat parallel to the MBT here (Fig. 8.3b). The thrusts, located up to 50 km from the active front, are believed to have accommodated out-of-sequence deformation during the Late Quaternary. JMT deforms fluvial terraces of approximately 10 ka with a mean slip rate of 7.5 mm/yr (Dey et al., 2016; Thakur et al., 2014; Yeats & Lillie, 1991). Long-term shortening of 14 ± 2 mm/yr, inferred from balanced cross-sections (Powers et al., 1998), matches the GPS-based slip rate of 14 ± 1 mm/yr along the ductile part of the MHT (Banerjee & Burgmann, 2002), supported also by the evidence of steady convergence between India and Eurasia over geological time scales (e.g., DeMets et al., 2017).

Seeber and Armbruster (1981) regarded the 1905 earthquake as originating on the MHT, and the absence of any surface ruptures led to an understanding that the rupture was on a blind thrust—a conclusion also reached by Yeats and Lillie (1991). A later model by Gahalaut et al. (1994) suggested thrust faulting with rupture terminating near the JMT. Wallace et al. (2005) used the Great Trigonometrical Survey of India (GTS) and 2001 GPS data to arrive at the same conclusion. However, Malik et al. (2015) and Sahoo and Malik (2017) argue that the earthquake occurred on the 60-km-long, right-lateral Kangra Valley Fault (KVF; see Fig. 8.3b). They further add that a similar earthquake occurred 1050 ± 150 years ago, an observation that has been challenged by Szeliga and Bilham (2017).

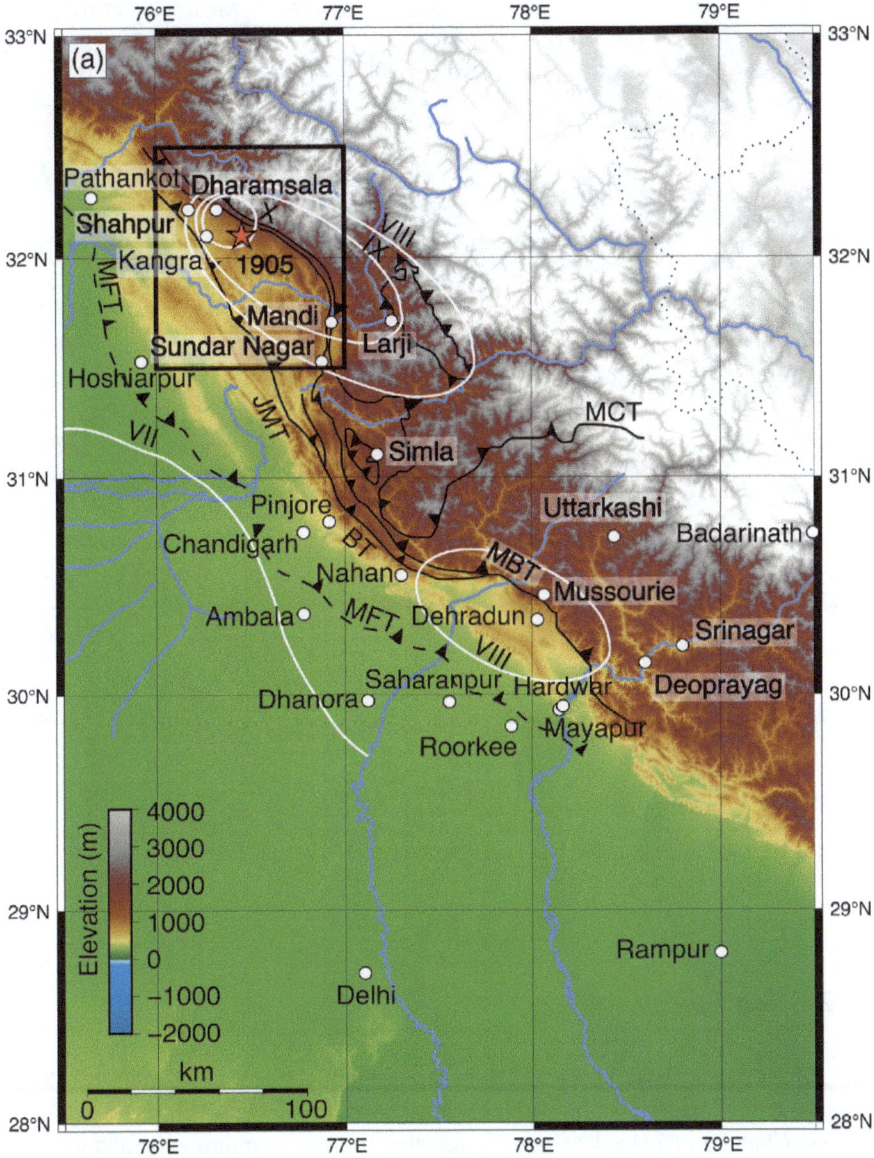

Fig. 8.3 **a** Area showing location of the Kangra earthquake and the isoseismal on the Rossi-Forel scale (after Middlemiss, 1910). Traces of the Jwalamukhi Thrust (JMT), Bilaspur Thrust (BT) are also shown along with other structures: MCT, MBT and MFT. The epicentre is in the highest-intensity zone. The area in the box is shown in Fig. 8.3b (after Rajendran and Rajendran, 2022). **b**. Location of the 1905 earthquake, trace of the Kangra Valley Fault (KVF) showing the GPS vector from Stevens (after Rajendran and Rajendran, 2022)

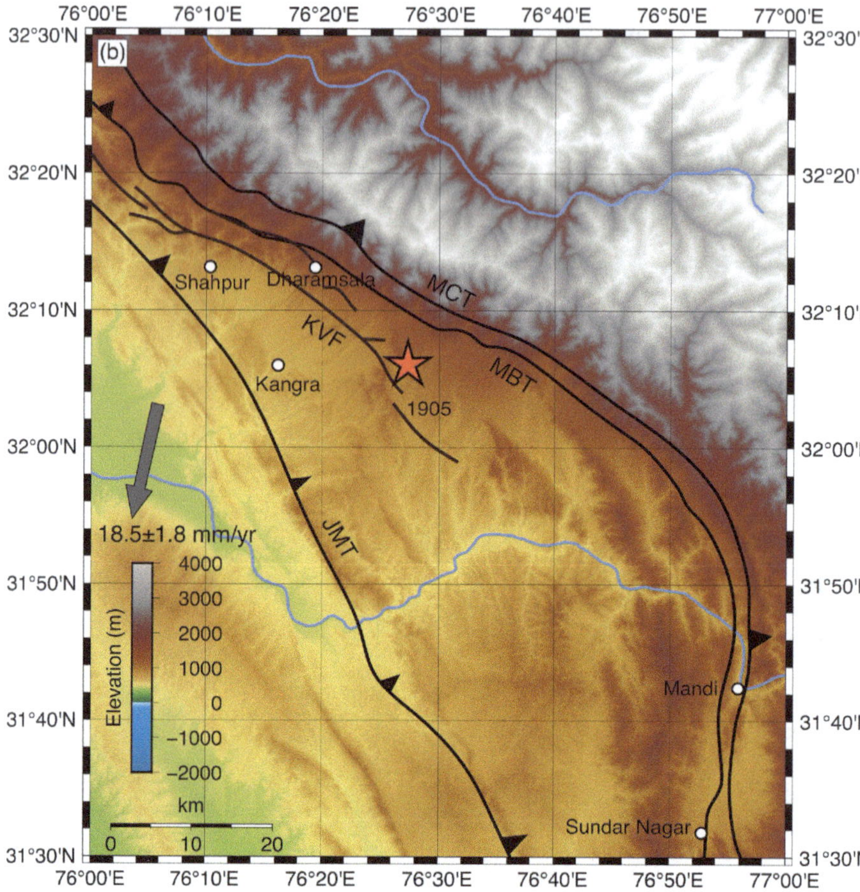

Fig. 8.3 (continued)

8.3.2 *Two Epicentral Tracts?*

The 1904 first-order geodetic levelling survey along the Sharanpur-Dehra Dun-Mussoorie line, resurveyed immediately after the earthquake, was utilised by Gahalaut and Chander (1992a, 1997) to suggest that the 1905 rupture extended within a few kilometres to Dehra Dun. Using re-evaluated raw data that reportedly contained many systematic errors, Bilham (2001) challenged this interpretation. From the re-assessed magnitude of Ms 7.8 ± 0.2 by Ambraseys and Bilham (2000), the inferred rupture length of 200 km, with no extension to Dehra Dun. Hough et al. (2005) identified two P-wave arrivals within the first two minutes on records from Colaba (Bombay/Mumbai) and interpreted them as independent earthquakes. They also analysed recordings from Göttingen and Leipzig (Germany), which showed S-wave arrivals separated by 7 to 8 min, and, given the longer sS-S travel time of the second event, it was considered triggered by a deeper source (Hough & Bilham, 2008; Hough et al., 2005).

8.3.3 Comments

Recent models suggest that the KVF does not appear to accommodate much slip (Szeliga & Bilham, 2017). The KVF's potential to generate future earthquakes should be assessed using more constrained GPS strain models. The fact that the 1905 earthquake triggered a second event raises the question of whether such dual events occur elsewhere along the Himalaya. Aside from the M_w 7.8 2015 Nepal earthquake (Gorkha), which triggered a second event (M_w 7.3) to its east, there are no such examples. However, this does not exclude the possibility that some prehistoric earthquakes occurred as twin events, which has implications for interpreting the length of paleo ruptures (and hence paleo-magnitudes). These have important implications for seismic hazard assessment.

8.4 1934 Bihar Nepal Earthquake

The Bihar earthquake of 15 January 1934 (M_w 8.4), one of the strongest among the twentieth-century Himalayan events, caused extensive death, damage and destruction in both central-eastern Nepal and the Bihar plains (e.g., Dunn et al., 1939) (Fig. 8.4). The official death toll was about 8000 in Nepal and 7000 in India, but the unofficial number touched 25,000. The earthquake was felt over almost the whole of northern and central India, demonstrating the power to deform a riverine landscape. Although its tectonic link to the Himalaya was not apparent at that time, it emerged as an illustration of the rupture mechanism of great Himalayan earthquakes in the later years. Thus, Seeber and Armbruster (1981) argued that these great earthquakes nucleated on the décollement, and their ruptures propagated southward toward the MFT, breaking the surface or terminating as a blind thrust (see Fig. 8.2).

According to the report by GSI (Dunn et al., 1939), ground shaking was most severe in an area of about 46,600 km^2 covering the alluvial plains of northern Bihar. The severely affected regions of Bettiah, Motihari, Sitamarhi, Madhubani, Muzaffarpur, Darbhanga and Monghyr (Munger) formed what was described as a 'slump belt', which also experienced severe liquefaction at some locations (Fig. 8.4). The epicentre was placed under the Gangetic alluvial plains of Bihar, based on the data from the Indian side, which was modified later using accounts from Nepal by Rana (1935). Seeber and Armbruster (1981) argued that the earthquake had originated on the detachment beneath the Lesser- and Sub-Himalaya. Chen and Molnar (1977) reviewed all the data to suggest its source at 27.55° N, 87.07° E, approximately 10 km south of Mount Everest (Fig. 8.4). In one of the earliest models, Singh and Gupta (1980) used a combination of thrust and strike-slip faulting for the 1934 earthquake.

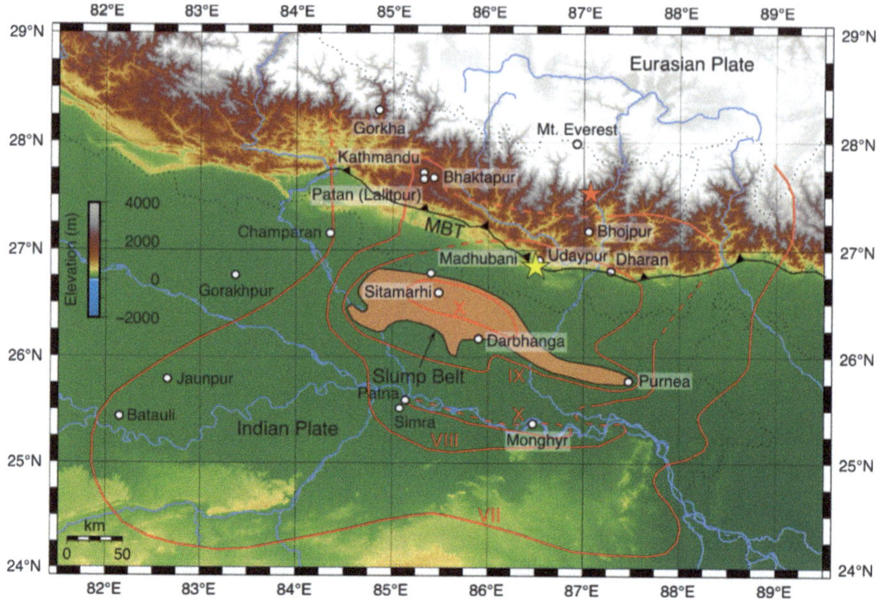

Fig. 8.4 Location of the 1934 earthquake and the intensity contours (Mercalli scale). The yellow and red stars denote the epicentral locations by Seeber and Armbruster (1981) and Chen and Molnar (1977), respectively; the slump belt, as reported in Dunn et al. (1939), is shown in beige (after Rajendran and Rajendran, 2022)

Several models discussed since then propose rather complex rupture processes, but none can explain all the features of the earthquake, including the observed damage. A model by Bilham (2019) proposed three possible scenarios. In one model, the rupture extent is bounded on the west by the 2015 earthquake rupture at 86.1°, whereas in an alternate scenario, the rupture overlaps it. In the third scenario, the southern edge of the rupture extends southward to the Himalaya Frontal Thrust, requiring the fault to be a blind thrust, an idea supported by many previous workers (e.g., Almeida et al., 2018; Duvall et al., 2020; Yeats & Thakur, 2008) (Fig. 8.5).

8.4.1 History of Earthquakes

Some of the most extensive trenching excavations in the search for evidence for past earthquakes were done on the MFT of central Nepal (e.g., Lavé et al., 2005; Kumar et al., 2001, 2006, 2010; Sapkota et al., 2013; Bollinger et al., 2014; Wesnousky et al., 2017a, b, 2019; Rizza et al., 2019; see Fig. 8.6 for the locations of trenches). The earliest was by Lavé et al. (2005), who reported a surface-breaking event with evidence for a 1100 CE event (M_W ~ 8.8), with likely surface displacement of ~17 (+ 5/– 3) metres with a lateral extent of ~ 240 km. This evidence ruled out any surface

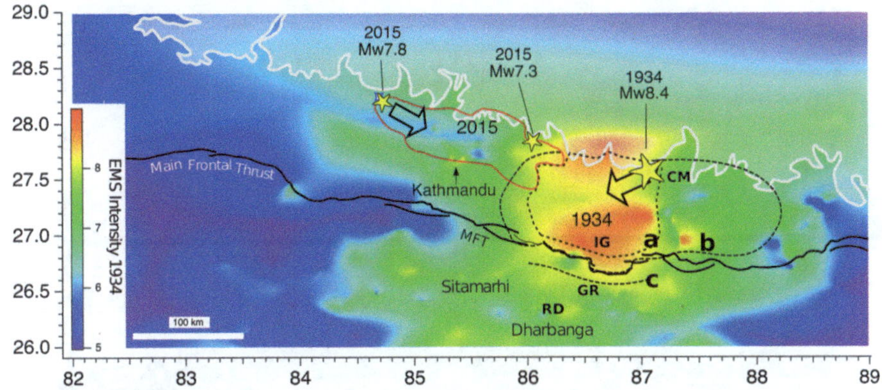

Fig. 8.5 Intensities observed in the 1934 earthquake (Ambraseys & Douglas, 2004; Martin & Szeliga, 2010). Epicentral locations by various authors are indicated as RD—Dunn et al. (1939), GR—Gutenberg and Richter (1949), IG-ISC/GEM, CM—Chen and Molnar (1977). Letters a, b and c denote rupture scenarios by Bilham (2019). Large unfilled arrows show inferred rupture propagation directions of the 1934 and 2015 earthquakes identified by Bilham (2019) (figure after Rajendran and Rajendran, 2022)

breaks from previous earthquakes during the past 1000 years, including the 1934 event. This proposition has been challenged by later studies (e.g., Bollinger et al., 2014; Sapkota et al., 2013) that reported earlier surface deformation and evidence for an event postdating 1021–1166 CE, attributed to the 1934 Bihar-Nepal and the 1255 CE events. Mishra et al. (2016) reported that the 1255 ruptured 800 km of the eastern Himalaya, reiterating the potential of the Himalayan plate boundary to generate future giant earthquakes of Mw ~ 9.0.

These observations have been challenged by Wesnousky et al., (2017a, b) after a re-examination of the radiocarbon age data from the Sir Khola trench. They argue that the temporal bounds overlap with an earthquake around 1100 CE, rather than the 1255 CE event. Rizza et al. (2019) also reported a surface scarp near the Sir Khola sites, but, as the authors admit, these features are obscured by post-earthquake aggradation processes, and the age constraints could be erroneous. They conclude that the penultimate rupture of the MFT must have occurred around ~ 1100 CE. Based on stratigraphy, structure and radiocarbon data from the Khayarmara trench, Wesnousky et al. (2019) concluded that ancient earthquakes occurred between 1146–1256, 1022–1102 and 1221–1262 CE on the eastern, central and western segments, respectively. By exploring the 1934 slump belt, Rajendran et al. (2016) reported multiple liquefaction episodes possibly generated by earthquakes during the intervals: 829–971, 886–1090, 907–1181, 1130–1376, 1112–1572, 1492–1672, 1733–1839 and 1814–1854 CE. The oldest event, earliest bracketed between 829 and 1181 CE, might be attributed to ~ 1100 CE. A more recent study supports the evidence for a great earthquake in 1100 CE, reported by Divyadarshini et al. (2023).

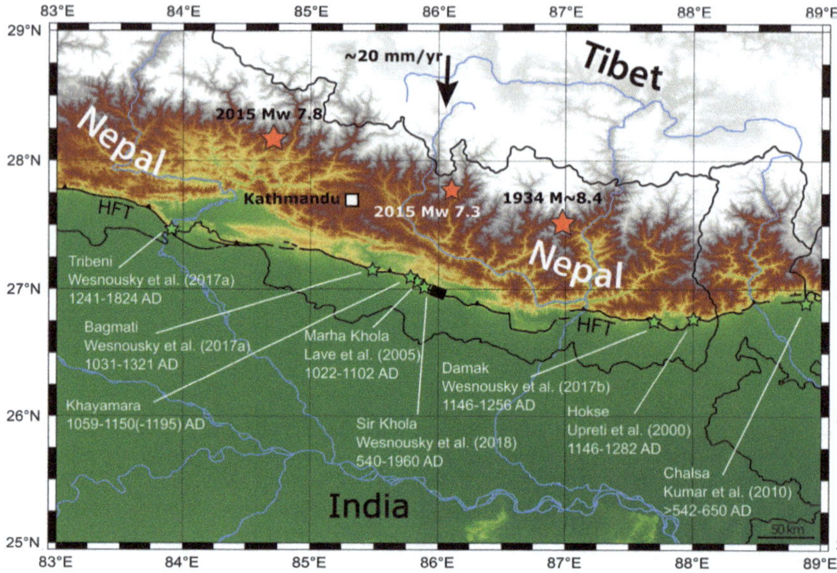

Fig. 8.6 Location of various trenches excavated along the MFT of the eastern Nepal Himalaya. Sites are tagged with the age of the last large surface displacement and citations. The red stars mark locations of the two large earthquakes of 2015. A filled black square near Sir Khola marks the location of the seismic lines of Almeida et al. (2018); the Charnath trench excavated by Rizza et al. (2019) lies to the east of Sir Khola. (Modified after Wesnousky et al., 2019)

8.4.2 Comments

The slump belt, the extensive liquefaction in the Ganga plains, and the missing rupture were key to the idea that the great Himalayan thrust earthquakes are blind thrust events (Seeber & Armbruster, 1981). On the lack of surface rupture, there was a change in perception when Sapkota et al. (2013) proposed otherwise, but later reinterpretations by Wesnousky et al., (2017a, b), from the same trench, ruled out that possibility. Further, using additional data from other sites, Wesnousky et al. (2019) argue that the trenches expose evidence for an earthquake between 1059 and 1195 CE, but none during the intervening period or from the great 1934 earthquake. The data on the chronology of past events from the source of the 1934 Bihar earthquake appear a bit incoherent, but that is the very nature of geologic evidence from a region of active tectonism and landscape alteration. As of now, the large penultimate earthquake associated with the 1934 source seems to have occurred around 1100 CE. Perhaps, more work from future events and application of techniques, especially seismic imaging, might help resolve some of these issues.

8.5 1950 Assam Earthquake

The 15 August 1950, M_W 8.6 earthquake originated in the rugged mountainous areas of the China-Tibet region, north of the village of Rima. It was felt over an area of 3 million km^2 in India, Myanmar (Burma), Bangladesh (then East Pakistan), Tibet and southern China. It caused extensive damage in many towns of Upper Assam and regions across the eastern Himalaya in areas including the Abor Hills, the Mishmi Hills and parts of the Assam Plains, with about 1500 casualties on the Indian side. Ground shaking was severe at Sadiya and the alluvial plain of the Lohit River, where fissuring and soil liquefaction occurred. Highest intensities were felt in the southwest quadrant of the Brahmaputra Valley, possibly due to site amplification (e.g., Mathur, 1953). (Poddar, 1950) (Fig. 8.7). Reports about ground shaking and landslides from the epicentral region come from Frank Kingdon-Ward (Kingdon-Ward, 1951).

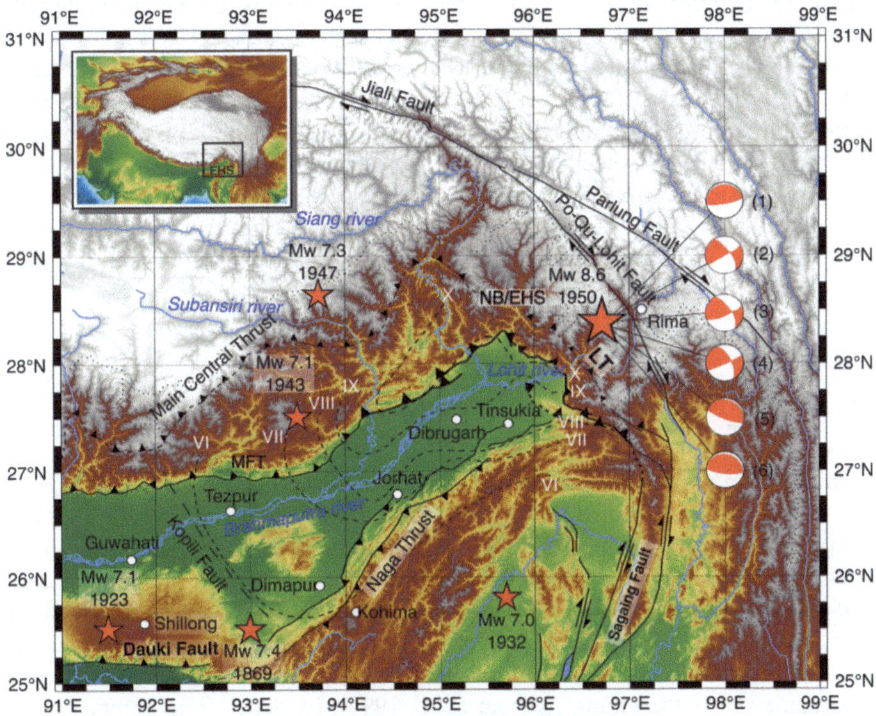

Fig. 8.7 Major structures and earthquakes in the eastern Himalaya: NB: Namche Barwa, EHS: the Eastern Himalayan Syntaxis. LT: Lohit Thrust. The focal mechanism solutions for the 1950 earthquake are from 1: Chen and Molnar (1977); 2: Tandon (1954); 3: Sylman et al. (1974); 4: Ben-Menahem et al. (1974); 5: Devachandra et al. (2014) and 6: Coudurier-Curveur et al. (2020). Dashed lines represent isoseismal for the 1950 earthquake (after Poddar, 1950; Figure after Rajendran and Rajendran, 2022)

8.5.1 Epicentral Location and Fault Models

The epicentre (28.6° N and 96.5° E), ~ 40 km west of Rima (Tillotson, 1951), was later relocated by Chen and Molnar (1977) as 28.36° N 96.45° E, at a focal depth of ~ 14 km with a magnitude of 8.6–8.7 (Richter, 1958). Diverse mechanisms have been proposed for the 1950 earthquake (Fig. 8.7). The initial solution by Ben-Menahem et al. (1974) proposed strike-slip faulting on a NNW-SSE-trending fault plane, consistent with the EHS. Chen and Molnar (1977) suggested thrust faulting on an easterly striking, gently north-dipping fault at the eastern end of the Himalayan plate boundary. According to Molnar (1990), there was a wide zone of strike-slip and shallow thrust faulting on multiple planes. Coudurier-Curveur et al. (2020) also suggest multiple ruptures on the MFT (Abor Hills) and the Mishmi Thrust (MT), based on tectonic scarps on Quaternary alluvial terraces. Simultaneous ruptures on two orthogonal planes with different dips (Main Frontal Thrust, 13–15°; Mishmi Thrust, 25–28°) appear very complex, but the model seems to be compatible with the composite main shock thrust fault mechanism. These authors also suggested that the earthquake had ruptured both the MFT (locally represented by the Abor Hills) and the Mishmi Thrust (MT), in a rather complex rupture model. They also used terrace uplifts on the frontal area of the Mishmi Hills and the ^{10}Be exposure ages from the terrace deposits to suggest multiple episodes of coseismic uplift, including that from the 1950 earthquake. A scarp with a cumulative height of ~ 3 m, identified near Pasighat (Priyanka et al., 2017), is another piece of evidence for coseismic uplift. Singh et al. (2021) also infer coseismic near-surface offsets due to the 1950 earthquake.

8.5.2 Past Earthquakes in the Region

Information on past earthquakes in the Assam Valley and the neighbouring regions is available in *Ahom* records and the colonial writings (Iyengar et al., 1999; Gait, 1906). Paleoliquefaction investigations around the Brahmaputra Valley and nearby areas provide geological evidence for earthquakes, dated in the age range of 1262–1635 CE, although the chronology needs further validation (Rajendran & Rajendran, 2011; Reddy et al., 2009) (see Fig. 8.8). The 1697 Sadiya earthquake, felt strongly in and around Sadiya Town, ~ 150 km southwest of the 1950 source is another important event in the region (Iyengar et al., 1999). Evidence from a trench in the central Bhutan Himalaya (Le Roux-Mallouf et al., 2016) suggests that an earthquake occurred between 1642 and 1836 CE with a likely magnitude of Ms 7. Hetényi et al. (2016) fine-tuned the date to 1714, based on excavations on the frontal thrust of Bhutan. This was further validated by Zhao et al. (2021), who used independent data to confirm that the faulting was associated with the 1714 CE earthquake. Evidence for a medieval offset from trenches excavated near Mymensingh, Bangladesh, for

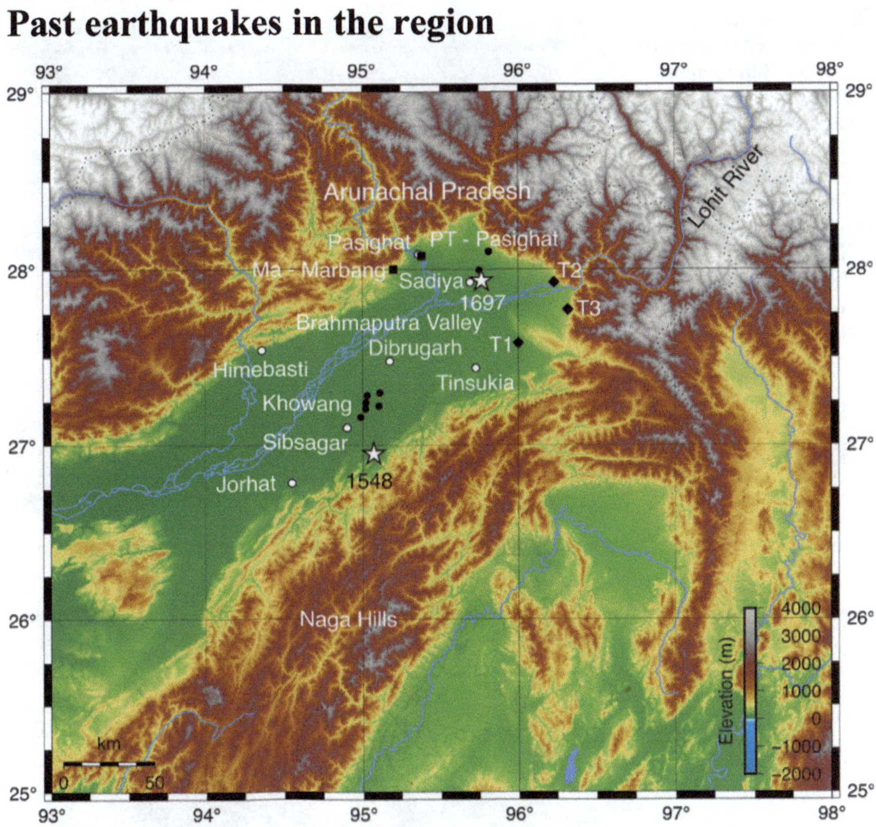

Fig. 8.8 Locations of historical earthquakes and selected sites of paleoseismological investigations in the northeast region (Arunachal Himalaya, Brahmaputra Valley and adjoining regions): PT: Pasighat; Ma: Marbang (Priyanka et al., 2017); T1 and T2 (Rajendran & Rajendran, 2011); T3 (Kamlang Nagar; Singh et al., 2021). Solid black circles mark the sites excavated by Reddy et al., (2009; sedimentary section from a site at Khowang is shown in Fig. 8.9) (after Rajendran and Rajendran, 2022)

the Garhgaon earthquake of 1548 CE, presented by Morino et al. (2009), is another significant event in this region.

8.5.3 Comments

It is not uncommon for earthquakes to lead to complex ruptures, sometimes on intersecting faults; still, they usually occur in a sequence rather than continuously, involving multiple faults (e.g., Li et al., 2020). The Arunachal MFT and the Mishmi

Fig. 8.9 Liqucfaction feature exposed on the left bank of the Burhi Dihing River at Khowang. The solid circle represents the organic material collected as an upper bound with its ^{14}C age. About 1.2 m high sand dyke (Dy) is seen with connectivity to the source sand and subsequent deposition over it (figure courtesy: D. V. Reddy)

Thrust are not known to have experienced any such previous ruptures (see Rajendran & Rajendran, 2021). Single and continuous rupture involving two orthogonal fault planes, as invoked by Coudurier-Curveur et al. (2020), must be a rarity and extremely complex. Although historical and geological records indicate a medieval earthquake in 1697, no earthquakes have mimicked the magnitude and rupture style of the 1950 earthquake in the last 1000 years. This premise, implying nearly bi-millennial return time for mega-quakes in the 1950 source region, is also supported by cosmogenic (^{10}Be) terrace ages (Coudurier-Curveur et al., 2020). However, the seismotectonic scenario raises the question of whether the complex rupture geometry assigned to the 1950 earthquake is repeatable in time in the same manner, or it is a rare one-time phenomenon.

8.6 1897 Shillong Earthquake

The 12 June 1897 (Mw 8.2) Shillong earthquake, one of the most damaging events in India, is among the earliest for which the instrumental magnitude of 8.7 was calculated (Gutenberg, 1956). There were widespread landslides and extensive coseismic liquefaction, mostly near Shillong and Guwahati, where most masonry buildings, including many bridges, stone houses and churches, were levelled. Liquefaction led to land subsidence in areas as far as Mymensingh, on the southern side of the Dauki fault, in Bangladesh. Oldham's report (1899), which provides the details of destruction, ground accelerations and results of a repeat triangulation survey initiated in 1860, is probably the most detailed documentation of a great earthquake of the nineteenth century. Oldham identified a 'hat-shaped' patch, with the highest intensity of X on the Rossi-Forel scale. In a later evaluation of the La Touche (Bilham, 2008) letters' transcripts, Ambraseys and Bilham (2003) revised the isoseismal map, location and magnitude to 8.2 (Fig. 8.10).

8.6.1 Epicentral Location and Fault Geometry

Oldham (1899) suggested a thrust fault below the plateau; Seeber and Armbruster (1981) argued for a north-dipping fault; and more recent studies refer to it as an 'intraplate' earthquake (e.g., Ambraseys & Bilham, 2003). The rupture of the 1897 earthquake was estimated to be about 200 km long, terminating near the Dauki fault in the south, but its northern limit remained uncertain (Molnar, 1990; Molnar & Pandey, 1989). Gahalaut and Chander (1992b) used Oldham's data to suggest a gently dipping thrust fault beneath the Shillong Plateau. These ideas were challenged by Bilham and England (2001), who used the pre- and post-earthquake triangulation survey data to propose a south-dipping reverse fault, named the Oldham Fault, spatially coinciding with the northern topographic front of the plateau (see Figs. 8.11 and 8.12).

Considering a similar geometry, Rajendran et al., (2004, 2006) suggested slip on the Brahmaputra fault, located north of the plateau's topographic front. These authors pointed out the instability of the 1859–1937 survey monuments, the absence of active tectonic expressions, microseismicity, or gravity anomaly signatures. However, England and Bilham (2015), after re-evaluation of the survey data, have reaffirmed the existence of the Oldham Fault. Contrasting views of the seismotectonics are further complicated by the paucity of surface expressions of the great earthquake (Srinivasan, 2003). Crustal structure derived from P-wave receiver function analysis by Mitra et al. (2018) implies faulting beneath the Brahmaputra rather than along the topographic front. In blind thrust scenarios, trenching may not provide much information on fault geometry; therefore, high-resolution seismic data may be highly useful for imaging the geometry of the faulted strata (Almeida et al., 2018). This debate might find a solution only through seismic imaging, which should be able to capture subsurface faults.

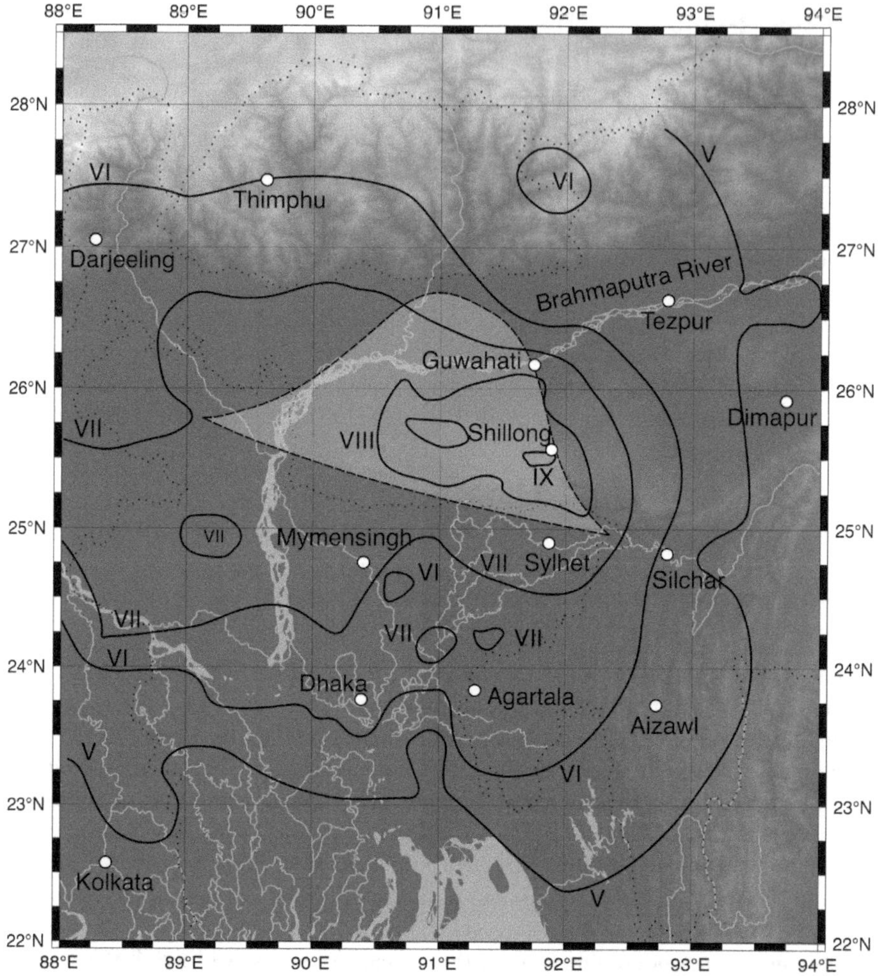

Fig. 8.10 Hat-shaped patch of Oldham (1899) overlapped on the isoseismal map of Ambraseys and Bilham (2003)

8.6.2 Historical Record of Earthquakes

Historical records on Assam provide information about its ancient temples—some more than 1500 years old—and their documented histories bear testimony to the ground-shaking effects of great earthquakes. Barua (1966) and Iyengar et al. (1999) refer to a wave of reconstruction activity around 835–860 CE, for example, the seventh–eighth century CE Kamakhya temple located about 8 km from the town of Guwahati (Rajendran et al., 2004). Excavations near the Kamakhya temple exposed evidence for two levels of destruction from the 1897 and ~ 1200 CE events, based on the age of pottery from the pre-1897 destruction and a similar penultimate level

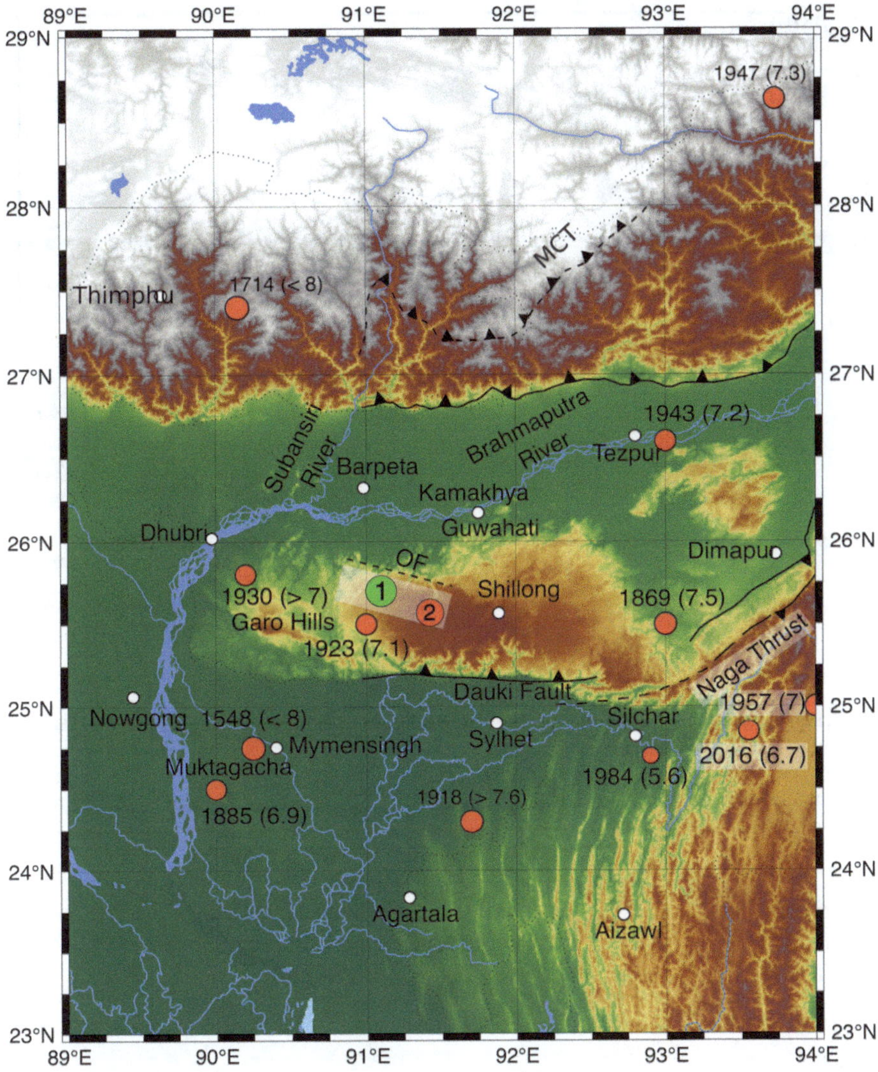

Fig. 8.11 Distribution of earthquakes near the Shillong Plateau (SP). Earthquake locations marked as 1 and 2 are from Ambraseys and Bilham (2003) and England and Bilham (2015), respectively. The geodetically constrained rupture zone of the 1897 earthquake (the Oldham fault; OF) is shown as a white transparent rectangle (after Rajendran & Rajendran, 2022)

(Fig. 8.13). A recurrence period of the order of 500 years for large earthquakes was suggested in the Brahmaputra Valley (Sukhija et al., 1999a, b), while Rajendran et al. (2004) consider that the predecessor of the 1897 occurred 1200 years ago.

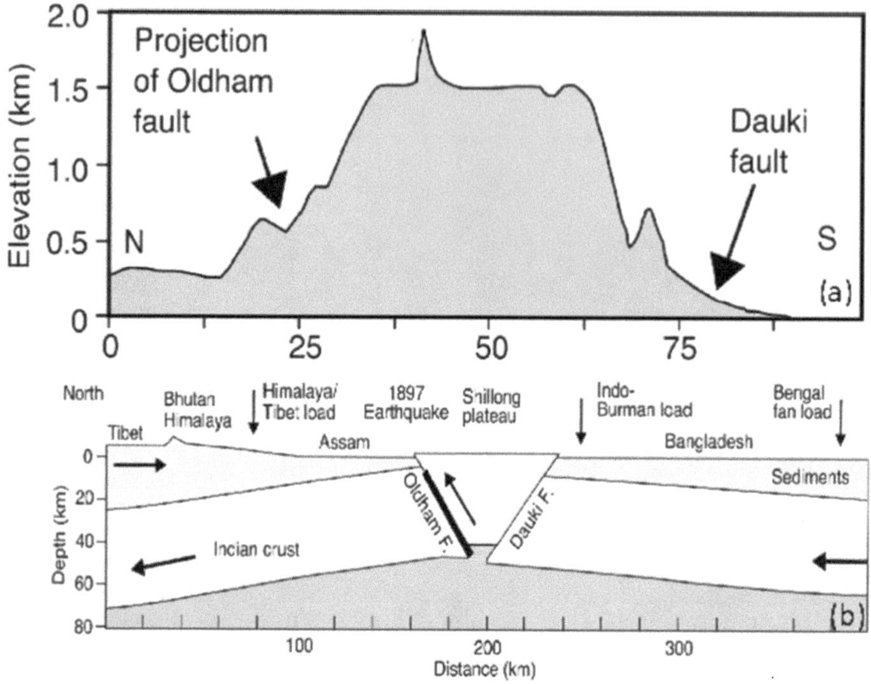

Fig. 8.12 **a** Topographic section through the central Shillong Plateau showing the projected surface location of the Oldham fault. **b** N/S section from Tibet to the Bay of Bengal showing schematic geometry of the plateau pop-up. The dip of the Dauki fault is conjectural, and it would intersect the 1897 rupture where it dips less than 40° (after Bilham & England, 2001) (figure after Rajendran and Rajendran, 2022)

Differing from these estimates, Bilham and England (2001) argued for a source-specific recurrence interval of 3000–8000 years, based on model-specific fault-slip data. The earthquake chronology needs validation from future work.

8.6.3 Concluding Remarks

The documented history of Himalayan earthquakes spans no more than two centuries. The oldest and best-documented historical great earthquake is the 1897 Shillong earthquake. The 1950 Assam earthquake is the most recent great earthquake to have originated in the Himalaya. Although the region has witnessed a few large events in recent history, none have had a magnitude > 8 in the Himalaya. Long quiescence along active plate boundaries is alarming, as it is a harbinger of future earthquakes. Many researchers have identified the central Himalaya as a region where the missing slip could be released during a future great earthquake. Over the last two decades, significant insights have been gained on each of the four earthquakes discussed here.

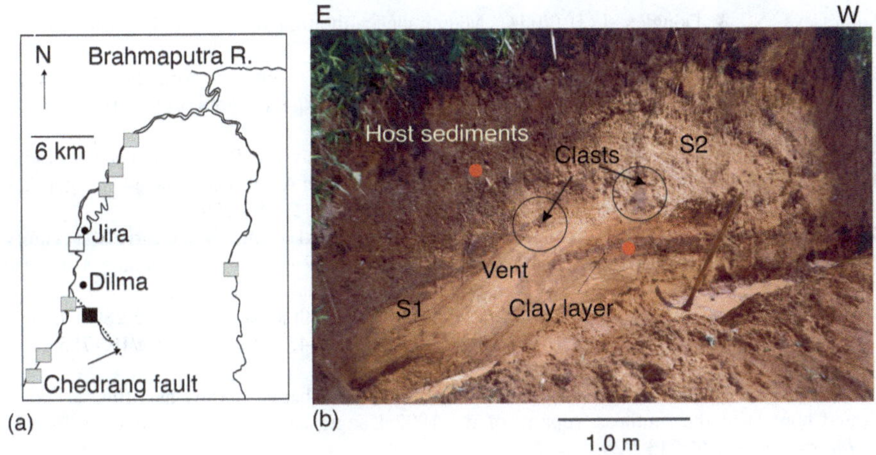

Fig. 8.13 **a** Locations of paleoliquefaction sites around the Chedrang River. Sites investigated (open circles: Rastogi et al. (1993); shaded circles: Sukhija et al. (1999a) and solid squares: Rajendran et al., 2004). **b** A view (facing north) showing the paleoliquefaction feature exposed at Dilma-1, located on the western bank of the Chedrang River. S1 is the older sill, and S2 is the younger sill. Samples collected from the ruptured clay layer and the host sediments (locations shown by red dots) yielded calibrated ages of 2170 ± 140 years B.P. and 1250 ± 80 years B.P., respectively. The section shown here is 1 m below the surface (figure after Rajendran and Rajendran, 2022)

However, more needs to be understood about their rupture mechanisms and past activity. Given the enormous progress made in the study of Himalayan earthquakes, the coming years will yield better answers to some of the outstanding questions raised here.

Acknowledgements KR thanks Prof. Debasis Roy, Department of Civil Engineering, Indian Institute of Science, and CPR thanks Dr Shailesh Nayak, Director, National Institute of Advanced Studies, for hosting. We thank the reviewers for their input in improving the text and Dr V.M. Tiwari for inviting this contribution.

References

Almeida, V., Hubbard, J., Liberty, L., Foster, A., & Sapkota, S. N. (2018). Seismic imaging of the main frontal thrust in Nepal reveals a shallow décollement and blind thrusting. *Earth and Planetary Science Letters, 494*, 216–225.

Ambraseys, N., & Bilham, R. (2003). Re-evaluated intensities for the great Assam earthquake of 12 June 1897, Shillong, India. *Bulletin of the Seismological Society of America, 93*, 655–673.

Ambraseys, N., & Jackson, D. (2003). A note on early earthquakes in northern India and southern Tibet. *Current Science, 84,* 570–582.

Ambraseys, N., & Bilham, R. (2000). A note on the Kangra Ms = 7.8 earthquake of 4 April 1905. *Current Science, 79*, 45–50.

Ambraseys, N., & Douglas, J. J. (2004). Magnitude calibration of north Indian earthquakes. *Geophysical Journal International, 159*, 165–206.

Banerjee, P., & Burgmann, R. (2002). Convergence across the northwest Himalaya from GPS measurements. *Geophysical Research Letters, 29*, 1652–1656. https://doi.org/10.1029/2002GL015184

Barua, R. K. L. (1966). *Early history of Kamarupa* (pp. 234). Lawyers Book Stall, Gauhati, India.

Ben-Menahem, A. E., & Schild, R. (1974). The source of the great Assam earthquake—an interplate wedge motion. *Physics of the Earth and Planetary Interiors, 9*, 265–289.

Bilham, R. (2019). Himalayan earthquakes: A review of historical seismicity and early 21st century slip potential. In P. J. Treloar, & M. P. Searle, (Eds.) *2019 Himalayan tectonics: A modern synthesis* (Vol. 483, pp. 423–482). Geological Society, Special Publications.

Bilham, R. (2008). Tom La Touche and the Great Assam Earthquake of 12 June 1897: Letters from the epicenter. *Seismological Research Letters, 9*, 426–437. https://doi.org/10.1785/gssrl.79.3.426

Bilham, R. (2001). Slow tilt reversal of the Lesser Himalaya between 1862 and 1992 at 78° E and bounds to the southeast rupture of the 1905 Kangra earthquake. *Geophysical Journal International, 144*, 713–728.

Bilham, R., & England, P. (2001). Plateau 'pop-up' in the great 1897 Assam earthquake. *Nature, 410*, 806–809.

Bollinger, L., Sapkota, S. N., Tapponnier, P., et al. (2014). Estimating the return times of great Himalayan earthquakes in eastern Nepal: Evidence from the Patu and Bardibas strands of the main frontal thrust. *Journal of Geophysical Research, 119*, 7123–7163. https://doi.org/10.1002/2014JB010970

Chen, W. P., & Molnar, P. (1977). Seismic moments of major earthquakes and the average rate of slip in central Asia. *Journal of Geophysical Research, 82*, 2945–2969.

Coudurier-Curveur, A., Tapponnier, P., Okal, E., et al. (2020). A composite rupture model for the great 1950 Assam earthquake across the cusp of the East Himalaya syntaxis. *Earth and Planetary Science Letters, 531*, 115928. https://doi.org/10.1016/j.epsl.2019.115928

DeMets, C., Calais, E., & Merkouriev, S. (2017). Reconciling geodetic and geological estimates of recent plate motion across the Southwest Indian ridge. *Geophysical Journal International, 208*, 118–133.

Dey, S., Thiede, R. C., Schildgen, T. F., et al. (2016). Holocene internal shortening within the northwest Sub-Himalaya: Out-of-sequence faulting of the Jwalamukhi Thrust, India. *Tectonics, 35*, 1–21.

Devachandra, M., Kundu, B., Catherine, J., et al. (2014). Global Positioning System (GPS) measurements of crustal deformation across the frontal eastern Himalayan syntaxis and seismic-hazard assessment. *Bulletin of the Seismological Society of America, 104*, 1518–1524.

Divyadarshini, A., Rajendran, C. P., John, B., Singh, V., Yadava, M. G., Sarkar, A., & Parida, S. (2023). Evidence of blind thrusting in the frontal part of the Central Himalaya and its significance for rupture evolution and size of a medieval earthquake. *Tectonophysics, 854*, 229798.

Dunn, J. A., Auden, J. B., Gosh, A. M. H., et al. (1939). *The Bihar–Nepal earthquake of 1934* (Vol. 73, p. 391). Memoirs of the geological society of India. (Survey of India, Calcutta).

Duvall, M., Waldron, J. W. F., Godin, L., & Najman, Y. (2020). Active strike-slip faults and an outer frontal thrust in the Himalayan foreland basin. *PNAS, 117*, 17615–17621.

England, P. C., & Bilham, R. (2015). The Shillong Plateau and the great 1897 Assam Earthquake. *Tectonics, 34*, 1792–1812. https://doi.org/10.1002/2015TC003902

Gahalaut, V. K., & Chander, R. (1992a). On the active tectonics of Dehradun region from observations of ground elevation changes. *Journal of the Geological Society of India, 39*, 61–68.

Gahalaut, V. K., & Chander, R. (1992b). A rupture model for the great earthquake of 1897, northeast India. *Tectonophysics, 204*, 163–174.

Gahalaut, V. K., Gupta, P. K., Chander, R., & Gaur, V. K. (1994) Minimum norm inversion of observed ground elevation changes for slips on the causative fault during the 1905 Kangra earthquake. *Proceedings of the Indian Academy of Sciences 103*, 401–411.

Gait, E. A. (1906). *A history of Assam* (p. 407). Thacker, Spink & Co.

Gutenberg, B., & Richter, C. F. (1949). *Seismicity of the earth and associated phenomena*. Princeton.

Hetényi, G., Le Roux-Mallouf, R., Berthet, T., et al. (2016). Joint approach combining damage and paleoseismology observations constrains the 1714 A.D. Bhutan earthquake at magnitude 8 ± 0.5. *Geophysical Research Letters, 43*, 10695–10702. https://doi.org/10.1002/2016GL071033

Hough, S., & Bilham, R. (2008). Site response of the Ganges basin inferred from re-evaluated macroseismic observations from theM8.1 Shillong 1897, M7.8 Kangra 1905 and 1934 Nepal M 8.1 earthquakes. *Journal of Earth System Science, 117*, 773–782.

Hough, S. E., Bilham, R., Ambraseys, N., & Feldl, N. (2005). Revisiting the 1897 Shillong and 1905 Kangra earthquakes in northern India: site response, Moho reflections and a triggered earthquake. *Currrent Science, 2*, 1632–1638.

Iyengar, R. N., Sharma, D., & Siddiqui, J. M. (1999). Earthquake history of India in medieval times. *Indian Journal of History of Science, 34*, 181–237.

Jayangondaperumal, R., Daniels, R. L., & Niemi, T. M. (2017). A paleoseismic age model for large-magnitude earthquakes on fault segments of the Himalayan Frontal Thrust in the Central Seismic Gap of northern India. *Quaternary International, 462*, 130–137. https://doi.org/10.1016/j.quaint.2017.04.008

Kingdon-Ward, F. (1951). Notes on the Assam earthquake. *Nature, 167*, 130–131.

Kumar, S., Wesnousky, S. G., Rockwell, T. K., et al. (2001). Earthquake recurrence and rupture dynamics of Himalayan frontal thrust, India. *Science, 294*, 2328–2331.

Kumar, S., Wesnousky, S. G., Rockwell, T. K., et al. (2006). Paleoseismic evidence of great surface rupture earthquakes along the Indian Himalaya. *Journal of Geophysical Research: Solid Earth 111*. https://doi.org/10.1029/2004JB003309

Kumar, S., Wesnousky, S. G., Jayangondaperumal, R., Nakata, T., Kumahara, Y., & Singh, V. (2010). Paleoseismological evidence of surface faulting along the northeastern Himalayan front, India: Timing, size, and spatial extent of great earthquakes. *Journal of Geophysical Research: Solid Earth, 115*. https://doi.org/10.1029/2009JB006789

Li, C., Zhang, G., Shan, X., et al. (2020). Surface rupture kinematics and coseismic slip distribution during the 2019 Mw7.1 Ridgecrest, California earthquake sequence revealed by SAR and optical images. *Remote Sensing 12*, 3883. https://doi.org/10.3390/rs12233883

Lave, J., Yule, D., Sapkota, S. N., et al. (2005). Evidence for a Great Medieval Earthquake (~ 1100 A.D.) in the Central Himalayas Nepal. *Science, 307*, 1302–1305. https://doi.org/10.1126/science.1104804

Le Fort, P. (1975). Himalayas: The collided range. Present knowledge of the continental arc. *American Journal of Science, 275*, 1–44.

Le Roux-Mallouf, R., Ferry, M., & Ritz, J.-F. (2016). First paleoseismic evidence for great surface-rupturing earthquakes in the Bhutan Himalayas. *Journal of Geophysical Research, 121*, 7271–7283. https://doi.org/10.1002/2015JB012733

Malik, J. N., Sahoo, S., Satuluri, S., & Okumura, K. (2015). Active fault and paleoseismic studies in Kangra valley: evidence of surface rupture of a great Himalayan 1905 Kangra earthquake (Mw 7.8), northwest Himalaya, India. *Bulletin of the Seismological Society of America 105*, 2325e2342. https://doi.org/10.1785/0120140304

Malik, J. N., Naik, S. P., Santiswarup, S., & Okumora, K. (2016). Paleoseismic evidence of the 1505 CE (?) and 1803 CE earthquakes from the foothill zone of the Kumaun Himalaya along the Himalayan frontal thrust (HFT) India. *Tectonophysics, 105*, 714–715. https://doi.org/10.1016/j.tecto.2016.07.026

Martin, S., & Szeliga, W. (2010). A catalog of felt intensity data for 589 earthquakes in India, 1636–2008. *Bulletin of the Seismological Society of America, 100*, 562–569.

Mathur, L. P. (1953). The Assam earthquake of 15 August 1950. In M. B. Rao (Ed.) *A compilation of papers on the Assam earthquake of August 15, 1950* (No. 1, pp. 56–60). Geology of Survery of India Publication.

Middlemiss, C. S. (1905). Preliminary account of the Kangra earthquake of 4 April 1905. *Geological Society of India Memoir, 32*(Pt 4), 258–294.

Middlemiss, C. S. (1910). The Kangra earthquake of 4th April 1905. *Memoirs of the Geological Survey of India, 38*, 409p.
Mishra, R. L., Singh, I., Pandey, A., et al. (2016). Paleoseismic evidence of a giant medieval earthquake in the eastern Himalaya. *Geophysical Research Letters, 43*, 5707–5715.
Mitra, S., Priestley, K. F., Borah, K., & Gaur, V. K. (2018). Crustal structure and evolution of the eastern Himalayan plate boundary system. *Journal of Geophysical Research: Solid Earth, 640*, 123621. https://doi.org/10.1002/2017JB014714
Molnar, P. (1990). A review of the seismicity and the rates of active underthrusting and deformation at the Himalaya. *Journal of Himalayan Geology, 1*, 131–154.
Molnar, P., & Deng, Q. (1984). Faulting associated with large earthquakes and the average rate of deformation in Central and Eastern Asia. *Journal of Geophysical Research, 89*, 6203–6227.
Molnar, P., & Pandey, M. R. (1989). Rupture zones of great earthquakes in the Himalayan region. *Proceedings of the Indian Academy of Sciences-Earth and Planetary Sciences, 98*, 61–70.
Morino, M., Kamal, A. S. M. M., & Muslim, D. (2009). Activity of dauki fault during the 1897 Ms 8.0 Great Assam earthquake confirmed by trench investigation at Gabrakahri Village, Haluaghat, Mymensingh, Bangladesh. *The Journal of the Geological Society of Japan, 115*, 11–12.
Morino, M., Kamal, A., Maksud, S. M., et al. (2014). A paleo-seismological study of the Dauki fault at Jaflong, Sylhet, Bangladesh: Historical seismic events and an attempted rupture segmentation model. *Journal of Asian Earth Sciences, 91*, 218–226. https://doi.org/10.1016/j.jseaes.2014.06.002
Mugnier, J. L., Gajurel, A., Huyghe, P., et al. (2013). Structural interpretation of the great earthquakes of the last millennium in the central Himalaya. *Earth-Science Reviews, 127*, 30–47.
Oldham, R. D. (1899). *Report of the great earthquake of 12th June 1897* (p. 379). Memoirs of the geological society of India. (reprinted: 1981, Geol Surv. India Calcutta).
Poddar, M. C. (1950). The Assam earthquake of 15th August 1950. *Indian Miner, 4*, 167–176.
Powers, P. M., Lillie, R. J., & Yeats, R. S. (1998). Structure and shortening of the Kangra and Dehra Dun Reentrants, Sub-Himalaya, India. *Geological Society of America Bulletin, 110*, 1010–1027.
Priyanka, R. S., Jayangondaperumal, R., Pandey, A., et al. (2017). Primary surface rupture of the 1950 Tibet-Assam great earthquake along the eastern Himalayan front India. *Science Report, 7*, 5433. https://doi.org/10.1038/s41598-017-05644-y
Rajendran, C. P., Rajendran, K., Duarah, B. P. et al. (2006). Reply to comment by R. Bilham on "Interpreting the style of faulting and paleoseismicity associated with the 1897 Shillong, northeast India, earthquake: Implications for regional tectonism". *Tectonics: 25*, TC2002. https://doi.org/10.1029/2005TC001902
Rajendran, C. P., Rajendran, K., Duarah, B. P. et al. (2004). Interpreting the style of faulting and paleoseismicity associated with the 1897 Shillong, northeast India, earthquake: Implications for regional tectonism. *Tectonics 23*, TC 4009. https://doi.org/10.1029/2003/TC001605
Rajendran, C. P., & Rajendran, K. (2005). The status of central seismic gap: A perspective based on the spatial and temporal aspects of the large Himalayan earthquakes. *Tectonophysics, 395*, 19–39.
Rajendran, K., & Rajendran, C. P. (2011). Revisiting some significant earthquake sources in the Himalaya: Perspectives on past seismicity. *Tectonophysics, 504*, 75–88.
Rajendran, C. P., Rajendran, K., Sanwal, J., & Sandiford, M. (2013). Archaeological and historical database on the medieval earthquakes of the central Himalaya: Ambiguities and inferences. *Seismological Research Letters, 87*, 1098–1108.
Rajendran, C. P., & Rajendran, K. (2021). Paleoseismic context of the 1950 earthquake: Implications for seismic gaps in the Eastern Himalaya. *Physics and Chemistry of the Earth, Parts A/B/C*.
Rajendran, C. P., Sanwal, J., Morell, K., et al. (2016). Stalagmite growth perturbations from the Kumaun Himalaya as potential earthquake recorders. *Journal of Seismology, 20*, 579–594.
Rajendran, C. P., John, B., & Rajendran, K. (2015). Medieval pulse of great earthquakes in the central Himalaya: Viewing past activities on the frontal thrust. *Journal of Geophysical Research, 120*, 1623–1641. https://doi.org/10.1002/2014JB011015

Rajendran, C. P., & Rajendran, K. (2022). *Earthquakes of the Indian Subcontinent: Seismotectonic perspectives* (pp. 253). Springer.

Rajendran, C. P., Singh, T., Mukul, M., Thakkar, M., Kothyari, G., John, B., & Rajendran, K. (2020). Paleoseismological studies in India (2016–2020): Status and prospects. *Proceedings of the Indian National Science Academy 86*(1), 585–607. https://doi.org/10.16943/ptinsa/2020/49787

Rana, B. S. (1935). Nepal Ko Maha Bhukampa (The Great Earthquake of Nepal). Jorganesh Press, Kathmandu (in Nepali), 250 pp.

Rana, B. S. (2013). *The great earthquake in Nepal* (First Published as Nepal Ko Maha Bhukampa in Nepali in 1934; Translated to English from the 2nd Nepali Edition (1935) by Kesar Lall) (pp, 136). Ratna Pustak Bhandar, Kathmandu, Nepal.

Rastogi, B. K., Chadha, R. K., & Rajagopalan, G. (1993). Palaeoseismicity Studies in Meghalaya. *Current Science, 64*, 933–935.

Reddy, D. V., Nagabhushanam, P., & Kumar, D. (2009). The great 1950 Assam earthquake revisited: Field evidence of liquefaction and search for paleoseismic events. *Tectonophysics, 474*, 463–472.

Richter, C. F. (1958). *Elementary seismology*. W.H. Freeman Company (p. 768). (Reprinted: S. Chand and Company, New Delhi).

Rizza, M., Bollinger, L., Sapkota, S. N., et al. (2019). Post-earthquake aggradation processes to hide surface ruptures in thrust systems: The M8.3, 1934, Bihar-Nepal earthquake ruptures at Charnath Khola (Eastern Nepal). *Journal of Geophysical Research, 124*, 9182–9207.

Sahoo, S., & Malik, J. N. (2017). Active fault topography along Kangra Valley fault in the epicentral zone of 1905 Mw 7.8 earthquake NW Himalaya India. *Quaternary International, 462*, 90–108.

Sapkota, S., Bollinger, L., Klinger, Y., et al. (2013). Primary surface ruptures of the great Himalayan earthquakes in 1934 and 1255. *Nature Geoscience, 6*, 71–76.

Seeber, L., Armbruster, J. (1981). Great detachment earthquakes along the Himalayan arc and long-term forecasting. In D. W. Simpson, & Richards) (Eds.) *Earthquake prediction-an international review* (Vol. 4, pp. 259–277). Maurice Ewing Series Am Geophys Union.

Singh, D. D., & Gupta, H. K. (1980). Source dynamics of two great earthquakes of the Indian subcontinent; The Bihar-Nepal earthquake of January 15, 1934 and the Quetta earthquake of May 30, 1935. *Bulletin of the Seismological Society of America, 70*, 757–773.

Singh, I., Pandey, A., & Mishra, R. L. (2021). Evidence of the 1950 great Assam earthquake surface break along the Mishmi thrust at Namche Barwa Himalayan syntaxis. *Geophysical Research Letters 48* e2020GL090893. https://doi.org/10.1029/2020GL090893

Srinivasan, V. (2003). Deciphering differential uplift in Shillong Plateau using remote sensing. *Journal of the Geological Society of India, 612*, 773–777.

Stevens, V. L., & Avouac, J.-P. (2015). Interseismic coupling on the main Himalayan thrust. *Geophysics Research Letter, 42*, 5828–5837. https://doi.org/10.1002/2015GL06484

Stevens, V., & Avouac, J.-P. (2021). On the relationship between strain rate and seismicity in the India-Asia collision zone: Implications for probabilistic seismic hazard. *Geophysical Journal International, 226*, 220–245.

Sukhija, B. S., Rao, M. N., Reddy, D. V., et al. (1999a). Paleoliquefaction evidence and periodicity of large prehistoric earthquakes in Shillong Plateau. *Indian Earth Planet Science Letter, 167*, 269–282.

Sukhija, B. S., Rao, M. N., Reddy, D. V., et al. (1999b). Timing and return period of major paleoseismic events in the Shillong Plateau, India. *Tectonophysics, 308*, 53–65.

Sylman, Y., Shimshonj, M., & Ben-Menahem, A. (1974). Theoretical amplitudes of body waves from a dislocation source in the earth, III. P, S, and surface reflections. *Physics of the Earth and Planetary Interiors, 8*, 130–147.

Szeliga, W., & Bilham, R. (2017). New constraints on the mechanism and rupture area for the 1905 Mw 7.8 Kangra earthquake, northwest Himalaya. *Bulletin of the Seismological Society of America, 107*, 2467–2479. https://doi.org/10.1785/0120160267

Tandon, A. N. (1954). Study of the great Assam earthquake of August 1950 and its aftershocks, Indian. *Journal of Meteorology & Geophysics, 5*, 95–137.

Thakur, V. C., Joshi, M., Sahoo, D., et al. (2014). Partitioning of convergence in Northwest Sub-Himalaya: Estimation of late quaternary uplift and convergence rates across the Kangra reentrant, North India. *International Journal of Earth Science, 103*, 1037–1056.

Tillotson, E. (1951). The great Assam earthquake of August 15, 1950. *Nature, 167*, 128–130.

Wallace, K., Bilham, R., Blume, F. et al. (2005). Surface deformation in the region of the 1905 Kangra Mw = 7.8 earthquake in the period 1846–2001. *Geophysical Research Letters 321*. https://doi.org/10.1029/2005GL022906

Wesnousky, S., Kumahara, Y., Chamlagain, D., et al. (2019). Large Himalayan Frontal Thrust Paleoearthquake at Khayarmara in Eastern Nepal. *Journal of Asian Earth Sciences, 174*, 346–351. https://doi.org/10.1016/j.jseaes.2019.01.008

Wesnousky, S. G., Kumahara, Y., Chamlagain, D., et al. (2017a). Geological observations on large earthquakes along the Himalayan frontal fault near Kathmandu, Nepal. *Earth and Planetary Science Letters, 457*, 366–375.

Wesnousky, S. G., Kumahara, Y., Chamlagain, D., et al. (2017b). Large paleoearthquake timing and displacement near Damak in eastern Nepal on the Himalayan frontal thrust. *Geophysical Research Letters, 44*, 8219–8226. https://doi.org/10.1002/2017GL074270

Yeats, R. S., & Thakur, V. C. (2008). Active faulting south of the Himalayan front: Establishing a new plate boundary. *Tectonophysics, 453*, 63–73.

Yeats, R. S., & Lillie, R. J. (1991). Contemporary tectonics of the Himalayan frontal fault system. Folds, blind thrusts and the 1905 Kangra earthquake. *Journal of Structural Geology, 13*, 227–233.

Zhao, W., & Nelson, K. D. et al. (1993). Deep seismic reflection evidence for continental underthrusting beneath southern Tibet. *Nature, 366*, 557–559.

Zhao, Y., Grujic, D., Baruah, S., et al. (2021). Paleoseismological findings at a new trench indicate the 1714 M8.1 earthquake ruptured the main frontal thrust over all the Bhutan Himalaya. *Frontiers in Earth Science*. https://doi.org/10.3389/feart.2021.689457

Chapter 9
Landslides in Himalaya

Khayingshing Luirei, Shubham K. Rajewar, Manish Mehta, Neeraj Gurjar, Anand Kumar Gupta, and V. K. Gahalaut

Abstract In the rugged terrains of geodynamically evolving Himalaya, the problems posed by landslides are acute. The present study aimed to integrate previous work and current findings to characterize the zone or area affected by landslides and its associated threats. The Main Central Thrust and Main Boundary Thrust zones define the zones severely affected by landslides. The thrust zones represent deformation planes along which rocks have moved repeatedly, causing the rock to be sheared, shattered and split by thrust into a multitude of rock-slabs. The bedrock condition with steep slopes is the most important factor, with triggering factors being rainfall, seismicity, tectonic movement and undermining of the slope due to rock cuts and toe erosion by rivers. Furthermore, numerous Himalayan townships in the MCT zones, such as Joshimath, Bhatwari, Dar, Loharkhet, etc. located over paleolandslide deposits, are facing a severe land subsidence problem. The Joshimath landslide is the most significant amongst them due to the crisis in December 2022–January 2023, when numerous cracks and fissures appeared on the ground and buildings. The GNSS data of Joshimath for the period between November 2022 and January 2023 suggested varying rate; till mid-November 2022, the site exhibited northward motion of ~ 1.5 mm/day with subsidence rate of ~ 0.8 mm/day while during January 2023 more intense movement of about ~ 15 mm/day towards north, with subsidence rate of ~ 11 mm/day occurred. The thrust zones are also susceptible to landslide damming and have preserved the historical records of several landslides that have dammed lakes/paleolakes with their cascading effects downstream after breaching.

K. Luirei · M. Mehta · A. K. Gupta · V. K. Gahalaut (✉)
Wadia Institute of Himalayan Geology, Dehradun, India
e-mail: vkgahalaut@yahoo.com

K. Luirei · S. K. Rajewar · M. Mehta · A. K. Gupta · V. K. Gahalaut
Academy of Scientific and Innovative Research (AcSIR), Ghaziabad, India

S. K. Rajewar · V. K. Gahalaut
CSIR—National Geophysical Research Institute, Hyderabad, India

N. Gurjar
Geodetic and Research Branch, Survey of India, Dehradun, India

V. M. Tiwari (ed.), *Himalayan Natural Hazards*, Indian National Science Academy,
https://doi.org/10.1007/978-981-92-3784-5_9

In many instances, these valley-blocking landslides led to the formation of epigenetic valleys and gorges. Considering the geomorphic changes brought by the landslides and incurred socio-economic losses, the present study documents key research works, including landslide susceptibility mapping and landslide risk assessment. It further highlights the need for more studies assessing the risk posed by landslides to support cost-effective resource allocation and strengthen disaster risk reduction and management strategies.

Keywords Landslides · Main central thrust · Main boundary thrust · Subsidence · GNSS · Landslide-dammed lakes

9.1 Introduction

The ongoing tectonic collision between the Indian and Eurasian (Tibetan) Plates (Dewey & Burke, 1973; Gansser, 1964; Heim & Gansser, 1939; LeFort, 1975; Molnar, 1984; Thakur et al., 2014; Yin, 2006) has given rise to several active thrusts and faults, including the Main Central Thrust (MCT), Main Boundary Thrust (MBT) and the Himalayan Frontal Thrust (HFT) (Fig. 9.1). These thrusts are splays to Main Himalayan Thrust (MHT) which is seismically active beneath the frontal Himalaya and hosts moderate to large-magnitude earthquakes (Seeber & Armbruster, 1981; Thakur, 2004; Valdiya, 1992). As a result of the collision, the rocks are folded, fractured and fragmented due to repeated movement and splitting along the fault planes (Valdiya, 1987), thereby making the MCT and MBT zones extremely vulnerable to landslides. The MCT does not only serve as lithological and structural boundaries but also acts as an orographic barrier to moisture-laden winds of the Indian summer monsoon (ISM) and western disturbance (WD), leading to enhanced rainfall in the windward side of the elevated terrain (Gadgil et al., 2007; Hunt et al., 2018; Sundriyal et al., 2024) and making the region more vulnerable to landslides. Each year, incessant rainfall and cloudburst cases are reported from various parts of the Himalaya. In addition to this, the Himalaya host various highly landslide susceptible rocks such as phyllite, shale, schist, poorly cemented sandstone, gneiss, quartzite, etc. (Gerrard, 1994; Luirei et al., 2023). Hence, the topography and orography process, geology and earthquake events in the Indian Himalaya facilitate a favourable condition for fatal landslide occurrences, and, consequently, landslides are widespread and recurrent across the region, making them a dominant natural hazard (Hukku & Jaitely, 1964; Ives & Messerli, 1981; Middlemiss, 1890; Oldham, 1880; Valdiya, 1985). A recent study by Froude & Petley, (2018) provides a global fatal landslide database, which suggests that between the years 2004 and 2016, a total of 4862 fatal landslides resulted in 55,997 deaths, with the highest (~ 75%) recorded for the Asian continent, and the Himalaya host the highest number of landslide events in the world. Besides the natural factors, anthropogenic pressures have significantly compounded the natural stressors on the Himalayan environment (Valdiya, 1998). Rapid population growth, deforestation, unsustainable agricultural practices and unplanned infrastructural development,

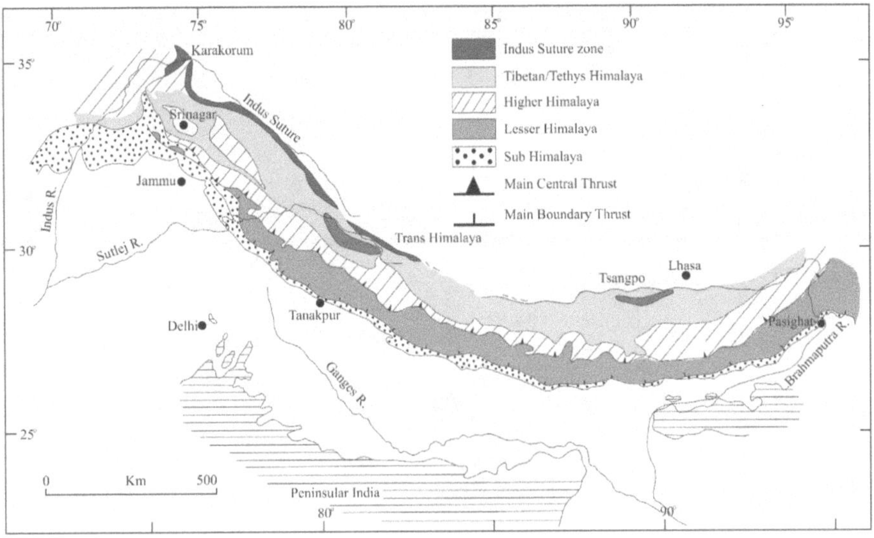

Fig. 9.1 Regional geology and major structure of the Himalaya (Gansser, 1964)

such as road construction, hydroelectric projects and urban expansion, have led to habitat fragmentation, soil erosion and biodiversity loss (Barnard et al., 2001; Dimri et al., 2021; Haigh, 1984; Haigh & Rawat, 2011; Kashyap et al., 2021; Lennartz, 2013; Negi et al., 2013; Petley et al., 2007; Shrestha et al., 2025; Sur et al., 2020; Valdiya, 1987). These human-induced activities not only amplify the intensity of natural hazards but also compromise the region's ecological resilience and enhance the risk associated with such hazards multifold (Highland & Bobrowsky, 2008). The problem posed by landslide occurrences is very severe and acute in the Himalaya, and as such, a holistic study of landslides by integrating geological, geomorphological, meteorological and anthropogenic factors is crucial for better planning and risk reduction, as well as from socio-economic and environmental perspectives. The present study documents several crucial research works, including landslide susceptibility mapping, landslide risk assessment and highlights the need for more studies assessing the risk posed by landslides in order to prioritize cost-effective resource allocation for better disaster risk reduction and management.

9.2 Landslides in the Indian Himalaya

The Indian Himalaya encompasses approximately 76% (~ 0.32 million km^2) of the India's total landslide-prone area (~ 0.42 million km^2) (Geological Survey of India) (Fig. 9.2). It suffers economic loss of nearly USD 1 billion and records more than 200 casualties each year (Naithani, 1999). In addition, landslide-dammed lake-related

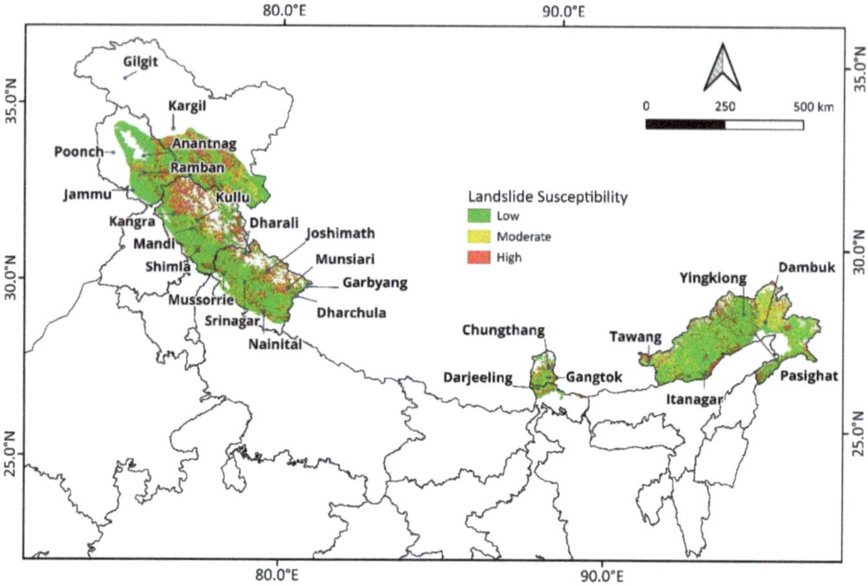

Fig. 9.2 Landslide susceptibility map of the Indian Himalaya (GSI, 2025a, b)

hazards are also quite common in the river valleys of the Himalaya (Joshi & Subramanian, 2025). In recent decades, with the growing frequency of meteorological extremes and enhanced anthropogenic disturbances, landslide cases in the Himalaya have increased multifold (Kaushal et al., 2025). Most of the landslides occur along the valleys; therefore, the landslides have been described on the basis of river valleys with a case study. Notable landslides occurring in river valleys in the MCT and MBT zones along with other significant landslide-prone and affected areas are described.

9.2.1 Landslides Along the River Valleys

Yamuna River Valley

In the Unata Gad Valley, a tributary of the Yamuna River, a massive landslide has been taking place since earlier than 1960 as evident from the topographical map. The landslide is located on northwest-facing slope and has affected an area of about 1.73 km^2 (Fig. 9.3). The landslide-affected slope forms part of the footwall block of the MCT. The bedrock is steeply dipping towards the northwest and is highly fractured, sheared and weathered, with little to no colluvial or debris cover (Fig. 9.3a–c). The landslides in this catchment are primarily governed by three factors, viz., highly sheared and weathered lithology, freeze and thaw and steep slopes. Kharsali is nestled over landslide debris, brought down and deposited by Unata Gad, and the sediments are about 60 m thick and rest over schistose rock (Fig. 9.3b–f). Jamir et al., (2017) have carried out potential slope instability modelling of Kharsali village using the finite element method and found that the slope section near the confluence of the Yamuna River and Unata Gad is the most vulnerable. Taking into consideration the thickness of the landslide debris, it is inferred that in one or more of the landslide events, the Yamuna River must have been blocked. The terminal part of the landslide debris fan is observed near Bapas, almost 2.3 km downstream of the confluence between the Yamuna River and Unata Gad, where the sediment thickness is about 40 m, and the strata become almost horizontal.

At Phoolchatti, a stretch of about 350 m long section of the road is being affected by subsidence (Fig. 9.3g). Both the bedrock made up of mica schists, and the overlying colluvial deposits have been affected by a slow-moving landslide with an apparent curve failure plane. The rockfall of Wazir Garh took place in the hanging wall block of the Munsiari thrust, made up of gneissic rock (Fig. 9.4a–d), and it resulted in changing the course of a high-gradient seasonal small stream. This landslide is the reactivation of an old landslide (Khanduri 2018) (Fig. 9.4a, b), and the bedrock exposed along the road cut has the attitude of 30° towards 70° and is traversed by two sets of joints, one set dips SW and the other trends parallel to the foliation but

Fig. 9.3 **a–c**. Google Earth images of Kharshali–Diph–Janki Chhati area, Yamuna River Valley, showing active landslide and landslide-debris fan. **d–f**. Kharshali settlement is on the top of a thick landslide-debris fan, which extends ~ 3 km downstream. The distal part of the fan shows grading of the sediments. **g**. Portion of the road section affected by landslide at Phool Chhati

Fig. 9.4 **a–d**. Google Earth images and field photographs of Wazirgarh in the Yamuna Valley indicating pre- and post-events. The rockfall took place in the hanging wall of the Munsiari Thrust, highly fractured gneissic rock constitutes the slope. **d–e**. Catchment area of Kanodia Gad, a tributary of Bhagirathi River experienced huge landslide in 1978 that blocked the Bhagirathi Valley, the landslide is still active as evident from Google Earth image and field observations. **f**. Bhatwari–Charethi area in Bhagirathi River Valley is prone to landslide as indicated by number of old landslide scars and landforms. **g–h**. Landslide at Pilang along Pilang Gad a tributary of Bhagirathi River, Google Earth Image showing the pre and post-events images

with a steeper dip (Fig. 9.4d). The SW dipping joints (48° towards 257°) are very important discontinuity planes as they are oriented towards the free face of the slope.

Bhagirathi River Valley

In the Bhagirathi Valley, the MCT zone is represented by a ~ 20 km wide zone with Ramgarh Thrust as the sole thrust and the Vaikrita Thrust as the roof thrust. Between these thrusts, a number of landslides are marked both recent and ancient and represent one of the valleys severely affected zones by landslides. Bhambri et al., (2017) mapped 2772 landslides, of which 1434 were new landslides, 533 landslides were reactivated and 451 old landslides; the maximum number of landslides occurred in the areas around Bhatwari, Pilang and Dharali during the extreme hydro-meteorological event of June-2013. The 1978 landslide of Kanodia Gad and the Varunawat Hill landslide of 2003 are the two most documented landslides. The landslide of Kanodia Gad was one of the largest landslides to have occurred in the twentieth century in the central Himalaya (Agarwal & Narain, 1991) (Fig. 9.4e, f). It occurred on 6th August 1978, and about 400 ha of land was destroyed. The debris generated blocked the Bhagirathi River for about 14 h, forming a lake which was 35 m high, 45 m wide and about three km long. The landslide lake was breached on 11th August 1978, causing immense destruction downstream. The 1991 Uttarkashi earthquake triggered many landslides around Uttarkashi, and some of the landslides triggered by the earthquake are still active (Gupta et al., 2025).

Bhatwari area is extremely vulnerable, as it is marked by both active and inactive landslides (Fig. 9.4g). Yadav et al. (2020) carried out geodetic monitoring of landslide movement at Raithal and Bhatwari, exhibiting a creep motion of the slope towards east since 2006 at a rate of 12 mm/year, which increased to 22 mm/year in 2016, with subsidence at a rate of ~ 6 mm/year during 2012–2016. Preceding studies and present observations suggest that the landslides at Bhatwari are recurrent. The east-facing slopes are more prone to landslides, and the area around Bhatwari–Charethi–Raithal is speckled by scars of old and active landslides. The landslide at Pilang is an ancient landslide that is still active, and the affected area measures about 0.32 km^2, and the landslide reactivated in the year 2013 (Fig. 9.4h, i) and affected both sides of the slopes of the Pilang Gad.

Alaknanda River Valley

In the Alaknanda River Valley, the valley section between Pipalkoti and Joshimath is marked by numerous landslides, both active and dormant. The Patalganga River watershed, one of the sub-basins in the Alaknanda Valley, represents a severely affected region by landslides. In the watershed, the bedrock is highly sheared, and as a result, numerous landslides occur mainly in monsoon season, and some of the landslides predate 1985, as evidenced by historical satellite imagery available on the Google Earth image. The landslide-affected bare slopes are extremely prone to severe erosion due to the highly crushed nature of the bedrock (Fig. 9.5a–c). On the opposite slope of Langsi, at Kimana, several landslides occur in the highly weathered and sheared bedrock of mica schist, where a section of slope is subsiding each monsoon. A few km away from Kimana, a portion of the slope is affected by landslides, which have been aggravated due to the construction of the road (Fig. 9.5d, e). The slope is made up of alternation of highly weathered schist and talc (Fig. 9.5d, e). The schistosity planes, along with the flakes of talc, are acting as the gliding planes for the slope movement. At Syun, the footwall block of the Munsiari Thrust is made up of moderately north-dipping quartzite. The bedrock is traversed by prominent, steeply south-dipping fractures. Further, the Maina Gad has steeply incised into the bedrock, resulting in steep to vertical slopes on its left valley with a number of waterfalls. The right bank is much gentler, but with the downcutting by Maina Gad, the base support has been removed, resulting in a number of landslides. The thin phyllitic layers separating the quartzite beds are acting as the gliding planes for the rock slides, while the undermined bedrock is falling due to wedges formed by intersecting discontinuity planes (Fig. 9.5f, g). Gohna Lake originated due to a landslide that resulted in the blocking of Birahi Ganga, a tributary of the Alaknanda River, in 1893. The lake was breached in 1894 and 1970. The 1970 event was caused by a cloudburst between Joshimath and Chamoli (Holland, 1984; Rana et al., 2013;

Fig. 9.5 **a–c**. Patalganga Valley, a sub-basin of the Alakhnanda River, is severely affected by landslides as imaged from Google Earth image. The bedrock is highly sheared due to tectonic activity and freeze–thaw action. **d–e**. Photographs showing hanging wall bedrock of the MCT comprising of highly sheared and weathered schist and alternation of talc and schist are very susceptible to landslides. **f–g**. The footwall block of the MCT is made up of thickly bedded quartzite cut across by steeply dipping joints, forming wedges resulting in wedge failure. The intervening thin phyllitic planes in the quartzite have initiated landslides

Weidinger, 1998). The event of 1970 caused unprecedented damage downstream (Agarwal & Narain, 1991; Rana et al., 2013). The landslides in the Gohna area are still active, affecting both the slopes of the Birahi Ganga.

Kali River Valley

In the northeastern part of the Uttarakhand Himalaya, both the Kali River and its tributary the Dhauli Ganga river Valley are severely affected by landslides. Along the Dhauli Ganga River Valley, a valley section between Tawaghat and Dar, stretching about 15 km in length, is affected by several old and active landslides (Fig. 9.6a–e). Some of the notable landslides along this stretch are the September 1977 landslide event near Sisna in the upslope of Tawaghat, the stray rockfalls at Tawaghat in August 1982 and the slide and rockfalls of Sobla in July 1993. At Tijam, massive paleolandslides are evident from the north-facing as well as the south-facing slopes, and one of these landslide events resulted in the damming of the Niangal Gad, a tributary of the Dhauli Ganga River (Luirei et al., 2006) (Fig. 9.6d). Tijam and Dar lie in the Vaikrita Thrust zone and also fall in the highly landslide susceptible zone (Luirei et al., 2006; Ram et al., 2024; Valdiya, 1985). Dar is nestled over a huge landslide debris, and the slope deformation predates 1998 and remains active. The old landslide landform measures about 2.37 km^2, and debris is deposited over highly sheared bedrock (Fig. 9.6a–c). The main cause of slope instability is the seepage of spring water through the loose debris, and the contact between the bedrock and the overlying debris is behaving as the slip plane. The head scarp developed between 2004 and 2012 as imaged from the Google Earth image. The scarp measured 67 m in length in 2012 and has grown up to 1158 m as of January 2024. Khet Village, located downstream of Dar, rests over more than 50 m thick landslide debris cover, comprising mainly of gneissic rock clasts. This landslide reactivates every monsoon. Futher, the valley section between Tawaghat and Chhirkhila also constitutes one of the areas extremely prone to landslides, the gneissic bedrock with foliation trending almost east–west. Down cutting by the Dhauli Ganga River has played a major role in the evolution of the valley by gradual erosion of the base of the slope, thereby initiating landslides from time to time. The debris over the slope contains rock boulders, some measuring more than 10 m in diameter.

In the Kali River Valley, the Garbyang paleolake resulted from the blocking of the Kali River at Chhialekh by moraine and landslide from the Api and Nampa glaciers in Nepal (Gansser, 1964; Heim & Gansser, 1939) in the Trans Himadri Fault zone. Juyal et al., (2004) have dated the varve sediments to be of 20 ka, which gives the age of the impounded sediments, suggesting that the damming event of the Kali River predated the oldest obtained sediment age. Malpa Village, situated along the Mansarovar pilgrimage track along this valley, has been experiencing landslides from time to time (Fig. 9.6f). The landslide of 18th August 1998 occurred in the crystalline rocks of the Vaikrita Group (Khanduri et al., 2018; Pant & Luirei, 1999; Paul et al., 2000). The rockfalls are predominantly controlled by joint sets 70° towards NW, 80° towards SE and 60° towards NNE with foliation plane dipping 35° towards NE, and the rocks are polyphasedly deformed (Pant& Luirei, 1999). On 14th August 2017, extreme precipitation in the catchment of Simkhola and Malpa Gad, small tributaries of the Kali River, led to flash flood in the Mangti and Malpa areas and triggered numerous small landslides. (Khanduri et al., 2018).

9.2.2 Landslides in the MBT Zone

The MBT zone of Uttarakhand and Himachal is prone to landslides (Valdiya, 1987, 1992). Google Earth Image demonstrates that the landslide activity in Amiya in the Gaula River Valley near Nainital predates 1985. The landslide occurs at the top of the footwall block of the MBT made up of sandstone and shales that dips 40°–48° towards NE and is traversed by five sets of intersecting joints (Pant & Luirei, 2005). This landslide has modified the slope and valley morphology, bringing significant changes in the land use and landform pattern (Bartarya & Valdiya, 1989; Pant &

Fig. 9.6 a–c. Vaikrita Thrust zone in the Dhauliganga River Valley is severely affected by slope deformation as evident from Dar and Tijam villages. Google Earth images show the precariously parched Dar village on south dipping slope with recent development of scars upslope of the village. **d**. Tijam village is settled on top of a landslide debris fan and paleolake. **e**. Field photograph showing reactivation of old landslide debris at Khet, Khet settlement is on the debris of an old landslide. **f**. Field photographs showing debris-filled narrow valley of Kali River at Malpa during the event of 1998

Luirei, 2005; Singh & Sarkar, 2005). During the landslide event of 1999, the Gaula River was temporarily blocked, and the water spilled into the paddy field at Amiya (Fig. 9.7a). The image also suggests modification of the slope to a larger extent as evident from the huge paleolandslide scar. In the MBT zone of the Dabka–Kosi River Valleys, landslides are mainly taking place in the hanging wall of the MBT (Fig. 9.7b, c). In Dehradun, the hanging wall of the MBT is characterized by severe erosion and landslides. The MBT zone is defined by highly sheared bedrock that easily gives way to gravitational slope deformation and erosion during the monsoon (Fig. 9.7d, e). Highly sheared and fractured bedrock is overlain by a thick deposit of fan constitutes the footwall block. The sediments are being supplied by landslides and severe erosion in the hinterland. A number of characteristic geomorphic features pertaining to landslides are observable in the region, an example of the present day being the landslide at Koti Village in the immediate hanging wall of the MBT, generating an enormous amount of debris material during the rainy season.

In the southeastern part of Western Himalaya in the Giri River Valley, the MBT, MBF and other splay thrusts are squeezed within a width of about 1 km. Here, a number of active fault traces are aligned parallel with the MBT and MBF and also cut across it. This zone is experiencing slope deformation by active landslides and faulting (Datta et al., 2024; Negi et al., 2021, 2023; Rautela & Lakhera, 2000). Four active landslides and a number of inactive landslides have been identified through detailed investigations in the MBT zone. The hanging wall block of the MBT is made up of thinly bedded and highly fractured slate and siltstone of Mandhali Formation; it dips moderately towards NE and is traversed by two important joint sets that dip S and E–SE. Near Malgi, a chronic landslide that reactivates almost every monsoon is responsible for frequent ground subsidence of a stretch of the road. The MBT (~ Murree Thrust) zone of the Chenab River Valley in the Western Himalaya is experiencing landslides from time to time, particularly during the monsoon. In the Ramban area, the Panjal Thrust and the MBT are closely spaced, and these thrust zones are severely affected by landslides (Aziz et al., 2023; Bahuguna & Khan, 2022; Khan et al., 2022; Mir et al., 2025). The Pernote landslide of 2024 occurred at the top of the footwall of the MBT and caused extensive damage, and many residents were displaced. The landslide was a complex debris slide and flow, involving mud, silt, clay and rock fragments. The factors responsible for the landslide include structural and lithological conditions (Mir et al., 2025). The slope is made up of alternation of mudstone, siltstone and sandstone. These intervening layers of mudstone between the lithologies make the slope prone to landslides. The prolonged rainfall was the triggering factor for the landslide, as the area received continuous rain before the landslide.

Arunachal Himalaya

Landslide studies carried out in the Arunachal Himalaya are still very sparse and poorly documented, except for a few studies (Chalpathi Rao, 1983; Luirei, 2007; Luirei & Bhakuni, 2008; Maurya et al., 2025; Parkash et al., 2002; Rawat & Joshi,

Fig. 9.7 **a**. Field photographs and Google Earth image illustrating the development of epigenetic river valley that resulted from the obstruction of Gaula River in the MBT zone due to landslides. Gaula River was dammed during the landslide event of 1999 (Inset photographs after Pant & Luirei, 2005). **b**. Huge thickness of landslide debris fan in the immediate footwall of the MBT in the Kosi River Valley at Ukhaldhunga. **c**. Highly sheared bedrock of Siwalik at the footwall of MBT in Kotadun, affected by landslide and erosion. **d–e**. The MBT zone of Dehradun is defined by a thick zone of fault gouge that is vulnerable to severe erosion and landslide. Huge amount of sediments is generated by landslides and erosion during the monsoon

2012; Rehman & Azhoni, 2023; Saha et al., 2023a, b; Singh & Devi, 2006). In the sub-Himalaya, not only are the geology and topography the main causative factors, but jhum cultivation is one of the contributing factors (Rawat & Joshi, 2012). Massive landslides occur near the exit of Dibang River into the alluvia of Brahmaputra River, generating an enormous amount of sediments. The quartzites thrust over the alluvium along the MBT and the hanging wall rocks are highly sheared and fractured with closely spaced intersecting joints. The discontinuity planes and steep slopes make the slope extremely susceptible to landslides and erosion during the monsoon. The hanging wall rocks of Mishmi Thrust are also highly sheared and crushed, causing many landslides (Fig. 9.8a–c). The footwall blocks of Mishmi Thrust and MBT are characterized by huge aprons of gigantic landslide debris fans. In the Itanagar–Banderdewa area, the lithology is the main factor, while the triggering factor is the rainfall (Parkash et al., 2002; Singh & Devi, 2006). The MBT zone between Pasighat and Rengging in the Siang River Valley is affected by landslides of anthropogenic and natural genesis. The Sillekorong Thrust, the MBT and the Sibo Korong Thrust are squeezed within a width of 1.5 km, where the rocks are highly sheared and crushed (Fig. 9.8d–g). The footwall blocks of the Sillekorong Thrust are made up of highly sheared carbonaceous shale of the Gondwana Group. A landslide that predates 1985 is still active and advancing (Luirei & Bhakuni, 2008). The continued slope deformation has exposed the MBT and the Sillekorong Thrust. A steeply NE-dipping Ranaghat fault is also one of the contributing factors for the instability of the slope. At Yagrung, situated west of Pasighat, a massive landslide occurs in the footwall block of the MBT predating 1985 and is still active (Luirei & Bhakuni, 2008) (Fig. 9.8h, i). The landslide is taking place in Upper Siwalik bedrock, generating a huge amount of debris each monsoon, destroying paddy fields downstream, and the landslide is advancing upslope.

9.3 Case Study of Joshimath Slide

In the past few years, the Joshimath slide has drawn the attention of several geoscientists because of the unique slow sliding, causing damage and threatening the population in a large region. Hence, in this paper, we have briefly discussed this particular case study. Joshimath town is nestled in the north-dipping slope made up of old landslide debris or glacial deposited material, and the area has been witnessing gradual sinking for a long time (Bisht & Rautela, 2010; Mishra Committee 1976). The slope deformation attained an alarming rate in December 2022, where numerous cracks appeared in the houses and slope (Fig. 9.9) and also oozing out of underground water along the fissures and cracks. Studies suggest that the construction of an underground tunnel disrupted the water-bearing horizon that supplies potable springs for many villages (Brandl et al., 2010; Millen et al., 2015; Nawani, 2015).

The first report of Interferometric Synthetic Aperture Radar (InSAR) analysis provided evidence of rapid sliding (https://blogs.agu.org/landslideblog/2023/01/18/joshimath-new-insar/ and https://blogs.agu.org/landslideblog/2023/01/23/joshimath-5/) of up to 50 cm since January 2018. Gahalaut et al., (2023) analysed Joshimath

Fig. 9.8 **a–c**. Arcuate mountain of the Arunachal Himalaya is affected by chronic landslides, generating an enormous amount of sediments forming gigantic debris fans. The highly crushed and fractured with steep slopes give way to slope movement during the onslaught of heavy rain in the months of May–September. **d–e**. The highly sheared bedrock in the thrust zones of Sillekorong Thrust, MBT and the Mingo Thrust gives way easily to toe erosion and landslides. **f–g**. The footwall block of the MBT, west of Pasighat, has been facing the problem of landslides since 2005. **h–i**. The channel of a small stream is strewn with rock fragments generated by a landslide in the upslope (h–upstream, i–downstream)

slide using the Global Navigation Satellite System (GNSS) site JOSH of SOI for a period of December 2021 to April 2023 and reported that till mid-November 2022, site exhibited towards northward motion of ~ 1.5 mm/day with subsidence rate of ~ 0.8 mm/day. Till end of December 2022, intense movement of ~ 7.4 mm/day with subsidence of ~ 4 mm/day occurred. Between January 1 and 4, 2023, more intense movement of about ~ 15 mm/day towards north, with subsidence rate of ~ 11 mm/day, was recorded. Based on the ratio of horizontal and vertical movement, they suggested a dip of the slip surface of about ~ 30°, which is quite reasonable considering the slope of the hillock itself is ~ 22° Also, they analysed the behaviour of this movement, whether it is stick–slip or viscous, by kinematic GNSS data processing at every 30 s with a focus on two episodes, i.e. 10–22 November 2022 and 25 December 2022 to 5 January 2023, and suggested the motion was smooth like viscous creep. Slope deformation investigations of Joshimath were also carried out by various workers through InSAR and GNSS revealing the kinematics, rate of deformations, zones of deformations and depth of the landslides (Dalal et al., 2025; Rather & Bukhari, 2024; Shankar et al., 2023; Sreejith et al., 2024; Wang et al., 2024). We have continued

Fig. 9.9 a–b. Photographs showing the crack/fissure near Manoharbhag, Joshimath have widened horizontally 63 cm at a rate of 5.2 cm/day between 31 December 2022 and 12 January 2023. Photographs also show that the land vertically shifted 46 cm in the same period

monitoring the GNSS site at Joshimath town since December 2021, the installation of the site, to June 2025 (Fig. 9.10a). After the intense sliding period of January 2023, the site still shows high speed with dominating northward motion (~ 154 mm/year). It also exhibits seasonal changes due to monsoon (July–October) which increases the sliding rate, as clearly seen in 2023, 2024 in the north and vertical components (Fig. 9.10b).

9.4 Landslide Susceptibility Mapping (LSM) and Landslide Risk Assessment

In the Himalaya, landslide susceptibility zoning was first carried out in Central Nepal along the Kathmandu–Kakani belt and developed prototype maps of landslide hazard (Ives & Messerli, 1981, 1984; Kienholz et al., 1984; Johnson et al., 1982). Valdiya (1985) attempted to identify broad belts in the Kumaun Himalaya vulnerable to mass movement and mapped the MCT and MBT zones as the zones of moderate to severe

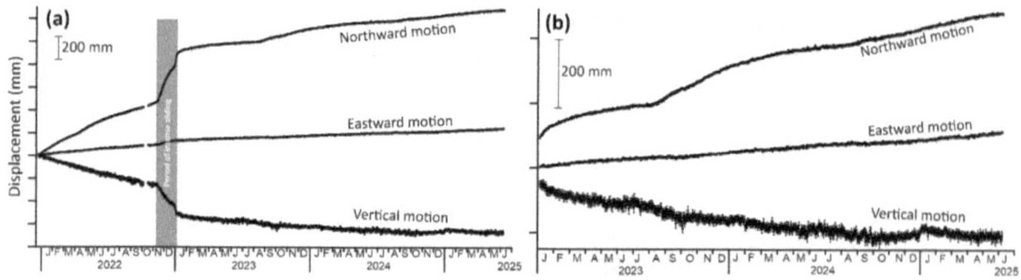

Fig. 9.10 Temporal variation of coordinate time-series in the three components of the GNSS site at JOSH (Joshimath town). **a**. Complete time series from December 2021 to June 2025. Grey shade highlights the period of intense sliding. **b**. A zoomed version of the period after the intense sliding to highlight the effect of the monsoon. Note a slight increase in the movement after the monsoon

mass movement and high erosion (Fig. 9.11a). Gupta et al., (2022) carried LSM for the states of Uttarakhand and Himachal Pradesh using bivariate methods (Fig. 9.11b, c) (Gupta et al., 2022, 2023). Recent landslide susceptibility mapping studies have greatly employed artificial intelligence and machine learning (AI/ML) methods, bivariate and multivariate statistical methods. Here, we present recent studies dealing with LSM at varying spatial scales, including regional scale, valley-specific regions, townships and critical segments of national highways.

In Jammu and Kashmir, LSM investigations have been conducted at both regional and corridor scales, including studies covering the regional scale (Khan et al., 2024), the Chenab Valley (Patil et al., 2020) and the stretches along NH-44 (Mir et al., 2024). In Himachal Pradesh, research spans regional scale (Gupta et al., 2023) as well as focused studies in the Giri Valley (Pareta & Pareta, 2012),the Kullu Valley (Meena et al., 2019), the Beas Valley (Kaur et al., 2024), Dharamshala township (Sweta et al., 2022) and vulnerable segments of the NH-154A (Singh & Kumar, 2017) and NH-3 (Kumar et al., 2018). In Uttarakhand, extensive work has addressed multiple river valley such as the Tons Valley (Raman & Punia, 2012; Srivastava et al., 2010), the Yamuna Valley (Kundu & Patel, 2019), the Bhagirathi Valley (Devi et al., 2024), the Alaknanda Valley (Pathak, 2016; Sangeeta & Maheshwari, 2022; Sundriyal et al., 2024), the Mandakini Valley (Gupta & Shukla, 2022), the Goriganga Valley (Kumar & Gupta, 2021), the Kali Valley (Ramakrishnan et al., 2013; Solanki et al., 2022), in addition with the township-level analysis in Mussorrie (Peethambaran et al., 2019; Ram & Gupta, 2022). Beyond northwest Himalaya, several studies have explored LSM in Sikkim Himalaya (Chowdhuri et al., 2020; Islam et al., 2022; Pal et al., 2019; Prakash et al., 2021; Sonker et al., 2024), while in the West Bengal, the Darjeeling and Kalimpong regions have received considerate attention (Basu & Pal, 2017; Chawla et al., 2018; Mondal & Mandal, 2019; Roy & Saha, 2019; Roy et al., 2019; Saha & Saha, 2020; Sarkar & Kanungo, 2004; Sarkar et al., 2013, 2016). In the northeast Himalayan states, research efforts include city-scale analysis in Guwahati (Bhusan et al., 2014; Phukon et al., 2012), regional assessment in Arunachal Pradesh (Rehman & Azhoni, 2023) and several studies across Manipur (Balamurugan et al., 2016; Chanu & Bakimchandra, 2022; Khusulio & Kumar, 2023; Singh et al., 2011),

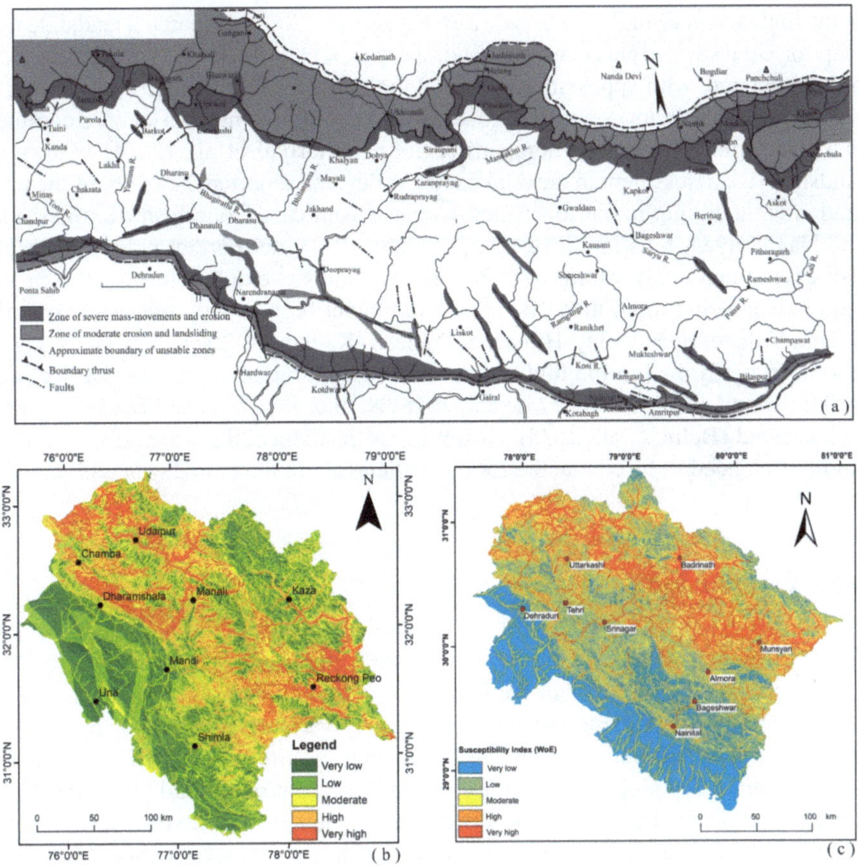

Fig. 9.11 **a**. Landslide-prone areas and zone of erosion of the Kumaun Himalaya (modified after Valdiya, 1985). **b–c**. Landslide susceptibility maps of the Himalayan states of Himachal Pradesh and Uttarakhand depicting very low, low, moderate, high and very high landslide susceptibility zones (Gupta et al., 2022, 2023)

Nagaland (Bilwa et al., 2015; Jamir et al., 2022; Sema et al., 2017; Yadav et al., 2023) and Tripura (Ghosh et al., 2013, 2017; Nath et al., 2024).

Landslide risk assessment is the quantification of hazard, exposure, vulnerability and consequently risk for an element in a landslide-affected area (Finlay & Fell, 1997). There are limited number of studies pertaining to landslide risk in the Indian Himalayan region. Hazard zoning map of Nainital township situated in the MBT zone was carried out on the basis of creep movement, past landslides, deformed civil structures and geological conditions into three zones—safe, moderately safe and very unsafe (Hukku & Jaitely, 1964; Sharma, 1981). Ram and Gupta (2022) conducted a risk assessment study for Mussorrie township in Uttarakhand and found that Nagar Palika Parisad ward has the largest part (~ 41%) of the area, with 37% buildings falling in the high and very high-risk zones. Sundriyal et al. (2024) prepared a landslide risk map for Uttarkashi, Tehri Garhwal, Rudraprayag, Chamoli and Pithoragarh districts of Uttarakhand, with approximately 5% of the total area falls in high and very high-risk zones, ~6% in moderate-risk zones, with very high to high-risk zones confined to Uttarkashi, Tehri and Pithoragarh regions. Sundriyal et al., (2024) performed landslide risk assessment in the Alaknanda Valley and reported that ~ 4% of the total land area, including major townships such as Joshimath, Gopeshwar, Karnaprayag and Nandprayag, falls into a very high to high-risk zone. Furthermore, a landslide risk assessment study, along with PS-InSAR analysis, exhibits that ~ 50% of the buildings housing more than 100 people are under high to very high-risk zones in Solang Village, Beas Valley, Himachal Pradesh (Kaur et al., 2024). Such works have also been performed in the highly landslide-susceptible areas such as Darjeeling–Sikkim Himalaya (Mondal & Mandal, 2018; Nath et al., 2021) and Kohima district of Nagaland (Belho et al., 2025). However, for resilience, there are still a lot more studies that need to be conducted for the settlements in the Himalayan region.

9.5 Conclusion

Results from the various studies and the present study reveal that the thrust zones, namely the MCT and MBT in the Himalaya, are highly susceptible to landslides. The susceptibility of these zones is governed by various factors, viz., state of the bedrock (highly sheared and fractured, split into multitudes of slabs), topography (steep—very steep slopes) and anthropogenic interventions. In addition, the MCT zone is the windward side of the orographic barrier that receives heavy rainfall, triggering numerous landslides, particularly during the monsoon season. Moreover, the narrow and steep valleys in these zones are also the sites for landslide-dammed lakes. Catastrophic bursting of landslide-dammed lakes is detrimental to the populace, as it not only causes debris flow but also initiates many landslides, as evident from the events of Gohna and Kanodia. Settlements over paleolandslide deposits in these zones are facing severe land subsidence and slope deformation in almost all the major river valleys. Numerous comprehensive research works have been carried out investigating the spatial susceptibility of the Himalaya to slope deformation, but there still persists a dearth of studies, such as estimating the risk posed by landslides and also site-specific studies primarily focusing on disaster risk management and reduction and development of early-warning systems.

Acknowledgements We thank Dr V M Tiwari for inviting us to contribute this article. We also thank an anonymous reviewer for his very constructive comments.

References

Agarwal, A., & Narain, S. (1991). 3rd State of India's environment, a citizen's report: floods, flood plains and environmental myths. In: Center for Science and the Environment, pp. 167.

Aziz, K., Sarkar, S., & Sahu, P. (2023). Geo-investigation and slope stability analysis of debris slides along Ramban-Gool road network, Jammu and Kashmir, India. *Journal of the Geological Society of India, 99*, 986–994. https://doi.org/10.1007/s12594-023-2419-5

Bahuguna, H., & Khan, I. (2022). Analyzing the genetic controls of landslide distributions in Jammu and Kashmir Himalaya. *Bulletin Series B, 74*, 1–178.

Balamurugan, G., Ramesh, V., & Touthang, M. (2016). Landslide susceptibility zonation mapping using frequency ratio and fuzzy gamma operator models in part of NH-39, Manipur, India. *Natural Hazards, 84*, 465–488.

Barnard, P. L., Owen, L. A., Sharma, M. C., & Finkel, R. C. (2001). Natural and human induced landsliding in the Garhwal Himalaya of Northern India. *Geomorphology, 40*, 21–35.

Bartarya, S. K., & Valdiya, K. S. (1989). Landslides and erosion in the catchment of the Ganla River, Kumaun Lesser Himalaya, India. *Mountain Research and Development, 9*, 405–419.

Basu, T., & Pal, S. (2017). Exploring landslide susceptible zones by analytic hierarchy process (AHP) for the Gish River Basin, West Bengal, India. *Spatial Information Research, 25*, 665–675.

Belho, K., Rawat, M. S., & Rawat, P. K. (2025). Threatening dynamics of landslide disaster risk in Himalaya region due to adverse climatic and anthropogenic changes: Geospatial approach. *Natural Hazards, 121*, 14091–14121.

Bhambri, R., Mehta, M., Singh, S., Jayangondaperumal, R., Gupta, A. K., & Srivastava, P. (2017). Landslide inventory and damage assessment in the Bhagirathi valley, Uttarakhand, during June 2013 flood. *Himalayan Geology, 38*, 193–224.

Bhusan, K., Kundu, S. S., Goswami, K., & Sudhakar, S. (2014). Susceptibility mapping and estimation of rainfall threshold using space based input for assessment of landslide hazard in Guwahati city in North East India. *The International Archives of the Photogrammetry, Remote Sensing and Spatial Information Sciences XL–8*, 15–19.

Bilwa, L. M., Yanthan, M. A., Madesh, P., & Hutti, B. (2015). Identification of potential landslide hazard zonation mapping using geoinformatics for Kohima region, Nagaland, India. *International Journal of Innovative Science and Research Technology* 2, 11. www.ijiset.com

Bisht, M. P. S., Rautela, P. (2010). Disaster looms large over Joshimath. *Current Science, 98*

Brandl, J., Gupta, V. K., Millen, B. (2010). Tapovan-vishnugad hydroelectric power project – experience with TBM excavation under high rock cover /. *Geomechanics and Tunnelling, 3*, 501–509. https://doi.org/10.1002/geot.201000048

Caine, N., & Mool, P. K. (1982). Landslides in the Kolpu khola Drainage, Middle Mountains, Nepal. *Mountain Research and Development, 2*, 157–173. https://doi.org/10.2307/3672961

Chalpathi Rao, R. V. (1983). Some major geotechnical aspects of environmental development of North Eastern Himalayas of Arunachal Pradesh: An appraisal. In: *Proceedings of Symposium on Geology Mineral Resources of Northeastern Himalaya*, 28–30 May 1976, Shillong (Vol. 43, pp. 29–33). GSI Misc Publication.

Chanu, M. L., & Bakimchandra, O. (2022). Landslide susceptibility assessment using AHP model and multi-resolution DEMs along a highway in Manipur India. *Environmental Earth Science, 81*, 5. https://doi.org/10.1007/S12665-022-10281-4

Chawla, A., Chawla, S., Pasupuleti, S., Rao, A. C. S., Sarkar, K., & Dwivedi, R. (2018). Landslide susceptibility mapping in Darjeeling Himalayas India. *Advances in Civil Engineering, 2018*, 6416492. https://doi.org/10.1155/2018/6416492

Chowdhuri, I., Pal, S. C., Arabameri, A., Ngo, P. T. T., Chakrabortty, R., Malik, S., Das, B., & Roy, P. (2020). Ensemble approach to develop landslide susceptibility map in landslide dominated Sikkim Himalayan region India. *Environmental Earth Science, 79*, 20. https://doi.org/10.1007/s12665-020-09224-9

Dalal, P., Hari, M., Xu, Y., Senapati, B., Gautam, P., Tiwari, D. K., et al. (2025). Deformation dynamics and hazard of slow-moving landslides: The 2023 Joshimath event Uttarakhand Himalaya. *Engineering Geology, 354*, 108201. https://doi.org/10.1016/j.enggeo.2025.108201
Datta, A., Luirei, K., Mehta, M. (2024). Slope instabilities and evolution of tectonic geomorphology along the strike of the Main Boundary Thrust zone in the western Himalaya India. *Natural Hazards Research, 4*(1), 118–133. https://doi.org/10.1016/j.nhres.2023.09.007
Devi, M., Gupta, V., & Sarkar, K. (2024). Landslide susceptibility zonation using integrated supervised and unsupervised machine learning techniques in the Bhagirathi Eco-Sensitive Zone (BESZ), Uttarakhand, Himalaya, India. *Journal of Earth System Science, 133*, 131. https://doi.org/10.1007/s12040-024-02344-w
Dewey, J. F., & Burke, K. C. A. (1973). Tibetan, Variscan and Precambrian basement reactivation products of continental collision. *The Journal of Geology, 81*, 683–692.
Dimri, A. P., Allen, S., Huggel, C., Ballesteros-Canovas, J. A., Rohrer, M., Shukla, A., et al. (2021). Climate change, cryosphere and impacts in the Indian Himalayan region. *Current Science, 120*, 774–790. https://doi.org/10.18520/cs/v120/i5/774-790
Finlay, P. J., & Fell, R. (1997). Landslides: Risk perception and acceptance. *Canadian Geotechnical Journal, 34*(2), 169–188.
Froude, M. J., & Petley, D. N. (2018). Global fatal landslide occurrence from 2004 to 2016. *Natural Hazards and Earth Systems Sciences, 18*, 2161–2181. https://doi.org/10.5194/nhess-18-2161-2018
Gadgil, S., Rajeevan, M., & Francis, P. A. (2007). Monsoon variability: Links to major oscillations over the equatorial Pacific and Indian oceans. *Current Science, 93*, 182–194.
Gahalaut, V. K., Gurjar, N., Kumar, A., Rajewar, S., Mohanty, A., Kumar, A., et al. (2023). Creeping slopes in NW Himalaya and Joshimath slide: Constraints from GPS measurements. *Geomatics Nat Hazards Risk, 14*, 2263622. https://doi.org/10.1080/19475705.2023.2263622
Gansser, A. (1964). *Geology of the Himalayas* (p. 289). Wiley Interscience.
Gerrard, J. (1994). The landslide hazard in the Himalayas: Geological control and human action. *Geomorphology, 10*, 221–230.
Geological Survey of India. (2025). Landslide hazards. *Geological Survey of India*. https://www.gsi.gov.in/webcenter/portal/OCBIS/pages_pageGeoInfo/pageLANDSLIDEHAZRD. Accessed on 02 July 2025
Geological Survey of India. (2025). Landslide Susceptibility Map. Bhukosh Portal. Retrieved 1 Sept 2025, from https://bhukosh.gsi.gov.in/Bhukosh/
Ghosh, K., Bandyopadhyay, S., & De, S. K. (2017). A comparative evaluation of weight-rating and analytical hierarchical (AHP) for landslide susceptibility mapping in Dhalai District, Tripura. In S. Hazra, A. Mukhopadhyay, A. R. Ghosh, D. Mitra, & V. K. Dadhwal (Eds.) *Environment and earth observation: case studies in India* (pp. 175–193). Springer International Publishing. https://doi.org/10.1007/978-3-319-46010-9_12
Ghosh, K., Bandyapadhayay, S., & De, S. K. (2013). Regional scale landslide risk assessment using fuzzy set approach: A case study on Dhalai District, Tripura. *J Indian Geomorphol, 2*, 82–90.
Gupta, S. K., & Shukla, D. P. (2022). Effect of scale and mapping unit on landslide susceptibility mapping of Mandakini River Basin, Uttarakhand, India. *Environmental Earth Science, 81*(14), 1–21.
Gupta, V., Kumar, S., Kaur, R., & Tandon, R. S. (2022). Regional-scale landslide susceptibility assessment for the hilly state of Uttarakhand, NW Himalaya, India. *Journal of Earth System Science, 131*, 1–18.
Gupta, V., Kaur, R., Kumar, S., Ram, B. K., & Tandon, R. S. (2023). Regional-scale landslide susceptibility assessment for the hilly state of Himachal Pradesh, NW Himalaya, India. *Current Science, 125*, 1369–1380.
Gupta, N., Das, J., Kanungo, D. P. (2025). Prediction of the spatial variability of rainfall- and human-induced landslides in the Bhagirathi Valley of the Indian NW Himalayas using machine learning techniques. *Environmental Earth Sciences, 84*(21). https://doi.org/10.1007/s12665-025-12568-8

Haigh, M. J. (1984). Landslide prediction and highway maintenance in the Lesser Himalaya, India. *Zeitschrift Für Geomorphologie Supplementband, 51*, 17–37.

Haigh, M., & Rawat, J. S. (2011). Landslide causes: Human impacts on a Himalayan landslide swarm. *Belgeo, 3–4*, 201–220. https://doi.org/10.4000/belgeo.6311

Heim, A., & Gansser, A. (1939). Central Himalaya. *Denkschr Schweiz Naturforsch Ges, 73*, 1–245.

Highland, L. M., & Bobrowsky, P. (2008). The landslide handbook—a guide to understanding landslides (Vol. 1325, pp. 129). U.S. Geological Survey Circular (1999).

Holland, T. H. (1984). Report on the Gohna landslip, Garhwal. In *Selections from the records of the Government of India in the Public Works Department* CCCXX–IV(324). Office of the Superintendent of Government Printing, Calcutta.

Hukku, B. M. & Jaitely, G. M. (1964). A geological report on the investigations of the stability of hill slopes around Nainital. *Geology Survey of India*, unpublished report

Johnson, K., Olson, E. A., & Manandhar, S. (1982). Environmental knowledge and response to natural hazards in mountainous Nepal. *Mountain Research and Development, 2*, 175–188.

Hunt, K. M. R., Turner, A. G., & Shaffrey, L. C. (2018). The evolution, seasonality and impacts of western disturbances. *Quarterly Journal Royal Meteorological Society, 144*, 278–290. https://doi.org/10.1002/qj.3200

Islam, A. R. M. T., Saha, A., Ghose, B., Pal, S. C., Chowdhuri, I., & Mallick, J. (2022). Landslide susceptibility modeling in a complex mountainous region of Sikkim Himalaya using new hybrid data mining approach. *Geocarto International, 37*, 9021–9046.

Ives, J. D., & Messerli, B. (1981). Mountain hazards mapping in Nepal: Introduction to an applied mountain research project. *Mountain Research and Development, 1*, 223–230.

Ives, J. D., & Messerli, B. (1984). Stability and instability of mountain ecosystems: Lessons learned and recommendations for the future. *Mountain Research and Development, 4*, 63–71.

Jamir, I., Gupta, V., Kumar, V., & Thong, G. T. (2017). Evaluation of potential surface instability using finite element method in Kharsali Village, Yamuna Valley, Northwest Himalaya. *J Mountain Sci, 14*, 1666–1676. https://doi.org/10.1007/s11629-017-4410-3

Jamir, M., Chang, C. N., Jamir, I., Thong, G. T., & Walling, T. (2022). Landslide susceptibility mapping of Noklak Town, Nagaland, North-east India using bivariate statistical method. *Geological Journal, 57*(12), 5250–5264.

Joshi, S., & Siva Subramanian, S. (2025). Landslide dam studies in Uttarakhand, India: Past, present and future. In B. Abolmasov, I. Alcántara-Ayala, Ž. Arbanas, D. Huntley, K. Konagai, M. Mikoš, K. Sassa, S. Sassa, & B. Tiwari (Eds.) *Progress in landslide research and technology* (Vol. 3, no. 2, pp. 199–206). Springer. https://doi.org/10.1007/978-3-031-72736-8_15

Juyal, N., Pant, R. K., Basavaiah, N., Yadava, M. G., Saini, N. K., & Singhvi, A. K. (2004). Climate and seismicity in the higher central Himalaya during the last 20 KYR: Evidences from Garbyang basin, Uttaranchal, India. *Palaeogeography, Palaeoclimatology, Palaeoecology, 213*, 315–330.

Kashyap, R., Pandey, A. C., & Parida, B. R. (2021). Spatio-temporal variability of monsoon precipitation and their effect on precipitation triggered landslides in relation to relief in Himalayas. *Spatial Information Research, 29*, 857–869. https://doi.org/10.1007/s41324-021-00392-8

Kaur, R., Gupta, V., & Chaudhary, B. S. (2024). Landslide susceptibility mapping and sensitivity analysis using various machine learning models: A case study of Beas valley, Indian Himalaya. *Bulletin of Engineering Geology and the Environment, 83*(6), 1–23.

Kaushal, S., Thanveer, J., Islam, S. M. ul. et al. (2025). Unveiling the amplifying impact: Anthropogenic activities and the two-fold surge in landslides in the Lesser Himalayas. *CATENA, 250*, 108771. https://doi.org/10.1016/j.catena.2025.108771

Khan, I., Bahuguna, H., & Kainthola, A. (2024). Regional landslide susceptibility zonation utilizing bivariate statistical techniques in the northwestern Himalayas, Jammu and Kashmir, India. *Journal of Earth System Science, 133*(3), 1–35.

Khan, I., Bahuguna, H., Kainthola, A., & Pandey, V. H. R. (2022). GIS-based multitemporal analysis of landslide susceptibility mapping along the Ramban-Banihal road section of National Highway-44, Jammu and Kashmir. In *National Conference on Landslide Risk Assessment and Mitigation in India.* Springer Nature Singapore, pp. 119–145.

Khanduri, S. (2018). *Hydro-meteorological disaster in Chamoli district of Uttarakhand Himalaya, causes and implications* (p. 45). Scholar's Press.

Khanduri, S., Sajwan, K. S., & Rawat, A. (2018). Disastrous events on Kelash-Mansarowar route, Dharchula Tehsil in Pithoragarh District, Uttarakhand in India. *Journal of Earth Science and Climatic Change, 9*(4), 1–4.

Khusulio, K., & Kumar, R. (2023). Feasibility assessment of multi-criteria decision making and quantitative landslide susceptibility methods: A case study of Mao-Maram Manipur. *Journal of Earth System Science, 132*(2), 1–27.

Kienholz, H., Schneider, G., Bichsel, M., et al. (1984). Mapping of Mountain Hazards and slope stability. *Mountain Research and Development, 4*, 247. https://doi.org/10.2307/3673145

Kumar, A., Sharma, R. K., & Bansal, V. K. (2018). Landslide hazard zonation using analytical hierarchy process along National Highway-3 in mid Himalayas of Himachal Pradesh, India. *Environmental Earth Science, 77*(20), 1–19.

Kumar, S., & Gupta, V. (2021). Evaluation of spatial probability of landslides using bivariate and multivariate approaches in the Goriganga valley, Kumaun Himalaya, India. *Natural Hazards, 109*(3), 2461–2488.

Kundu, V., & Patel, R. C. (2019). Susceptibility status of landslides in Yamuna valley, Uttarakhand, NW-Himalaya, India. *Himalayan Geology, 40*(1), 30–49.

Le Fort, P. (1975). Himalayas: The collided range, present knowledge of the continental arc. *American Journal of Science, 275A*, 1–44.

Lennartz, T. (2013). Constructing roads—constructing risks? Settlement decisions in view of landslide risk and economic opportunities in western Nepal. *Mountain Research and Development, 33*, 364–371.

Luirei, K. (2007). Slope failures in the Main Boundary Thrust Zone along Siang valley between Pasighat and Rengging, East Siang district, Arunachal Pradesh. *Journal of the Geological Society of India, 69*, 858–861.

Luirei, K., & Bhakuni, S. S. (2008). Landslides along frontal part of Eastern Himalaya in East Siang and Lower Dibang Districts, Arunachal Pradesh, India. *Journal of Geological Society of India, 71*(3), 321–330.

Luirei, K., Mehta, M., Iqbal, A., Nazir, S., & Kothyari, G. C. (2023). Factors influencing the slope instability of the Mussoorie-Bhitarli area, Garhwal Himalaya. *Geosciences Journal, 27*(3), 337–350.

Luirei, K., Pant, P. D., & Kothyari, G. C. (2006). Geomorphic evidence of neotectonic movements in Dharchula area, Northeast Kumaun: A prospective of the recent tectonic activity. *Journal of the Geological Society of India, 67*, 92–100.

Maurya, P., Gupta, V., Luirei, K. (2025). Evaluation of relative active tectonics and the spatial distribution of landslides in the Yamne Valley (Arunachal Pradesh), Northeastern Himalaya, India. *Geological Journal, 60*, 2984–3015. https://doi.org/10.1002/gj.5197

Meena, S. R., Mishra, B. K., & Piralilou, S. T. (2019). A hybrid spatial multi-criteria evaluation method for mapping landslide susceptible areas in Kullu Valley, Himalayas. *Geosciences, 9*(4), 156.

Middlemiss, C. S. (1890). Geological sketch of Nainital with some remarks on natural conditions governing the mountain slopes. *Records of the Geological Survey of India, 21*, 213–234.

Millen, B., Höfer-Öllinger, G., Brandl, J. (2015). Change in hydraulic properties of rock mass due to tunnelling. In: G. Lollino, D. Giordan, K. Thuro et al. (Eds.), *Engineering geology for society and territory* (Vol. 6, pp. 937–940). Springer.

Mir, R. A., Aziz, K., Singh, S., Sharma, I., Shah Za, A., & Ahmed, R. (2025). Pre and post landslide phenomena in Pernote area, Ramban District, Jammu and Kashmir, India: Implications for early warning, mitigation, and preparedness. *Landslides, 22*, 1577–1596. https://doi.org/10.1007/s10346-025-02469-z

Mir, R. A., Habib, Z., Kumar, A., & Bhat, N. A. (2024). Landslide susceptibility mapping and risk assessment using total estimated susceptibility values along NH44 in Jammu and Kashmir, Western Himalaya. *Natural Hazards, 120*(5), 4257–4296.

Mishra Commission. (1964). Report of the commission set up by the Government of India vide letter No. 142/23-5/44/76 dated 08.04.1976.

Molnar, P. (1984). Structure and tectonics of the Himalaya: Constraints and implications of geophysical data. *Annual Review of Earth and Planetary Sciences, 12*, 489–518.

Mondal, S., & Mandal, S. (2018). Landslide susceptibility and risk: A micro level study from the Balason River basin in Darjeeling Himalaya. *Arabian Journal of Geosciences, 11*(9), 1–15.

Mondal, S., & Mandal, S. (2019). Landslide susceptibility mapping of Darjeeling Himalaya, India using index of entropy (IOE) model. *Appl Geomatics, 11*(2), 129–146. https://doi.org/10.1007/s12518-018-0248-9

Naithani, A. K. (1999). The Himalayan landslides. *Employ News 23*(47), 1–2, 20–26.

Nath, N. K., Gautam, V. K., Pande, C. B., Mishra, L. R., Raju, J. T., Moharir, K. N., & Rane, N. L. (2024). Development of landslide susceptibility maps of Tripura, India using GIS and analytical hierarchy process (AHP). *Environmental Science and Pollution Research, 31*(5), 7481–7497.

Nath, S. K., Sengupta, A., & Srivastava, A. (2021). Remote sensing GIS-based landslide susceptibility and risk modeling in Darjeeling-Sikkim Himalaya together with FEM-based slope stability analysis of the terrain. *Natural Hazards, 108*(3), 3271–3304.

Nawani, P. C. (2015). Groundwater ingress in head race tunnel of tapovan: Vishnugad hydroelectric project in higher Himalaya, India. In: G. Lollino, D. Giordan, K. Thuro et al. (Eds.), *Engineering geology for society and territory* (Vol. 6, pp. 941–945). Springer

Negi, I. S., Kumar, K., Kathait, A., & Prasad, P. S. (2013). Cost assessment of losses due to recent reactivation of Kaliasaur landslide on National Highway 58 in Garhwal Himalaya. *Natural Hazards, 68*, 901–914. https://doi.org/10.1007/s11069-013-0663-5

Negi, R., Sati, S. P., Kumar, D., & Rana, S. S. (2021). Assessment of landslides susceptibility in Giri watershed, Northwest Himalaya, Himachal Pradesh, India. *Journal of Mountain Research 16*(1), 45–59. https://doi.org/10.51220/jmr.v16i1.5

Negi, R., Sati, S. P., Puniya, M. K., Biswas, M., Jayal, T., Rawat, A., Rana, S. S., & Sharma, V. (2023). Assessment of active tectonics of Giri Valley, NW Himalaya: Insights from geomorphic signature using remote sensing and GIS. In *Disaster develop* (p. 173).

Oldham, R. D. (1880). Note on the Nainital landslide 18th September 1880. *Records of the Geological Survey of India, 13*, 277–281.

Pal, S. C., Das, B., & Malik, S. (2019). Potential landslide vulnerability zonation using integrated analytic hierarchy process and GIS technique of Upper Rangit catchment area, West Sikkim, India. *Journal of the Indian Society of Remote Sensing, 47*(10), 1643–1655.

Pant, P. D., & Luirei, K. (1999). Malparockfalls of 18 August 1998 in the northeastern Kumaun Himalaya. *Journal of the Geological Society of India, 54*(4), 415–420.

Pant, P. D., & Luirei, K. (2005). Amiya landslide in the catchment of Gaula River, Southern Kumaun, Uttaranchal. *Journal of the Geological Society of India, 65*(3), 291–295.

Pareta, K., & Pareta, U. (2012). Landslide modeling and susceptibility mapping of Giri River watershed, Himachal Pradesh (India). *International Journal on Science and Technology, 1*(2), 91–100.

Parkash, S., Sah, M. P., Virdi, N. S. (2002). Inventory of slope failures along National Highway-52A between Banderdewa and Itanagar, Arunachal Pradesh. *Bulletin of Indian Geologist Association, 35*, 49–60

Pathak, D. (2016). Knowledge based landslide susceptibility mapping in the Himalayas. *Geoenvironmental Disasters, 3*(1), 1–11.

Patil, A. S., Bhadra, B. K., Panhalkar, S., & Patil, P. T. (2020). Landslide susceptibility mapping using landslide numerical risk factor model and landslide inventory prepared through OBIA in Chenab Valley, Jammu and Kashmir (India). *Journal of the Indian Society of Remote Sensing, 48*(3), 431–449.

Paul, S. K., Bartarya, S. K., Rautela, P., Mahajan, A. K. (2000). Catastrophic mass movement of 1998 monsoons at Malpa in Kali Valley Kumaun Himalaya (India). *Geomorphology, 35*(3–4), 169–180. https://doi.org/10.1016/S0169-555X(00)00032-5

Peethambaran, B., Anbalagan, R., Shihabudheen, K. V., & Goswami, A. (2019). Robustness evaluation of fuzzy expert system and extreme learning machine for geographic information system-based landslide susceptibility zonation: A case study from Indian Himalaya. *Environment and Earth Science, 78*(6), 1–20.

Phukon, P., Chetia, D., & Das, P. (2012). Landslide susceptibility assessment in Guwahati city, Assam using analytic hierarchy process (AHP) and geographic information system (GIS). *International Journal of Computer Applications, 2*(1), 2231–4946.

Petley, D. N., Hearn, G. J., Hart, A., Rosser, N. J., Dunning, S. A., Oven, K., & Mitchell, W. A. (2007). Trends in landslide occurrence in Nepal. *Natural Hazards, 43*, 23–44. https://doi.org/10.1007/s11069-006-9100-3

Prakash, B., Kush, K., Varun, J., & Deepesh, G. (2021). Causes of the triggering of Chamoli glacier burst of 7 February 2021 in Uttarakhand, India. *Disaster Advances, 14*(7), 60–67.

Ram, P., & Gupta, V. (2022). Landslide hazard, vulnerability, and risk assessment (HVRA), Mussoorie township, Lesser Himalaya, India. *Environment, Development and Sustainability, 24*(1), 473–501.

Ram, B. K., Shawez, M., Gupta, V., & Rawat, G. (2024). Ground subsidence in Dar village (Darma valley), Pithoragarh district, Kumaun Himalaya, India: A Himalayan disaster in waiting. *Journal of Earth System Science, 133*(1), 34.

Ramakrishnan, D., Singh, T. N., Verma, A. K., Gulati, A., & Tiwari, K. C. (2013). Soft computing and GIS for landslide susceptibility assessment in Tawaghat area, Kumaon Himalaya, India. *Naturtal Hazards, 65*(1), 315–330.

Raman, R., & Punia, M. (2012). The application of GIS-based bivariate statistical methods for landslide hazards assessment in the upper Tons River valley, Western Himalaya, India. *Georisk, 6*(3), 145–161.

Rana, N., Singh, S., Sundriyal, Y. P., & Juyal, N. (2013). Recent and past floods in the Alaknanda valley: Causes and consequences. *Current Science, 105*(9), 1209–1212.

Rather, A. A., & Bukhari, S. K. (2024). Understanding Joshimath landslide using PS interferometry and PSDS InSAR. *Journal of Earth System Science, 133*(2), 93. https://doi.org/10.1007/s12040-024-02312-4

Rautela, P., & Lakhera, R. C. (2000). Landslide risk analysis between Giri and Tons rivers in Himachal Himalaya (India). *International Journal of Applied Earth Observation and Geoinformation, 2*, 153–160. https://doi.org/10.1016/S0303-2434(00)85009-6

Rawat, J. S., & Joshi, R. C. (2012). Remote-sensing and GIS-based landslide-susceptibility zonation using the landslide index method in Igo River Basin, Eastern Himalaya, India. *International Journal of Remote Sensing, 33*, 3751–3767.

Rehman, S., & Azhoni, A. (2023). Analyzing landslide susceptibility, health vulnerability and risk using multi-criteria decision-making analysis in Arunachal Pradesh, India. *Acta Geophysica, 71*, 101–128. https://doi.org/10.1007/s11600-022-00943-z

Roy, J., & Saha, S. (2019). Landslide susceptibility mapping using knowledge driven statistical models in Darjeeling District, West Bengal, India. *Geoenvironmental Disasters, 6*(1), 1–18.

Roy, J., Saha, S., Arabameri, A., Blaschke, T., & Bui, D. T. (2019). A novel ensemble approach for landslide susceptibility mapping (LSM) in Darjeeling and Kalimpong Districts, West Bengal, India. *Remote Sensing, 11*(23), 2866.

Saha, A., & Saha, S. (2020). Comparing the efficiency of weight of evidence, support vector machine and their ensemble approaches in landslide susceptibility modelling: A study on Kurseong region of Darjeeling Himalaya India. *Remote Sensing Applications: Society and Environment, 19*, 100323.

Saha, S., Bera, B., Shit, P. K., Sengupta, D., Bhattacharjee, S., Sengupta, N., Majumdar, P., & Adhikary, P. P. (2023a). Modelling and predicting of landslide in Western Arunachal Himalaya, India. *Geosystems and Geoenvironment, 2*(2), 100158.

Saha, S., Bera, B., Shit, P. K., Sengupta, D., et al., (2023b) Modelling and predicting of landslide in Western Arunachal Himalaya, India. *Geosystems and Geoenvironment* 2(2). https://doi.org/10.1016/j.geogeo.2022.100158

Sangeeta, & Maheshwari, B. K. (2022). Spatial predictive modelling of rainfall-and earthquake-induced landslide susceptibility in the Himalaya region of Uttarakhand, India. *Environmental Earth Sciences, 81*(8), 237.

Sarkar, S., & Kanungo, D. P. (2004). An integrated approach for landslide susceptibility mapping using remote sensing and GIS. *Photogrammetric Engineering and Remote Sensing, 70*(5), 617–625.

Sarkar, S., Roy, A. K., & Martha, T. R. (2013). Landslide susceptibility assessment using information value method in parts of the Darjeeling Himalayas. *Journal of the Geological Society of India, 82*(4), 351–362.

Sarkar, S., Roy, A. K., & Raha, P. (2016). Deterministic approach for susceptibility assessment of shallow debris slide in the Darjeeling Himalayas, India. *CATENA, 142*, 36–46.

Seeber, L., & Armbuster, J. (1981). Great detachment earthquakes along the Himalayan arc and the long-term forecasts. In E. W. Simpson & P. G. Richards (Eds.), *Earthquake prediction: An international review, Maurice Ewing Series 4* (pp. 259–277). American Geophysical Union.

Sema, H. V., Guru, B., & Veerappan, R. (2017). Fuzzy gamma operator model for preparing landslide susceptibility zonation mapping in parts of Kohima Town, Nagaland, India. *Modeling Earth Systems and Environment, 3*(2), 499–514.

Sharma, A. K. (1981). Structural study of Nainital with special reference to the hillside instability (p. 126). Unpublished Ph.D. Thesis, Kumaun University, Nainital.

Shankar, H., Chauhan, P., Singh, D., Bhandari, R., Bhatt, C. M., Roy, A., et al. (2023). Multi-temporal InSAR and Sentinel-1 for assessing land surface movement of Joshimath town, India. *Geomatics Natural Hazards Risk, 14*(1), 2253972. https://doi.org/10.1080/19475705.2023.2253972

Shrestha, M., Sharma, S., & Shrestha, R. P. (2025). Landslides in the Himalayas: A comprehensive review of hazards, impacts, and adaptive strategies. *Rural and Regional Development, 3*, 10002. https://doi.org/10.70322/rrd.2025.10002

Singh, T., & Devi, M. (2006). Landslide occurrences and risk assessment in Itanagar Capital Complex, Arunachal Himalaya. *Himalayan Geology, 27*, 145–162.

Singh, C. D., Behera, K., & Rocky, W. S. (2011). Landslide susceptibility along NH-39 between Karong and Mao, Senapati district, Manipur. *Journal of the Geological Society of India, 78*(6), 559–570.

Singh, T. N. & Sarkar, K. (2005). Geotechnical investigation of Amiyan landslide hazard zone in Himalayan region, Uttaranchal, India. In *Proceedings of Geotechnical Engineering for Disaster Mitigation and Rehabilitation, the 1st International Conference* (pp. 355–360).

Singh, K., & Kumar, V. (2017). Landslide hazard mapping along national highway-154A in Himachal Pradesh, India using information value and frequency ratio. *Arabian Journal of Geosciences, 10*(24), 1–18.

Solanki, A., Gupta, V., & Joshi, M. (2022). Application of machine learning algorithms in landslide susceptibility mapping, Kali Valley, Kumaun Himalaya, India. *Geocarto International, 37*(27), 16846–16871.

Sonker, I., Tripathi, J. N., & Swarnim. (2024). Landslide susceptibility mapping using morphological and hydrological parameters in Sikkim Himalaya: frequency ratio model and geospatial technologies. *Natural Hazards, 120*(7), 6797-6832.

Sreejith, K. M., Jasir, M. C. M., Sunil, P. S., Rose, M. S., Saji, A. P., Agrawal, R., et al., (2024). Geodetic evidence for cascading landslide motion triggered by extreme rain events at Joshimath, NW Himalaya. *Geophysical Research Letters 51*(9) e2023GL106427. https://doi.org/10.1029/2023GL106427

Srivastava, V., Srivastava, H. B., & Lakhera, R. C. (2010). Fuzzy gamma based geomatic modelling for landslide hazard susceptibility in a part of Tons river valley, northwest Himalaya, India. *Geomatics Natural Hazards Risk, 1*(3), 225–242.

Sundriyal, Y., Kumar, S., Chauhan, N., Kaushik, S., Kumar, V., Rana, N., & Wasson, R. (2024). An integrated approach of machine learning and remote sensing for evaluating landslide hazards and risk hotspots, NW Himalaya. *Remote Sensing Applications: Society and Environment, 33*, 101140. https://doi.org/10.1016/j.rsase.2024.101140

Sur, U., Singh, P., & Meena, S. R. (2020). Landslide susceptibility assessment in a Lesser Himalayan road corridor (India) applying fuzzy AHP technique and Earth-observation data. *Geomatics Nat Hazards Risk, 11*, 2176–2209. https://doi.org/10.1080/19475705.2020.1836038

Sweta, K., Goswami, A., Peethambaran, B., Bahuguna, I. M., & Rajawat, A. S. (2022). Landslide susceptibility zonation around Dharamshala, Himachal Pradesh, India: An artificial intelligence model–based assessment. *Bulletin of Engineering Geology and the Environment, 81*(8), 1–24.

Thakur, V. C. (2004). Active tectonics of Himalayan Frontal Thrust and seismic hazard to Ganga plain. *Current Science, 86*, 1554–1560.

Thakur, V. C., Joshi, M., Sahu, D., Suresh, N., Jayangondaperumal, R., & Singh, A. (2014). Partitioning of convergence in Northwest Sub-Himalaya: Estimation of late Quaternary uplift and convergence rates across the Kangra reentrant, North India. *International Journal of Earth Sciences, 103*, 1037–1056.

Valdiya, K. S. (1998). Catastrophic landslides in Uttaranchal, Central Himalaya. *Journal of the Geological Society of India, 52*(4), 483–486.

Valdiya, K. S. (1992). Main Boundary Thrust Zone in Himalaya. *Ann Tectonicae, 6*, 54–84.

Valdiya, K. S. (1985). Accelerated erosion and landslide-prone zones in the central Himalaya. In *Concepts and strategies* (pp. 122–138). Gyanodaya Prakashan, Nainital.

Valdiya, K. S. (1987). *Environmental geology: Indian context*. Tata McGraw-Hill Pub. Co., Ltd.

Wang, W., Motagh, M., Xia, Z., Lu, Z., Karimzadeh, S., Zhou, C., et al., (2024). Characterization of transient movements within the Joshimath hillslope complex: Results from multi-sensor InSAR observations. *PFG–Journal of Photogrammetry, Remote Sensing and Geoinformation Science, 92*(5) 629–648. https://doi.org/10.1007/s41064-024-00315-w

Weidinger, T. J. (1998). Case history and hazard analysis of two lake-damming landslides in the Himalayas. *Journal of Asian Earth Sciences, 16*, 323–331.

Yadav, R. K., Gahalaut, V. K., Gautam, P. K., Jayangondaperumal, R., Sreejith, K. M., Singh, I., Kumar, A., Vivek, V. J., Agrawal, R., Catherine, J. K., et al. (2020). Geodetic monitoring of landslide movement in the Garhwal region of NW Himalaya. *Himalayan Geology, 41*, 126–138.

Yadav, M., Pal, S. K., Singh, P. K., & Gupta, N. (2023). Landslide susceptibility zonation mapping using frequency ratio, information value model, and logistic regression model: A case study of Kohima district in Nagaland, India. In P. Thambidurai, & T. N. Singh, (Eds.), *Landslides: Detection, prediction and monitoring: technological developments* (pp. 333–363). Springer International Publishing. https://doi.org/10.1007/978-3-031-23859-8_17

Yin, A. (2006). Cenozoic tectonic evolution of the Himalayan orogen as constrained by along-strike variation of structural geometry, exhumation history, and foreland sedimentation. *Earth Science Reviews, 76*, 1–131.

Chapter 10
Structural Controls on Seismicity in the Himalaya

M. Ravi Kumar and Jyotima Kanaujia

Abstract This contribution synthesizes the relation between crustal architecture and seismogenesis in the Himalaya. Collation of the knowledge of historical earthquakes and present day earthquake locations suggests that seismicity along the Himalayan arc is variable, influenced by the geometry, nature of coupling along the Main Himalayan Thrust (MHT), and segmentation controlled by structural features indenting the orogen. Also, the nature of the Himalayan seismicity belt, coinciding with the transition from a locked to creeping portion of the MHT, is variable among different segments. The Sikkim Himalaya stands out as a distinct segment revealing a dominance of transverse tectonics, evidenced by strike-slip earthquake mechanisms, the occurrence of lower crustal, uppermost mantle earthquakes due to an imbricated crust that is accommodating the Indian plate convergence. Preliminary estimates of the maximum earthquake magnitudes, based on the current knowledge of locking width, average slip, and rupture dimensions, are 7.0–7.3 for Arunachal, 7.6–7.8 for Sikkim–Bhutan, 8.1–8.2 for Garhwal–Kumaun, and 8.2–8.3 for Kashmir–Himalaya segments. However, these estimates need to be validated with improved knowledge of the geometry of the MHT, locking width, slip rates, and nature of segmentation.

Keywords Himalaya · Segmentation · Main Himalayan Thrust · Seismicity

Jyotima Kanaujia contributed equally to this work.

M. Ravi Kumar (✉)
CSIR-National Geophysical Research Institute, Hyderabad, Telangana, India
e-mail: mravingri@gmail.com

J. Kanaujia
Department of Earth and Environmental Sciences, Indian Institute of Science Education and Research, Mohali, Punjab, India
e-mail: jyotima@iisermohali.ac.in

V. M. Tiwari (ed.), *Himalayan Natural Hazards*, Indian National Science Academy,
https://doi.org/10.1007/978-981-92-3784-5_10

10.1 Introduction

The Himalayan orogen, nearly spanning a 2400 km stretch from Kashmir in the west to Arunachal Pradesh in the east, is a result of the collision between the Indian and Eurasian plates ongoing since about 50 Ma (Molnar & Tapponnier, 1977). The convergence is primarily accommodated along the Main Himalayan Thrust (MHT), a shallow-dipping décollement that separates the underthrusting Indian plate from the overlying Himalayan wedge. It is advocated that the MHT and the associated ramp structures play a pivotal role in the nucleation of large and great earthquakes in the region (Stevens & Avouac, 2015). Geodetic observations of crustal strain across the Himalayan orogen indicate that approximately 2 cm/year of the ongoing convergence between the Indian and Eurasian tectonic plates (Bilham, 1995; Molnar, 1990) is accommodated within the mountain belt via episodic slip along a seismically active décollement. This deformation occurs through a stick–slip mechanism, characteristic of continental collision zones, where interseismic strain accumulation is periodically released during seismic events (Molnar, 1990; Seeber & Armbruster, 1981). Historically, the Himalaya has experienced a large number of devastating earthquakes like the 1555 Kashmir, 1714 Bhutan, 1803 Kumaun–Garhwal, 1905 Kangra, 1934 Bihar–Nepal, and the 1950 Assam earthquakes, each producing widespread damage and significant loss of life (Bilham, 1995; Chen & Molnar, 1977). In the last 20 years, two large earthquakes, namely the 2005 Muzaffarabad (M_w 7.6) and the 2015 Gorkha (M_w 7.8) earthquakes, occurred along the Himalayan belt (Fig. 10.1).

10.2 Panoptic View of Seismogenesis

The steady state and evolutionary models are the most popular ones to describe the seismotectonics of the Himalaya (Seeber & Armbruster, 1981; Ni & Barazangi, 1984). They hypothesize that the great and major earthquakes occur on a northward, gently dipping ($\sim$ 5–10°) detachment under the Outer and Lesser Himalaya. Also a significant characteristic of small and moderate-magnitude events is that they are confined within a narrow band adjacent to the down-dip margin of the detachment plane (Seeber & Armbruster, 1981; Ni & Barazangi, 1984; Molnar, 1990; Sapkota et al., 2013). This seismic activity is localized in a transitional zone between the actively slipping detachment to the south and the aseismically deforming segment to the north. The spatial distribution of these events, which projects onto the surface across the Main Central Thrust (MCT) within a 20–30 km wide corridor, defines the Himalayan Seismic Belt (HSB). A bimodal nature of Himalayan seismicity has been widely observed (Baranowski et al., 1984; Ni & Barazangi, 1984; Pandey et al., 1995), characterized by distinct behaviors. Partial-rupture earthquakes (reaching magnitudes up to M_w 7.8) repeatedly affect the deeper section of the MHT but rarely propagate to the surface. In contrast, major earthquakes (M_w 8) occur infrequently and extensively rupture the entire seismogenic depth (Fig. 10.1). The shallower focal

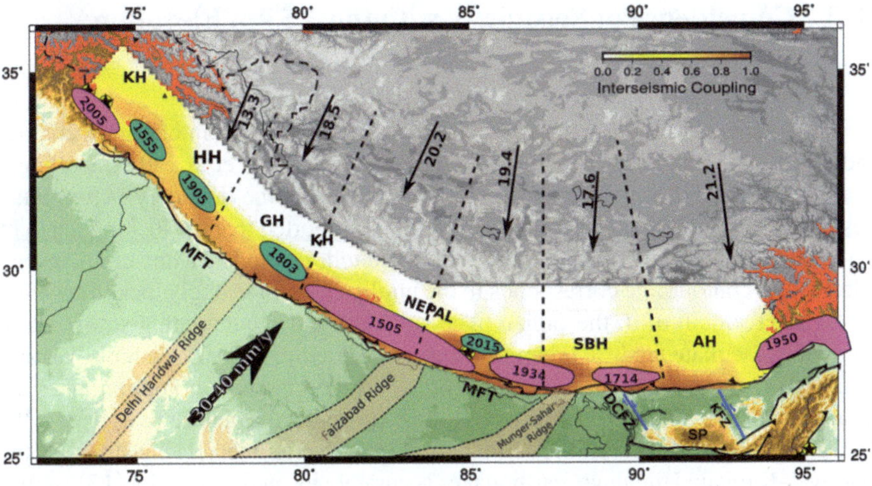

Fig. 10.1 Topographic map of the Himalaya with the interseismic coupling values (Stevens & Avouac, 2015) and rupture zones of previous significant earthquakes. Green colored regions denote ruptures of large earthquakes of magnitude < 8, and pink ones are those of great earthquakes (Mw ≥ 8), Transverse ridges (transparent patches) and fault zones (blue lines) are plotted. Black Solid Arrow: the Indian plate motion rate. Thin Black Arrows: convergence rates in different Himalayan segments (Stevens & Avouac, 2015). KH: Kashmir Himalaya, HH: Himachal Himlaya, GH: Garhwal Himalaya, KH: Kumaun Himalaya, SBH: Sikkim-Bhutan Himalaya, AH: Arunachal Himalaya, MFT: Main Frontal Thrust, DCFZ: Dhubiri-Chungthang Fault Zone, KFZ: Kopili Fault Zone, SP: Shillong Plateau

depths (10–15 km) of the partial ruptures typically cluster around a mid-crustal structural ramp (Pandey et al., 1995). Such ramps along the MHT, normally associated with the high topographic front (Lavé and Avouac 2001; Pandey et al., 1995), are inferred from clustering of seismicity, imaging of the subsurface using geophysical data and thermomechanical modeling. They control both the along-strike and down-dip segmentation of rupture (Lavé and Avouac 2001; Pandey et al., 1995). In some regions, out-of-sequence thrusting at the front of the Higher Himalaya plays a significant role in accommodating shortening and influencing the earthquake cycle (Whipple et al., 2016).The 2015 Gorkha earthquake (M_w 7.8) is the largest recorded event in this category. In contrast, evidence for surface ruptures at the Main Frontal Thrust (MFT), probably associated with great ($M_w \geq 8$) events (Bollinger et al., 2016; Hetényi et al. 2016; Kumar et al., 2010; Lavé et al. 2005; Sapkota et al., 2013), has been found from paleoseismological studies. The 1934 M_w 8.4 Bihar–Nepal (Bollinger et al., 2016; Sapkota et al., 2013) and the 1950 M_w 8.7 Assam (the largest earthquake ever recorded on land) earthquakes (Chen & Molnar, 1977) probably fall in that category.

10.3 Coincidence of Seismicity with the 3.5 km Elevation Front

Several studies reported a spatial association between the 3.5 km elevation contour and small-magnitude earthquakes in the Himalaya (Avouac et al., 2015; Bollinger et al., 2004) (Fig. 10.2). This elevation generally marks the boundary between the Lesser and Higher Himalaya. Bilham et al. (2001) reported that the locking depth of the MHT closely follows the 3.5 km elevation contour marked by a line of microseismicity. This contour line corresponds to the up-dip end of a strongly locked segment of the MHT. South of it, the Indian plate is locked and accumulating strain, while north of it, the plate tends to creep or deform in a ductile, aseismic fashion. Vernant et al. (2014) emphasized that in central Nepal, the 3.5 km contour approximates the location of the locking line, below which, at 18 km depth, the plate slips aseismically. To the south of the locking line, earthquakes occur due to interseismic stress. In the Garhwal–Kumaun Himalaya, earthquakes accurately located using a local 3D tomographic model reveal that seismicity is concentrated along the HSB, coinciding with the 3.5 km elevation contour (Kanaujia et al., 2025). In Nepal Himalaya, aftershocks of the April 25, 2015, M_w7.8 earthquake were mostly located north of this contour (Avouac et al., 2015). In the Himachal–Jammu Himalaya, the western segment of Himachal has very less seismicity, and in the eastern segment, it is clustered along the 3.5 km contour (locking line) (Fig. 10.2). In the Eastern Himalaya (Sikkim and Bhutan), west of Bhutan, seismicity still follows the 3.5 km contour. However, to the east of Bhutan, seismicity does not follow this trend. Figure 10.2 shows the Himalayan seismicity from 1977 to date, with the 3.5 elevation contour. Regions where the elevation contour does not relate with seismicity warrant further investigation. For instance, a widely scattered distribution of moderate, minor, and microseismic events within the Himalayan Seismic Belt (HSB) is prominently observed in the central Garhwal Himalaya region. This pattern of reduced seismic activity in the HSB suggests the potential absence of a structural ramp beneath sections of the Himalaya where subsurface ridges in the foreland enhance the curvature of the orogenic arc (Gahalaut & Arora, 2012).

10.4 Variable Coupling, Segmentation, and Seismicity

Although the Indian plate convergence along the arc appears continuous, seismicity is not uniformly distributed (Fig. 10.2). Instead, the Himalaya is partitioned into distinct segments characterized by varied rupture histories and coupling behavior (Figs. 10.1, 10.2 and 10.3). Geodetic observations from GPS and InSAR, combined with seismological data, enabled estimation of interseismic coupling along the MHT (Duputel et al., 2016; Elliott et al., 2016). Results from inversion of geodetic data reveal patches of highly coupled segments separated by low-coupling barriers, which in some cases correspond to known geological features such as transverse ridges or

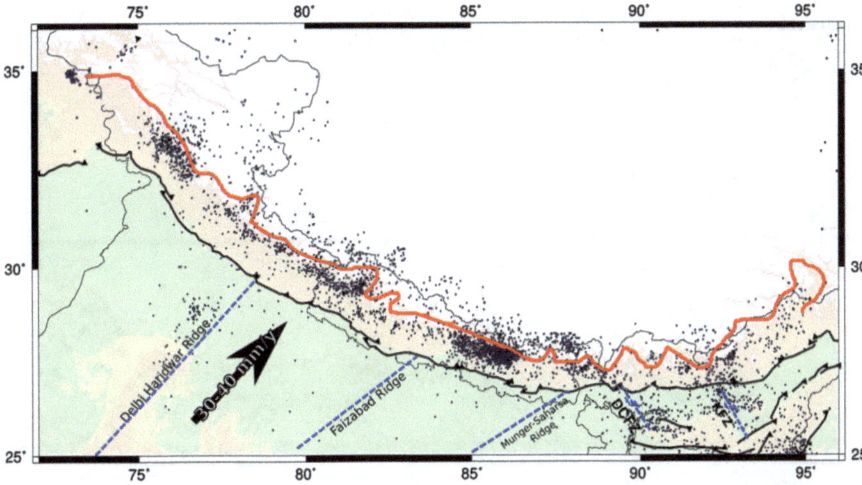

Fig. 10.2 Epicentral locations of M ≥ 3 earthquakes in the Himalayan arc (black dots) during 1977–2025, sourced from the ISC catalog. All other details are the same as in Fig. 10.1. KH: Kashmir Himalaya, HH: Himachal Himalaya, GKH: Garhwal–Kumaun Himalaya, SH: Sikkim Himalaya, BH: Bhutan Himalaya, AH: Arunachal Himalaya, MFT: Main Frontal Thrust, DCFZ: Dhubiri–Chungthang Fault Zone, KFZ: Kopili Fault Zone. Red line: ~ 3.5 elevation contour

crustal heterogeneities in the foreland basin (Stevens & Avouac, 2015; Dal Zilio et al., 2020). Interestingly, the highly coupled patches coincide with rupture areas of major historical earthquakes, implying that they represent zones of significant strain accumulation. In contrast, the lowly coupled segments may be experiencing aseismic slip or lower stress buildup. Along-strike variations in seismic moment release corroborate segmented strain accumulation, with certain portions of the arc consistently releasing small seismic energy (Dal Zilio et al., 2020).

Rupture propagation along the MHT is influenced by both arc-normal and arc-parallel geological structures (Fig. 10.1). The transverse ridges in the Indian plate, such as the Delhi Hardwar Ridge (DHR) and Faizabad Ridge, act as physical and rheological barriers to rupture growth. For instance, the eastern extent of the 1905 Kangra earthquake and the western limit of the 1803 Kumaun–Garhwal earthquake coincide with the extension of DHR into Himalaya (Fig. 10.1). Further, the western terminations of the 1934 Bihar–Nepal and 2015 Gorkha–Nepal earthquakes align with the Faizabad Ridge (Gahalaut & Kundu, 2012). These features likely represent regions of reduced coupling, where strain is accommodated aseismically or by smaller events, inhibiting rupture propagation of great earthquakes. Segmentation of the MHT limits the rupture length, thereby constraining the maximum magnitude of earthquakes in each segment. While full-width ruptures in single segments can produce earthquakes of magnitude 8, multi-segment ruptures can result in even larger ones. Determining whether such multi-segment ruptures are feasible requires integrating GPS slip rates, seismic imaging, and rheological modeling to estimate realistic upper bounds on earthquake magnitudes (Avouac et al., 2015).

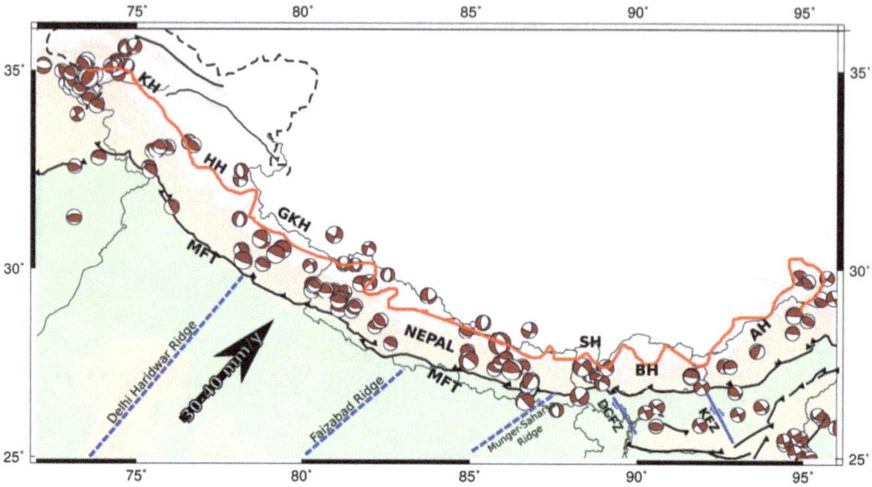

Fig. 10.3 Focal mechanisms derived from the Global CMT catalog of M $\geq$ 5 earthquakes in the Himalayan arc. All other details are the same as in Fig. 10.2. Red line: $\sim$ 3.5 km elevation contour

In northwestern Himalaya, a seismic gap between the 1905 Kangra and the 2005 Muzaffarabad earthquakes, called the Kashmir gap, has been identified (Mitra et al., 2023). Despite strong interseismic locking inferred from modeling of GPS data (Avouac et al., 2015; Gahalaut et al., 2013), this segment has not experienced a great event in over five centuries. The Beas–Sutlej Ridge in Himachal and the associated lineaments likely define a transverse segmentation boundary, evidenced by abrupt changes in the MHT geometry and rupture terminations (Rajendran & Rajendran, 2021). Seismic tomographic models show steep MHT segments and a high-velocity crust in these zones (Gupta et al., 2022). The 1905 Kangra earthquake (M_w $\sim$ 7.8) might have been terminated along one such boundary, underscoring the role of transverse features in limiting rupture lengths and segmentation (Fig. 10.1).

In the Garhwal–Kumaun region, seismicity is concentrated along the ramp and flat portions of the MHT (Figs. 10.2 and 10.4a). High-resolution seismic tomography reveals a steep ramp in Garhwal that transitions into a gentler dip in Kumaun, probably indicating a lateral ramp separating the two segments (Kanaujia et al., 2025). The width of MHT, measured from the Main Frontal Thrust (MFT) to the Lesser Himalayan ramp, is approximately 115 km in Garhwal and 130 km in Kumaun. Moment tensor solutions of events greater than M_w 3.8 track the ramp-flat structure, with most earthquakes occurring above 20 km depth. The Uttarkashi (1991, M_w 6.8) and Chamoli (1999, M_w 6.2) earthquakes occurred near the ramp region. Tomographic evidence also reveals the Delhi–Haridwar Ridge as a high-velocity anomaly influencing the MHT geometry and probably acting as a segmentation boundary.

In central Himalaya, Nepal is one of the most extensively studied segments, particularly in the aftermath of the 2015 Gorkha earthquake. The rupture of this M_w 7.8 earthquake initiated near the mid-crustal ramp of the MHT and propagated eastward

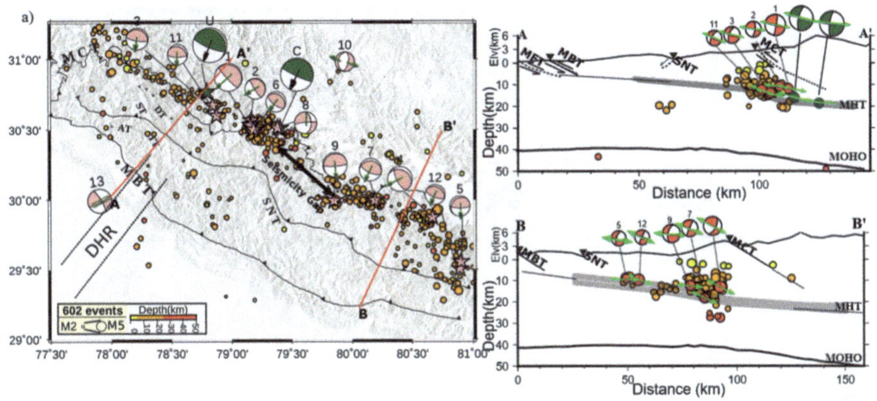

Fig. 10.4 **a** Overview of seismicity (colored coded circles) of Garhwal–Kumaun Himalaya segment. Epicenters as well as the depth sections along cross-sections in Garhwal (AA') and Kumaun (BB') are presented. The focal mechanisms of earthquakes (M $\geq$ 3.8) are plotted in pink and two moderate-magnitude earthquakes, namely the 1991 Uttarkashi (Mw 6.8) and 1999 Chamoli (Mw 6.6) are plotted in green colors. The gray patch in depth sections (right hand sides) is the MHT configuration taken from Kanaujia et al. (2025). **b** Overview of seismicity (color coded circles) of Sikkim Himalaya segment. Epicenters as well as depth sections along the cross-sections are presented. Red lines: $\sim$ 3.5 km elevation contour. **c** Regional seismicity (color coded circles) of Arunachal Himalaya. The epicenters are color coded according to their respective hypocenter depths. Red lines: $\sim$ 3.5 km elevation contour

for over 150 km, remaining blind, confined between 10 and 25 km depth (Avouac et al., 2015; Elliott et al., 2016). Seismic and geodetic studies indicate that this rupture occurred on a well-locked patch of the MHT (Pandey et al., 1995; Stevens & Avouac, 2015). Aftershock patterns and GPS models show stress release localized near the up-dip limit of a mid-crustal low-velocity layer (Elliott et al., 2016).

The eastern Himalaya exhibits a contrasting tectonic behaviour compared to its central and western counterparts. In Sikkim, the 2011 M_w 6.9 earthquake and its aftershocks revealed complex rupture mechanisms involving shallow strike-slip, mid-crustal oblique-slip, and deeper thrust faulting (Hazarika et al., 2010) (Fig. 10.3). Seismicity is distributed from the upper crust ($\sim$ 10 km) down to the lower crust (> 30 km), with significant events associated with the NW–SE trending structures such as the Darjeeling–Chungthang Fault Zone (DCFZ) and Kopili Fault Zone (KFZ) (Figs. 10.2, 10.3). Crustal imaging studies in Sikkim using receiver functions and local earthquake tomography reveal low-velocity anomalies, anisotropy, and high Vp/Vs zones, interpreted as fluid-rich and structurally complex domains (Singh et al., 2024). These features are linked to earthquakes at varied depths and suggest partial decoupling of crustal layers. In eastern Arunachal Pradesh, seismicity is less frequent, but deeper (30–50 km) (Fig. 10.4c), reflecting underthrusting of the Indian plate beneath the Mishmi Hills and Lohit ranges. The MHT geometry is poorly resolved here, as revealed by receiver function imaging (Singh et al., 2021). The eastern

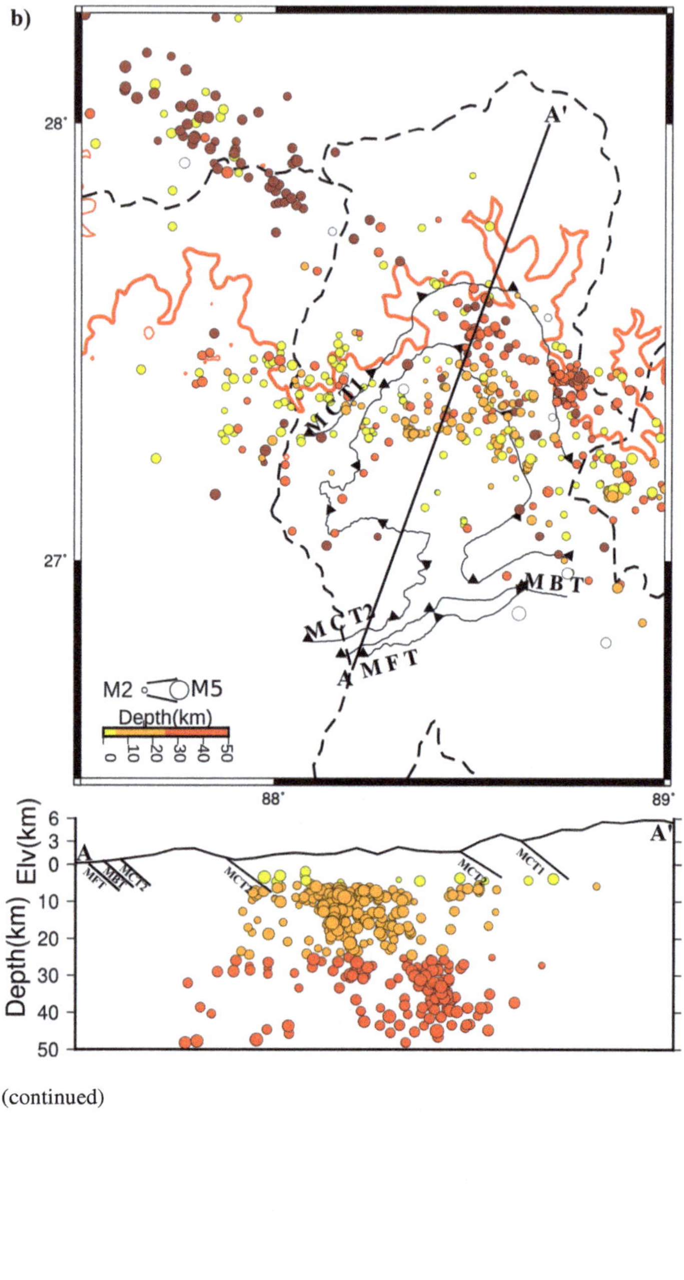

Fig. 10.4 (continued)

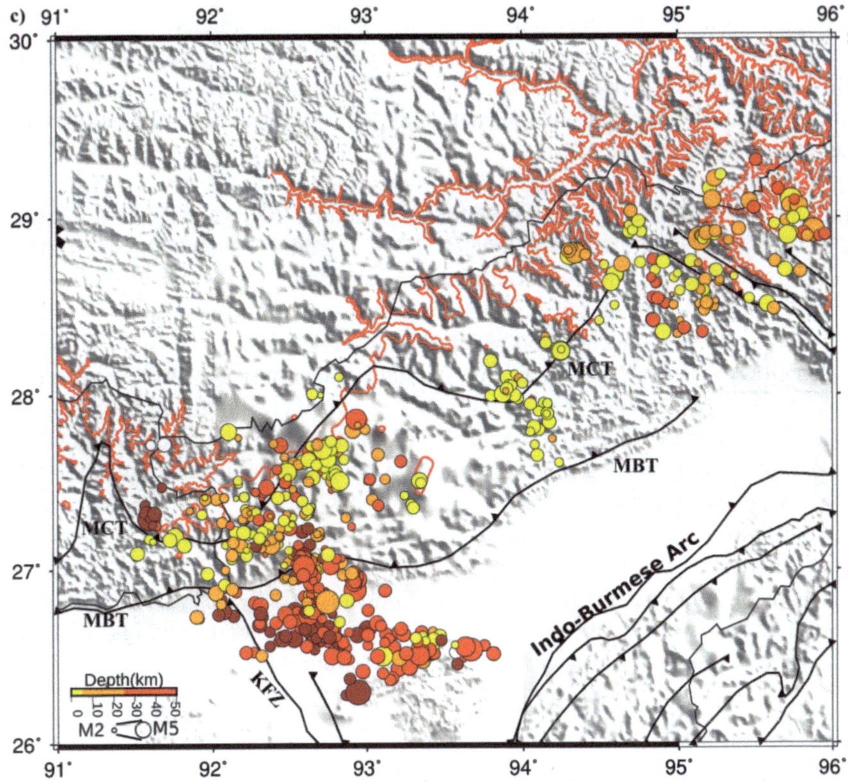

Fig. 10.4 (continued)

syntaxis exhibits highly oblique convergence, with seismic energy distributed across multiple fault strands.

10.5 Regional Seismicity Patterns

10.5.1 Kashmir and Himachal Himalaya

Modeling of crustal structure including the geometry of the Main Himalayan Thrust (MHT) using the Jammu and Kashmir Seismological Network data revealed a low-velocity layer coinciding with the MHT having a flat-ramp geometry (Mitra et al., 2023). The flat portion beneath the Sub-to-Lesser Himalaya lies between 6 and 10 km depth with a dip of $\sim 4°$. The mid-crustal frontal ramp found beneath the Kishtwar Higher Himalaya and Zanskar Range is between 10 and 16 km depth, having a dip of 13 to 17°. Also, a Lesser Himalayan duplex (LHD) bound by the MHT and the Main Central Thrust is documented beneath the Kishtwar Window. This feature contains high-angle imbricate structures associated with moderate seismicity. In the Kashmir Valley, the MHT is shallower, at ~ 10 km depth. Also, a lateral ramp causes westward shallowing of the MHT. Aftershocks of the 2013 Kishtwar earthquake are located at the intersection of the frontal–lateral ramp and mark the transition from a down-dip locked to a creeping segment.

In a detailed investigation using continuous seismic waveform data from the Jammu and Kashmir Seismological Network, spanning 2015–2017, Shamim et al. (2024) identified and tentatively located 1064 seismic events with magnitudes ranging from 0.6 to 4.7 and hypocentral depths extending to approximately 60 km. Events with focal depths less than 20 km are predominantly situated at or above the MHT, with a notable concentration beneath the Kishtwar Higher–Himalaya region. These shallow clusters align with the mid-crustal frontal ramp of the MHT and define a 50 km-wide transition zone between a fully locked and creeping fault behavior. An up-dip, frictionally locked segment of the MHT is estimated to span 100 km in width, suggesting the potential for a fully ruptured maximum magnitude earthquake of 8.4. Enhanced seismic activity southwest and east of the Kishtwar Window delineates the lateral ramps of the MHT, which may act as structural barriers influencing the extent and propagation dynamics of future ruptures (Shamim et al., 2024). In Himachal Pradesh, the background seismicity remains comparatively low west of the DHR but exhibits a gradual increase towards the Beas–Sutlej river system (see Figs. 10.1 and 10.2), highlighting regional variations in tectonic strain release.

10.5.2 Garhwal and Kumaun Himalaya

Seismicity patterns in this region exhibit conspicuous differences between Garhwal and Kumaun, reflecting variations in the geometry, locking width and structural complexity of the MHT. In the Garhwal Himalaya, seismicity is concentrated along a ~ 50 km wide zone straddling the MHT ramp. A notable seismic gap exists within this zone, which has been interpreted in terms of a lateral ramp separating the Garhwal and Kumaun segments. Relocated seismicity and focal mechanism solutions reveal that the MHT here exhibits a flat-ramp-flat structure, with earthquakes closely following the geometry of the ramp-flat transition. Moment tensor solutions of moderate earthquakes ($M_w \geq 3.8$) show predominantly thrust faulting, consistent with the ongoing plate convergence and closely track the steep ramp-flat geometry of the MHT (Fig. 10.4a) (Kanaujia et al., 2024, 2025). Moderate-magnitude events like the 1991 M_w 6.8 Uttarkashi earthquake and the 1999 M_w 6.2 Chamoli earthquake occurred along the MHT ramp between 12 and 15 km depth. Seismicity is observed along the Main Central Thrust (MCT), with 96% of the events occurring above 20 km depth, indicating that the seismogenic zone is concentrated in the upper and mid-crust. Seismic tomography in Garhwal reveals the MHT as a 2–3% low-velocity layer (LVL) with elevated Vp/Vs ratios, suggestive of fluid presence. The width of MHT from the Main Frontal Thrust (MFT) to the Higher Himalaya varies from 115 km in Garhwal to 130 km in Kumaun. In the latter, the MHT dips more gently, and seismicity is more diffuse, possibly due to differences in the rheology and/or fluid content.

10.5.3 Nepal Himalaya

In Nepal, seismic activity is largely concentrated along the frontal zone of the Higher Himalaya, with the majority of the seismic moment release occurring at crustal depths ranging from 10–20 km (Pandey et al., 1995). This pattern of earthquake distribution is indicative of interseismic stress buildup in the upper crust, coupled with ductile deformation (creep) in the lower crust beneath the Higher Himalayan region. Enhanced seismicity is observed near 82 °E in western Nepal and around 87 °E in the east. Notably, a segment between 82.5 °E and 85 °E in western Nepal exhibits a relatively subdued seismic behavior, forming a zone of lower seismic activity (Fig. 10.2). The lateral variations in seismicity are related to segmentation of the Main Himalayan Thrust (Pandey et al., 1995). Also, a correlation with lateral variations of geological structures suggests that segmentation is an ancient feature. In addition, Pandey et al. (1995) inferred four 250–400 km long segments capable of producing earthquakes like the $\sim$ M8.4 1934 Bihar–Nepal earthquake that occurred in eastern Nepal. The recurrence interval between two such earthquakes on a given segment is estimated to be between 130 and 260 years, invoking the characteristic earthquake model.

Lateral heterogeneities in mid-crustal seismic activity suggest along-strike variations in deep fault behavior, potentially influenced by fluid distribution, local structural configurations of ramp-flat systems, and differential stress accumulation across fault segments. These spatial variations may play a critical role in modulating the propagation and extent of intermediate-to-large-magnitude earthquakes, including great events such as the historically documented 1505 CE earthquake in far-western Nepal. This region exhibits distinct spatial and temporal variations in mid-crustal seismicity patterns. Seismic events are concentrated within a depth range of 5–20 km, forming three discrete, sub-parallel belts aligned with the Himalayan frontal thrust system. High-precision hypocenter locations highlight a strong correlation between seismicity clusters and specific structural elements such as mid-crustal ramps and tectonically bounded slivers. Laporte et al. (2021) proposed that the MHT comprises multiple fault segments, interconnected yet bounded by structural ramps and transverse (tear) faults. This segmentation is hypothesized to influence coseismic rupture propagation, either by limiting rupture extent or by enabling complete failure of locked zones. Consequently, this geometric complexity likely contributes to lateral variations in seismic coupling during the interseismic phase, thereby shaping regional seismic hazard potential.

The largest instrumentally recorded earthquake in the Himalaya occurred in April 2015 near Kathmandu, in a region where the pattern of fault locking was previously well documented (Ader et al., 2012). Local earthquake tomographic studies clearly imaged the MHT and its ramp as an LVL. The main shock of 2015 earthquake was located at the upper corner of the ramp. Analysis of teleseismic earthquake waveforms and Synthetic Aperture Radar images suggests that this earthquake originated northwest of Kathmandu and lies within a cluster of background seismicity fringing the bottom part of the locked portion of the MHT (Avouac et al., 2015). A 140 km rupture propagated eastward at about 2.8 km/s and unzipped the lower edge of the locked portion of the fault. Further, the eastward unzipping of the fault happened again during the M_w7.3 aftershock on May 12, 2015. Correlation of the seismic structure with the locations of the main shock and aftershocks reveals that the rupture extended for > 40 km to the south of the source zone originating in the MHT ramp. Also, a two-stage rupture towards the eastern margin of the rupture zone was suggested, with the rupture getting terminated on the floor thrust at the footwall imbricate below the region of maximum coseismic uplift (Pandey et al., 2017). Avouac et al. (2015) suggested that the transfer of stress to neighboring regions during the Gorkha earthquake should facilitate future rupture of the areas of the MHT adjacent and up-dip of its rupture zone. According to a study that used aftershock data from the Gorkha earthquake, the MHT is comprised of a duplex system of steeply descending faults rather than a lower ramp (Mendoza et al., 2019). The duplex system consists of three high-angle active faults between 10 and 20 km, inferred from a diffuse area of dense aftershock distribution east of the Kathmandu region (Mendoza et al., 2019). The highest rates of surface uplift and horizontal convergence between southern Tibet and India are also identified vertically above this duplex region. This is envisaged to form as scraped-off material from the underthrusting Indian plate migrates southward along the MHT, creating a series of narrower ramps in contrast to a much wider and shallow-dipping one.

10.5.4 Sikkim Himalaya

The eastern Nepal, Sikkim, and western Bhutan regions exhibit a transverse tectonic behavior in contrast to the rest of the Himalaya. They are characterized by deep crustal seismicity with strike-slip source mechanisms having fault planes oblique to the Indian plate convergence direction (Hazarika et al., 2010; Pavankumar & Manglik, 2021) (Figs. 10.3 and 10.4b). The Dhubri–Chungthang Fault Zone (DCFZ) is hypothesized to transfer strain from the Higher Himalaya in northwest Sikkim to the northwestern part of Shillong. Strain partitioning across the orogen led to crustal scale fault systems causing segmentation (Murphy et al., 2014; Vernant et al., 2014). The 2011 M_w 6.9 Sikkim earthquake exemplifies transverse tectonic deformation along the Himalayan front. A combination of shallow and deep seismic events characterizes tectonic activity in this region (Fig. 10.4b), with deeper occurrences predominantly localized in the northern segment of Sikkim (De & Kayal, 2003) (Fig. 10.3). Notably, the majority of these earthquakes exhibits strike-slip characteristics rather than thrust-type mechanisms, linked to the influence of transcurrent fault systems situated between the MBT and MCT (Hazarika et al., 2010). Seismic events are primarily confined to the upper to middle crustal depths, though a subset extends into the lower crust. However, precise identification of these focal regions remains challenging due to the linear alignment of seismic stations across the Himalayan strike, resulting in limited azimuthal distribution and spatial resolution (Hazarika et al., 2010).

It is speculated that the Dhubri–Chungthang Fault Zone (DCFZ) extends northward into Sikkim (Figs. 10.1 and 10.3). Uthaman et al. (2024) mapped a 20 km-wide mid-crustal seismogenic zone corresponding to DCFZ at Chungthang and its northward termination in the Higher Himalaya, affirmed by a marked variation in focal mechanisms. In addition, they propose that earthquakes along imbricated segments in the lower crust, previously imaged by receiver functions, possibly originate in response to crustal shortening and seismicity to the northwest of Sikkim is triggered by shearing along the Moho (Uthaman et al., 2024).

The M_w 6.9 magnitude earthquake of Sept. 18, 2011 in Sikkim and its 150 aftershocks revealed that the main shock was caused by a dextral motion along a NW–SE fault plane and dominant clustering in the NW–SE direction. The focal mechanisms of most aftershocks are found to be similar to that of the main shock (Kumar et al., 2012) with the NW–SE fault plane coinciding with the aftershock trend. Aftershock patterns and mechanisms suggest accommodation of Indian plate convergence by dextral motion along steeply dipping fault planes. These results speculate the possibility of transverse tectonics due to differential block motion within the Indian crust. New insights of complex multi-fault dynamics in Sikkim Himalaya are revealed from hypocenters and mechanisms of local earthquakes (Chaudhuri et al., 2024). Earthquakes are found to occur from shallow to deeper levels in the crust. In the upper crust, near MFT, the mechanism is strike-slip with a thrust component. For mid-crustal events below the MHT ramp (20–35 km depth), the focal mechanisms are dominantly strike-slip with a dextral sense of motion. Events in the lower crust (35–50 km depth) are located above the Moho and align along north-dipping imbricated segments with a strike-slip plus thrust component (Chaudhuri et al., 2024). Distributed seismicity indicates a heterogeneous crust in Sikkim, with imbrication in the lower crust causing earthquakes at those depths.

The Sikkim and Bhutan Himalaya regions are also characterised by deep crustal/uppermost mantle earthquakes. Modeling of the global, regional and local waveforms of the September 18, 2011 (M_w 6.9) Sikkim main shock and its major aftershocks revealed that the main shock occurred at 53 ± 4 km depth with the rupture extent being at least 20 km in the Indian lower crust. The source mechanism revealed motion on a near vertical NW–SE dextral strike-slip fault, transverse to the local strike of the Himalayan arc (Paul et al., 2015). The rupture was initiated at the SE end of the fault and then propagated northwest. A total of twenty small to moderate aftershocks ($M_b \geq 3.0$) followed. All of them were located SE of the main shock, between 12 and 50 km depth, with a predominance of strike-slip mechanisms. Seismicity and earthquake source mechanisms in Sikkim and Bhutan Himalaya reveal that the entire underthrust Indian crust is seismogenic and deforms by dextral strike-slip motion on oblique structures (Paul et al., 2015).

10.5.5 Bhutan Himalaya

The Sikkim–Bhutan seismic gap experienced a long earthquake quiescence, since the ~ M8.1 earthquake in 1714. The low seismicity in Bhutan compared to its adjoining segments is attributed to a low-convergence rate, nature of subsurface structures, aseismic creep, or a stress shadow caused by the great 1897 Shillong Plateau earthquake. Within the northeastern part of the Indian Plate, the Shillong Plateau has significant seismic activity, south of the Himalayan belt. The apparent reduced seismicity in Bhutan is attributed to this cause (Grujic et al. 2018). Also, it is suggested that a thicker crust in western Bhutan causes lateral changes in stresses on the Main Himalayan Thrust (MHT) and potentially affects the interseismic coupling and deformation regime (Diehl et al., 2017). The tectonic framework of Sikkim, western Bhutan and their adjacent foreland areas is dominated by strike-slip deformation within the Indian subcontinental crust. Pronounced strain accommodation occurs along a NW–SE oriented mid-crustal fault system, which extends from Chungthang in northeastern Sikkim to Dhubri near the Shillong Plateau's western margin. This dextral-slip Dhubri–Chungthang Fault Zone (DCFZ) serves as a critical structural element, partitioning the Indian basement and the MHT between eastern Nepal and western Bhutan. Functioning as the western boundary of the Shillong block, the DCFZ effectively links the active Himalayan deformation front with the Shillong Plateau. To the east, the Kopili fault, posited as the Shillong block's eastern margin, exhibits characteristics of a broad, mid-crustal seismicity zone within the foreland region. Geological imaging in eastern Bhutan reveals a seismically active, planar segment of the MHT, which may correlate with either a partially creeping fault section or elevated seismicity linked to the 2009 Mw 6.1 event. In the western-central Bhutan region, concentrated microseismicity at the High Himalaya' frontal zone suggests the development of a mid-crustal ramp structure, alongside accumulating stress on a fully locked MHT segment.

A comprehensive analysis of GPS-derived kinematic data across the Sikkim–Bhutan Himalayan region reveals interseismic convergence rates ranging between 16.2 and 18.5 mm/y. These rates align with geodetic observations documented in other Himalayan segments (Li et al., 2022). Notably, the slip deficit zone in the western Bhutan portion exhibits a lateral dimension of approximately ~ 100 km, significantly exceeding the 30–40% narrower widths observed in Sikkim and eastern Bhutan. A topographic embayment within the decoupling interface zone near the Sikkim–Bhutan border correlates spatially with the down-dip extents of the 1934 M_w 8.2 and 1714 M_w 8.1 megathrust ruptures. This geometric correlation suggests that the segmentation and recurrence patterns of major thrust earthquakes in this Himalayan domain are governed by heterogeneous frictional characteristics along the MHT. Furthermore, the distribution of seismic events in the Bhutan Himalaya reveals a spatially diffuse pattern, extending beyond the lower boundary of the locked zone. This observation challenges existing models that solely attribute the seismicity of Bhutan to stress buildup at the basal limit of the seismogenic portion of the MHT (Li et al., 2022).

10.5.6 Arunachal Himalaya

The seismotectonic framework of Arunachal Pradesh and the adjoining Eastern Himalayan Syntaxial (EHS) zone displays distinctive features compared to central and western Himalaya. Seismicity in this sector shows strong along-strike variability: While the western Arunachal Himalaya and the syntaxial bend are highly active, central Arunachal records relatively fewer earthquakes, mostly confined to the upper 20 km (Mahesh et al., 2023) (Fig. 10.4c). In contrast, the entire crust appears seismically active within the syntaxial zone, pointing to enhanced deformation where the Indian, Eurasian, and Burmese plates interact. Western Arunachal is further influenced by the extension of Kopili Fault, which cuts across the Main Himalayan Thrust and generates deeper earthquakes that extend into the foreland near the Mikir Hills. Notably, seismic clustering beneath the Higher Himalaya, often observed in association with locked MHT patches elsewhere, is absent in this region, suggesting that strain accumulation mechanisms differ fundamentally from those in Nepal or Garhwal–Kumaun.

Passive seismological imaging reveals that the crust in Arunachal is more transparent, with weaker reflectivity than other Himalayan segments (Singh et al., 2021). Mahesh et al. (2023) reported that both velocity models and seismicity distribution indicate a gently dipping MHT in Arunachal, with no robust evidence of a mid-crustal ramp; only weak ramp-like features are tentatively observed in the eastern part. The seismically less active, flat portion of the MHT beneath central Arunachal could represent a partially creeping segment or a structurally less deforming zone at the Himalayan front. Together, these observations emphasize that the Arunachal and EHS regions are not simply a continuation of the central Himalayan seismotectonic regime but represent a structurally and mechanically distinct domain, shaped by lithospheric heterogeneity, plate interactions, and possible transitions from locked to creeping behavior along the MHT.

10.6 Concluding Remarks

The anatomy of the Indian crust varies significantly along the strike of the Himalayan arc. The Main Himalayan Thrust (MHT) manifests as a low-velocity layer (LVL) in Garhwal–Kumaun and Nepal, whereas in Arunachal, it is imaged as a high-velocity layer (HVL), indicating sharp differences in interface properties. Seismicity, locking width and seismic coupling along the MHT are likewise variable and punctuated by transverse features, seismicity gaps, and possible slab tears. In Garhwal–Kumaun and Nepal, seismicity is largely controlled by the MHT and its associated ramps, in contrast to Arunachal where such clustering is not observed. The Sikkim Himalaya stands out as distinct, dominated by strike-slip tectonics, lower-crustal to uppermost mantle earthquakes, and an imbricated crustal structure. The maximum earthquake magnitudes along the Himalayan arc are potentially constrained by the width and

Table 10.1 Estimated maximum earthquake magnitudes (M_w) for different Himalayan segments

Segment	Locked width (km)	Rupture length (km)	Avg. slip (m)	Rupture area (km^2)	Estimated M_w (max)	Key references
Kashmir Himachal Himalaya	120–140	250–300	6–7	~ 35,000	8.2–8.3	Bilham (2004), Rajendran and Rajendran (2021), Mitra et al. (2023)
Garhwal–Kumaun Himalaya	115–130	220–250	~ 6	~ 32,000	8.1–8.2	Kanaujia et al. (2024), Kanaujia et al. (2025), Gupta et al. (2022)
Central Nepal Himalaya	120–130	250–300	6–7	~ 36,000	8.2–8.3	Pandey et al. (1995), Avouac et al. (2015), Elliott et al. (2016)
Sikkim–Bhutan Himalaya	100–110	150–180	~ 5	~ 18,000	7.6–7.8	Hazarika et al. (2010), Chaudhuri et al. (2024)
Arunachal Himalaya	≤ 100 (weakly locked)	≤ 120	3–4	~ 10,000	7.0–7.3	Mahesh et al. (2023), Singh et al. (2021)

depth of the seismogenic zone, slip deficit rates, rupture extents of historical earthquakes, paleoseismological evidence, structural segmentation and rheological properties of the crust. Using known information, first-order estimates of the maximum expected earthquake magnitude (M_w) for the major arc segments are calculated and listed in Table 10.1. However, these preliminary values need to be refined with more precise knowledge of these parameters.

Acknowledgements We thank V.M. Tiwari for inviting us to write this article. M.R.K acknowledges the support of CSIR through an Emeritus Scientist fellowship. JK acknowledges the support from DST Inspire faculty research grant (DST/INSPIRE/04/2023/001762).

References

Ader, T., Avouac, J. P., Liu-Zeng, J., et al. (2012). Convergence rate across the Nepal Himalaya and interseismic coupling on the Main Himalayan thrust: Implications for seismic hazard. *Journal of Geophysical Research: Solid Earth 117*(B4).

Avouac, J. P., Meng, L., Wei, S., et al. (2015). Lower edge of locked main Himalayan thrust unzipped by the 2015 Gorkha earthquake. *Nature Geoscience, 8*(9), 708–711.

Baranowski, J., Armbruster, J., Seeber, L., et al. (1984). Focal depths and fault plane solutions of earthquakes and active tectonics of the Himalaya. *Journal of Geophysical Research: Solid Earth, 89*(B8), 6918–6928.

Bilham, R. (1995). Location and magnitude of the 1833 Nepal earthquake and its relation to the rupture zones of contiguous great Himalayan earthquakes. *Current Science, 69*(2), 101–128.
Bilham, R. (2004). Earthquakes in India and the Himalaya: Tectonics, geodesy and history. *Annals of GEOPHYSICS, 47*, 2–3.
Bilham, R., Gaur, V. K., & Molnar, P. (2001). Himalayan seismic hazard. *Science, 293*(5534), 1442–1444.
Bollinger L, Avouac JP, Cattin R, et al. (2004) Stress buildup in the Himalaya. *Journal of Geophysical Research: Solid Earth 109*(B11).
Bollinger, L., Tapponnier, P., Sapkota, S. N., et al. (2016). Slip deficit in central Nepal: Omen for a repeat of the 1344 AD earthquake? *Earth, Planets and Space, 68*(1), 12.
Chaudhuri, D., Banerjee, R., Kumar, A., et al. (2024). Role of intraplate strike-slip earthquakes in accommodating convergence across the Eastern Himalayan plate boundary system. *Lithosphere 2024*(Special 15).
Chen, W. P., & Molnar, P. (1977). Seismic moments of major earthquakes and the average rate of slip in central Asia. *Journal of Geophysical Research, 82*(20), 2945–2969.
Dal Zilio, L., Jolivet, R., & van Dinther, Y. (2020). Segmentation of the main Himalayan thrust illuminated by Bayesian inference of interseismic coupling. *Geophysical Research Letters 47*(4), Article e2019GL086424.
De, R., & Kayal, J. R. (2003). Seismotectonic model of the Sikkim Himalaya: Constraint from microearthquake surveys. *Bulletin of the Seismological Society of America, 93*(3), 1395–1400.
Diehl, T., Singer, J., Hetényi, G., et al. (2017). Seismotectonics of Bhutan: Evidence for segmentation of the Eastern Himalayas and link to foreland deformation. *Earth and Planetary Science Letters, 471*, 54–64.
Duputel, Z., Vergne, J., Rivera, L., et al. (2016). The 2015 gorkha earthquake: A large event illuminating the Main Himalayan thrust fault. *Geophysical Research Letters, 43*(6), 2517–2525.
Elliott, J. R., Jolivet, R., González, P. J., et al. (2016). Himalayan megathrust geometry and relation to topography revealed by the gorkha earthquake. *Nature Geoscience, 9*(2), 174–180.
Gahalaut, V. K., & Arora, B. R. (2012). Segmentation of seismicity along the Himalayan arc due to structural heterogeneities in the underthrusting Indian plate and overriding Himalayan wedge. *Episodes Journal of International Geoscience, 35*(4), 493–500.
Gahalaut, V. K., & Kundu, B. (2012). Possible influence of subducting ridges on the Himalayan arc and on the ruptures of great and major Himalayan earthquakes. *Gondwana Research, 21*(4), 1080–1088.
Gahalaut, V. K., Kundu, B., Laishram, S. S., et al. (2013). Aseismic plate boundary in the Indo-Burmese wedge, northwest Sunda Arc. *Geology, 41*(2), 235–238.
Grujic, D., Hetényi, G., Cattin, R., et al. (2018). Stress transfer and connectivity between the Bhutan Himalaya and the Shillong plateau. *Tectonophysics, 744*, 322–332.
Gupta, S., Mahesh, P., Kanna, N., et al. (2022). 3-d seismic velocity structure of the Kumaun-Garhwal (Central) Himalaya: Insight into the Main Himalayan Thrust and earthquake occurrence. *Geophysical Journal International, 229*(1), 138–149.
Hazarika, P., Kumar, M. R., Srijayanthi, G., et al. (2010). Transverse tectonics in the Sikkim Himalaya: Evidence from seismicity and focal-mechanism data. *Bulletin of the Seismological Society of America, 100*(4), 1816–1822.
Hetényi, G., Le Roux-Mallouf, R., Berthet, T., et al. (2016). Joint approach combining damage and paleoseismology observations constrains the 1714 ad Bhutan earthquake at magnitude 80.5. *Geophysical Research Letters, 43*(20), 10–695.
Kanaujia, J., Ravi Kumar, M., Sunilkumar, T. C., et al. (2024). Nature of the Himalayan seismicity belt in the Garhwal-Kumaun segment and its tectonic implications. *Seismological Research Letters, 95*(4), 2386–2398.
Kanaujia, J., Kumar, M. R., Bandari, N., et al. (2025). Tomographic evidence for segmentation of the main Himalayan thrust in the Uttarakhand Himalaya. *Tectonophysics, 900*, 230655.

Kumar, S., Wesnousky, S. G., & Jayangondaperumal, R., et al. (2010). Paleoseismological evidence of surface faulting along the northeastern Himalayan front, India: Timing, size, and spatial extent of great earthquakes. *Journal of Geophysical Research: Solid Earth 115*(B12).

Kumar, M. R., Hazarika, P., Prasad, G. S., Singh, A., Saha, S. (2012). Tectonic implications of the September 2011 Sikkim Earthquake and its aftershocks. *Current Science, 102,* 5, 788–792.

Laporte, M., Bollinger, L., Lyon-Caen, H., et al. (2021). Seismicity in far western Nepal reveals flats and ramps along the Main Himalayan thrust. *Geophysical Journal International, 226*(3), 1747–1763.

Lavé, J., & Avouac, J. P. (2001). Fluvial incision and tectonic uplift across the Himalayas of central Nepal. *Journal of Geophysical Research: Solid Earth, 106*(B11), 26561–26591.

Lavé, J., Yule, D., Sapkota, S., et al. (2005). Evidence for a great medieval earthquake (1100 AD) in the central Himalayas, Nepal. *Science, 307*(5713), 1302–1305.

Li, S., Tao, T., Gao, F., et al. (2022). Present-day fault kinematic around the eastern Himalayan Syntaxis and probable viscoelastic relaxation perturbation following the 1950 Mw 8.7 Assam earthquake. *Journal of Asian Earth Sciences, 238*, 105396.

Mahesh, P., Srijayanthi, G., & Kumar, M. R. (2023). Seismotectonics and 3-D seismic velocity structure of the Arunachal Himalaya. *Tectonophysics, 869*, 230098.

Mendoza, M. M., Ghosh, A., Karplus, M. S., et al. (2019). Duplex in the main Himalayan thrust illuminated by aftershocks of the 2015 Mw 7.8 Gorkha earthquake. *Nature Geoscience, 12*(12), 1018–1022.

Mitra, S., Sharma, S., Wanchoo, S. K., et al. (2023). 3D crustal structure of the Jammu and Kashmir Himalaya: Signatures of mid-crustal ramp and lesser Himalayan duplex. *Journal of Geophysical Research: Solid Earth 128*(11):e2023JB027350.

Molnar, P. (1990). A review of seismicity and the rates of active underthrusting and deformation at the Himalaya. *Journal of Himalayan Geology, 1*, 131–154.

Molnar, P., & Tapponnier, P. (1977). The collision between India and Eurasia. *Scientific American, 236*(4), 30–41.

Murphy, M. A., Taylor, M. H., Gosse, J., et al. (2014). Limit of strain partitioning in the Himalaya marked by large earthquakes in western Nepal. *Nature Geoscience, 7*(1), 38–42.

Ni, J., & Barazangi, M. (1984). Seismotectonics of the Himalayan collision zone: Geometry of the underthrusting Indian plate beneath the Himalaya. *Journal of Geophysical Research: Solid Earth, 89*(B2), 1147–1163.

Pandey, A. K., Saikia, D., & Kumar, M. R. (2017). Earthquake genesis in Nepal Himalaya: A perspective from imaging of the 25th April 2015 mw 7.8 earthquake source zone. *Journal of Asian Earth Sciences, 141*, 259–267.

Pandey, M. R., Tandukar, R. P., Avouac, J. P., et al. (1995). Interseismic strain accumulation on the Himalayan crustal ramp (Nepal). *Geophysical Research Letters, 22*(7), 751–754.

Paul, H., Mitra, S., Bhattacharya, S. N., et al. (2015). Active transverse faulting within underthrust Indian crust beneath the Sikkim Himalaya. *Geophysical Journal International, 201*(2), 1072–1083.

Pavankumar, G., & Manglik, A. (2021). Complex tectonic setting and deep crustal seismicity of the Sikkim Himalaya: An electrical resistivity perspective. *Physics and Chemistry of the Earth, Parts a/b/c, 124*, 103077.

Rajendran, C. P., & Rajendran, K. (2021). Paleoseismic context of the 1950 earthquake: Implications for seismic gaps in the Eastern Himalaya. *Physics and Chemistry of the Earth, Parts a/b/c, 124*, 103055.

Sapkota, S. N., Bollinger, L., Klinger, Y., et al. (2013). Primary surface ruptures of the great Himalayan earthquakes in 1934 and 1255. *Nature Geoscience, 6*, 71–76.

Seeber, L., & Armbruster, J. G. (1981). Great detachment earthquakes along the Himalayan arc and long-term forecasting. *Earthquake Prediction: An International Review, 4*, 259–277.

Shamim, S. K., Ghosh, A., Mitra, S., Priestley, K., Sharma, S., & Wanchoo, S. K. (2024). Locked frontal and lateral ramps on the main Himalayan thrust beneath NW Himalaya illuminated by precisely located seismicity. *Geophysical Research Letters 51*(21), e2024GL110160.

Singh, A., Saikia, D., & Kumar, M. R. (2021). Seismic imaging of the crust beneath Arunachal Himalaya. *Journal of Geophysical Research: Solid Earth 126*(3):e2020JB020616.

Singh, A., Kumar, G., Singh, C., et al. (2024). Evidence for an imbricated lower crust and its role in accommodating crustal shortening in Sikkim Himalaya. *Tectonics 43*(6):e2023TC008044. (E2023TC008044 2023TC008044).

Stevens, V. L., & Avouac, J. P. (2015). Interseismic coupling on the main Himalayan thrust. *Geophysical Research Letters, 42*(14), 5828–5837.

Uthaman, M., Singh, C., Singh, A., et al. (2024). Complex multi-fault dynamics in Sikkim Himalaya: New insights from local earthquake analysis. *Geochemistry, Geophysics, Geosystems, 25*(6):e2023GC011363.

Vernant, P., Bilham, R., Szeliga, W., et al. (2014). Clockwise rotation of the Brahmaputra valley relative to India: Tectonic convergence in the eastern Himalaya, Naga Hills, and Shillong plateau. *Journal of Geophysical Research: Solid Earth, 119*(8), 6558–6571.

Whipple, K. X., Shirzaei, M., Hodges, K. V., et al. (2016). Active shortening within the Himalayan orogenic wedge implied by the 2015 Gorkha earthquake. *Nature Geoscience, 9*(9), 711–716.

Part II
Hydro-meteorological Hazards

Chapter 11
Monsoon Hydroclimatic Hazards in the Himalaya: Patterns, Drivers and Impacts

Raju Attada, Rohtash Saini, Ankita, Ali P. Yunus, and A. P. Dimri

Abstract This study investigates monthly and seasonal-scale precipitation variability, long-term trends, extreme precipitation events (EPEs) and their underlying atmospheric drivers over the Western Himalaya (WH) during the ISM from 1979 to 2020 using the ERA5 reanalysis dataset. The WH region exhibits highly variable monthly precipitation patterns during the ISM, driven by complex interactions between local topography and large-scale atmospheric dynamics. Our results reveal that July and August are the dominant monsoon months, with August exhibiting the most pronounced increases in both mean precipitation and extreme precipitation event's frequency, mainly over the Himalayan foothills belt of WH. Spatial patterns highlight that the southern slopes of Uttarakhand and Himachal Pradesh exceed 120 mm/day. Trend analysis indicates a general intensification of EPEs, particularly concentrated in August, alongside complex shifts in mean precipitation patterns on a monthly scale during ISM. Thermodynamic diagnostics show that precipitation distributions are associated with high vertically integrated moisture transport (VIMT exceeding 400 kg m^{-1} s^{-1}), high moist static energy (MSE) and high cloud cover, all supporting deep convection. Findings also reveal that El Niño-Southern Oscillation (ENSO) is a significant modulator for Himalayan precipitation, with most El Niño and La Niña years associated with dry and wet conditions, respectively. Furthermore, findings show a negative correlation of WH precipitation with sea surface temperatures over the Niño 3.4 region, suggesting ENSO as a key driver of precipitation variability in the Himalaya during the monsoon season. Orography-amplified extreme monsoon rainfall drives landslides in the Western Himalaya, necessitating integrated forecasting and climate-resilient infrastructure.

Keywords Western Himalaya · Precipitation · Variability · Extreme events · landslides

R. Attada (✉) · R. Saini · Ankita · A. P. Yunus
Department of Earth and Environmental Sciences, Indian Institute of Science Education and Research Mohali, Mohali, Punjab, India
e-mail: rajuattada@iisermohali.ac.in

A. P. Dimri
School of Environmental Sciences, Jawaharlal Nehru University, New Delhi, India

V. M. Tiwari (ed.), *Himalayan Natural Hazards*, Indian National Science Academy,
https://doi.org/10.1007/978-981-92-3784-5_11

11.1 Introduction

The Western Himalaya (WH) are the strongest modulator of South Asian weather and climate system, significantly influenced by the Indian Summer Monsoon (ISM) from June to September (JJAS). WH received approximately 65–80% of the annual precipitation across the Himalayan region during monsoon (Saini & Attada, 2023; Singh & Bharti, 2019). The complex topography of the WH, with elevations varying from downstream plain areas to peaks exceeding 7000 m, acts as an orographic barrier. This barrier intercepts moisture-laden southwest monsoonal winds, leading to enhanced precipitation through forced ascent and condensation of air masses over the windward side of WH (Barros et al., 2004; Dimri et al., 2016; Houze et al., 2017; Kulkarni et al., 2013; Rasmussen & Houze, 2012). However, this highly heterogeneous terrain and its elevation-dependent weather and climate dynamics make the region highly prone to hydroclimatic extremes, including flash floods, landslides and cloudbursts (Dimri & Niyogi, 2013; Rasmussen & Houze, 2012; Saini & Attada, 2023, 2025; Vellore et al., 2020).

In recent decades, the WH has experienced an increase in the frequency and intensity of extreme precipitation events (EPEs) during the ISM (Chevuturi & Dimri, 2016; Negi et al., 2021; Priya et al., 2017; Riley et al., 2021). The EPEs, such as the Kedarnath flood in 2013, the Leh floods in 2010, the Chamoli river floods in 2016 and the July flood of 2023, among others, have devastating effects on the WH and the nearby plain area. Such EPEs in the WH are not only becoming more intense but also more widespread, often affecting densely populated and ecologically fragile areas of the upstream and downstream of the Himalaya (Hunt et al., 2018a; Jena & Azad, 2022; Nischal & Attada, 2025; Saini & Attada, 2025). Several further studies report that changes in monsoon rainfall characteristics, particularly increasing rainfall intensity during summer, are strongly linked to disaster occurrence, even in regions where seasonal mean rainfall shows weak or mixed trends (Guhathakurta et al., 2011; Karki et al., 2017). The 2021 Chamoli disaster in Himalaya highlights the cascading impacts of hydro-meteorological extremes (Sain et al., 2021; Shugar et al., 2021). This event was initiated by a slope failure and ice–rock avalanche. Antecedent wet conditions associated with changing monsoon–cryosphere interactions played a critical role in destabilizing high-altitude slopes and amplifying downstream flood impacts (Dash et al., 2023; Shugar et al., 2021). More recently, the 2023 Himachal Pradesh floods and landslides were directly linked to persistent and extreme monsoon rainfall during July–August, characterized by prolonged wet spells and multiple cloudburst events (Gowhar et al., 2024; Raghuvanshi & Agarwal, 2024; Saini and Attada, 2025). These findings are consistent with recent evidence showing enhanced moisture transport, stronger convective activity and increased frequency of extreme precipitation events under a warming climate, which significantly elevates geohazard risk in steep Himalayan terrain (Dash et al., 2023; Joseph et al., 2015; Kumari et al., 2023; Ranalkar et al., 2016; Sana et al., 2024; Vellore et al., 2016).

The consequences are severe, leading to significant loss of life, damage to infrastructure, agricultural failures, glacial hazards, landslides and disruptions to hydropower production and water availability downstream (Bohlinger et al., 2019; Cook et al., 2013; Dimri et al., 2023).

Understanding the occurrence, drivers and consequences of these precipitation characteristics and EPEs is paramount for several climatic and atmospheric settings. Firstly, the WH is the source of major river systems such as the Ganga, Indus and Brahmaputra, which sustain millions of people in northern India and neighbouring countries (Bookhagen & Burbank, 2010; Chawla et al., 2018; Cook et al., 2013; Hunt et al., 2018b; Ikram et al., 2023; Sinha & Tandon, 2014; Suhas et al., 2023). Secondly, the region is characterized by a highly heterogeneous precipitation regime, influenced by mesoscale and synoptic-scale systems, including monsoon lows or monsoon depressions (MDs), western disturbances (WDs) and the Intertropical Convergence Zone (ITCZ) (Das et al., 2006; Hunt et al., 2019; Saini & Attada, 2023, 2024, 2025; Vellore et al., 2020). Thirdly, there is a prominent scarcity of high-resolution and consistent observational data due to the sparse distribution of rain gauges and the remoteness of mountainous terrain (Baudouin et al., 2020; Hunt et al., 2018a; Li et al., 2018; Saini et al., 2023). To address these knowledge gaps, recent studies have utilized high-resolution reanalysis (ERA5) and satellite-based (IMERG). The dynamical and thermodynamical process along with the Himalayan orography leads to hydroclimatic extremes in the WH. Dynamically, large-scale circulations, such as Rossby wave breaking, mid to upper-tropospheric troughs and upper-tropospheric divergent, facilitate moisture transport and vertical lifting over the Himalaya (Saini et al., 2026a, 2026b; Chevuturi & Dimri, 2016; Kornhuber et al., 2017; Nischal et al., 2024; Shekhar et al., 2015). These atmospheric settings are also linked to quasi-resonant amplification (QRA) and baroclinic wave patterns that enhance synoptic-scale convergence (Saini & Attada, 2023, 2024, 2025; Athira et al., 2024). Thermodynamically, increased atmospheric water vapour due to warming, enhanced diabatic heating, and moist static energy gradients are crucial for the initiation and intensification of deep convection (Chansaengkrachang et al., 2018; Dube et al., 2014; Khan et al., 2023; Zheng et al., 2020). Furthermore, global warming has led to pronounced temperature increases over the WH, with a warming rate nearly double that of the Indian plains (Nischal et al., 2022; Sabin et al., 2020; Saini et al., 2023). This elevation-dependent warming alters the snowline, shifts precipitation phase from snow to rain and changes the dynamics of runoff generation and flood risk (Dimri et al., 2016; Liu et al., 2022; Singh & Kumar, 1997; Swain et al., 2022).

Consequently, these extreme precipitation events trigger devastating floods and landslides, particularly when they occur over fragile and geomorphologically sensitive landscapes, thereby severely impacting livelihoods (Dixit et al., 2024). A global analysis by Froude and Petley (2018) reported a general increase in recorded landslide occurrences between 2004 and 2010, followed by a decline until April 2015, after which the incidence has shown a renewed upward trend, particularly pronounced along the Himalayan Arc. Their study revealed a strong correlation between mean monthly rainfall and landslide occurrences in South Asia ($R^2 = 0.99$), underscoring the dominant role of seasonal rainfall as a trigger. For instance, in July 2023, Himachal Pradesh experienced an unprecedented 436% increase in rainfall relative to the normal average, which led to more than 4000 landslides across the region (Kaushal et al., 2025). Based on this data, a threshold between triggering and non-triggering rainfall events was distinguished using an exponential rainfall-duration relationship: $I = 37.844D^{-0.95}$.

Despite these scientific studies conducted over the WH, significant gaps persist in our understanding of the spatial variability and associated EPEs at a monthly scale across the Himalaya. Much of the existing research has concentrated on individual case studies, leaving a lack of comprehensive and seasonal progression of the precipitation variability over the Himalayan arc during the ISM. Addressing this gap, the present study investigates precipitation variability and hydroclimatic extremes across the WH during the monsoon season on a monthly scale from June to September. The primary objective is to examine the spatiotemporal variability of precipitation and its extremes during the ISM. Secondly, we investigate the physical mechanisms and drivers that trigger EPEs. Finally, we assess the geomorphic and hydrological impacts of extreme precipitation in the complex orographic regions of the Himalaya.

This study is organized as follows. Section 2 elucidates the data and methodology. Section 3 evaluates the ISM mean precipitation, EPEs, trends, along with their potential drivers and impacts. Section 4 offers the summary and conclusions of the study.

11.2 Data and Methodology

11.2.1 Study Area

The WH lies between 25.5° N and 37.5° N latitude and 72° E and 84° E longitude of the northwestern region of the Indian Himalaya. The elevation of WH varies from hundreds of metres near sea level to over 6500 m. It encompasses the Himalayan foothills, the Shivalik ranges and the Siwalik Hills, with the WH, in particular, demonstrating a high degree of topographical complexity shown in Fig. 11.1, which marks it as a severe and complex landscape compared to the global average.

11.2.2 Methodology

This study utilizes the European Centre for Medium-Range Weather Forecasts (ECMWF) developed reanalysis ERA5 (Hersbach & Dee, 2016) data to examine precipitation distribution and hydroclimate extremes during the ISM over the WH. ERA5 provides high-resolution data using a 4D-Var data assimilation system within the integrated forecasting system (IFS). The dataset spans from 1979 to 2020 with a spatial resolution of 0.25° × 0.25° and hourly temporal resolution, which was aggregated into daily and monthly means for this analysis. Key ERA5 variables employed in the study include total cloud cover, relative humidity, vertical pressure

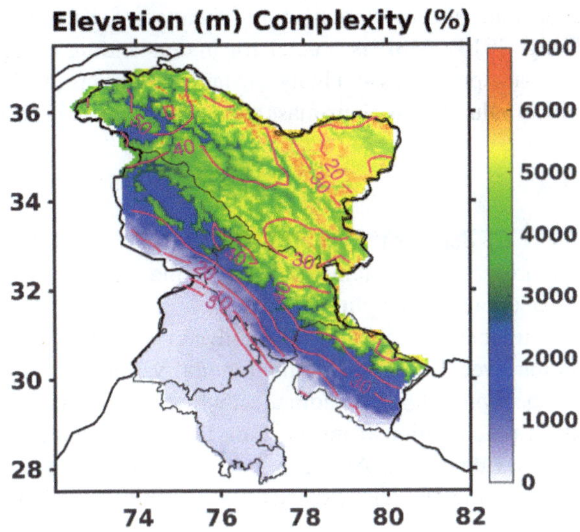

Fig. 11.1 **a** Topography (shaded; units in metres) map of Western Himalaya. Magenta counters represent the topography complexity (%) derived from GTOP030 data

velocity (omega), geopotential height, zonal (*u*) and meridional (*v*) winds, as well as derived variables moist static energy (MSE) and vertically integrated moisture transport (VIMT).

In this study, we analysed the long-term means and trends of precipitation on the monthly scale (June to September, JJAS) during the monsoon season. Further, EPEs were identified on a grid-by-grid basis using the 95th percentile threshold of daily precipitation derived from the full 42-year JJAS dataset. Subsequently, intensity and frequency of the EPEs were calculated for the Uttarakhand (UK), Himachal Pradesh (HP) and J&K. To investigate associated physical mechanisms and the drivers behind the precipitation variability and EPEs, VIMT and MSE were analysed.

Vertically integrated moisture transport (VIMT) accounts for the total amount of moisture present in the atmosphere (Fasullo & Webster, 2003; Zhang et al., 2001). It is calculated by vertically integrating the zonal (*u*) and meridional (*v*) moisture fluxes from the surface up to 300 hPa, considering specific humidity (*q*) and gravitational acceleration (*g*). This provides a comprehensive measure of atmospheric moisture transport to the study area. To obtain insights into the fundamental physical mechanisms driving changes in precipitation distribution and EPEs, a monthly-scale map is generated spanning the period from 1979 to 2020. The following equation is used to calculate the VIMT.

$$\mathrm{VIMT} = -\frac{1}{g}\int_{1000\ \mathrm{hPa}}^{300\ \mathrm{hPa}} qV\,\mathrm{d}p \tag{11.1}$$

Moist static energy (MSE) is a thermodynamic parameter that accounts for the latent heat released during water vapour condensation and the potential energy associated with the air parcel's height. MSE comprises three main components: dry-air enthalpy (heat content), specific gravitational potential energy and a term representing the contribution from latent heating (Andersen & Kuang, 2012; Dube et al., 2014; Wu & Deng, 2013). MSE is crucial for understanding energy conservation during moist adiabatic processes and in hydrostatic conditions (Maloney 2009; Dube et al., 2014). MSE (value per total air mass) is defined by Eq. (11.2).

$$\mathrm{MSE} = C_P T + gz + L_v q, \tag{11.2}$$

where C_P denotes the specific heat capacity at constant pressure (1004 J kg^{-1} K^{-1}), T represents the air temperature, g is the gravitational acceleration (9.8 m s^{-2}), z is the geopotential height.

Interannual variability was analysed to identify wet and dry years of ISM precipitation over the WH. This interannual variability was calculated using normalised seasonal precipitation anomalies, which were calculated by subtracting the seasonal mean precipitation from the long-term precipitation and dividing by the seasonal standard deviation. A wet and dry year was defined as one in which the summer precipitation anomaly exceeds $\pm$ 0.5 standard deviations (SD) of the mean, respectively. Furthermore, an empirical orthogonal function (EOF) analysis was conducted to explore the influence of large-scale climatic drivers (Barros et al., 2004; Kar & Rana, 2014). Sea surface temperature (SST) over the Niño 3.4 index was obtained from the National Oceanic and Atmospheric Administration (NOAA). A 5-year running correlation analysis was performed to evaluate the relationship between the El Niño-Southern Oscillation (ENSO) and the ISM Precipitation.

11.3 Results and Discussions

11.3.1 Distribution of Precipitation over WH

Precipitation distribution over the WH reveals distinct seasonal and spatial variability on a monthly scale during ISM. The Himalayan foothill belt receives the maximum precipitation during the month of July, followed by August, as shown in Fig. 11.2a. Findings show that the monsoon onset starts in June, intensifies in July and August and begins to retreat in September (Fig. 11.2a). The months of July and August represent the peak monsoon period, with both ERA5 and IMERG showing the highest precipitation over the southern slopes of the Himalaya, majorly in region of UK and HP during these months. The spatial distribution of monthly mean precipitation shows strong orographic influence, with the highest accumulation along south-facing slopes, where moist monsoon winds are forced to ascend, resulting in heavy precipitation (Das et al., 2006; Hunt et al., 2019; Saini & Attada, 2023, 2024, 2025; Vellore et al., 2020). In contrast, the higher-altitude region of WH receives less precipitation due to the prevailing rain-shadow effect. We have further investigated the temporal trends of mean precipitation through an area-averaged time series at monthly and seasonal scale in UK, HP and J&K states of WH (Fig. 11.2b).

The time series of daily mean precipitation shows increasing trend in the long term (in ERA5) as well as recent decades (in IMERG) during the JJAS seasonal scale; however, shows mixed trends in monthly scale. Peak monsoon month August shows statistically significant increasing trends for all three Himalayan states (Fig. 11.2b), while decreasing long-term trends during the month of July. These findings align with previous studies reporting spatiotemporal heterogeneity in monsoon rainfall trends across the Himalaya (Ménégoz et al. 2013; Nischal et al., 2022; Saini and Attada 2025).

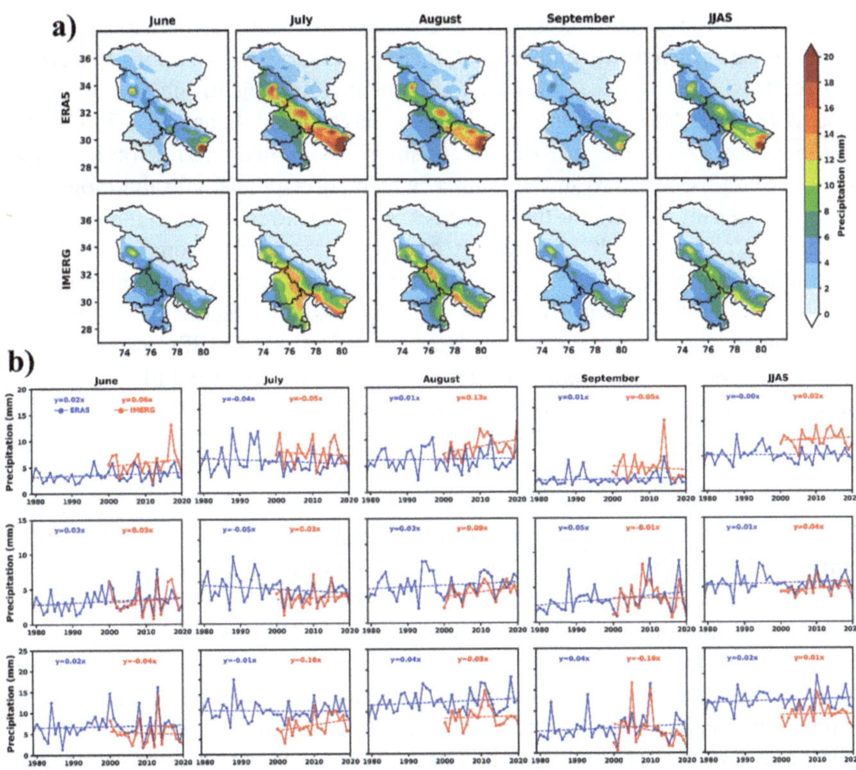

Fig. 11.2 **a** Spatial distribution of precipitation climatology (mm) for June, July, August, September and the seasonal mean (JJAS). The first row represents the ERA5 reanalysis data (1979–2020), while the second row represents the IMERG observations (2000–2020). b) Area-averaged precipitation trends over the Himalayan states (third row: J&K, fourth row: Himachal Pradesh, and fifth row: Uttarakhand) for the same months, and the seasonal mean (JJAS)

11.3.2 Spatiotemporal Trend of Extreme Precipitation Intensity and Frequency

We analyzed the spatiotemporal variability in the intensity and frequency of EPEs over the WH. Figure 11.3 presents the spatial distribution of precipitation intensity exceeding the 95th percentile threshold at each grid point during the ISM. At the monthly scale, spatial patterns highlight a progressive intensification of extreme precipitation throughout the monsoon period, as shown in the first panel (ERA5) and second panel (IMERG). In June, precipitation intensity remains low across WH, consistent with the mean condition of the precipitation during the early stages of monsoon onset. During July and August, there is a notable rise in intensity, particularly over the southern slopes of HP and UK, where orographic lifting boosts convective activity (Fig. 11.3). Hotspots in the Himalayan foothill belt of HP, UK and J&K frequently experience intense precipitation. A significant reduction in precipitation intensity is observed in September, which corresponds with the monsoon's withdrawal. During the month of September, significant reduction in the precipitation intensity, aligning with the monsoon's retreat phase. The seasonal (JJAS) mean precipitation from both ERA5 and IMERG confirms a spatial gradient in precipitation intensity, with low precipitation over high-altitude regions in the northern WH and higher precipitation along the Himalayan foothills. IMERG reports higher precipitation intensities compared to ERA5, especially in the foothill regions. The spatial trends in 95th percentile precipitation intensity (shown in the third and fourth panels of Fig. 11.3) reveal considerable heterogeneity in extreme precipitation over the WH. The month of June shows a mostly increasing trend in both ERA5 and IMERG. During July, ERA5 indicates a decreasing trend over the Himalayan foothill belt; however, IMERG shows a mixed trend during recent decades. It highlights specific hotspots within the WH where statistically significant increases in extreme precipitation intensity are observed, reaching up to 12 mm day^{-1} per decade. While available precipitation datasets reasonably capture the broad spatial distribution of precipitation events, most exhibit substantial limitations over complex terrain in the northwestern Indian Himalayan region, particularly in quantifying high-altitude precipitation (Nischal et al., 2022; Palazzi et al., 2013; Saini & Attada, 2023). ERA5 suffers from model-related precipitation biases associated with convective parameterization schemes and an inadequate representation of regional orography. Similarly, IMERG shows weak agreement with in-situ rain-gauge observations, often misclassifying warm orographic clouds as non-precipitating due to the use of fixed infrared precipitation and cold-cloud duration thresholds (Paredes Trejo et al., 2016). These limitations have been consistently documented through station-based evaluations in several studies (e.g. Nischal et al., 2022; Saini & Attada, 2023; Ullah et al., 2018).

Further, we have investigated the trends for area-averaged precipitation extremes at the monthly and seasonal scale for the Himalayan states, UK, J&K and HP (Fig. 11.4). The findings show heterogeneity in spatial distribution in the precipitation trends for the Himalayan states. J&K experiences a rise in the intensity of extreme precipitation. In contrast, the UK is seeing a decrease, while HP displays a

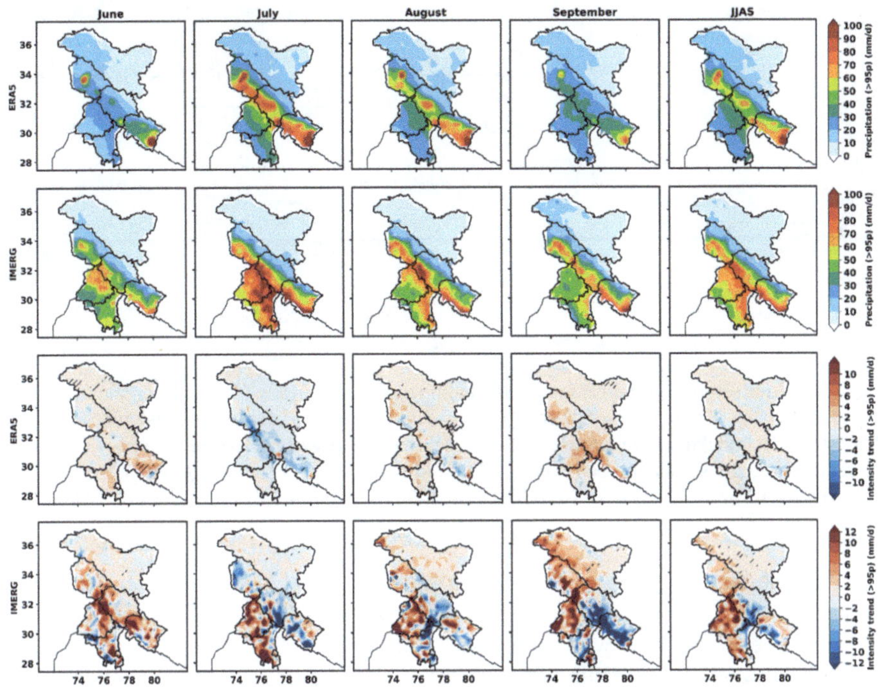

Fig. 11.3 Spatial distribution of mean precipitation intensity exceeding the 95th percentile (mm day^{-1}) for June, July, August, September, and the JJAS season, derived from ERA5 reanalysis (first row; 1979–2020) and IMERG satellite observations (second row; 2000–2020). Trends in extreme precipitation intensity are shown for ERA5 (third row; 1979–2020) and IMERG (fourth row; 2000–2020). Statistically significant trends ($p < 0.05$) are indicated by hatching

more complex pattern with mixed positive and negative trends on a monthly scale. The JJAS seasonal mean highlights a positive trend in ERA5, while IMERG shows modest declines over the UK.

Figure 11.5 demonstrates the spatial and temporal trends in the frequency of EPEs. Panel (5a) shows the spatial distribution of EPE trends (events per decade), while panel (5b) represents the time series of area-averaged frequency of EPE. In June, findings show a notable increase in EPE frequency over large parts of J&K and HP, particularly across the Pir Panjal and Dhauladhar ranges, with IMERG capturing a more pronounced trend. In July, trends are mixed across all three states, with ERA5 showing decreases in EPEs over J&K. August also shows scattered patterns, with weak trends in ERA5 and localized increases in IMERG, particularly over J&K and HP. September exhibits clearer increasing trends, especially over northern J&K and parts of central and eastern HP and UK, with ERA5 capturing stronger signals than IMERG. On the seasonal scale (JJAS), IMERG highlights strong and widespread increases in EPE frequency across most of J&K and northern HP, while ERA5 shows similar but comparatively weaker trends. Uttarakhand exhibits more localized increases, particularly in the IMERG dataset.

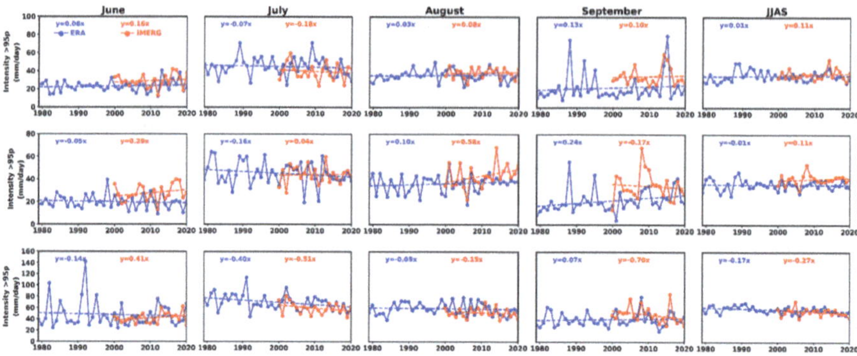

Fig. 11.4 Time series of area-averaged precipitation intensity (mm day^{-1}; > 95th percentile) for the Himalayan states. The first, second, and third rows correspond to Jammu and Kashmir, Himachal Pradesh, and Uttarakhand, respectively., while columns represent months of June, July, August, September, and the seasonal mean (JJAS). The blue and red lines correspond to data from ERA5 and IMERG, respectively, with dashed lines indicating the linear trend

The time series in panel (5b) supports these spatial patterns. Over J&K, the frequency of EPEs in June has increased sharply in recent decades, with IMERG showing a steeper rise (18 events mm day^{-1}/decade) than ERA5 (3 events mm day^{-1}/decade). July and August exhibit higher year-to-year variability, but September shows a consistent rising trend in ERA5, while IMERG remains more stable. For the full JJAS season, IMERG indicates a significant increase in EPE frequency over J&K (8 events mm day^{-1}/decade), with a sharp rise in the most recent decades. In HP, both datasets show increasing EPE trends in June and September, with ERA5 indicating a strong rise in September (11 events mm day^{-1}/decade). July shows almost no change, while August displays modest increases in both datasets. For JJAS, both datasets indicate a gradual rise (2–3 events mm day^{-1}/decade). Uttarakhand, in contrast, shows relatively subdued trends across all months, though IMERG captures increasing EPEs in September (0.13 mm day^{-1}/decade) and a modest seasonal rise (0.02 mm day^{-1}/decade). Results show that the months of June and September are emerging as periods of significant EPE increases, possibly linked to shifting monsoon onset and retreat. While ERA5 broadly agrees with the direction of change, it generally reports a lower magnitude of increase. These results highlight the rising risk of EPEs, especially in J&K and HP, which may have serious implications for flood hazards, landslides and infrastructure resilience in this ecologically fragile region.

Figure S1 illustrates the spatial distribution of low-level wind speed and relative vorticity during the ISM using ERA5 reanalysis data. Figure (S1.a) shows the mean wind speed at the 850 hPa, revealing a well-developed southwesterly low-level jet (LLJ) over the Arabian Sea (AS), with maximum wind speeds exceeding 16 m/s. As the monsoon winds travel across the Indian land mass, wind speeds weaken but transporting moisture across peninsular India and into the Himalayan foothills, where orographic convergence plays a key role in triggering heavy precipitation events. Figure (S1.b) depicts the mean relative vorticity, highlighting regions of cyclonic

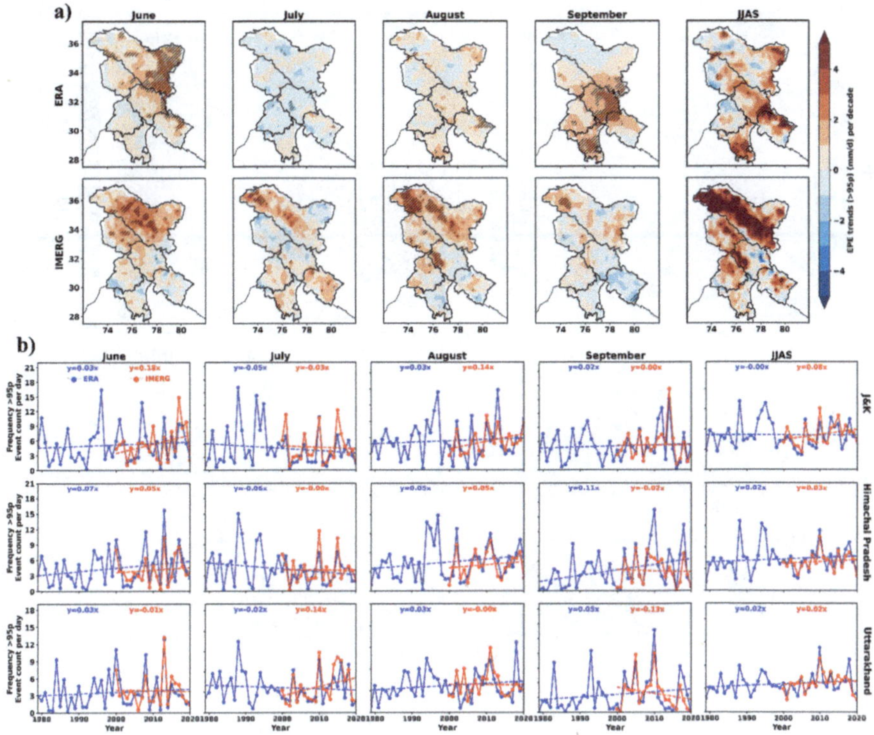

Fig. 11.5 **a** Spatial distribution of extreme precipitation frequency exceeding the 95th percentile (mm day^{-1} per decade) for June, July, August, September, and the JJAS season, derived from ERA5 reanalysis (first row; 1979–2020) and IMERG satellite observations (second row; 2000–2020). **b** Area-averaged trends in extreme precipitation frequency (event count day^{-1}) over the Himalayan states (third row: Jammu & Kashmir; fourth row: Himachal Pradesh; fifth row: Uttarakhand), shown by the blue line (ERA5: 1979–2020) and the red line (IMERG: 2000–2020)

(positive) and anticyclonic (negative) circulation. Positive vorticity (in blue) dominates over the northern Bay of Bengal, central India and along the Himalayan foothills regions, often associated with enhanced atmospheric instability and upward motion conducive to deep convection. Intensified low-level wind dynamics and vorticity play a key role in modulating moisture transport, atmospheric instability and the spatial distribution of hydroclimate extremes over the WH during the monsoon season (Saini and Attada; Hunt et al., 2018a).

Further, to understand the atmospheric dynamics and thermodynamics on precipitation variability, VIMT has been calculated on a monthly scale during ISM as shown in Fig. 11.6. The finding clearly illustrates the progression of moisture transport from June through September. The southwesterly moisture flow from the AS and the southeasterly flow from the Bay of Bengal (BoB) are well depicted in the month of June, indicating the arrival of the monsoon and the transport of moisture towards the Himalaya (Fig. 11.6a). This moisture transport intensifies in July and

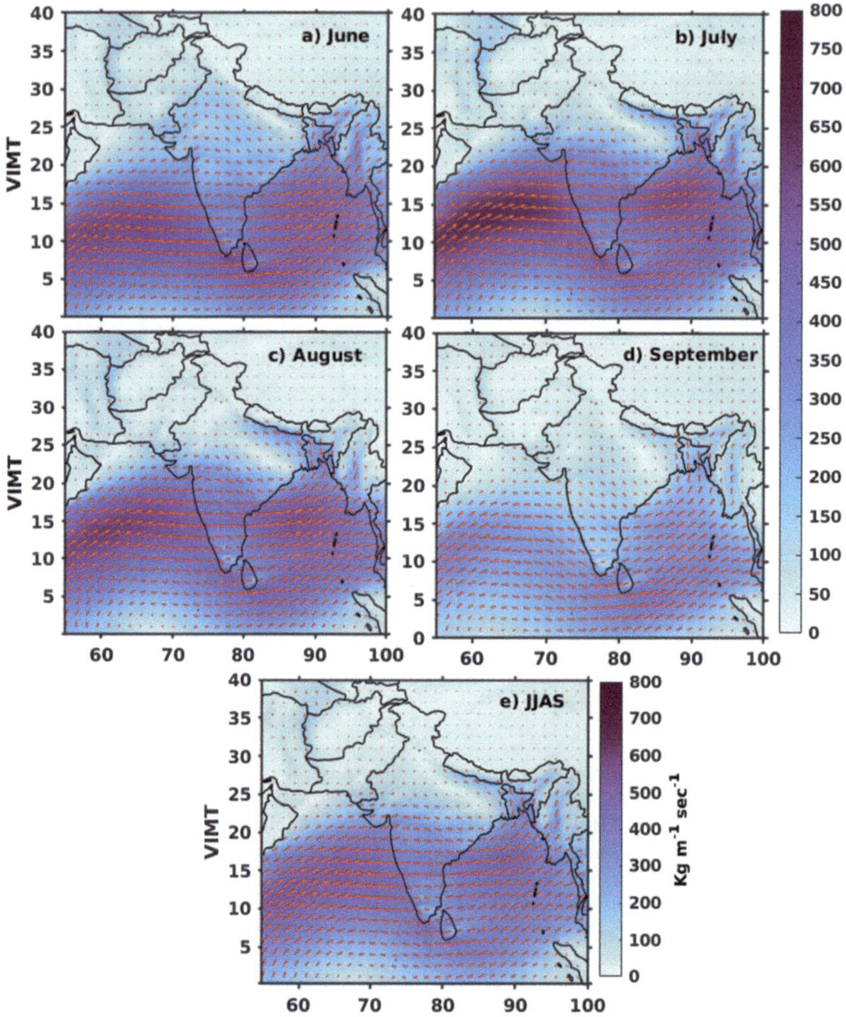

Figure S1 Climatological spatial distribution of **a** wind speed (m/s) and **b** relative vorticity (s^{-1}) at 850 hPa during JJAS, based on ERA5 reanalysis (1979–2020)

August, indicating the peak of the monsoon season. During these months, the VIMT magnitude reaches its maximum (> 600 Kg m^{-1} sec^{-1}) as indicated by the deep red and purple shading, signifying an influential and sustained influx of moisture carried by the strong southwesterly winds from the AS (Figs. 11.6b-c). This intensified VIMT, combined with moisture drawn from the BoB, plays a key role in driving precipitation over the WH. Monsoon's withdrawal becomes evident (see Fig. 11.6d) during September, as the strength of the VIMT weakens. Panel (e) of Fig. 11.6 shows the mean conditions during the JJAS period, emphasizing the strong southwesterly VIMT originating from the AS which is the dominant feature. Overall results show

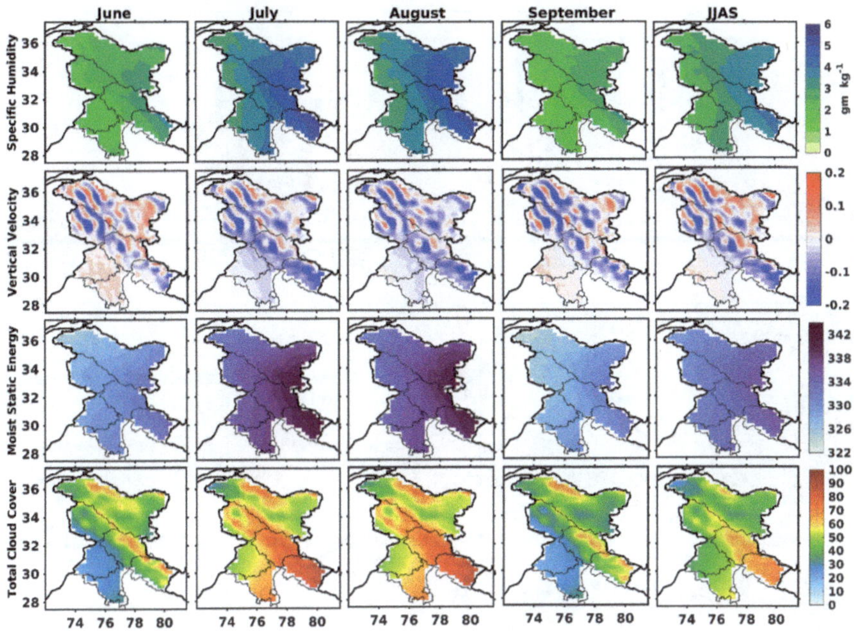

Fig. 11.6 Climatological spatial distribution of vertically integrated moisture transport (VIMT; kg m^{-1} s^{-1}) for June, July, August, September and the JJAS season, based on ERA5 reanalysis (1979–2020)

that AS and BoB are the major moisture sources, with their seasonal cycle of moisture transport directly influencing the timing and intensity of monsoon precipitation.

Further, a comprehensive analysis of the monsoon dynamics over the WH region reveals a spatiotemporal evolution of key atmospheric parameters from June to September shown in Fig. 11.7. The monthly progression of the monsoon onset, peak monsoon and withdrawal conditions is well depicted in Fig. 11.7. June exhibits relatively low specific humidity (first row), weak vertical velocity (indicating less upward motion) and scattered cloud cover (fourth row), while July and August show a high available moisture over the foothills along with a strong negative vertical velocity, signifying vigorous and widespread upward motion of air, which fuels the build-up of MSE, a key indicator of atmospheric instability. This atmospheric instability leads to more cloud formation, signalling the peak of convective activity and the establishment of the monsoon's core phase. These high-intensity monsoon atmospheric settings persist through August, with both July and August showing peak values for specific humidity, vertical velocity, MSE and cloud cover, over the foothills where orographic lift enhances convection. As the monsoonal season progressed, a decline in specific humidity, a weakening of upward air movement and a reduction in overall cloud cover is exhibited during September. At the seasonal scale, high moisture, strong ascent with high MSE and widespread cloud cover over the Himalayan foothills highlight the region's vital influence on monsoon behaviour.

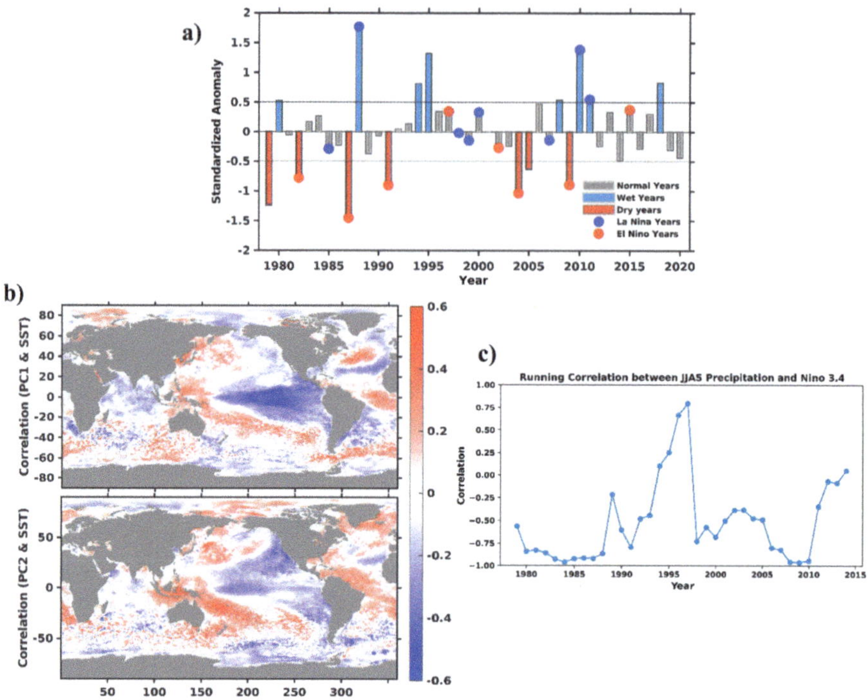

Fig. 11.7 Climatological spatial distribution of (first row) specific humidity (g/kg), (second row) vertical velocity (Pa/s), (third row) moist static energy (kJ/kg), and (fourth row) total cloud cover (%) for June, July, August, September and the seasonal mean (JJAS, rightmost column), based on ERA5 reanalysis (1979–2020)

Figure 11.8 shows the relationship between the ENSO and precipitation variability over the WH. Figure 11.8a reveals that most dry years are associated with El Niño, characterized by warmer SSTs over Niño 3.4, while most wet years agree with La Niña, defined by cooler SSTs over the Niño 3.4 region. This finding emphasizes that ENSO is the stronger modulator of the interannual variability of monsoon precipitation in the WH region. This relationship is further explored in Fig. 11.8 (b), which presents a spatial correlation map between the dominant modes of precipitation variability (represented by Principal Components 1 and 2) and global SSTs. The finding shows a strong negative correlation between the PC1 and SSTs over the Niño 3.4 region. This signifies that colder-than-average SSTs in this region (La Niña condition) are directly linked to an increase in precipitation, thereby reinforcing the findings from panel (a). The secondary mode (PC2) also exhibits a similar, albeit weaker, negative correlation, confirming that both primary and secondary modes of precipitation variability are strongly influenced by SSTs in the Pacific. Finally, ENSO recurrence is further explored, as shown in Fig. 11.8 (c), during the JJAS time period. A 5-year running window is then applied to compute the correlation between Pacific SST anomalies and Himalayan precipitation, as this timescale aligns with ENSO's typical recurrence period (Chen & Wallace, 2015; Tyrrell et al., 2015). Findings show that 5-year running correlations vary over time (interannual scale), signifying that ENSO influences on WH precipitation variability. These results are consistent with previous studies (Bharati et al., 2025; Kamil et al., 2019) which demonstrate that ENSO exerts a strong but seasonally dependent influence on Himalayan hydroclimate through large-scale circulation adjustments, particularly via modulation of the subtropical jet and associated synoptic systems. During El Niño phases, a southward displacement of the subtropical jet enhances moisture transport from the Arabian Sea, leading to more intense winter precipitation (Bharati et al., 2025). While during monsoon season extreme precipitation frequency substantially increases over the western Himalaya during El Niño compared to La Niña years. Proxy-based and observational studies further support the existence of long-standing ENSO teleconnections with Himalayan precipitation and snow accumulation at interannual timescales (Islam et al., 2025; Sahu et al., 2025).

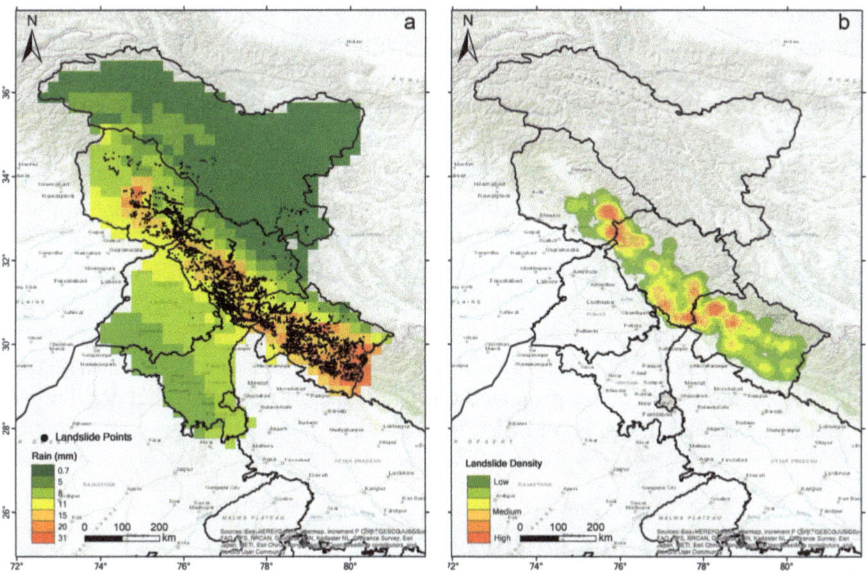

Fig. 11.8 **a** Time series of standardised precipitation anomalies over the Western Himalaya during the Indian Summer Monsoon (ISM), with bars indicating wet (blue), dry (red), and normal (grey) years. El Niño and La Niña events are marked with red and blue dots, respectively. Panel **b** illustrates the spatial correlation between global sea surface temperature (SST) and the time series of the first two principal components (PC1 and PC2) during the ISM. Panel **c** displays the 5-year running correlation between mean rainfall in the Northwest (NW) region and the Niño 3.4 SST index

11.3.3 Geomorphic and Hydrological Impacts of Extreme Precipitation in the Western Himalaya

Heavy rainfall often occurs over the Himalayan foothills, frequently triggering landslides in the region (Hunt & Dimri, 2021). We assessed the spatiotemporal variability of landslides triggered during the peak monsoon precipitation period across the WH. Figure 11.9 provides a combined analysis of monsoon rainfall and landslide distribution. Panel (a) overlays landslide occurrences, sourced from the Geological Survey of India, on average July precipitation from ERA5. A dense concentration of landslides is evident along the southern slopes of Uttarakhand and Himachal Pradesh, where mean daily rainfall often exceeds 15 mm and reaches over 30 mm in localized zones. Panel (b) further supports this pattern, depicting high landslide point density along the same orographic corridor, particularly across the Lesser Himalaya. The spatial overlap of intense rainfall and high landslide density aligns with the seasonal evolution of the ISM, which typically begins in June, intensifies through July and August and retreats by September. Both ERA5 and IMERG datasets indicate maximum rainfall accumulation during July along the south-facing slopes of the Himalaya, where moist monsoonal winds are forced to ascend steep terrain, enhancing orographic precipitation and convective activity. Thermodynamic diagnostics further confirm the presence of elevated moisture, high moist static energy and extensive cloud cover during these periods. In addition, the steep topographic gradients, complex lithology and active tectonic regime exacerbate slope instability here. Rapid infiltration of intense rainfall elevates pore-water pressure and increases saturation loading, reducing slope strength. The resulting clustering of landslides in both space and time reflects the strong control of extreme precipitation on geomorphic hazards. This cascading relationship between atmospheric forcing and terrain response highlights the growing vulnerability of the WH region to monsoon-triggered landslides, particularly under projected climate warming scenarios that are likely to enhance monsoonal intensity and EPE frequency.

11.4 Summary and Conclusion

This research offers an in-depth evaluation of precipitation variability and hydroclimatic extremes over the WH during the ISM, utilizing high-resolution ERA5 reanalysis and IMERG satellite data.

- The analysis highlights a distinct spatial and temporal pattern in precipitation, with July and August emerging as peak precipitation months, over the WH.
- EPEs, identified using the 95th percentile threshold, show an increase in frequency and intensity, especially in J&K and HP. Significant increasing trends are observed during the monsoon onset month (June) and monsoon withdrawal (September) months. IMERG tends to capture stronger EPE than ERA5, though both datasets align on overall trends.

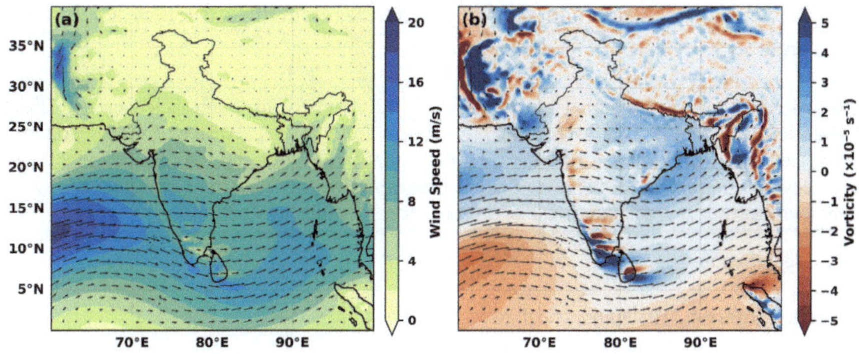

Fig. 11.9 **a** Landslide points (sourced from Geological Survey of India) overlaid on average July precipitation (ERA5). **b** Landslide point density (magnitude-per-unit area) plotted for the Western Himalaya

- WH are experiencing the intensified transport of moisture from the AS and BoB, coupled with increased specific humidity, stronger upward air movement and higher moist static energy. These factors play a significant role in driving precipitation variability and the occurrence of EPEs. Additionally, orographic and dynamic atmospheric processes, such as low-level jets and cyclonic vorticity, play a role in enhancing localized convective activity.
- The study also confirms a strong teleconnection between ENSO phases and monsoonal precipitation over the WH, with La Niña years associated with wetter conditions and El Niño with drier conditions over the WH. However, the running correlation between ENSO and Himalayan precipitation shows varying correlations over time, indicating modulation by additional climatic factors.
- Extreme precipitation events during the peak monsoon months, amplified by orographic forcing and thermodynamic conditions, are the primary drivers of rainfall-induced landslides in the WH, highlighting the urgent need for integrated hazard forecasting and climate-resilient infrastructure planning in this vulnerable region.

Overall, the findings highlight rising monsoon variability and an increasing risk of hydroclimatic extremes in the WH. These trends have significant implications for water resource management, disaster preparedness and regional climate adaptation in this ecologically and socioeconomically sensitive region. The increasing frequency and intensity of extreme precipitation events highlight the urgent need to improve predictive capabilities to mitigate flood hazards and associated geomorphic and hydrological impacts. Accurate prediction of the location, intensity, frequency and duration of heavy precipitation remains a major challenge for effective flood forecasting in the Himalaya (Nanditha & Mishra, 2021). Improving weather and climate models to better represent monsoon lows, Western Disturbances and their interactions, combined with high-resolution simulations using the WRF model, can support early-warning systems, adaptation strategies and resilient infrastructure development in the region.

Acknowledgements The authors are thankful to the Ministry of Earth Sciences (MoES/16/03/2021-RDESS/NARM-3), Government of India for funding this research. Authors gratefully acknowledge the IISER Mohali for providing the essential facilities and resources that facilitated the successful completion of this research. We thank the IISER Super Computational facility. We extend our heartfelt appreciation to the reviewers for their valuable feedback.

Ethics Declarations

Competing Interest The authors declare no competing interests.

Ethical Approval All authors have read and understood the "Ethical Responsibilities of Authors" and have complied with these guidelines where applicable, as outlined in the Instructions for Authors.

Consent to Participate Not applicable.

Consent to Publish All authors consent to the publication of this research.

Data Availability The datasets utilized in this study are publicly accessible and can be obtained from their respective sources.

References

Andersen, J. A., & Kuang, Z. (2012). Moist static energy budget of MJO-like disturbances in the atmosphere of a zonally symmetric aquaplanet. *Journal of Climate, 25*, 2782–2804. https://doi.org/10.1175/JCLI-D-11-00168.1

Athira, K. S., Attada, R., & Rao, V. B. (2024). Synoptic dynamics of cold waves over North India: Underlying mechanisms of distinct cold wave conditions. *Weather Clim Extrem, 43*, 100641. https://doi.org/10.1016/j.wace.2024.100641

Barros, A. P., Kim, G., Williams, E., & Nesbitt, S. W. (2004). Probing orographic controls in the Himalayas during the monsoon using satellite imagery. *Natural Hazards and Earth System Sciences, 4*, 29–51. https://doi.org/10.5194/nhess-4-29-2004

Baudouin, J. P., Herzog, M., & Petrie, C. A. (2020). Cross-validating precipitation datasets in the Indus River basin. *Hydrology and Earth System Sciences, 24*, 427–450. https://doi.org/10.5194/hess-24-427-2020

Bharati, P., Deb, P., Hunt, K. M., Orr, A., & Dash, M. K. (2025). ENSO-induced latitudinal variation of the subtropical jet modulates extreme winter precipitation over the Western Himalaya. *Advances in Atmospheric Sciences, 42*(3), 427–437. https://doi.org/10.1007/s00376-024-4057-2.

Bohlinger, P., Sorteberg, A., Liu, C., et al. (2019). Multiscale characteristics of an extreme precipitation event over Nepal. *Quarterly Journal of the Royal Meteorological Society, 145*, 179–196. https://doi.org/10.1002/qj.3418

Bookhagen, B., & Burbank, D. W. (2010). Toward a complete Himalayan hydrological budget: Spatiotemporal distribution of snowmelt and rainfall and their impact on river discharge. *Journal of Geophysical Research—Earth Surface, 115*. https://doi.org/10.1029/2009JF001426

Chansaengkrachang, K., Luadsong, A., & Aschariyaphotha, N. (2018). Vertically integrated moisture flux convergence over Southeast Asia and its relation to rainfall over Thailand. *Pertanika Journal of Science and Technology, 26*, 235–246.

Chawla, I., Osuri, K. K., Mujumdar, P. P., & Niyogi, D. (2018). Assessment of the weather research and forecasting (WRF) model for simulation of extreme rainfall events in the upper Ganga Basin. *Hydrology and Earth System Sciences, 22*, 1095–1117. https://doi.org/10.5194/hess-22-1095-2018

Chen, X., & Wallace, J. M. (2015). ENSO-like variability: 1900–2013. Journal of climate, *28*(24), 9623-9641

Chevuturi, A., & Dimri, A. P. (2016). Investigation of Uttarakhand (India) disaster-2013 using weather research and forecasting model. *Natural Hazards, 82*(3), 1703–1726. https://doi.org/10.1007/s11069-016-2264-6

Cook, E. R., Palmer, J. G., Ahmed, M., Woodhouse, C. A., Fenwick, P., Zafar, M. U., Wahab, M., & Khan, N. (2013). Five centuries of upper Indus River flow from tree rings. *Journal of Hydrology, 486*, 365–375. https://doi.org/10.1016/j.jhydrol.2013.02.004

Das, S., Ashrit, R., & Moncrieff, M. W. (2006). Simulation of a Himalayan cloudburst event. *Journal of Earth System Science, 115*, 299–313. https://doi.org/10.1007/BF02702044

Dash, S., Shekhar, M. S., Bhardwaj, P., Swain, M., Satyawali, P. K., & Mohanty, U. C. (2023). Chamoli disaster 2021, Uttarakhand: A study on the role of a hidden meteorological parameter. *Journal of Earth System Science, 132*(4), 193. https://doi.org/10.1007/s12040-023-02207-w

Dimri, A. P., & Niyogi, D. (2013). Regional climate model application at subgrid scale on Indian winter monsoon over the western Himalayas. *International Journal of Climatology, 33*, 2185–2205. https://doi.org/10.1002/joc.3584

Dimri, A. P., Pooja, Jeelani, G., & Mohanty, U. C. (2023). Western disturbances vs non-western disturbances days winter precipitation. *Climate Dynamics, 61*(9), 4825–4847.

Dimri, A. P., Thayyen, R. J., Kibler, K., Stanton, A., Jain, S. K., Tullos, D., & Singh, V. P. (2016). A review of atmospheric and land surface processes with emphasis on flood generation in the southern Himalayan rivers. *Science of the Total Environment, 556*, 98–115. https://doi.org/10.1016/j.scitotenv.2016.02.206

Dixit, S., Siva Subramanian, S., Srivastava, P., Yunus, A. P., Martha, T. R., & Sen, S. (2024). Numerical-model-derived intensity–duration thresholds for early warning of rainfall-induced debris flows in a Himalayan catchment. *Natural Hazards and Earth System Sciences., 13;24*(2), 465–480.

Dube, A., Ashrit, R., Ashish, A., Sharma, K., Iyengar, G. R., Rajagopal, E. N., & Basu, S. (2014). Forecasting the heavy rainfall during Himalayan flooding-June 2013. *Weather Clim Extrem, 4*, 22–34. https://doi.org/10.1016/j.wace.2014.03.004

Fasullo, J., & Webster, P. J. (2003). A hydrological definition of Indian monsoon onset and withdrawal. *Journal of Climate, 16*, 3200–3211. https://doi.org/10.1175/1520-0442(2003)016<3200a:AHDOIM>2.0.CO;2

Froude, M. J., & Petley, D. N. (2018). Global fatal landslide occurrence from 2004 to 2016. *Natural Hazards and Earth System Sciences., 18*(8), 2161–2181.

Gowhar, S., Hussain, M., & Srivastava, A. (2024). Stability assessment of slopes under normal weather and heavy rainfall conditions in the Himalayan region. *Transportation Infrastructure Geotechnology, 11*(3), 978–1007. https://doi.org/10.1007/S40515-023-00314-1/METRICS

Guhathakurta, P., Sreejith, O. P., & Menon, P. A. (2011). Impact of climate change on extreme rainfall events and flood risk in India. *Theoretical and Applied Climatology, 104*, 133–148. https://doi.org/10.1007/s00704-010-0342-5

Hersbach, H., & Dee, D. (2016). *ERA5 reanalysis is in production, ECMWF newsletter 147* (Vol. 147, p. 7). ECMWF. ECMWF Newsl.

Houze, R. A., Jr., McMurdie, L. A., Rasmussen, K. L., Kumar, A., & Chaplin, M. M. (2017). Multiscale aspects of the storm producing the June 2013 flooding in Uttarakhand, India. *Monthly Weather Review, 145*, 4447–4466. https://doi.org/10.1175/MWR-D-17-0004.1

Hunt, K. M. R., & Dimri, A. P. (2021). Synoptic-scale precursors of landslides in the western Himalaya and Karakoram. *Science of the Total Environment, 776*, 145895. https://doi.org/10.1016/j.scitotenv.2021.145895

Hunt, K. M. R., Turner, A. G., & Shaffrey, L. C. (2018a). Extreme daily rainfall in Pakistan and North India: Scale interactions, mechanisms, and precursors. *Monthly Weather Review, 146*, 1005–1022. https://doi.org/10.1175/MWR-D-17-0258.1

Hunt, K. M. R., Turner, A. G., & Shaffrey, L. C. (2018b). The evolution, seasonality and impacts of western disturbances. *Quarterly Journal of the Royal Meteorological Society, 144*, 278–290. https://doi.org/10.1002/qj.3200

Hunt, K. M. R., Turner, A. G., & Shaffrey, L. C. (2019). Falling trend of western disturbances in future climate simulations. *Journal of Climate, 32*, 5037–5051. https://doi.org/10.1175/JCLI-D-18-0601.1

Ikram, R. M., Hazarika, B. B., Gupta, D., Heddam, S., & Kisi, O. (2023). Streamflow prediction in mountainous region using new machine learning and data preprocessing methods: A case study. *Neural Computing and Applications, 35*, 9053–9070. https://doi.org/10.1007/s00521-022-08163-8

Islam, N., Vennemann, T., Büntgen, U., Krusic, P. J., Shah, S. K., Lane, S. N. (2025). Tree-ring based May-June streamflow reconstruction of Zemu River in the Eastern Himalaya. *Journal of Hydrology: Regional Studies, 60*, 102508. https://doi.org/10.1016/j.ejrh.2025.102508

Jena, P., & Azad, S. (2022). Identification of wet-prone regions over northwest Himalaya using high-resolution satellite seasonal estimates. *Natural Hazards, 112*, 1727–1748. https://doi.org/10.1007/s11069-022-05246-6

Joseph, S., Sahai, A. K., Sharmila, S., Abhilash, S., Borah, N., Chattopadhyay, R., Pillai, P. A., Rajeevan, M., & Kumar, A. (2015 Apr). (20115) north Indian heavy rainfall event during June 2013: Diagnostics and extended range prediction. *Climate Dynamics, 44*(7), 2049–2065. https://doi.org/10.1007/s00382-014-2291-5

Kamil, S., Almazroui, M., Kang, I. S., Hanif, M., Kucharski, F., Abid, M. A., & Saeed, F. (2019). Long-term ENSO relationship to precipitation and storm frequency over western Himalaya–Karakoram–Hindukush region during the winter season. *Climate Dynamics, 53*(9), 5265–5278.

Kar, S. C., & Rana, S. (2014). Interannual variability of winter precipitation over Northwest India and adjoining region: Impact of global forcings. *Theoretical and Applied Climatology, 116*, 609–623. https://doi.org/10.1007/s00704-013-0968-z

Karki, R., Hasson, S. U., Schickhoff, U., Scholten, T., & Böhner, J. (2017). Rising precipitation extremes across Nepal. *Climate, 5*(1), 4.

Kaushal, S., Thanveer, J., ul Islam, S. M., Negi, A., Dhanshyan, A., Beetan, Y., & Yunus, A. P. (2025). Unveiling the amplifying impact: Anthropogenic activities and the two-fold surge in landslides in the lesser Himalayas. *Catena, 250*, 108771.

Khan, M., Khan, A. U., Khan, S., & Khan, F. A. (2023). Assessing the impacts of climate change on streamflow dynamics: A machine learning perspective. *Water Science and Technology, 88*, 2309–2331. https://doi.org/10.2166/wst.2023.340

Kornhuber, K., Petoukhov, V., Karoly, D., Petri, S., Rahmstorf, S., & Coumou, D. (2017). Summertime planetary wave resonance in the northern and southern hemispheres. *Journal of Climate, 30*, 6133–6150. https://doi.org/10.1175/JCLI-D-16-0703.1

Kulkarni, A., Patwardhan, S., Kumar, K. K., Ashok, K., & Krishnan, R. (2013). Projected climate change in the Hindu Kush–Himalayan region by using the high-resolution regional climate model PRECIS. *Mountain Research and Development The Mayan, 33*(2), 142–151.

Kumari, S., Chaudhary, A., & Shankar, V. (2023). Modelling of rainfall threshold for the initiation of landslides in lesser Himalayan region using THRESH. *Modeling Earth Systems and Environment, 9*, 3207–3215. https://doi.org/10.1007/s40808-022-01660-8

Li, H., Haugen, J. E., & Xu, C. Y. (2018). Precipitation pattern in the Western Himalayas revealed by four datasets. *Hydrology and Earth System Sciences, 22*, 5097–5110. https://doi.org/10.5194/hess-22-5097-2018

Liu, J., Yuan, X., Zeng, J., Jiao, Y., Li, Y., Zhong, L., & Yao, L. (2022). Ensemble streamflow forecasting over a cascade reservoir catchment with integrated hydrometeorological modeling and machine learning. *Hydrology and Earth System Sciences, 26*, 265–278. https://doi.org/10.5194/hess-26-265-2022

Maloney, E. D. (2009). The moist static energy budget of a composite tropical intraseasonal oscillation in a climate model. *Journal of Climate, 22*, 711–729. https://doi.org/10.1175/2008JCLI2542.1

Ménégoz, M., Gallée, H., & Jacobi, H. W. (2013). Precipitation and snow cover in the Himalaya: From reanalysis to regional climate simulations. *Hydrology and Earth System Sciences, 17*, 3921–3936. https://doi.org/10.5194/hess-17-3921-2013

Nanditha, J. S., & Mishra, V. (2021). On the need of ensemble flood forecast in India. *Water Security, 12*, 100086. https://doi.org/10.1016/j.wasec.2021.100086

Negi, V. S., Thakur, S., Dhyani, R., et al. (2021). Climate change observations of indigenous communities in the Indian Himalaya. *Weather, Climate, and Society, 13*, 245–257. https://doi.org/10.1175/WCAS-D-20-0077.1

Nischal, & Attada, R. (2025). Multiphysics ensemble framework for representing Western Himalayan precipitation climatology and extremes: An assessment from the WRF model. *Weather and Forecasting, 40*, 149–172. https://doi.org/10.1175/WAF-D-24-0057.1

Nischal, Attada, R., & Hunt, K. M. (2022). Evaluating winter precipitation over the western himalayas in high resolution Indian regional reanalysis using multi-source climate datasets. *Journal of Applied Meteorology and Climataology, 61*(11), 1607–1627. https://doi.org/10.1175/JAMC-D-21-0172.1

Nischal, Attada, R., Hunt, K. M. R., & Barlow, M. (2024). Underlying physical mechanisms of winter precipitation extremes over India's high mountain region. *Quarterly Journal of the Royal Meteorological Society*, 1–23. https://doi.org/10.1002/qj.4661

Palazzi, E., Von Hardenberg, J., & Provenzale, A. (2013). Precipitation in the Hindu-Kush Karakoram Himalaya. *Journal of Geophysical Research: Atmospheres, 118*, 85–100. https://doi.org/10.1029/2012JD018697

Paredes Trejo, F. J., Alves Barbosa, H., Peñaloza-Murillo, M. A., Moreno, M. A., & Farias, A. (2016). Intercomparison of improved satellite rainfall estimation with CHIRPS gridded product and rain gauge data over Venezuela. *Atmósfera, 29*(4), 323–342.

Priya, P., Krishnan, R., Mujumdar, M., & Houze, R. A. (2017). Changing monsoon and midlatitude circulation interactions over the Western Himalayas and possible links to occurrences of extreme precipitation. *Climate Dynamics, 49*, 2351–2364. https://doi.org/10.1007/s00382-016-3458-z

Raghuvanshi, A. S., & Ankit, A. (2024). Multiscale dynamics of transient merging between western disturbances and monsoonal lows: Connections to the July 2023 flood in Himachal Pradesh. *Atmospheric Research, 107401*. https://doi.org/10.1016/j.atmosres.2024.107401

Ranalkar, M. R., Chaudhari, H. S., Hazra, Sawaisarje, G. K., & Pokhrel, R. (2016). Dynamical features of incessant heavy rainfall event of June 2013 over Uttarakhand, India. *Natural Hazards*, 1579–1601. https://doi.org/10.1007/s11069-015-2040-z

Rasmussen, K. L., & Houze, R. A. (2012). A flash-flooding storm at the steep edge of high terrain. *Bulletin of the American Meteorological Society, 93*, 1713–1724. https://doi.org/10.1175/BAMS-D-11-00236.1

Riley, C., Rupper, S., Steenburgh, J. W., Strong, C., Kochanski, A. K., & Wolvin, S. (2021). Characteristics of historical precipitation in high mountain asia based on a 15-year high resolution dynamical downscaling. *Atmosphere (Basel), 12*. https://doi.org/10.3390/atmos12030355

Sabin, T. P., Krishnan, R., Vellore, R., et al. (2020). Climate change over the Himalayas. In *Assessment of climate change over the Indian region* (pp. 207–222). Springer Singapore.

Sahu, R., Kumar, P., Gupta, R., & Ahirwar, S. (2025). Teleconnections and long-term precipitation trends in the Alaknanda River basin, Uttarakhand, India. *Earth Systems and Environment, 9*(4), 3043–3058. https://doi.org/10.1007/s41748-024-00536-4

Sain, K., Kumar, A., Mehta, M., et al. (2021). A perspective on Rishiganga-Dhauliganga flash flood in the Nanda Devi biosphere reserve, Garhwal Himalaya, India. *Journal of the Geological Society of India*, 335–338. https://doi.org/10.1007/s12594-021-1691-5

Saini, R., & Attada, R. (2023). Analysis of Himalayan summer monsoon rainfall characteristics using Indian high-resolution regional reanalysis. *International Journal of Climatology*, 1–22. https://doi.org/10.1002/joc.8087

Saini, R., & Attada, R. (2024). Deciphering the characteristics and drivers of the summer monsoon precipitation extremes over the Indian Himalayas. *Journal of Geophysical Research: Atmospheres., 129*(20), e2024JD042226.

Saini, R., & Attada, R. (2025). Indian summer monsoon precipitation and extremes over the Indian Himalayas in WRF dynamically downscaled (HARv2) reanalysis. *Climate Dynamics, 63.* https://doi.org/10.1007/s00382-025-07648-1

Saini, R., Attada, R., Hunt, K. M., Kizhuveettil, S., & Pandidurai, D. (2026a). Interaction of the monsoon trough and western disturbance ignites multiday extreme rainfall event in July 2023 over North India. *Monthly Weather Review, 154*(2), 183–201. https://doi.org/10.1175/MWR-D-25-0147.1

Saini, R., Attada, R., Hale, N., Kumar, P., Srivastava, A., & Yesubabu, V. (2026b). Cloud microphysics sensitivity and predictability of a multi-day extreme rainfall event over the Indian Himalayas in convective-permitting simulations. *Meteorology and Atmospheric Physics, 138*(4), 64. https://doi.org/10.1007/s00703-026-01169-0.

Saini, R., Sharma, N., & Attada, R. (2023). Delving into recent changes in precipitation patterns in the Western Himalayas under global warming. *IntechOpen.* https://doi.org/10.5772/intechopen.1002028

Sana, E., Kumar, A., Robson, E., Prasanna, R., Kala, U., & Toll, D. G. (2024). Preliminary assessment of series of land-slides and related damage by heavy rainfall in Himachal Pradesh, India, during July 2023. *Landslides, 21*(4), 919–931. https://doi.org/10.1007/S10346-023-02209-1/METRICS

Sharma, N., Saini, R., Sreehari, K., Pathaikara, A., Punde, P., & Attada, R. (2023). Hydrological extremes in Western Himalayas-trends and their physical factors. In *Natural hazards-new insights*. IntechOpen. https://doi.org/10.5772/intechopen.109445

Shekhar, M. S., Pattanayak, S., Mohanty, U. C., Paul, S., & Kumar, M. S. (2015). A study on the heavy rainfall event around Kedarnath area (Uttarakhand) on 16 June 2013. *Journal of Earth System Science, 124*, 1531–1544. https://doi.org/10.1007/s12040-015-0621-6

Shugar, D. H., Jacquemart, M., Shean, D., Bhushan, S., Upadhyay, K., Sattar, A., Schwanghart, W., McBride, S., De Vries, M. V. W., Mergili, M., & Emmer, A. (2021). A massive rock and ice avalanche caused the 2021 disaster at Chamoli, Indian Himalaya. *Science, 373*(6552), 300–306.

Singh, C., & Bharti, V. (2019). Rainfall characteristics over the northwest Himalayan region. *Remote Sens Northwest Himal Ecosyst*, 171–194. https://doi.org/10.1007/978-981-13-2128-3_8

Singh, P., & Kumar, N. (1997). Impact assessment of climate change on the hydrological response of a snow and glacier melt runoff dominated Himalayan river. *Journal of Hydrology, 193*, 316–350. https://doi.org/10.1016/S0022-1694(96)03142-3

Sinha, R., & Tandon, S. K. (2014). Indus-ganga-Brahmaputra Plains: The alluvial landscape. *World Geomorphol Landscapes, 53–63.* https://doi.org/10.1007/978-94-017-8029-2_5

Suhas, D. L., Ramesh, N., Kripa, R. M., & Boos, W. R. (2023). Influence of monsoon low pressure systems on south Asian disasters and implications for disaster prediction. *npj Climate and Atmospheric Science, 6*, 1–7. https://doi.org/10.1038/s41612-023-00376-5

Swain, S., Mishra, S. K., Pandey, A., Pandey, A. C., Jain, A., Chauhan, S. K., & Badoni, A. K. (2022). Hydrological modelling through SWAT over a Himalayan catchment using high-resolution geospatial inputs. *Environmental Challenges, 8*, 100579. https://doi.org/10.1016/j.envc.2022.100579

Tyrrell, N. L., Dietmar, D., Claudia, F., Scott, W., & Mike, R. (2015). The influence of global sea surface temperature variability on the large-scale land surface temperature. *Climate Dynamics*, 2159–2176.

Ullah, W., Wang, G., Ali, G., Hagan, D. F. T., Bhatti, A. S., & Lou, D. (2018). Comparing multiple precipitation products against in situ observations over different climate regions of Pakistan. *Remote Sensing, 11*, 628.

Vellore, R. K. (2016). Monsoon-extratropical circulation interactions in Himalayan extreme rainfall. *Climate Dynamics, 46*, 3517–3546. https://doi.org/10.1007/s00382-015-2784-x

Vellore, R. K., Bisht, J. S., Krishnan, R., Uppara, U., Di Capua, G., & Coumou, D. (2020). *Sub-synoptic circulation variability in the Himalayan extreme precipitation event during June 2013.* Springer Vienna.

Wu, X., & Deng, L. (2013). Comparison of moist static energy and budget between the GCM-simulated madden-julian oscillation and observations over the Indian Ocean and western Pacific. *Journal of Climate, 26*, 4981–4993. https://doi.org/10.1175/JCLI-D-12-00607.1

Zhang, H., Henderson-Sellers, A., & Mcguffie, K. (2001). The compounding effects of tropical deforestation and greenhouse warming on climate. *Climatic Change, 49*, 309–338. https://doi.org/10.1023/A:1010662425950

Zheng, T., Feng, T., Xu, K., & Cheng, X. (2020). Precipitation and the associated moist static energy budget off Western Australia in conjunction with Ningaloo Niño. *Frontiers in Earth Science, 8*, 1–13. https://doi.org/10.3389/feart.2020.597915

Chapter 12
Snow Avalanche Hazard in Indian Himalayan Region: A Review, Challenges and Way Forward

Ashwagosha Ganju, Amreek Singh, Praveen Kumar Srivastava, Harendra Singh Negi, and MS Shekhar

Abstract Snow avalanches are common in the high-altitude, snow-bound regions of the Himalaya. Following a sharp rise in avalanche-related fatalities during the mid to late 1960's, systematic studies in the Indian Himalayan Region (IHR) began with the collection of reliable data from 1971–72. While avalanches took place at many locations across the snow-covered Himalaya, early efforts focused on the western Himalaya due to frequent human presence. Initial research helped to reduce casualties substantially by mid-nineties. With rapid infrastructure development in the IHR, the need for accurate warnings and effective control measures grew, prompting studies in snow material behaviour, avalanche terrain, modelling and remote sensing. Additionally, the effect of climate change in IHR, which delayed the onset of winter precipitation, increased liquid precipitation, made winter relatively warmer, has added new dimensions in combating snow avalanche-related problems, necessitating renewed research on wet snow avalanches. These efforts provided valuable insights and prompted the pursuit of solutions to complex challenges. The paper outlines the emerging challenges and recommends future strategies for managing avalanche risks in IHR.

Keywords Snow · Avalanche · Fatalities · Climatology · Mitigation · Bulletins · Remote sensing

A. Ganju (✉) · A. Singh · P. K. Srivastava · H. S. Negi · M. Shekhar
Defence Geoinformatics Research Establishment, Chandigarh, India
e-mail: ashwagosha@gmail.com

A. Singh
e-mail: amreek.dgre@gov.in

P. K. Srivastava
e-mail: pk.srivastava.dgre@gov.in

H. S. Negi
e-mail: hs.negi.dgre@gov.in

M. Shekhar
e-mail: Sudhanshu.dgre@gov.in

V. M. Tiwari (ed.), *Himalayan Natural Hazards*, Indian National Science Academy,
https://doi.org/10.1007/978-981-92-3784-5_12

12.1 Introduction

Snow avalanches are among the most hazardous natural events in high mountain regions of Himalaya, posing serious risks to life, infrastructure and ecosystems. Natural triggers like heavy snowfall, glacial breakoffs, weak layer failure and seismicity, along with human activity such as mountaineering expeditions, often contribute to avalanche release with harmful impacts. Avalanche activity in the IHR may have peaked during the Little Ice Age (LIA), which coincided with glacier advances in the northern hemisphere. Although winter precipitation declined and temperatures rose after 1900, avalanches likely persisted albeit with moderate intensity. Historical records of avalanche accidents are sparse and so are the occurrence records.

Defence Geoinformatics

Research Establishment (DGRE), formerly Snow & Avalanche Study Esablishment (SASE), is the only institute in India actively engaged in avalanche research and associated snow-meteorological data collection and other field-oriented activities in IHR. (Ganju et al., 2002) conducted the first comprehensive analysis of avalanche accidents up to 2000. Foundational snow climatology studies (Sharma & Ganju, 2000); (Gusain et al., 2009) laid the groundwork for further research. Much of this foundational work is archived in the Annual Reports of the Defence Geoinformatics Research Establishment (DGRE), SASE, Chandigarh.

By late-nineties, the nation had developed strong capabilities in avalanche forecasting and control in IHR (Rao, 1985). However, growing demands for precise predictions in the twenty-first century highlighted gaps in routine data collection. This spurred advancements in numerical modelling and deployment of machine learning (ML) tools to enhance forecasting accuracy. Recent efforts integrate remote sensing, snow mechanics and past avalanche data to refine prediction tools and control strategies. With climate change intensifying risks, there is a push to unify models and technologies for reliable, area-specific alerts, alongside public awareness initiatives to bolster community preparedness.

12.2 Snow Avalanche Data

Avalanche research relies on multidisciplinary datasets, including meteorological, geospatial and field observations. DGRE maintains nearly 50 years of data from 63 manual observatories and 26 Automatic Weather Stations (AWS), 5 Research and Development (R&D) Centres, covering mainly the north-west and central Himalaya. Yet, compared to nations like Switzerland's SLF—with 100 years of data from ~ 200 observatories, India's dataset remains limited.

12.2.1 Meteorological and Snowpack Data

The initial operational constraints created data gaps in the past, which was partially substituted with the numerical techniques, at a few stations. To gain a better appreciation of spatial spread of this data, interpolation techniques with additional satellite inputs helped in bridging the gaps. Machine learning models, (Kaur et al., 2022), using multivariate imputation and nearest neighbour techniques, have improved forecasting while handling missing data. Manually collected field data—Shoval/Block/Collapse tests, stratigraphy, shear tests etc. are vital for validating models.

12.2.2 Remote Sensing and GIS Data

Satellites like IRS AWiFs, LISS-III, MODIS, Sentinel-1/-2 and SAR, Hyperspectral have been used to provide additional data on snow cover and terrain in IHR. High-resolution DEMs, combined with terrain variables, are now widely used for GIS-based avalanche susceptibility mapping.

12.2.3 Historical Avalanche Records

The briefly analysed snow avalanche accident records (Ganju et al., 2002) offer preliminary insights on the magnitude of the problem. This needs further revision to draw a comprehensive picture of the problem of avalanching in IHR. A compilation of fatalities due to snow avalanches in Indian Himalaya over the years is presented in Fig. 12.1.

12.3 Snow Avalanche Distribution, Topography and Climatology

The region-wise distribution of snow, hazard peaks and avalanche types of middle and inner Himalaya are given in Table 12.1. The Himalaya have an alpine climate with sharp temperature variations by altitude and contrasting wet southern and dry northern slopes. The region was first classified into three snow climatic zones (Sharma & Ganju, 2000) which was further analysed and improved (Gusain et al., 2009).

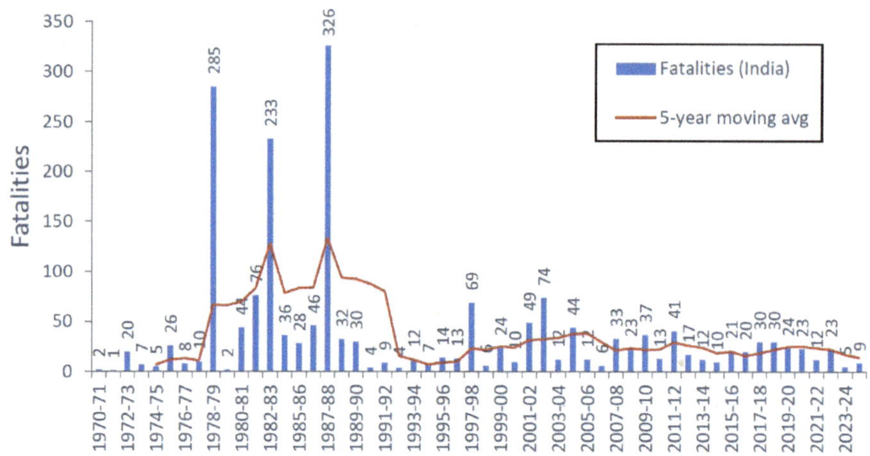

Fig. 12.1 Fatalities due to snow avalanches in Indian Himalaya

Table 12.1 Region-wise distribution of snow, hazard peaks and avalanche types in IHR

Himalayan region	Snow avalanche distribution (Himalayan ranges)			
	Snow regime	Altitude range	Avalanche type	
			Middle Himalaya	Inner (great) Himalaya
Eastern	• Monsoonal, • Moist wet snowpacks	• 2500–4000 m	• Forested slopes • Occasional wet slab • Rain-on-snow avalanches	• Wet slab avalanches from maritime snowpacks
Central	• Winter westerlies fed. • Mixed spatio-temporal snow • Heavy snowfall • Rapid accumulation	• 3000–4,500 m • Passes/ valleys • Steep orography	-Shallow slab, avalanche chains -Regular avalanches	-Higher frequency -Deep slab large avalanches -Weak layer failure
Western	• Winter westerlies • Arid, cold, continental deep snow	• 3,500–5,500 m, • Steep complex terrain, • Peaks and passes	• Dry/wet slabs, • Cornice collapses • Uneven snow accumulation	• Dry slab avalanches • Failure of wind loaded slopes • Basel facets failure

12.4 Snow Avalanche Mitigation

Snow avalanche risks can be addressed through permanent (active) methods or avoidance-based (passive) methods, depending on the area's strategic importance. While some key locations have been permanently secured and others are underway, passive mitigation remains the primary approach across most of the Himalayan region.

12.4.1 Active Methods: Structural Control/Afforestation/ Artificial Triggering

The active methods include permanent solution to the avalanche problem such as construction of avalanche (sheds) galleries over which avalanche flows, leaving the road protected all the time, mounds and dams, etc. At a few places in IHR (HP), avalanche galleries have protected roads while snow nets and bridges at other places in Jammu and Kashmir (J&K) have helped to prevent avalanche initiation. Near Shri Badrinath Temple in Uttarakhand (UK), diversion dams and earthen mounds deflect and reduce the run-out distance of avalanches. Tree planting and controlled blasting in avalanche-prone regions have shown limited and unstandardized success in IHR.

12.4.2 Passive Methods: Snow Avalanche Forecasting

Avalanche forecasting is aimed at identifying the likely location and time of avalanche occurrence with its type (dry, wet, slab, powder, etc.) and size (amount of snow involved) in the area of interest. Avalanche release involves complex interactions among snowpack, weather and terrain, requiring forecasting across three spatial scales: microscale (~ 1–10 km^2), mesoscale (~ 10–100 km^2) and macroscale (~ 100–1000 km^2). Microscale forecasting depends on in-situ stability tests, though these are limited, especially during storms. Mesoscale forecasts use detailed snowpack profiles from manual snow pits, which are time intensive and subjective. Nowadays physical models offer an alternative by simulating snowpack evolution. Macroscale forecasting relies on statistical models using historical weather and avalanche records.

Avalanche forecasting in the IHR evolved from early reliance on contributory snow, terrain and weather parameters. Subsequently discriminant analysis and nearest neighbour models (Buser et al., 1987) were improved (Ganju & Singh, 2004) for forecasting and issuing probability-based forecasts. Further advancements with expert systems, neural networks (Pant & Ganju, 2006) (Singh & Ganju, 2008) improved results. The applications were linked with numerical weather prediction (NWP) forecasts for longer lead periods (Singh et al., 2005). Random Forest (RF) is being

Degree of Danger	Danger Level	Snow Condition	Avalanche likelihood	Preferred action
1	Green	Generally safe condition. Snowpack on slopes, if any, is generally stable with isolated instability.	Avalanche activity is possible with high external loading e.g., seismic tremors, explosives or movement in formation zones.	• Valley movement is generally safe. • Movement on snow-loaded slopes with *care* only after ascertaining its stability. • Explore slope stabilization by Artificial Triggering. • Watch/prepare for higher danger level.
2	Yellow	Partly unsafe condition.	Small size triggering is possible on few avalanche paths.	• Valley movements with *care*. • Avoid movement on snow-loaded slopes. • Explore slope stabilization by Artificial Triggering. • Watch/prepare for higher danger level.
3	Orange	Unsafe condition	Triggering is possible from the most avalanche paths and may reach the valley in medium size.	• Restrict movements to carefully selected safer routes through valley only and with *extreme care*. • No movement on snow-loaded slopes. • Evacuate from unprotected settlements on/near the avalanche paths. • Watch/prepare for higher danger level.
4	Red	Highly unsafe condition.	Triggering is possible from all avalanche paths and may reach the valley bottom in large sizes. Airborne avalanches are likely.	• Suspend all movements. • Evacuate from all settlements on/near the avalanche paths. • Watch/prepare for higher danger level.
5	Black	Extremely unsafe condition.	Large size avalanches are likely from all possible avalanche paths even from moderately steep terrain. Avalanches may follow unexpected flowpaths.	Evacuate from avalanche prone areas.

- **Movement with care:** All safety measures to be observed while crossing suspected avalanche paths
- **Movement with extreme care:** Rescue party shall stand by in addition to above.

Fig. 12.2 Description of avalanche danger levels issued at DGRE

employed to assess avalanche potential by analysing snow-meteorological conditions (Chawla & Singh, 2021), offering probabilistic guidance and key feature bounds to support field-based forecasting decisions. However, limited and imbalanced avalanche occurrence data hindered model improvement (Kala et al., 2025) addressed this issue using multi-classification and resampling techniques, achieving POD scores of 0.83 (RF), 0.65 (SVM) and 0.75 (Logistic Regression). Despite this progress, expert judgement remained central, with guidance from AI-integrated, weather, remote sensing and GIS models.

The Avalanche Warning Bulletin (AWB), issued daily by DGRE, forecasts avalanche danger across different sectors. It supports planning for civil and defence authorities. AWB provides regional risk levels and snowpack details; forecasts may be imprecise. A description of avalanche danger levels included in a typical AWB is given in Fig. 12.2.

The overall framework of operational avalanche forecasting is given in Fig. 12.3.

12.5 Technological Breakthrough

12.5.1 Equipment and Instrumentation

The avalanche detection technologies used in IHR are as follows.

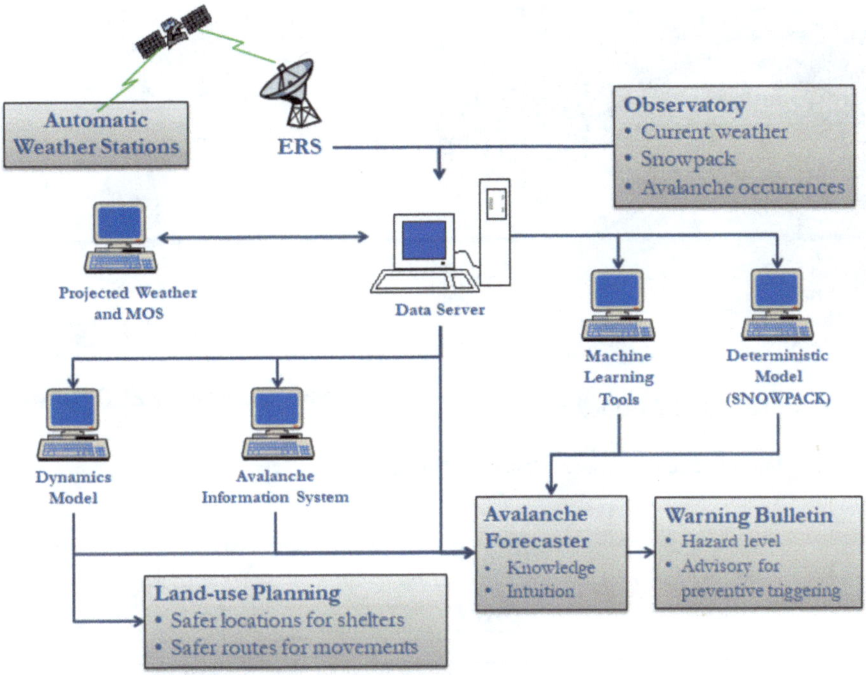

Fig. 12.3 Overall framework of operational avalanche early warning system at DGRE for IHR

Real-Time Automated Detection of Avalanches

Three sensor technologies are currently used for real-time avalanche detection: infrasonic sensor array, seismic sensor array-based and radar-based detection. Doppler radar technology is limited to specific avalanche paths. Infrasound arrays detect avalanches from all directions within a valley bowl. Seismic sensors placed along avalanche tracks capture ground vibrations. These systems are complementary and often used in combination for specific operational needs.

Infrasonic Sensor Array-Based Detection of Avalanches

Infrasound technology detects low-frequency pressure waves (1.0–20.0 Hz) from avalanches. As wind noise also falls in this range, robust processing is critical. Advanced techniques have enabled flow regime identification avalanche velocity estimation (Havens et al., 2014) and automated detection (Mayer et al., 2020). Infrasonic sensors were deployed at remote Himalayan sites Dhundi (HP) (Fig. 12.4) and Patsio (J&K) using 4-element arrays. The indigenous 'Infra-Locator' tool, developed using these signals, applies PMCC and TDDF algorithms for avalanche detection

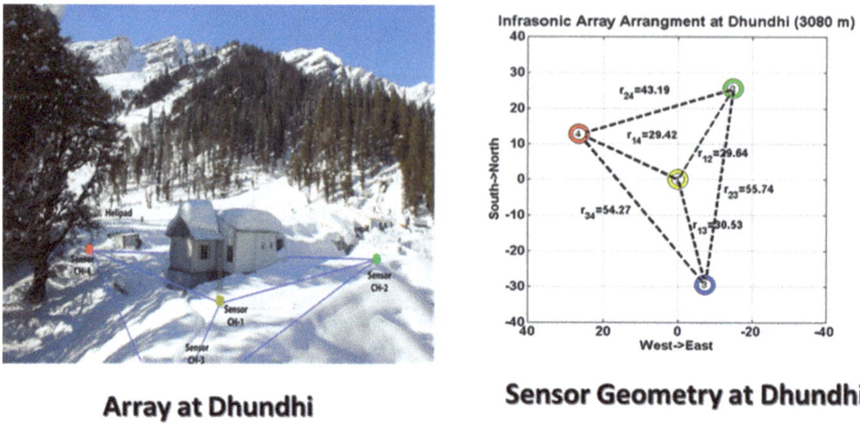

Fig. 12.4 Infrasonic sensor array at Dhundi (3080 m)

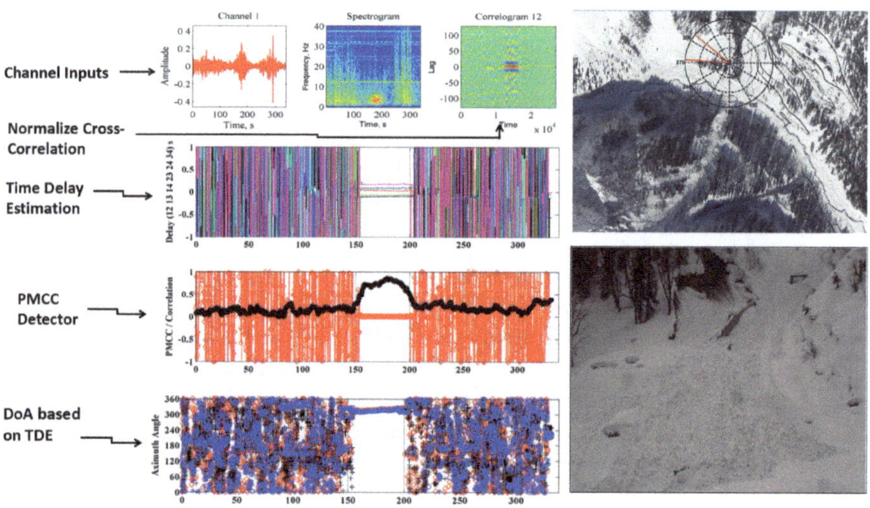

Fig. 12.5 Example of avalanche detection using infrasonic array technology at Dhundi (HP)

and localization (Fig. 12.5), offering reliable event tracking, even in snowstorms, and supporting real-time forecasting and corridor safety.

Seismic Sensor Array-Based Detection of Avalanches

Avalanches generate seismic vibrations detectable through signal processing (Surinach, 2004), linking signals to avalanche dynamics. Amplitude and spectral content

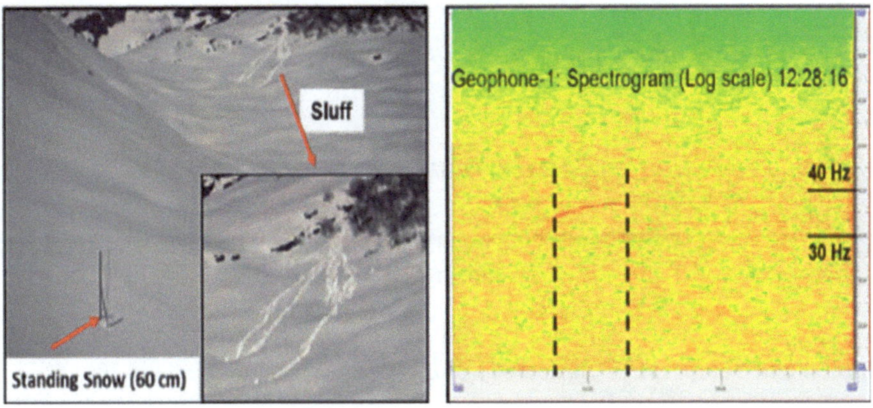

Fig. 12.6 Sluff event captured on 14 Feb 2018 by geophone and corresponding spectrogram

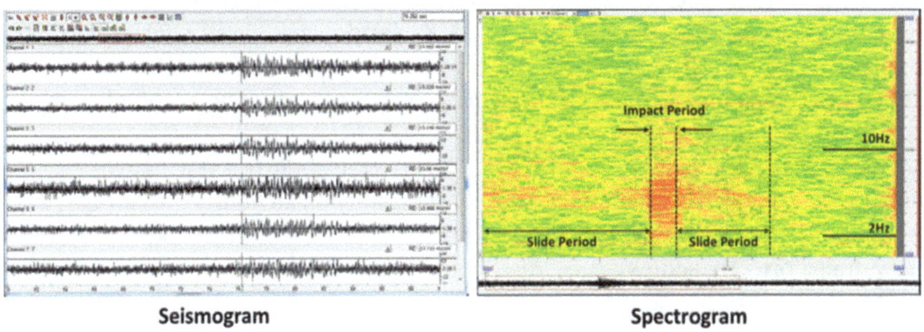

Fig. 12.7 Seismogram and spectrogram of a wet snow avalanche event of 23 Mar 2018

depend more on distance than avalanche size, with large avalanches showing triangular spectrograms. (Heck et al., 2018) applied hidden Markov models for automated detection, identifying signatures like > 10 s duration, > 1 Hz frequency and spindle-shaped seismograms. DGRE deployed geophones near Dhundi (Das et al., 2019). A sluff event on 14 Feb 2018 showed weak 33–37 Hz signals for ~ 50 s (Fig. 12.6), while a wet avalanche on 23 Mar 2018 produced stronger 2–10 Hz signals over 110 s with distinct phases (Fig. 12.7).

Radar-Based Detection of Avalanches

Active systems like Frequency Modulated Continuous Wave (FMCW) and Pulse Doppler Radars in C/X Band offer more precise avalanche detection than infrasound or seismic methods but are limited to predefined paths due to their narrow field of view. All these radars cover up to 10 km^2 within a 5 km range. Pulse Doppler Radars, such as those by IBTP Koschuch, use range-gating for better flow analysis

(Koschuch, 2018). DGRE has deployed such a radar in Sikkim with video and alarm integration for real-time monitoring and traffic control.

12.5.2 *Modelling and Simulations*

Snowpack, mass movement and weather simulation remained the forefront approaches in the study of snow avalanches.

Spatial and Temporal Evolution of Mountain Snowpack

Physical process-based snowpack models simulate mass and energy exchanges between snow, atmosphere and terrain. In 1D models, the snowpack is treated as layered, evolving under conservation laws. Prominent 1D models include Crocus (France; Brun et al., 1992), SNOWPACK (https://models.slf.ch/p/snowpack/), (Switzerland; Lehning et al., 1999), seNorge (Norway; Saloranta, 2012) and SNOWGRID (Olefs et al., 2013), each differing in numerical schemes, meteorological inputs and operational integration.

DGRE has implemented SNOWPACK 1D, driven by hourly AWS data (Fig. 12.8), offering forecasters high-resolution snow stratigraphy on an hourly basis (Fig. 12.9). However, 1D models lack capabilities to simulate spatial processes like wind-driven snow transport. To address this, distributed approaches as given in subsequent paragraphs have been adopted.

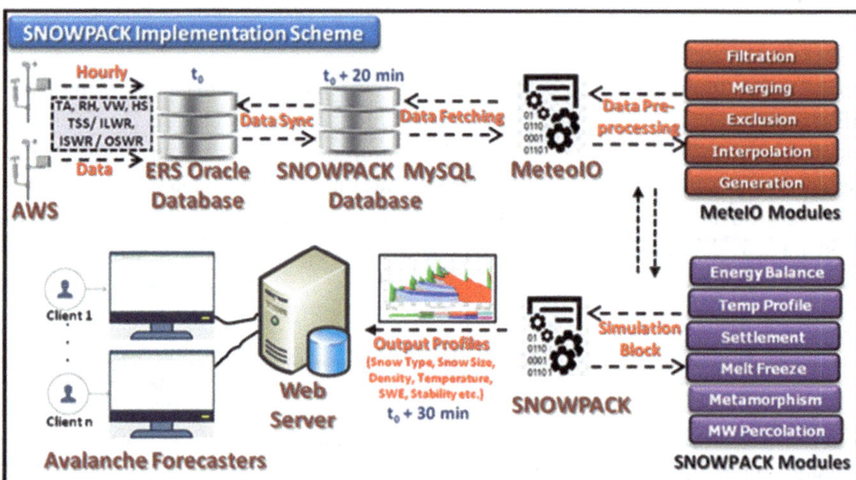

Fig. 12.8 Implementation of SNOWPACK simulations at DGRE. Process chain to run SNOWPACK simulation has been made fully automatic

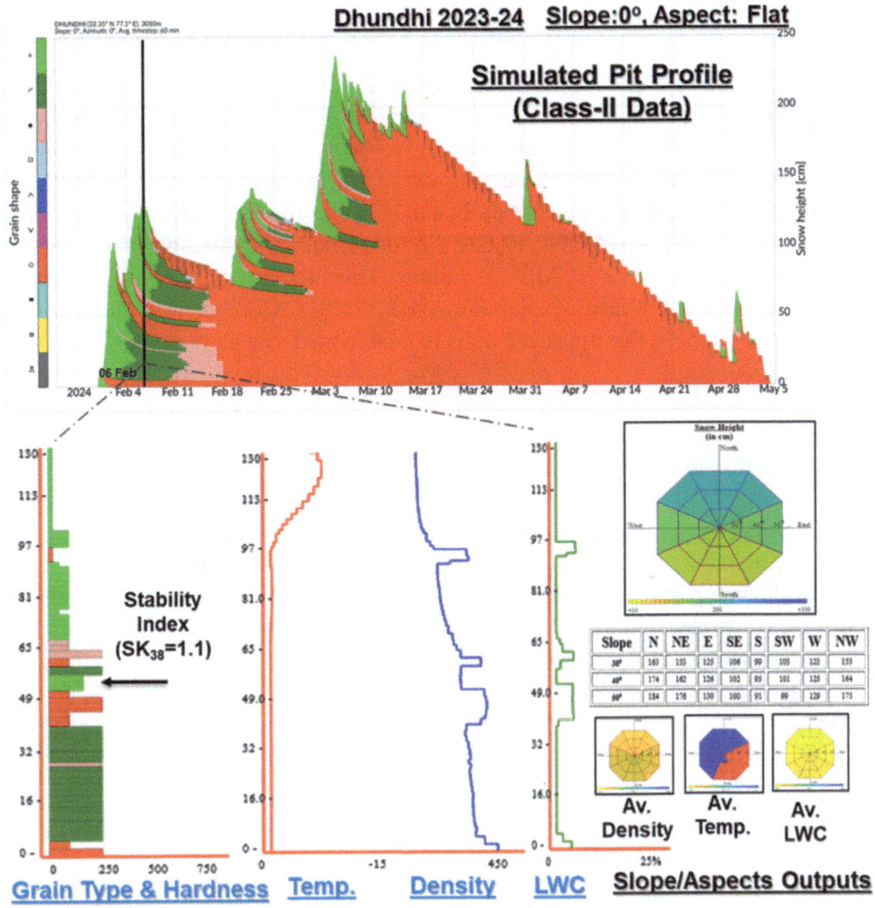

Fig. 12.9 SNOWPACK 1D model output in terms of simulated snowpack stratigraphy

Topographic Class-Based Approach

The topographic class-based approach divides terrain into representative units to capture surface heterogeneity. Meteo France's CEN applies SAFRAN (Durand et al., 1998), which segments the French Alps into ~ 1000 km^2 'Massifs' with assumed elevation-based meteorological homogeneity. SAFRAN merges ground observations with coarse NWP data to produce hourly meteorological inputs for the 1D Crocus model, stratified in 300 m elevation bands and 17 slope/orientation configurations. The SAFRAN + Crocus-like system has not yet been implemented in IHR.

Grid-Based Approach

Grid-based snowpack modelling relies on NWP outputs. In Canada, a simplified surface energy balance model predicts surface-hoar formation across 25,000 grid points (2.5 km spacing), while full 1D SNOWPACK runs at 3000 selected points. In the USA, CAIC uses WRF model outputs (2 km spacing) to drive SNOWPACK at each grid point, updating profiles daily using 24-h forecasts. DGRE is developing an operational grid-based SNOWPACK 3D model using high-resolution meteorological fields from AWS or NCMRWF outputs. This includes building an automated toolchain to generate hourly spatial snowpack profiles. Combined with avalanche data and AI-ML tools, this approach aims to enable next-gen autonomous avalanche forecasting.

Rapid Mass Movement Simulation

The RAMMS model (https://ramms.slf.ch/ramms/) has been used to simulate extreme avalanche scenarios—flow height, velocity and impact pressure—along major avalanche paths. Its depth-averaged avalanche flow equations using ASTER GDEM v2 (30 m resolution) while estimating release areas with fracture height linked to three-day snowfall and friction coefficients derived from detailed terrain analysis, have helped in planning road network and new habitations in avalanche-prone areas in IHR.

Microstructure-Based Computational Snow Mechanics

The elastic and multi-axial strength properties of snow layers are crucial for slab avalanche release models. These properties, closely linked to snow density and microstructure, help explain heterogeneity in macroscopic failure behaviour. (Srivastava et al., 2016) developed a fabric tensor and density-based orthotropic elasticity and strength model that extends failure analysis to multi-axial loading conditions beyond typical laboratory capabilities.

Numerical Weather Simulation

The use of weather forecast model MM5 (Fifth-Generation Penn State/NCAR Mesoscale Model) for avalanche forecasting in IHR started in 2004–05 (A. Singh et al., 2005). The model output was further downscaled to observatory stations from 10 km x 10 km grid resolution (Srinivasan et al., 2010) to gain improvement in avalanche forecast in IHR. Recently DGRE adopted the Weather Research and Forecast (WRF) model, using inputs from National Centre for Medium Range Weather Forecast (NCMRWF) (https://www.ncmrwf.gov.in) to generate 3 km grid weather forecasts up to 6 days in advance for the IHR. The location-specific precipitation forecast using Model Output Statistics (MOS) is also attempted for improved avalanche forecasting.

12.5.3 Remote Sensing

It is difficult to obtain snow cover information on repetitive basis from vast snow-covered mountains of IHR using ground conventional techniques, i.e. snow-meteorological observatories, etc. These have several limitations due to inaccessibility, ruggedness, lack of infrastructure facilities and moreover provide limited regional information. Satellite and air-borne remote sensing integrated with GIS provides vital spatial and temporal inputs for forecasting and hazard mitigation in the IHR. Multi-sensor approaches supported snow cover mapping and derivation of key parameters such as snow cover area (SCA), snow surface temperature (SST), albedo, snow depth (SD) and snow water equivalent (SWE). A high-resolution satellite data and digital elevation model (DEM) have also supported in terrain characterization and avalanche hazard mapping in IHR region (Bühler, 2012). Avalanche occurrence detection using multi-sensor satellite data at different spatial resolution is under implementation in IHR region.

Snow Cover Variability

Although the gross variability of snow cover across different snow climatic zones in the IHR was known (Gusain et al. 2009; Sharma & Ganju, 2000), further investigations (Negi et al., 2009) have explored seasonal and inter-annual changes in snow cover affect avalanche activity. Climate-induced reductions in snow cover have been examined in IHR and globally (Almagioni et al., 2025). Satellite sensors like Sentinel-1 (SAR) and MODIS were used to map regional snow cover (Singh et al., 2018). The reduced snow does not necessarily imply fewer avalanches, as shallow snowpacks can still avalanche under suitable conditions.

Avalanche Mapping

Avalanche registration in the IHR began with manually prepared avalanche atlases for various Himalayan Road corridors, documenting terrain features, past events and accidents, etc. While hard-bound ten such atlases exist, the growing need for updated atlases made them unsustainable. They have been now replaced using remote sensing and DEM-derived parameters for automated avalanche susceptibility and risk mapping (Gusain et al., 2019). Susceptibility and exposure maps with satellite-derived products, e.g. high-resolution DEMs with gravity flow simulation models like

RAMMS (Bühler et al., 2022) have shown promise and have also been implemented in the IHR.

Snow Depth Detection

Snow depth, vital for avalanche mitigation, was initially measured via field observatories and post-storm surveys along key road axes. To enhance spatial coverage, (Gusain et al., 2016) proposed integrating in-situ data with MODIS and DEM for 500 m resolution snow depth maps. Recently, (Singh et al., 2024) created multifactor models estimating snow depth at 500 m resolution across Western Himalayan zones. Though errors persist, particularly under non-dry snow conditions, yet this marks a major progress. Structure-from-motion photogrammetry (Buhler et al., 2016), SFM photogrammetry (Dewali et al., 2022) improved preparation of snow depth maps.

Avalanche Debris Detection

Avalanche behaviour—determined by type, release mechanism and snow volume, is also identified by travel distance and deposition patterns. In IHR, (Singh et al., 2020) tested methods including the Ratio Method, GLCM, Avalanche Debris Index and OBIA using Sentinel-2 data for debris detection, which has been found effective for medium-resolution debris mapping. More recently, (Abhinav et al., 2025) applied Fuzzy Random Forest and Fuzzy Support Vector Machine (FSVM) models with RGB and YCbCr colour spaces for debris detection via Sentinel-2, finding FSVM with YCbCr to yield the highest accuracy. The operational implementation of the developed techniques is underway in IHR.

Ice Avalanches Monitoring

Climate change has intensified the threat of ice avalanches in the Himalaya due to rapid glacier retreat, causing tributary glaciers to detach from trunk glaciers and form unstable hanging glaciers (Bhardwaj & Sam, 2022). In the IHR, research on hanging glacier mechanics remains limited. In the absence of a regional hanging glacier inventory large-scale risk modelling is in preliminary stage (Laha, 2024).

12.6 Climate Change and Snow Avalanche

Over recent decades, Himalayan snow climatology has shown signs of significant alteration. Rising temperatures, particularly above 4000 m, have led to decreased snow accumulation, earlier onset of snowmelt and reduced snow persistence (Pepin et al., 2015). The snow cover area and the variability that one finds in different snow

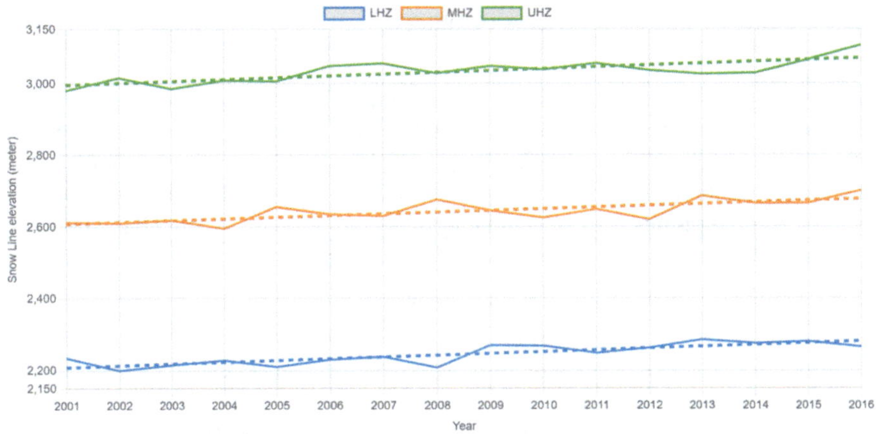

Fig. 12.10 Temporal variation of snow line elevation (SLE) in the Lower Himalayan Zone (LHZ), Middle Himalayan Zone (MHZ) and Upper Himalayan Zone (UHZ) across the Western Himalaya during 2001–2016. The graphs illustrate inter-annual variability and an overall upward shift in snow line elevation (Singh et al., 2018)

climatic zones (Singh et al., 2018) bring further insight in the shift in snow climatic zonation of western Himalaya. Based on the data from 2001 to 2016, all three zones show a gradual increase in snow line elevation (Fig. 12.10). Upper Himalayan Zone (UHZ) at the highest elevation range shows marginal shift in snow line. MHZ shows a steady upward trend starting near 2600 m and ending closer to 2700 m. LHZ also shows a steady rising trend from approximately 2200 m.

Rising snow avalanche incidences in IHR have been linked to climate change and regional warming (Ballesteros-Cánovas et al., 2018). Similar trends have been reported globally (Komac & Zorn, 2023) and within the IHR (Bansal et al., 2025). Given the vastness of snow reserves in Himalaya and faster-than-global-average warming (Shekhar & Ganju, 2022), avalanche risks are projected to rise. Tree-ring studies near Solang–Dhundi (Ballesteros-Cánovas et al., 2018) showed a surge in avalanche activity since the 1970’s, correlating with warmer winters and earlier springs. While direct action avalanches may increase due to increase in extreme events, studies on contrasting snowfall trends: a general decrease in seasonal snow cover (Naegeli et al., 2022), reduced snowfall days, increasing instability and even variable increase in certain regions (Choudhury et al., 2021), highlight the possibility of non-uniform nature of increasing (decreasing) trend of avalanching, which needs further investigation. However, higher temperatures, reduced snow and more rainfall are altering snowpack dynamics, making avalanches more frequent and hazardous. Previously low-risk areas may now be at risk, with rain-on-snow events increasing wet slab avalanche potential. The upslope retreat of avalanche zones (Giacona et al., 2021) and emergence of wet avalanches in previously dry regions suggests significant changes underway. Limited studies, hindered by poor winter mobility, prevent broad generalization on the effect of climate change on avalanching in IHR.

12.7 Challenges and Way Forward

Snow avalanches in the IHR result from a complex interaction of snow evolution, weather, terrain and human activity. Seasonal snow cover is not just a climatic feature but vital to ecosystem health, water security, defence and safety. While forecasting and control methods are improving through advanced modelling and observation, climate change presents ongoing challenges. Some of the important challenges and way forward points that can be immediately taken up in IHR are as follows:

- The ruggedness of the terrain posed a huge challenge in the past despite humongous efforts by DGRE/SASE in establishing a network of high-altitude observatories. This network needs now to be strengthened, especially in the eastern and central Himalaya.
- Weather modelling has made significant advances world over and now there is a need to further improve on optimization for the complex Himalayan terrain, to further improve avalanche forecasting.
- Avalanche data collection is hindered by inaccessibility, poor connectivity, cloud cover and infrequent satellite passes. Having proven their efficacy at select places in Himalaya, a wider deployment of seismic sensors, infrasonic devices and radars is essential.
- Several attempts in the past have demonstrated the efficacy of bringing various stakeholders together in combating hazards rampant in Himalaya. Enhance collaboration among researchers, local authorities and armed forces to develop valley and district-specific early warning systems is the need of the hour.
- The human efforts of data collection in the past by SASE/DGRE have their own limitations and already reached a saturation level. With the successes seen elsewhere in the world there is a need to adopt automated, AI-driven avalanche forecasting using remote sensing from optical and microwave satellites, reducing human dependence.
- The Unmanned Aerial Systems (UAS) have proved very successful in different aspects and reduced manual effort. We must now ensure year-round access to key Himalayan Road corridors through permanent UAV deployment for controlled triggering and guiding dedicated snow removal machinery.
- DGRE in collaboration with other institutions like India Meteorological Department (IMD), National Disaster Management Authority (NDMA) and academia is already on the way to develop AI/ML-based mobile alert systems for timely avalanche warnings and faster disaster response, which needs to be expedited.
- The initial experiments on fusing various satellite data have shown promising results. The academia needs to be involved fully to fuse data from multifrequency SAR (e.g. Sentinel-1, NISAR) and optical sensors (e.g. Sentinel-2) for real-time onboard satellite processing and faster product delivery.
- Build local capacity by integrating traditional knowledge with GIS and promoting inter-agency collaboration (e.g. GSI, DGRE, NDMA, ICIMOD) for effective risk management.

With a multidisciplinary approach that integrates field science, modelling and policy, nation can strengthen its resilience against one of its most formidable mountain hazards. As Himalaya continue to evolve, proactive snow and avalanche management will be essential not just for survival, but for sustainable prosperity in these magnificent but fragile mountains.

References

Abhinav, A., Sharma, A., Dhamija, S., & Negi, H. S. (2025). Avalanche debris detection from sentinel-2 data using fuzzy machine learning and colour spaces for the Indian Himalaya. *Remote Sensing Letters, 16*(6), 606–618. https://doi.org/10.1080/2150704X.2025.2488532

Almagioni, C. D., Manara, V., Diolaiuti, G. A., Maugeri, M., Spezza, A., & Fugazza, D. (2025). Snow cover variability and trends over Karakoram, Western Himalaya and Kunlun Mountains during the MODIS era (2001–2024). *Remote Sensing, 17*(5). https://doi.org/10.3390/rs1705 0914

Singh, A., Srinivasan, K., & Ganju, A. (2005). Avalanche forecast using numerical weather prediction in Indian Himalaya. *Cold Regions Science and Technology, 43*(1–2), 83–92., ISSN 0165-232X. https://doi.org/10.1016/j.coldregions.2005.05.009

Ballesteros-Cánovas, J. A., Trappmann, D., Madrigal-González, J., Eckert, N., & Stoffel, M. (2018). Climate warming enhances snow avalanche risk in the Western Himalayas. *Proceedings of the National Academy of Sciences of the United States of America, 115*(13), 3410–3415. https://doi.org/10.1073/PNAS.1716913115

Bansal, J. K., Goswami, A., & Snehmani, & Roy, A. (2025). Snow avalanche in the Indian Himalayas: Hazard zonation and climate change trends in Kullu region of Himachal Pradesh, India. *Physics and Chemistry of the Earth, 138*. https://doi.org/10.1016/j.pce.2025.103882

Bhardwaj, A., & Sam, L. (2022). Reconstruction and characterisation of past and the most recent slope failure events at the 2021 rock-ice avalanche site in Chamoli, Indian Himalaya. *Remote Sensing, 14*(4). https://doi.org/10.3390/rs14040949

Brun, E., David, P., Sudul, M., & Brunot, G. (1992). A numerical model to simulate snow-cover stratigraphy for operational avalanche forecasting. *Journal of Glaciology, 38*(128), 13–22. https://doi.org/10.3189/s0022143000009552

Bühler, Y. (2012). Remote sensing tools for snow and avalanche research. *International Snow Science Workshop, Anchorage, Alaska*, 264–268. https://www.researchgate.net/publication/233 324781

Buhler, Y., Adams, M. S., Bosch, R., & Stoffel, A. (2016). Mapping snow depth in alpine terrain with unmanned aerial systems (UASs): Potential and limitations. *The Cryosphere, 10*(3), 1075–1088. https://doi.org/10.5194/tc-10-1075-2016

Bühler, Y., Bebi, P., Christen, M., Margreth, S., Stoffel, L., Stoffel, A., Marty, C., Schmucki, G., Caviezel, A., Kühne, R., Wohlwend, S., & Bartelt, P. (2022). Automated avalanche hazard indication mapping on a statewide scale. *Natural Hazards and Earth System Sciences, 22*(6), 1825–1843. https://doi.org/10.5194/nhess-22-1825-2022

Buser, Q., Butler, M., & Good, W. (1987). Avalanche forecast by nearest neighbour method. *International Association of Hydrological Sciences Publication, 162*, 557–569.

Chawla, M., & Singh, A. (2021). A data efficient machine learning model for autonomous operational avalanche forecasting. *Natural Hazards and Earth System Sciences Discussions*. https://doi.org/10.5194/nhess-2021-106

Choudhury, A., Yadav, A. C., & Bonafoni, S. (2021). A response of snow cover to the climate in the northwest Himalaya (NWH) using satellite products. *Remote Sensing, 13*(4), 1–22. https://doi.org/10.3390/rs13040655

Das, R. K., Sharma, H., Chouhan, R., Garg, R. K., & Acharya, A. (2019). Preliminary results of seismic monitoring of avalanches in Pir Panjal ranges of Himalaya. *Int. Symp. on Snow Avalanches & Mitigation Strategies*, 195–196.

Dewali, S., Jain, K., Dhamija, S., & Singh, D. K. (2022). Mapping snow depth and spatial variability using SFM photogrammetry of UAV images over rugged mountainous regions of the Western Himalaya. *Geocarto International, 37*(5).

Durand, Y., Giraud, G., & Mérindol, L. (1998). Short-term numerical avalanche forecast used operationally at Météo-France over the Alps and Pyrenees. *Annals of Glaciology, 26*, 357–366.

Ganju, A., & Singh, A. (2004). *A new approach to avalanche forecasting and disaster management for western Himalaya.* World Congress on National Disaster Mitigation.

Ganju, A., Thakur, N. K., & Rana, V. (2002). *Characteristics of avalanche accidents in Western Himalayan region, India* (pp. 200–207). *International Snow Science Workshop.*

Giacona, F., Eckert, N., Corona, C., Mainieri, R., Morin, S., Stoffel, M., Martin, B., & Naaim, M. (2021). Upslope migration of snow avalanches in a warming climate. https://doi.org/10.1073/pnas.2107306118/-/DCSupplemental

Gusain, H. S., Mishra, V. D., Arora, M. K., Mamgain, S., & Singh, D. K. (2016). Operational algorithm for generation of snow depth maps from discrete data in Indian Western Himalaya. *Cold Regions Science and Technology, 126*, 22–29.

Gusain, H. S., Negi, H. S., Dhamija, S., & Mishra, V. D. (2019). Development of avalanche information system using remote sensing and GIS technology in the Indian Karakoram Himalaya. In Current science (Vol. 117, Issue 1).

Gusain, S. H., Chand, D., Thakur, N., Singh, A., & Ganju, A. (2009). Snow avalanche climatology of Indian Western Himalaya. *International Symposium on Snow and Avalanches*, 85–93. https://www.researchgate.net/publication/249008685

Havens, S., Marshall, H. P., Johnson, J. B. and, & Nicholson, B. (2014). Calculating the velocity of a fast-moving snow avalanche using an infrasound array. *Geophysical Research Letters, 41*, 6191–6198.

Heck, M., Hammer, C., van Herwijnen, A., Schweizer, J., & Fah, D. (2018). Automatic detection of snow avalanches in continuous seismic data using hidden Markov models. *Natural Hazards and Earth System Sciences, 18*, 383–396.

Kala, M., Jain, S., Singh, A., & Krishnan, N. C. (2025). Addressing class imbalance in avalanche forecasting. *Cold Regions Science and Technology, 231*. https://doi.org/10.1016/j.coldregions.2024.104411

Kaur, P., Joshi, J. C., & Aggarwal, P. (2022). A multi-model decision support system (MM-DSS) for avalanche hazard prediction over north-west Himalaya. *Natural Hazards, 110*(1), 563–585. https://doi.org/10.1007/s11069-021-04958-5

Komac, B., & Zorn, M. (2023). Impact of climate change on snowpack and avalanches in Slovenia: The SOČA Valley case study. *Geographia Polonica, 96*(1), 29–51. https://doi.org/10.7163/GPol.0244

Koschuch, R. (2018). 8 years' experience in avalanche detection by using a pulse Doppler radar. In J.-T. Fischer (Ed.), *International snow science workshop, Innsbruck* (pp. 636–639).

Laha, S., Majeed, U., Banerjee, A., Rashid, I., Steiner, J., & Vijay, S. (2024). Assessing potential risk of glacier avalanches to hydropower infrastructure in the Himalayan region. *Natural Hazards, 120*(5), 1–26.

Lehning, M., Bartelt, P., Brown, B., Russi, T., Stockli, U., & Zimmerli, M. (1999). SNOWPACK model calculations for avalanche warning based upon a new network of weather and snow stations. *Cold Regions Science and Technology, 30*. www.elsevier.comrlocatercoldregions

Mayer, S., van Herwijnena, A., Ulivierib, G., & Schweizer, J. (2020). Evaluating the performance of an operational infrasound avalanche detection system at three locations in the Swiss Alps during two winter seasons. *Cold Regions Science and Technology, 173*, 1–10.

Naegeli, K., Franke, J., Neuhaus, C., Rietze, N., Stengel, M., Wu, X., & Wunderle, S. (2022). Revealing four decades of snow cover dynamics in the Hindu Kush Himalaya. *Scientific Reports, 12*(1). https://doi.org/10.1038/s41598-022-17575-4

Negi, H. S., Thakur, N. K., Kumar, R., & Kumar, M. (2009). Monitoring and evaluation of seasonal snow cover in Kashmir valley using remote sensing, GIS and ancillary data. *Journal of Earth System Science, 118*(6), 711–720. https://doi.org/10.1007/s12040-009-0057-y

Olefs, M., Schoner, W., Suklitsch, M., Wittmann, C., Niedermoser, B., Neururer, A., & Wurzer, A. (2013). SNOWGRID—A new operational snow cover model in Austria. In *International Snow Science Workshop Grenoble-Chamonix Mont-Blanc* (pp. 38–45).

Pant, L. M., & Ganju, A. (2006). Application of neural networks for avalanche forecasting. *Defence Science Journal, 56*(4), 559–567. https://doi.org/10.14429/dsj.56.1921

Pepin, N., Bradley, R. S., Diaz, H. F., Baraer, M., Caceres, E. B., Forsythe, N., Fowler, H., Greenwood, G., Hashmi, M. Z., Liu, X. D., Miller, J. R., Ning, L., Ohmura, A., Palazzi, E., Rangwala, I., Schöner, W., Severskiy, I., Shahgedanova, M., Wang, M. B., Yang, D. Q. (2015). Elevation-dependent warming in mountain regions of the world. In Nature climate change (Vol. 5(5), pp. 424–430). Nature Publishing Group. https://doi.org/10.1038/nclimate2563

Rao, N. M. (1985). Avalanche protection and control in the Himalayas. *Defence Science Journal, 35*(2).

Saloranta, T. M. (2012). Simulating snow maps for Norway: Description and statistical evaluation of the seNorge snow model. *The Cryosphere, 6*(6), 1323–2012.

Sharma, S. S., & Ganju, A. (2000). Complexities of avalanche forecasting in Western Himalaya—an overview. *Cold Regions Science and Technology, 31*(2), 95–102. https://doi.org/10.1016/S0165-232X

Shekhar, M. S., & Ganju, A. (2022). Avalanche phenomena in northwest Himalaya: A review. In *VayuMandal* (Vol. 48, 2).

Singh, A., & Ganju, A. (2008). A new approach to analyze nearest Neighbors for avalanche forecasting. *National Snow Science Workshop*, 271–276. https://www.researchgate.net/publication/318593731

Singh, D. K., Gusain, H. S., Mishra, V., & Gupta, N. (2018). Snow cover variability in north-west Himalaya during last decade. *Arabian Journal of Geosciences, 11*(19). https://doi.org/10.1007/s12517-018-3926-3

Singh, D. K., Tanniru, S., Singh, K. K., Negi, H. S., & Ramsankaran, R. (2024). Passive microwave remote-sensing-based high-resolution snow depth mapping for Western Himalayan zones using multifactor modeling approach. *The Cryosphere, 18*(1), 451–474. https://doi.org/10.5194/tc-18-451-2024

Singh, K. K., Singh, D. K., Thakur, N. K., Dewali, S. K., Negi, H. S., & Snehmani, & Mishra, V. D. (2020). Detection and mapping of snow avalanche debris from Western Himalaya, India using remote sensing satellite images. *Geocarto International, 37*(9), 2561–2579. https://doi.org/10.1080/10106049.2020.1762762

Srinivasan, K., Kumar, A., Verma, J., & Ganju, A. (2010). Statistical downscaling of MM5 model output to better assess avalanche threats. *Annals of Glaciology, 51*(54), 14–18. https://doi.org/10.3189/172756410791386472

Srivastava, P. K., Chandel, C., Mahajan, P., & Pankaj, P. (2016). Prediction of anisotropic elastic properties of snow from its microstructure. *Cold Regions Science and Technology, 125*, 85–100.

Surinach, E. (2004). Spanish avalanche research: Experimental sites and seismic measurements, in: Snow and avalanche test sites. In M. Naaim & F. Naaim Bouvet (Eds.), (pp. 71–183).

Chapter 13
Glacial Lake and Outburst Floods in the Indian Himalaya: Challenges and Way Forward

Suraj Mal, Arshad Reza, Atul Kumar, and A. P. Dimri

Abstract Growing glacial lakes, related floods and potential damages to the infrastructure and human population of the Indian Himalayan Region (IHR) have been a great concern to the scientific community in the twenty-first century. Consequently, there is a significant increase in the large and regional-scale as well as case-based glacial lake outburst flood (GLOF) hazard assessment studies using satellite data and modelling approaches. Yet, the GLOF prediction and hence the timely risk mitigation and management remain a big challenge for the scientific community and management authorities. Therefore, the present study reviews past studies on potential GLOF assessments and documented GLOF events to address gaps and challenges in GLOF studies and a way forward that can aid effective GLOF risk mitigation and management. The IHR has experienced > 200 GLOF events in the last century. The majority of the documented GLOFs have been reported to be caused by ice-dam failures and are primarily concentrated in the Ladakh region. In contrast, the central Himalayan GLOFs have been mostly caused by moraine-dam failures. The IHR has experienced several GLOF events over the twenty-first century, which caused substantial damage to critical infrastructure. Currently, overall GLOF danger is highest in the western Himalayan region, while a significant threat is also posed in the eastern Himalaya (Sikkim). It is observed that the current GLOF models are not yet capable of predicting the GLOF events, as the role of complex interplay of geophysical mechanisms, atmospheric and socio-economic processes is not fully understood and hence not yet well-integrated in the current GLOF modelling frameworks. Therefore, the present study emphasizes on an urgent need to utilize machine-learning-based modelling approaches to successfully predict GLOF events for timely mitigation and management interventions.

S. Mal (✉) · A. Reza
Centre for the Study of Regional Development, Jawaharlal Nehru University, Delhi, India
e-mail: surajmal@jnu.ac.in

A. Kumar
Department of Civil Engineering, Indian Institute of Technology (BHU), Varanasi, India

A. P. Dimri
School of Environmental Sciences, Jawaharlal Nehru University, Delhi, India

V. M. Tiwari (ed.), *Himalayan Natural Hazards*, Indian National Science Academy,
https://doi.org/10.1007/978-981-92-3784-5_13

Keywords Glacial lakes · Outburst floods · Risk · Climate change · Indian Himalayan Region (IHR)

13.1 Introduction

Glacial lakes and related outburst floods in the Himalayan mountains have recently received significant attention among the global scientific community, policy-makers and management authorities (Emmer et al., 2022; Mondal et al., 2026; Ravindra et al., 2024). In the last decades, increased glacial lake numbers, surface area coverage, water volume and potential flood risks (Kumar et al., 2025a, 2025c; Sahu et al., 2023) to the infrastructure and communities have been observed, which is likely to increase manyfold in the twenty-first century (Mal et al., 2021; Sattar et al., 2021; Zheng et al., 2021a). Assessing the critically dangerous glacial lakes and related flood risks primarily requires a detailed understanding of the spatial distribution of glacial lakes and growth patterns, including types, areal coverage, potential flood volumes, adjacent terrain characteristics and exposed infrastructure and human population (Shugar et al., 2020; Wang et al., 2020; Zhang et al., 2022a; Zheng et al., 2021a).

The first detailed inventory of glacial lakes (*n*: 8790, > 0.01 km^2) was conducted across the broader Hindu Kush Himalaya (HKH) between 1999 and 2005 (Ives et al., 2010). Since then, numerous studies have been conducted at a larger HKH and/or High Mountain Asia (HMA), yielding varying results, from ~ 1300 (Nie et al., 2013) to more than 20,000 glacial lakes (Chen et al., 2021; Kumar et al., 2025c; Wang et al., 2020). Recent studies reveal significantly different results of glacial lake inventories in the Indian Himalayan Region (IHR), e.g. 329 (> 0.05 km^2) (Dubey & Goyal, 2020), 4418 (> 0.01 km^2) (Mal et al., 2021; Zheng et al., 2021a) and 5094 (> 0.01 km^2) (Kumar et al., 2025c). Differences in the extents of study domains, datasets and methodology used limit the inter-comparison across the studies. For instance, previous studies used different minimum mapping areas of a lake (0.001–0.05 km^2), elevation thresholds (e.g. > 3500 m asl), timing and spatial resolution of satellite data and mapping techniques (Kumar et al., 2025c).

These glacial lakes are highly dynamic and have shown significantly enhanced growth in numbers (up to 20%), surface area (up to 23%) (Bolch et al., 2019; Chen et al., 2021; Kumar et al., 2025c; Zhang et al., 2022a), upper elevation (up to 6200 m asl) (Kumar et al., 2025c; Mohanty et al., 2023; Nie et al., 2017) and water volumes (up to 45%) since the 1990s (Shugar et al., 2020). Growth of glacial lakes exhibits an east–west pattern in the Himalaya, where the southern slopes of the central and eastern Himalaya particularly reveal high expansion, in contrast to a decline in the northwestern Himalaya since the 1990s (Gardelle et al., 2011; Kumar et al., 2025c, 2025b; Nie et al., 2017; Zhang et al., 2022b). In addition, the water volume of these lakes has also increased (Fujita et al., 2013; Patel et al., 2017; Sharma et al., 2018). Consequently, associated outburst flood risk in the HKH region has increased (Zheng et al., 2021a), which is attributed to enhanced warming (up to 0.2 and 0.3 °C/decade)

(Dimri et al., 2021; Krishnan et al., 2019) and accelerated glacier recession (Bolch et al., 2019; Brun et al., 2017).

A glacial lake is prone to failure and further to flooding, when (a) the lake volume increases to a level that it exceeds the shear strength of its moraine-dam, (b) degradation (failure and piping) of moraine-dam, (c) permafrost thawing (Byers et al., 2022), (d) lake upslope failures (e.g. avalanches, landslides, and rock fall) (Allen et al., 2016b), (e) glacier calving (Harrison et al., 2018; Jain et al., 2025; Nie et al., 2018; Westoby et al., 2014) and (f) seismic events (Kargel et al., 2016). The glacial lakes connected and/or fed by the parent glaciers have particularly grown in the last decades and are responsible for devastating floods, while those are disconnected and dammed by lateral moraines (Emmer et al., 2022; Gupta et al., 2022), have remained relatively stable but can still pose risks in case of extreme precipitation events, e.g. Chorabari in 2013 (Allen et al., 2016a; Bhambri et al., 2016; Raj & Kumar, 2016). These GLOF events in HKH typically happen in monsoon months, as the prolonged precipitation coupled with higher temperatures saturate, destabilize and mobilize the terrain, leading to overflow and/or lake failures (Richardson & Reynolds, 2000; Westoby et al., 2014). The risk of potential GLOF events has further increased (Ravindra et al., 2024) due to increasing exposure of the human population and critical infrastructure, e.g. hydropower projects, located close to the existing glacial lakes (Allen et al., 2016a; Li et al., 2022; Zheng et al., 2021a). GLOF frequency in the HKH has been observed to increase particularly in the twenty-first century (Dimri et al., 2021; Harrison et al., 2018; Hock et al., 2019; Jain et al., 2025; Veh et al., 2019). Some of the notable GLOF events in the IHR include Chong Kumdan (1929–32) in the Shyok valley, NW Himalaya (Hewitt & Liu, 2010), Chorabari/Kedarnath floods (2013) in Uttarakhand (Allen et al., 2016a; Bhambri et al., 2016) and South Lhonak flood (2023) in Sikkim (Sattar et al., 2025).

Although the GLOF studies in the IHR have been conducted for over the last century (Shrestha et al., 2023), with a significant surge in the last decades (Mondal et al., 2026), major research gaps and challenges still persist. These are related to the complex interplay of climatological, meteorological and geophysical drivers, compounded with substantially increasing exposure of population and infrastructure growth in a changing climate. Lake volume data, an important and critical input of GLOF modelling, remain largely limited and flood scenarios as well as associated potential damage from the majority of the lakes are yet to be accurately estimated (Allen et al., 2022; Kumar et al., 2025a). Current studies estimating damage potentials from these lakes vary widely in terms of data and methodologies, limiting their comparison and posing a major challenge in GLOF forecasting. The study, therefore, synthesizes the existing studies on potential GLOFs and documented GLOF events, highlighting the major research gaps, as well as current and future challenges in the GLOF research. The study further highlights possible way forward to address the challenges related to forecasting, mitigation and management of GLOFs in the Indian Himalayan states and union territories, including (a) Ladakh: LA, (b) Jammu and Kashmir: JK, (c) Himachal Pradesh: HP, (d) Uttarakhand: UK, (e) Sikkim: SK and (f) Arunachal Pradesh: AP (Fig. 13.1).

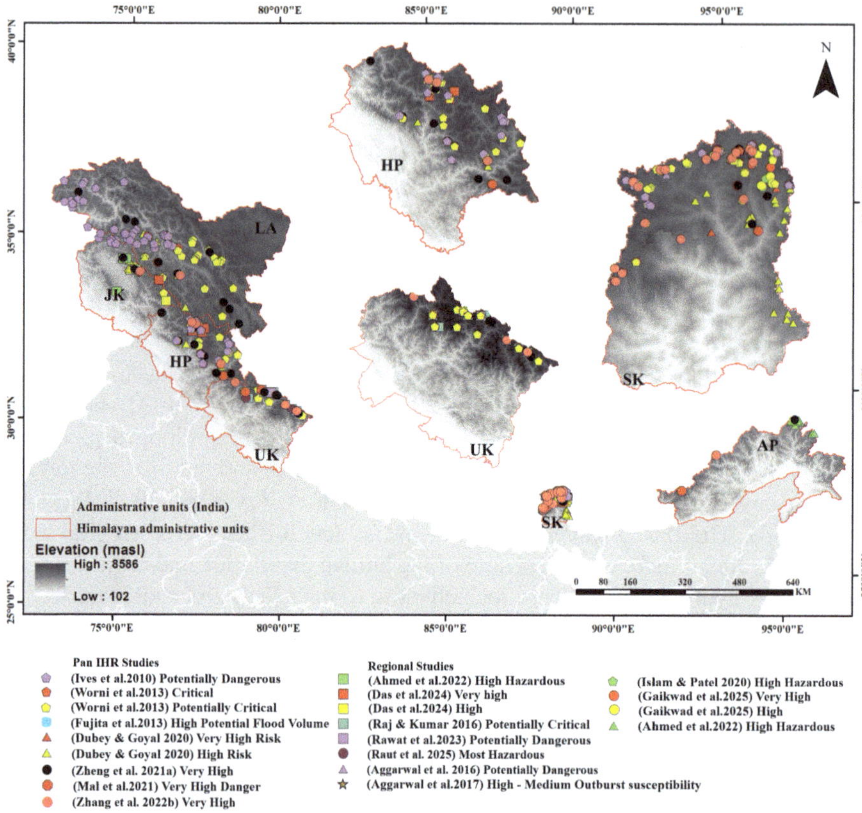

Fig. 13.1 Comparison of existing GLOF hazard assessments conducted across the IHR, including both pan-IHR studies (Dubey & Goyal, 2020; Fujita et al., 2013; Ives et al., 2010; Mal et al., 2021; Worni et al., 2013; Zhang et al., 2022b; Zheng et al., 2021a) and regional assessments (>10 lakes considered in the assessment), JK (Ahmed et al., 2022a), HP (Das et al., 2024), UK (Raj & Kumar, 2016; Raut et al., 2025; Rawat et al., 2023), SK (Aggarwal et al., 2016, 2017; Gaikwad et al., 2025; Islam & Patel, 2022) and AP (Ahmed et al., 2022b). Zheng et al. (2021a) has been used here as the baseline dataset for comparison with the pan-IHR studies. Only highly dangerous lakes, as classified in the studies, are considered here for the comparison. A shapefile of the administrative boundaries was obtained from the Survey of India (https://onlinemaps.surveyofindia.gov.in)

13.2 Database and Methods

The present study uses publicly available comprehensive database of potentially dangerous glacial lakes compiled from previous studies, integrating both pan-HKH and regional-scale (> 10 glacial lakes) studies (Fig. 13.1 and references cited therein). In some of the studies, coordinate information of the assessed hazardous lakes was not available; therefore they are not considered in the present study. The highly dangerous/critical lakes referred to by different terms in different studies were integrated within a GIS environment to compare and identify high-risk regions across the IHR.

Similarly, documented GLOF events in the IHR have been compiled from the previous studies conducted over the larger HKH and HMA region (Figs. 13.2 and 13.3), using various sources, such as peer-reviewed literature. These studies/datasets provide comprehensive information on the lake type, timing, and triggers since 1533 and are available in the public domain (Lützow et al., 2023; Shrestha et al., 2023; Veh et al., 2022). Information on physical processes, hydrological characteristics and consequent damages is rarely available, except in some cases, e.g. Chong Kumdan (1929–1933). As of 2024, > 700 GLOF events have been reported from across the HMA region (Shrestha et al., 2023), out of which ~ 235 events occurred in the IHR. Based on the compiled inventory, the GLOF density has also been estimated as the ratio of number of GLOF events and total glacial lake area (km^2) (Zheng et al., 2021b).

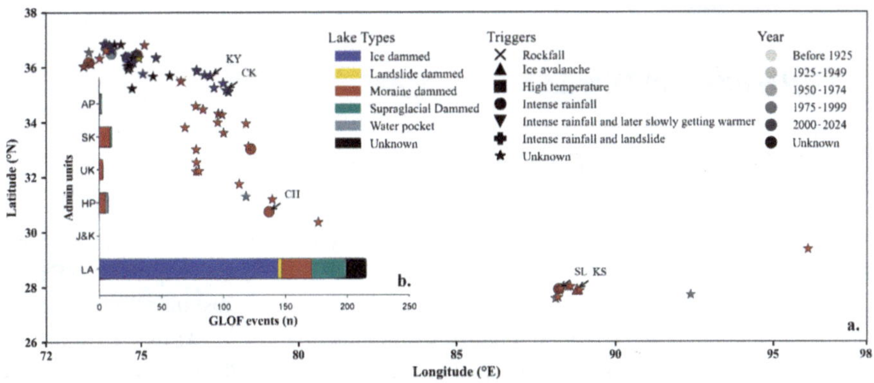

Fig. 13.2 **a** Documented GLOFs events, as classified based on lake types (different colours), triggers (different markers) (Zheng et al., 2021b) and temporal occurrence (grey to black) across the Indian Himalayan Region (IHR), compiled from previous studies (Bhambri et al., 2019; Carrivick & Tweed, 2016; Nie et al., 2018; Shrestha et al., 2023; Veh et al., 2019, 2022; Zheng et al., 2021a, 2021b). Some historically significant and high-impact GLOF events have been specifically marked as KY—Kyagar and CK—Chong Kumdan (JK), CH—Chorabari (UK), SL—South Lhonak and KS-Kongyangmi La Tsho (SK). **b** GLOF typology across IHR based on the lake type

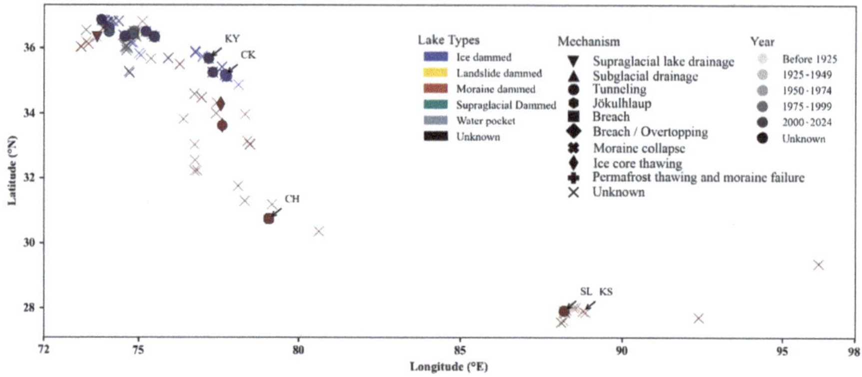

Fig. 13.3 Documented GLOFs events, classified based on lake types (different colours), mechanisms (different markers) and temporal occurrence (grey to black) across the Indian Himalayan Region (IHR) based on previous studies (Bhambri et al., 2019; Lützow et al., 2023; Nie et al., 2018; Shrestha et al., 2023; Veh et al., 2019; Zheng et al., 2021a, 2021b). Some historically significant and high-impact GLOF events: same as Fig. 13.2

13.3 Results and Discussion

13.3.1 Potential GLOFs Across the IHR

Identification and assessment studies on potentially dangerous glacial lakes, both at Pan-HKH and regional scales, are a relatively recent research priority (Dimri et al., 2021; Jain et al., 2025; Ravindra et al., 2024), but have substantially increased, especially after the Chorabari/Mandakini event (16/17 June 2013) in Uttarakhand. Earlier studies largely focused on the hazard aspect of the GLOFs (Fujita et al., 2013; Ives et al., 2010; Worni et al., 2013), while the recent studies incorporate exposure, vulnerability, susceptibility and risk parameters as well (Dubey & Goyal, 2020; Mal et al., 2021; Rinzin et al., 2023; Zheng et al., 2021a). Recently, substantial methodological improvements to address the gaps in the GLOF studies have been observed, from simple hazards to risk assessments (Allen et al., 2022; Rinzin et al., 2023). Currently, the northwestern Himalaya (LA and HP) have the highest concentration of lakes (120) susceptible to downstream floods. The concentration of potentially dangerous lakes decreases/increases towards the central (UK: 22)/ eastern Himalaya (SK) (101) indicating a west–east gradient. In the easternmost state (AP), a total of 15 high-risk lakes have been reported in various studies (Fig. 13.1 and references cited therein).

The first GLOF hazard assessment found 71 potentially dangerous lakes in the IHR, including 41 in J&K (LA: 36, JK: 5), 16 in HP and 14 in SK and none in the UK (Ives et al., 2010). In contrast, 104 critical and potentially critical lakes were identified in the IHR by another study (Worni et al., 2013), while Fujita et al. (2013) suggest only four critical lakes. In the recent studies (post 2020), number of potentially dangerous lakes varies between 24 and 73 (Dubey & Goyal, 2020; Mal et al., 2021; Zheng et al., 2021a), owing to notable differences in the methods. At the regional scale, it is observed that most of the studies are focused on SK and UK, while other Indian Himalayan states have fewer studies, for instance, Ladakh (Fig. 13.1). Most of the studies, both pan-HKH and regional, consider SK as the most susceptible to potential GLOFs, followed by UK, HP, AP and JK. AP remains the least studied region in the IHR with regard to potentially dangerous glacial lakes.

A number of regional-scale studies applied the analytical hierarchy process (AHP) method, using various criteria (e.g. characteristics of lakes, dam types and geometry, mass movement, seismic activities) for assessing the GLOF susceptibility levels (Prakash & Nagarajan, 2017). AHP is a statistical model and subject to individual bias and limited scale, and does not capture underlying physical processes. Therefore, its ability to capture the Chorabari lake flood-type case (2013), which was a very small right-lateral moraine-dammed lake (< 0.025 km^2), and triggered by an extreme rainfall event, is very limited. Hence, Chorabari lake was not identified as dangerous by any of the studies prior to its breach (Table 13.1). On the contrary, the South Lhonak lake (a larger proglacial lake) that breached in 2023 was identified as highly dangerous by at least five regional-scale studies before its failure (Table 13.1) and has been subject to detailed investigations (Sattar et al., 2019, 2021, 2025).

Significant differences in the results of potential GLOFs studies are seen (Fig. 13.1 and references cited therein), which is attributed to variations in methodologies and variables considered. A recent study (Mal et al., 2021) further improved upon the results obtained from large-scale GLOF assessment by incorporating an independent expert-inspection approach in the Google Earth environment and identified 25 high-priority lakes for in-depth risk management and mitigation. The regular and detailed GLOF risk assessments are essentially required to be carried out and updated on the dangerous lakes that have been commonly identified by different studies, e.g. South Lhonak lake in SK (Aggarwal et al., 2017; Sattar et al., 2019, 2021; Worni et al., 2013). The GLOF mitigation and management accordingly are to be prioritized to minimize any damage.

13.3.2 Spatio-temporal Distribution of GLOFs in the IHR

In total, 237 GLOF events have been reported to occur across the IHR since 1533 (Figs. 13.2 and 13.3). It is observed that GLOF events are spatially concentrated in the LA region (215), hence considered a GLOF hotspot (Fig. 13.4), followed by SK (10), HP (7), UK (3) and AP (2). No GLOF event has been reported so far in the JK region. Of the total 215 GLOF events in LA, 144 were caused by failure of ice-dammed lakes, followed by 27 from supraglacial and 24 from moraine-dammed lakes. Most of the reported GLOF events in SK and HP were triggered by moraine-dam failures. It is also observed that all the ice-dam failure-related GLOF events occurred in LA (Northwestern Himalayan and Karakoram), a phenomenon that is typically absent in the other parts of the HKH region (Bhambri et al., 2019; Hewitt & Liu, 2010; Shrestha

Table 13.1 List of GLOF events that occurred in the IHR, and the studies identified them as potential GLOFs

Admin units	Name	Years	Lat Long	Ives et al. (2010)	Worni et al. (2013)	Dubey and Goyal (2020)	Mal et al. (2021)	Zheng et al. (2021a, 2021b)	Raj and Kumar (2016)	Aggarwal et al. (2016)	Aggarwal et al. (2017)	Islam and Patel (2022)	Ahmed et al. (2022a, 2022b)	Gaikwad et al. (2025)	Trigger and References
LA	Unnamed	Unknown	33.13 78.41					VH							(1)* Moraine-dammed (Shrestha et al., 2023; Zheng et al., 2021a, 2021b)
LA	Passu	6/01/ 2008	36.46 74.88	PD											(1) Triggered by an intense rainfall event (Shrestha et al., 2023)
UK	Chorabari	16/17 June 2013	30.75 79.06						PCL						(1)Triggered by an intense rainfall event (Allen et al., 2016a; Shrestha et al., 2023)
SK	Goecha Lake	Unknown	27.60 88.19		PCL	HR		VH						VHH	(4)*Moraine-dammed, (Shrestha et al., 2023; Zheng et al., 2021a, 2021b)
SK	Unnamed	Unknown	27.87 88.86	PD		VHR	VHD	VH		PD					(5)*Moraine-dammed, (Zheng et al., 2021a, 2021b) (Shrestha et al., 2023)
SK	Kongyangmi La Tsho	Apr-Oct 1997	27.90 88.78			HR								HH	(1)* Moraine-dammed, (Nie et al., 2018; Shrestha et al., 2023; Zheng et al., 2021a)

(continued)

Table 13.1 (continued)

Admin units	Name	Years	Lat Long	Ives et al. (2010)	Worni et al. (2013)	Dubey and Goyal (2020)	Mal et al. (2021)	Zheng et al. (2021a, 2021b)	Raj and Kumar (2016)	Aggarwal et al. (2016)	Aggarwal et al. (2017)	Islam and Patel (2022)	Ahmed et al. (2022a, 2022b)	Gaikwad et al. (2025)	Trigger and References
SK	**South Lhonak**	**3/10/ 2023**	**27.91 88.20**	**PD**	**PCL**					**PD**	**HMS**	**HHZ**			(5) **Triggered by intense rainfall, landslide, permafrost thaw and moraine failure** (Shrestha et al., 2023)
SK	Unnamed	Unknown	28.01 88.49	PD	PCL					PD	HMS				(4) Moraine-dammed, (Shrestha et al., 2023; Zheng et al., 2021a, 2021b)
SK	Om Mane Lake	Unknown	28.01 88.58							PD					(1) Moraine-dammed, (Shrestha et al., 2023; Zheng et al., 2021a, 2021b)
AP	Unnamed	Unknown	29.36 96.12										HHZ		(1) Moraine-dammed, (Shrestha et al., 2023; Zheng et al., 2021a, 2021b)

*Numbers mentioned in brackets indicate studies (*n*) that identified the lake as Potentially Dangerous, though the nomenclature varies among different studies. Only South Lhonak, (3/10/ 2023) GLOF event was identified as potentially dangerous before its failure, while other GLOF events in the region were not identified by any of the previous studies. PD (Potentially dangerous), PCL (Potentially Critical), CL (Critical lakes) VHR (Very High Risk), HR (High Risk), VHD (Very High Danger), VH (Very High), HMS (High-Medium Outburst Sus), HHZ (High Hazardous), VHH (Very High Hazardous), HH (High Hazardous) et al., 2023). The GLOF events from supraglacial lakes are mostly reported from LA, while one each in SK and AP. Water pockets, also known as subglacial tunnelling and piping, are highly uncommon in the IHR, causing only two GLOF events in HP and one in LA. All the unknown GLOF events for lake type (16) occurred in LA, which are mostly confined to the pre-satellite data era (Fig. 13.4). It is interesting to note that LA has experienced a large number of GLOFs in the past, but the studies on potential GLOFs are rare (Gupta et al., 2022), which can be linked to a long gap in these events post-1940s.

The GLOFs/year have been increasing since the last century in the IHR, with the first quarter of the twenty-first century recording the highest GLOF events each year (2.4). Before 1925, GLOFs/year remained significantly low (0.18) and are largely unknown for their timings and physical mechanisms, which is linked to their under-reporting amid poor technology and communication networks in the pre-satellite era. The database on GLOF events during the pre-satellite era has been compiled by previous studies (Bhambri et al., 2019; Chen et al., 2010; Feng, 1991; Hewitt & Liu, 2010; Mason, 1929; Shrestha et al., 2023; Shroder et al., 1998) based on old research papers, regional newspapers and other regional reports. These data base not only systematically catalogue the past GLOFs, but also offer critical insights into GLOF occurrences, triggering events and physical mechanisms in some cases.

Even during the satellite image era (Post-1970s), under-reporting of GLOF events remains a big challenge in the IHR due to their remote locations and limited communication (Lützow et al., 2023; Veh et al., 2020; Zheng et al., 2021a). It indicates possible major gaps in the historical GLOF database, highlighting the need for further improvements based on remote sensing images and modelling approaches (Veh et al., 2019). Some of the unknown GLOF events post-1970s were recently identified using

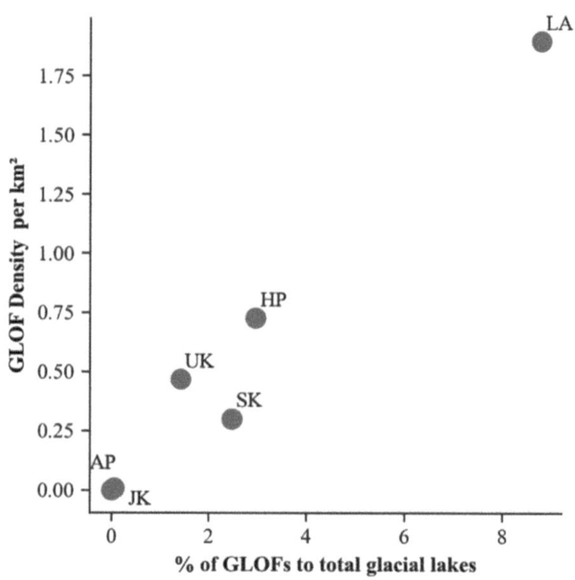

Fig. 13.4 Scatter plot of GLOF density and % of actual GLOF events to total glacial lakes in the IHR since 1533, indicating their hotspot

modelling approaches and updated in the GLOF database (Lützow et al., 2023; Veh et al., 2019; Zheng et al., 2021b).

Moraine-dammed lakes failure-related GLOFs display a more recent increase (Emmer et al., 2022; Harrison et al., 2018), especially post-2000, which are triggered by intense rainfall activities, prolonged extreme temperature conditions and associated rockfall and ice avalanches (Jain et al., 2025; Sattar et al., 2025). The complex interrelations of these forces and their role in GLOF triggers are yet to be fully understood and quantified (Prakash & Nagarajan, 2017). In most of the historical GLOF events, triggers are known; however, their mechanisms, e.g. strength and activities of moraines and surrounding slopes, remain largely unknown (Fig. 13.3). Besides, the degradation of ground ice within glacier foreland and glacial lake moraine, along with their material composition and contributions to GLOF triggers and processes, remain largely unaccounted for (Sattar et al., 2025; Shugar et al., 2021).

13.3.3 Gaps and Limitations in GLOF Studies and Scope for Further Research

The present study identified some important gaps related to the inventories of glacial lakes, parameters/factors considered to evaluate dangerous lakes and significant variations in the methodological frameworks across the studies (Jain et al., 2025; Kumar et al., 2025c). As a result, the findings obtained from across the studies offer limited cross-comparison (Kumar et al., 2025c). The latest and updated glacial lakes inventories are the most important input parameters for their hazard assessment, which are mostly generated based on satellite datasets with different spatial resolutions (Landsat: 30 m, Sentinel: 10 m). Mapping of lakes based on higher/medium resolution satellite data, for instance, Resourcesat-2 (5.8 m)/Landsat (30 m) can map/omit smaller glacial lakes (supraglacial), hence yielding significantly higher/lower lake numbers (Bhambri et al., 2018). Different minimum lake area thresholds, e.g. $\geq 0.0036\ km^2$ (Kumar et al., 2025c; Zheng et al., 2021a), ≥ 0.03–$0.05\ km^2$ (Dubey & Goyal, 2020; Nie et al., 2017), also produce significantly different results (number and area). Besides, different elevational thresholds (e.g. $\geq$ 2900 and $\geq$ 3500 m asl) (Kumar et al., 2025c; Maharjan et al., 2018), as observed in different studies, also significantly influence the inventory results. Glacial lakes mapping using satellite data is an extremely time-consuming and challenging task (Mal et al., 2020). Therefore, there is a need to develop and deploy machine-learning techniques with consistent methodologies for lake mapping, leading to large-scale lake inventories, and thereby aiding improved and updated GLOF modelling as well as timely hazard and risk management (Sahu & Singh, 2025; Sahu et al., 2023).

For identifying potentially dangerous glacial lakes at a relatively larger spatial scale (e.g. state or river basin), different methodologies have been deployed. These approaches can be grouped in two broader categories, viz., (a) AHP or multi-criteria decision-based methods, which are generally applied over relatively smaller regions

(Aggarwal et al., 2017; Prakash & Nagarajan, 2017) and (b) large-scale assessment studies (Mal et al., 2021; Zheng et al., 2021a). AHP or multi-criteria decision-based studies rely on statistical models that generally consider parameters related to (1) lake characteristics, (2) dam and glacier conditions, (3) physical settings of the adjacent region, (4) seismic activity and (5) climatic factors (Aggarwal et al., 2017; Prakash & Nagarajan, 2017). However, these methods are often influenced by subjectivity and tend to overlook the complex interactions of related forces. In addition, these methods are generally applied over relatively smaller regions, which limits the comparability with other studies and substantially affect the efficiency to predict the potential GLOF events, e.g. Chorabari Lake breach (2013) was not identified as potentially dangerous in any of the prior assessments (Table 13.1).

Large-scale GLOF assessments have considered (1) lake area (proxy to lake volume), (2) watershed area (proxy to precipitation amount), (3) dam conditions and (4) topographic potential to evaluate the hazard level of the lakes (Allen et al., 2016b; Mal et al., 2021; Zheng et al., 2021a). Furthermore, exposure of different sectors and population vulnerabilities has been assessed for the worst-case GLOF risk scenarios using flood trajectories, estimated using a modified single-flow model (Huggel et al., 2004), which has effectively captured numerous documented GLOF events (Allen et al., 2016b, 2019; Zheng et al., 2021a). However, most of these studies/models are also statistical in nature and do not consider the dynamic nature of underlying factors. Overall, the complex and interrelated mechanisms responsible for GLOF events are not yet sufficiently addressed in automated large-scale assessments. Consequently, the accuracy and reliability of GLOF prediction models remain a major limitation of large-scale assessments, as evidenced by the South Lhonak GLOF event (2023) that was not considered highly dangerous in earlier studies (Table 13.1).

Despite significant advances in hydrodynamic modelling and computational efficiencies in recent decades, large-scale simulations of GLOFs scenarios, including inundation depth and extents, flood arrival time, flow velocities and discharge, debris flow extent and volumes, remain subject to several limitations that affect the reliability of assessment results. Constraints of major case-based modelling frameworks (e.g. HEC-RAS) include (1) unaccounted dynamic nature of precipitation and temperature events, (2) under-capture of complex and dynamic interactions between surrounding terrain, glacier ice, moraine materials, and permafrost degradation, (3) availability of high-resolution digital elevation models and (4) lake bathymetric information. In addition, glacier-lake interactions and highly dynamic triggers, viz., subglacial tunnelling and other related physical processes and permafrost thaw, are yet to be incorporated in the latest models (e.g. BASEMENT, HEC-RAS, and r.avaflow). Latest modelling frameworks (i.e. r.avaflow) incorporate topographic potential (e.g. secondary events, such as landslides). However, it still does not account for subglacial processes and permafrost thaw, and is limited to individual lake case studies due to high computational requirements. Moreover, socio-economic-environmental factors are either excluded or oversimplified in the existing models, which leaves a great scope for wider improvements in the modelling environments and consequent GLOF modelling results. Therefore, reliable prediction of GLOF events remains a critical challenge due to the highly dynamic and complex interactions among geophysical, climatological, meteorological and hydrological processes, which are yet to be fully integrated in the modelling frameworks. Overall, major gaps in glacial lake inventories, methods of hazard assessments and GLOF modelling approaches and frameworks have been observed. It limits the reliability and applicability in GLOF event prediction and hence the timely interventions related to its mitigation and related risk management.

13.4 Conclusion

The comprehensive review of potential and historically recorded GLOF studies across the IHR underscores the escalating risk posed by existing glacial lakes. While the triggering factors of the GLOFs are usually well known and understood, the underlying long and short-term dynamic forces, particularly geophysical, hydrological, meteorological and social processes, are not sufficiently integrated in the modern modelling approaches and frameworks, thereby limiting their applicability for large-scale GLOF assessments. Consequently, there is an urgent need for more integrated and regionally tailored GLOF modelling approaches, integrating complex interactions of related processes. The state-of-the-art machine-learning modelling approaches using high-resolution data are critically required to solve the larger issue of GLOF forecasting. This will improve GLOF risk assessments as well as risk communication and also help pre- and post-GLOF event management, including preparedness and adaptive strategies for vulnerable communities and critical infrastructure in the downstream regions.

Acknowledgements The authors are thankful to the working groups and individual researchers for making GLOF (potential and historical events) data publicly available. We also thank the reviewer for constructive suggestions, which helped us greatly to improve the overall quality of the manuscript.

References

Aggarwal, A., Jain, S. K., Lohani, A. K., & Jain, N. (2016). Glacial lake outburst flood risk assessment using combined approaches of remote sensing, GIS and dam break modelling. *Geomatics, Natural Hazards Risk, 7*, 18–36. https://doi.org/10.1080/19475705.2013.862573

Aggarwal, S., Rai, S. C., Thakur, P. K., & Emmer, A. (2017). Inventory and recently increasing GLOF susceptibility of glacial lakes in Sikkim, Eastern Himalaya. *Geomorphology, 295*, 39–54. https://doi.org/10.1016/j.geomorph.2017.06.014

Ahmed, R., Ahmad, S. T., Wani, G. F., et al. (2022a). High resolution inventory and hazard assessment of potentially dangerous glacial lakes in upper Jhelum basin, Kashmir Himalaya, India. *Geocarto International, 37*, 10681–10712. https://doi.org/10.1080/10106049.2022.2038693

Ahmed, R., Ahmad, S. T., Wani, G. F., et al. (2022b). Glacial lake changes and the identification of potentially dangerous glacial lakes (PDGLs) under warming climate in the Dibang River Basin, Eastern Himalaya, India. *Geocarto International, 37*, 17659–17685. https://doi.org/10.1080/10106049.2022.2134461

Allen, S. K., Linsbauer, A., Randhawa, S. S., et al. (2016b). Glacial lake outburst flood risk in Himachal Pradesh, India: An integrative and anticipatory approach considering current and future threats. *Natural Hazards, 84*, 1741–1763. https://doi.org/10.1007/s11069-016-2511-x

Allen, S., Rastner, P., Arora, M., et al. (2016a). Lake outburst and debris flow disaster at Kedarnath, June 2013: Hydrometeorological triggering and topographic predisposition. *Landslides*. https://doi.org/10.1007/s10346-015-0584-3

Allen, S. K., Sattar, A., King, O., et al. (2022). Glacial lake outburst flood hazard under current and future conditions: Worst-case scenarios in a transboundary Himalayan basin. *Natural Hazards and Earth Systems Sciences, 22*, 3765–3785. https://doi.org/10.5194/nhess-22-3765-2022

Allen, S. K., Zhang, G., Wang, W., et al. (2019). Potentially dangerous glacial lakes across the Tibetan Plateau revealed using a large-scale automated assessment approach. *Sci Bull, 64*, 435–445. https://doi.org/10.1016/j.scib.2019.03.011

Bhambri, R., Hewitt, K., Kawishwar, P., et al. (2019). Ice-dams, outburst floods, and movement heterogeneity of glaciers, Karakoram. *Global and Planetary Change, 180*, 100–116. https://doi.org/10.1016/j.gloplacha.2019.05.004

Bhambri, R., Mehta, M., Dobhal, D. P., et al. (2016). Devastation in the Kedarnath (Mandakini) Valley, Garhwal Himalaya, during 16–17 June 2013: A remote sensing and ground-based assessment. *Natural Hazards, 80*, 1801–1822. https://doi.org/10.1007/s11069-015-2033-y

Bhambri, R., Misra, A., Kumar, A., et al. (2018). Glacier Lake inventory of Himachal Pradesh. *Himalayan Geology, 39*, 1–32.

Bolch, T., Shea, J. M., Liu, S., et al. (2019). Status and change of the cryosphere in the extended Hindu Kush Himalaya Region. In P. Wester, A. Mishra, A. Mukherji, & A. B. Shrestha (Eds.), *The Hindu Kush himalaya assessment: Mountains, climate change, sustainability and people* (pp. 209–255). Springer, Cham.

Brun, F., Berthier, E., Wagnon, P., et al. (2017). A spatially resolved estimate of High Mountain Asia glacier mass balances from 2000 to 2016. *Nature Geoscience, 10*, 668–673. https://doi.org/10.1038/ngeo2999

Byers, A. C., Shugar, D. H., Chand, M. B., et al. (2022). Three recent and lesser-known glacier-related flood mechanisms in high mountain environments. *Mountain Research and Development, 42*, A12–A22. https://doi.org/10.1659/MRD-JOURNAL-D-21-00045.1

Carrivick, J. L., & Tweed, F. S. (2016). A global assessment of the societal impacts of glacier outburst floods. *Global and Planetary Change, 144*, 1–16. https://doi.org/10.1016/j.gloplacha.2016.07.001

Chen, F., Zhang, M., Guo, H., et al. (2021). Annual 30 m dataset for glacial lakes in high mountain asia from 2008 to 2017. *Earth System Science Data, 13*, 741–766. https://doi.org/10.5194/essd-13-741-2021

Chen, Y., Xu, C., Chen, Y., et al. (2010). Response of glacial-lake outburst floods to climate change in the Yarkant River basin on northern slope of Karakoram Mountains, China. *Quaternary International, 226*, 75–81. https://doi.org/10.1016/j.quaint.2010.01.003

Das, S., Das, S., Mandal, S. T., et al. (2024). Inventory and GLOF susceptibility of glacial lakes in Chenab basin, Western Himalaya. *Geomatics, Natural Hazards Risk, 15*. https://doi.org/10.1080/19475705.2024.2356216

Dimri, A. P., Allen, S., Huggel, C., et al. (2021). Climate change, cryosphere and impacts in the Indian Himalayan Region. *Current Science, 120*, 1012–1012.

Dubey, S., & Goyal, M. K. (2020). Glacial lake outburst flood hazard, downstream impact, and risk over the Indian Himalayas. *Water Resources Research, 56*, 1–21. https://doi.org/10.1029/2019WR026533

Emmer, A., Allen, S. K., Carey, M., et al. (2022). Progress and challenges in glacial lake outburst flood research (2017–2021): A research community perspective. *Natural Hazards and Earth Systems Sciences, 22*, 3041–3061. https://doi.org/10.5194/nhess-22-3041-2022

Feng, Q. (1991). Characteristics of glacier outburst flood in the Yarkant river, Karakorum mountains. *GeoJournal, 25*, 255–263. https://doi.org/10.1007/BF02682195

Fujita, K., Sakai, A., Takenaka, S., et al. (2013). Potential flood volume of Himalayan glacial lakes. *Natural Hazards and Earth Systems Sciences, 13*, 1827–1839. https://doi.org/10.5194/nhess-13-1827-2013

Gaikwad, D., Tiwari, R. K., Kumar, M., & Guha, S. (2025). Hazard, and risk modelling of glacial lakes in the Sikkim Himalaya: Integrating qualitative and quantitative approaches for hazard assessment. *Geomorphology, 471*, 109577. https://doi.org/10.1016/j.geomorph.2024.109577

Gardelle, J., Arnaud, Y., & Berthier, E. (2011). Contrasted evolution of glacial lakes along the Hindu Kush Himalaya mountain range between 1990 and 2009. *Global and Planetary Change, 75*, 47–55. https://doi.org/10.1016/j.gloplacha.2010.10.003

Gupta, A., Maheshwari, R., Guru, N., et al. (2022). Quantitative prioritization of potentially critical glacial Lakes in the Indus River basin using satellite derived parameters. *Geocarto International, 37*, 7508–7530. https://doi.org/10.1080/10106049.2021.1975831

Harrison, S., Kargel, J. S., Huggel, C., et al. (2018). Climate change and the global pattern of moraine-dammed glacial lake outburst floods. *The Cryosphere, 12*, 1195–1209. https://doi.org/10.5194/tc-12-1195-2018

Hewitt, K., & Liu, J. (2010). Ice-Dammed lakes and outburst floods, Karakoram Himalaya: Historical perspectives on emerging threats. *Physical Geography, 31*, 528–551. https://doi.org/10.2747/0272-3646.31.6.528

Hock, R., Rasul, G., Adler, C., et al. (2019). High mountain areas. In *IPCC special report on the ocean and cryosphere in a changing climate, Geneva* (pp. 131–202).

Huggel, C., Haeberli, W., Kääb, A., et al. (2004). An assessment procedure for glacial hazards in the Swiss Alps. *Canadian Geotechnical Journal, 41*, 1068–1083. https://doi.org/10.1139/T04-053

Islam, N., & Patel, P. P. (2022). Inventory and GLOF hazard assessment of glacial lakes in the Sikkim Himalayas, India. *Geocarto International, 37*, 3840–3876. https://doi.org/10.1080/10106049.2020.1869332

Ives, J. D., Shrestha, R. B., Mool, P. K. (2010). *Formation of glacial lakes in the Hindu Kush-Himalayas and GLOF risk assessment, Lalitpur, Nepal.*

Jain, S. K., Jain, S. K., & Ahmed, R. (2025). Glacial lake outburst floods (GLOFs): Causes, modeling, and mitigation in Indian Himalayan region. *GeoJournal, 90*, 85. https://doi.org/10.1007/s10708-025-11326-4

Kargel, J. S., Leonard, G. J., Shugar, D. H., et al. (2016). Geomorphic and geologic controls of geohazards induced by Nepal's 2015 Gorkha earthquake. *Science*, 80–351, aac8353. https://doi.org/10.1126/science.aac8353

Krishnan, R., Shrestha, A. B., Ren, G., et al. (2019). Unravelling climate change in the Hindu Kush Himalaya: Rapid warming in the mountains and increasing extremes BT—The Hindu Kush Himalaya assessment: Mountains, climate change, sustainability and people. In: P. Wester, A. Mishra, A. Mukherji, A.B. Shrestha (Eds) (pp. 57–97). Springer, Cham.

Kumar, A., Mal, S., Schickhoff, U., & Dimri, A. P. (2025c). Basin-scale spatio-temporal development of glacial lakes in the Hindukush-Karakoram-Himalayas. *Global and Planetary Change, 245*, 104656. https://doi.org/10.1016/j.gloplacha.2024.104656

Kumar, A., Mal, S., & Schickhoff, U. (2025b). Spatio-temporal evolution of glacial lakes in the upper Ganga Basin, Central Himalayas. In S.C. Rai (Ed.) *Himalaya: Mountains of destiny* (pp. 59–82). Springer, Cham.

Kumar, A., Mal, S., Schickhoff, U., et al. (2025a). Assessing the role of regional characteristics in estimating the volume of glacial lakes in the upper Indus-Ganga-Brahmaputra basins, Hindu Kush Himalaya. *Journal of Hydrology, 657*, 133016. https://doi.org/10.1016/j.jhydrol.2025.133016

Li, D., Lu, X., Walling, D. E., et al. (2022). High Mountain Asia hydropower systems threatened by climate-driven landscape instability. *Nature Geoscience, 15*, 520–530. https://doi.org/10.1038/s41561-022-00953-y

Lützow, N., Veh, G., & Korup, O. (2023). A global database of historic glacier lake outburst floods. *Earth System Science Data, 15*, 2983–3000. https://doi.org/10.5194/essd-15-2983-2023

Maharjan, S. B., Mool, P. K., Lizong, W., et al. (2018). The status of glacial lakes in the Hindu Kush Himalaya. *ICIMOD Research Report, 70.*

Mal, S., Allen, S. K., Frey, H., et al. (2021). Sectorwise assessment of glacial lake outburst flood danger in the Indian Himalayan Region. *Mountain Research and Development, 41*, R1–R12. https://doi.org/10.1659/MRD-JOURNAL-D-20-00043.1

Mal, S., Kumar, A., Bhambri, R., et al. (2020). Inventory and spatial distribution of glacial lakes in Arunachal Pradesh, Eastern Himalaya, India. *Journal of the Geological Society of India, 96*, 609–615. https://doi.org/10.1007/s12594-020-1610-1

Mason, K. (1929). Indus floods and Shyok glaciers. *Himalayan Journal, 1*, 10–29.

Mohanty, L., Maiti, S., & Dixit, A. (2023). Spatio-temporal assessment of regional scale evolution and distribution of glacial lakes in Himalaya. *Frontiers in Earth Science, 10*, 1–15. https://doi.org/10.3389/feart.2022.1038777

Mondal, A., Pandey, A., Mondal, S., & Kundu, S. (2026). Impact assessment of Glacial Lake Outburst Floods (GLOFs) in the Himalaya using satellite imageries over the last 25 years (2000–2024): A comprehensive review. *Remote Sensing Applications: Society and Environment, 41*, 101845. https://doi.org/10.1016/j.rsase.2025.101845

Nie, Y., Liu, Q., & Liu, S. (2013). Glacial lake expansion in the Central Himalayas by landsat images, 1990–2010. *PLoS ONE, 8*, 1–8. https://doi.org/10.1371/journal.pone.0083973

Nie, Y., Liu, Q., Wang, J., et al. (2018). An inventory of historical glacial lake outburst floods in the Himalayas based on remote sensing observations and geomorphological analysis. *Geomorphology, 308*, 91–106. https://doi.org/10.1016/j.geomorph.2018.02.002

Nie, Y., Sheng, Y., Liu, Q., et al. (2017). A regional-scale assessment of Himalayan glacial lake changes using satellite observations from 1990 to 2015. *Remote Sensing of Environment, 189*, 1–13. https://doi.org/10.1016/j.rse.2016.11.008

Patel, L. K., Sharma, P., Laluraj, C. M., et al. (2017). A geospatial analysis of Samudra Tapu and Gepang Gath glacial lakes in the Chandra Basin, Western Himalaya. *Natural Hazards, 86*, 1275–1290. https://doi.org/10.1007/s11069-017-2743-4

Prakash, C., & Nagarajan, R. (2017). Outburst susceptibility assessment of moraine-dammed lakes in Western Himalaya using an analytic hierarchy process. *Earth Surface Processes and Landforms, 42*, 2306–2321. https://doi.org/10.1002/esp.4185

Raj, K. B. G., & Kumar, K. V. (2016). Inventory of glacial lakes and its evolution in Uttarakhand himalaya using time series satellite data. *Journal of the Indian Society of Remote Sensing, 44*, 959–976. https://doi.org/10.1007/s12524-016-0560-y

Raut, M., Pant, S., Kunmar, P., et al. (2025). Identification and assessment of potentially dangerous glacial lakes in Uttarakhand, India. Natural Hazards. https://doi.org/10.1007/s11069-025-07481-z

Ravindra, R., Kulkarni, A. V., Dimri, A. P., et al. (2024). Recent Indian studies in Himalayan cryosphere. *Proceedings of the Indian National Science Academy, 90*, 415–425. https://doi.org/10.1007/s43538-024-00237-6

Rawat, M., Jain, S. K., Ahmed, R., & Lohani, A. K. (2023). Glacial lake outburst flood risk assessment using remote sensing and hydrodynamic modeling: A case study of Satluj basin, Western Himalayas, India. *Environmental Science and Pollution Research, 30*, 41591–41608. https://doi.org/10.1007/s11356-023-25134-1

Richardson, S. D., & Reynolds, J. M. (2000). An overview of glacial hazards in the Himalayas. *Quaternary International, 65–66*, 31–47. https://doi.org/10.1016/S1040-6182(99)00035-X

Rinzin, S., Zhang, G., Sattar, A., et al. (2023). GLOF hazard, exposure, vulnerability, and risk assessment of potentially dangerous glacial lakes in the Bhutan Himalaya. *Journal of Hydrology, 619*, 129311. https://doi.org/10.1016/j.jhydrol.2023.129311

Sahu, R., Ramsankaran, R., Bhambri, R., et al. (2023). Evolution of supraglacial lakes from 1990 to 2020 in the Himalaya-Karakoram Region using cloud-based google earth engine platform. *Journal of the Indian Society of Remote Sensing, 51*, 2379–2390. https://doi.org/10.1007/s12524-023-01773-2

Sahu, R., & Singh, D. P. (2025). Conventional and deep learning approaches for glacial lake mapping using remote sensing data: A comprehensive review. *International Journal of Remote Sensing, 46*, 3992–4020. https://doi.org/10.1080/01431161.2025.2496001

Sattar, A., Cook, K. L., Rai, S. K., et al. (2025). The Sikkim flood of October 2023: Drivers, causes, and impacts of a multihazard cascade. *Science*, 80–387, eads2659. https://doi.org/10.1126/science.ads2659

Sattar, A., Goswami, A., & Kulkarni, A. V. (2019). Hydrodynamic moraine-breach modeling and outburst flood routing—A hazard assessment of the South Lhonak lake, Sikkim. *Science of the Total Environment, 668*, 362–378. https://doi.org/10.1016/j.scitotenv.2019.02.388

Sattar, A., Goswami, A., Kulkarni, A. V., et al. (2021). Future glacial lake outburst flood (GLOF) hazard of the South Lhonak Lake, Sikkim Himalaya. *Geomorphology, 388*, 107783. https://doi.org/10.1016/j.geomorph.2021.107783

Sharma, R. K., Pradhan, P., Sharma, N. P., & Shrestha, D. G. (2018). Remote sensing and in situ-based assessment of rapidly growing South Lhonak glacial lake in eastern Himalaya, India. *Natural Hazards, 93*, 393–409. https://doi.org/10.1007/s11069-018-3305-0

Shrestha, F., Steiner, J. F., Shrestha, R., et al. (2023). A comprehensive and version-controlled database of glacial lake outburst floods in High Mountain Asia. *Earth System Science Data, 15*, 3941–3961. https://doi.org/10.5194/essd-15-3941-2023

Shroder, J. F., Bishop, M. P., & Scheppy, R. (1998). Catastrophic flood flushing of sediment, Western Himalaya, Pakistan. In J. Kalvoda & C. L. Rosenfeld (Eds.), *Geomorphological hazards in high mountain areas* (pp. 27–48). Springer, Cham.

Shugar, D. H., Burr, A., Haritashya, U. K., et al. (2020). Rapid worldwide growth of glacial lakes since 1990. *Nature Climate Change, 10*, 939–945. https://doi.org/10.1038/s41558-020-0855-4

Shugar, D.H., Jacquemart, M., Shean, D., et al. (2021). A massive rock and ice avalanche caused the 2021 disaster at Chamoli, Indian Himalaya. *Science, 80*. https://doi.org/10.1126/science.abh4455

Veh, G., Korup, O., von Specht, S., et al. (2019). Unchanged frequency of moraine-dammed glacial lake outburst floods in the Himalaya. *Nature Climate Change, 9*, 379–383. https://doi.org/10.1038/s41558-019-0437-5

Veh, G., Lützow, N., Kharlamova, V., et al. (2022). Trends, breaks, and biases in the frequency of reported glacier lake outburst floods. *Earth's Future, 10*. https://doi.org/10.1029/2021EF002426

Veh, G., Korup, O., & Walz, A. (2020). Hazard from Himalayan glacier lake outburst floods. *Proceedings of the National Academy of Sciences of the United States of America, 117*, 907–912. https://doi.org/10.1073/pnas.1914898117

Wang, X., Guo, X., Yang, C., et al. (2020). Glacial lake inventory of high-mountain Asia in 1990 and 2018 derived from Landsat images. *Earth System Science Data, 12*, 2169–2182. https://doi.org/10.5194/essd-12-2169-2020

Westoby, M., Westoby, M. J., Glasser, N. F., et al. (2014). Modelling outburst floods from moraine-dammed glacial lakes. *Earth-Science Reviews*. https://doi.org/10.1016/j.earscirev.2014.03.009

Worni, R., Huggel, C., & Stoffel, M. (2013). Glacial lakes in the Indian Himalayas—From an area-wide glacial lake inventory to on-site and modeling based risk assessment of critical glacial lakes. *Science of the Total Environment, 468–469*, 71–84. https://doi.org/10.1016/j.scitotenv.2012.11.043

Zhang, M., Chen, F., Guo, H., et al. (2022a). Glacial lake area changes in high mountain Asia during 1990–2020 using satellite remote sensing. *Research 2022*. https://doi.org/10.34133/2022/9821275

Zhang, T., Wang, W., Gao, T., et al. (2022b). An integrative method for identifying potentially dangerous glacial lakes in the Himalayas. *Science of the Total Environment, 806*, 150442. https://doi.org/10.1016/j.scitotenv.2021.150442

Zheng, G., Allen, S. K., Bao, A., et al. (2021a). Increasing risk of glacial lake outburst floods from future third pole deglaciation. *Nature Climate Change, 11*, 411–417. https://doi.org/10.1038/s41558-021-01028-3

Zheng, G., Bao, A., Allen, S., et al. (2021b). Numerous unreported glacial lake outburst floods in the third pole revealed by high-resolution satellite data and geomorphological evidence. *Science Bulletin, 66*, 1270–1273. https://doi.org/10.1016/j.scib.2021.01.014

Part III
Forest Fire

Chapter 14
Forest Fire in the Himalayan Region—Status and Forecast

Vipin Upadhyay, Sunil Kumar, Archana Sharma, and Amit Kumar

Abstract Fire is one of the five elements discussed in the Indian philosophy of life. When it comes to the forest, an occasional, limited and low-intensity fire can be beneficial for forest clearance and regeneration. But the fires that have occurred in recent times are of high intensity and frequency, spread over a large geographical area, mostly in the foothills and lower Himalayan regions, where surface temperatures are high during the pre-monsoon season. Climate change and anthropogenic activities have further enhanced the probability of fire. Initial studies focused on post-fire assessment of extent and loss, which have now been transformed into machine learning models for predicting high-risk areas, fire forecasting and near-real-time monitoring. This advancement becomes possible due to technological advancements in the field of geospatial science, i.e. development of high spatial and temporal satellite systems and fire-sensitive satellite products such as MODIS, Sentinel and Thermal remote sensing, etc. In this review, our focus is on how forest fire research has evolved through technological advancement and how forest fire itself has evolved as a hazard. We identify various aspects of forest fires and discuss them here. We found that forest fires have become a major concern in recent times and have the potential to raise multiple-fold in the future. This threat is not only for forest ecology, but also for atmospheric pollution, soil health, human health, climate change and the socioeconomic functions of society.

Keywords Forest fire · Hazard · Climate change · Himalayan · Machine learning · Forecasting

---

V. Upadhyay · S. Kumar · A. Sharma · A. Kumar (✉)
Environmental Technology Division, CSIR-Institute of Himalayan Bioresource Technology, Palampur, Himachal Pradesh, India
e-mail: amit.kumar07@csir.res.in

S. Kumar · A. Sharma · A. Kumar
Academy of Scientific and Innovation Research (AcSIR), Ghaziabad, Uttar Pradesh, India

V. Upadhyay
Central University of Himachal Pradesh, Kangra, Himachal Pradesh, India

V. M. Tiwari (ed.), *Himalayan Natural Hazards*, Indian National Science Academy,
https://doi.org/10.1007/978-981-92-3784-5_14

14.1 Introduction

The Himalayan forests are one of the ecologically rich and sensitive mountain ranges in the world. These forests spread over the subtropical lowland (Shivalik-Tarai) region of Jammu-Kashmir, Himachal, Uttarakhand, Sikkim and Assam states of India, Nepal and Bhutan to the alpine grasslands of Hindukush Himalaya to Arunachal Pradesh. These forests provide habitat for a high number of flora and fauna (Biswas et al., 2025). The importance of Himalayan forests is not limited to biodiversity (Sati, 2024), but also serves as a source of origin for major river systems (Immerzeel et al., 2010), protects against soil erosion and slopes from falling, recharges groundwater and controls the flow of streams (Kumar & Kumar, 2022). Additionally, the Himalayan Forest also provides the potential to mitigate the adverse effects of climate change by capturing billions of tonnes of carbon and millions of people get their livelihoods (Chaudhary et al., 2023). However, this resource is now faced with a serious threat of forest fires (Bhattacharya et al., 2025). Fires are part of nature and were rare and limited earlier (Sati, 2024), but they have now become more frequent and seem like a disaster in this region (Sharma et al., 2023). In the western Himalaya, like Jammu and Kashmir, Uttarakhand and Himachal Pradesh, a large area has been identified as a burned area every year between 2001 and 2021, which was comparatively very high in years like 2016 and 2024 (Gupta et al., 2024). This is not limited to the Indian Himalaya, but the Himalaya in Nepal and Bhutan have also reported hectares of forests affected by fire. For example, in the state of India, fire incidents increased from 922 in 2002 to 41,600 in 2019 (Mina et al., 2023). In the Almora district of Uttarakhand itself, 7380 fire incidents occurred from 2001 to 2018, spread over 3150 km^2 and were more frequent and severe in 2016 and 2018 (Rawat et al., 2022). Similarly, a loss of 17 ha of tree cover in Pauri Garhwal district of Uttarakhand due to forest fires between 2020 and 2024 (Saini et al., 2024) caused significant loss of life, property and ecosystems. The impacts of these fires are not limited to the western Himalaya but have a far-reaching impact, extending to the central Himalaya and eastern Himalayan regions, such as in Nepal, an average of 3098 fire incidences occur every year, accounting for the destruction of 172,040.65 hectares (ha) of forest (Bhujel et al., 2021). In Madhesh Province, Nepal, alone assessed 6368 fire incidents, and 6158.22 km^2 of total burnt area were reported during the 22 year research period from 2001 to 2023, and 1050 forest fires in the period of 22 years during 2000–2022, damaging about 7932.42 hectares of forest land in the eastern Himalayan state of India, i.e. Sikkim (Dahal, 2024).

14.2 Ecological Implications and Soil Dynamics of Forest Fire

The Himalayan forests are considered the lungs of the Indian subcontinent region (Mina et al., 2023). Thus, disturbances in the forest's structure, forest regeneration (Sapkota et al., 2025) and native species composition (Payra, 2022) become matters of concern. The effects of fire can range from mild to drastic changes in species composition (Sugandha et al., 2025), soil quality (Banerjee & Sati, 2025) and forest regeneration (Sapkota et al., 2025). The ecological process has long-term effects (Dhar et al., 2023), such as shrinking forest cover to decline in diversity (Sahana et al., 2025). Different ecological zones of the Himalaya show significantly higher densities of saplings and seedlings in non-fire regions compared to the fire-affected region (Sapkota et al., 2025). In Fire-affected areas, fire-tolerant invasive species begin to grow and outcompete less-tolerant plants (Kaushal & Aggarwal, 2025), ultimately resulting in changing the ecological setup of the forest (Sahana et al., 2025). These kinds of observations are common in the pine forest, where fire not only promotes regeneration of *Pinusroxburghii* and *Myricaesculenta* but also suppresses the regeneration of sensitive species like *Quercusleucotrichophora* (Dobhal et al., 2025). Soil microorganisms, which are essential for nutrient cycling, are also affected (Payra, 2022). Occasional mild fires can increase nutrient availability and seed germination (Kaiser et al., 2025), but frequent or intense fires disrupt natural regeneration cycles and have long-term consequences for forest growth (Sahana et al., 2025). The chemical composition of the soil also changes (Bargali et al., 2024). After a fire, soil pH increases (Dobhal et al., 2025), and nutrients such as nitrogen, phosphorus and potassium become more abundant (Sharma et al., 2023). However, this also leads to a decrease in soil organic carbon (Banerjee & Sati, 2025), which is essential for soil fertility and moisture retention. The immediate increase in soil nutrients may lead to a temporary increase in vegetation, but over time, vegetation diversity declines (Sharma et al., 2023), and the ecosystem's ability to recover is impaired (Aryalakshmi et al., 2025). The most serious concern is the spread of non-native invasive species (Joshi et al, 2024) and the struggle for native species to regenerate. These plants threaten local biodiversity and further undermine ecosystem stability (Aryalakshmi et al., 2025).

14.3 Socioeconomic Aspects and Role of Community Participation in Himalayan Forest Fire

Forest fires are not only an ecological concern but also a socioeconomic crisis with a deep impact on the lives and livelihoods of communities. The increasing fire incidents in the Himalaya affect food security (Mamgain et al., 2023), livelihood (Chaudhary et al., 2023), health and social harmony (Wang et al., 2018). During the 2020 COVID-19 lockdown, it was particularly observed that western Himalayan states reported a temporary reduction in fire incidents due to a decline in human activities, while no such effect was seen in Northeast India (Puri, 2024). This makes it clear that the role of human activities varies by region, and human factors are the main cause of fires (Payra, 2022). Surveys conducted in fire-prone areas revealed that most communities are not adequately prepared to deal with fires (Parkash, 2010). Most people had a sense of participation (Sapkota et al., 2015), but there was a severe lack of resources, equipment (Parajuli et al., 2022), policy and awareness (Bhujel et al., 2021). However, there were examples in some places where communities themselves created effective models (Banerjee & Sati, 2025). Community fire management in Nepal's Shivalik hills is an encouraging example of how collective efforts of the community can prevent fires by taking collective decisions and resource sharing (Sapkota et al., 2015). At the same time, traditional ecological knowledge is still relevant today (Banerjee & Sati, 2025). Practices such as controlled burning, removal of accumulated understory biomass and maintaining fire lines (Sati, 2024), in association with scientific methods, are extremely effective (Rawat et al., 2022). One major challenge is accumulation of dry leaves, especially pine needles (Payra, 2022) that act as fuel, but can be transformed into a resource by timely collection and removal. Now, some innovative initiatives use pine needles to make products (Malik et al., 2020), like bio-briquettes (Sengar, 2022), PINEPEAT (Gupta et al., 2024), electricity generation (Kala & Subbarao, 2017), bioethanol (Dwivedi et al., 2024) and activated carbon (Asrani et al., 2024), etc. This reduces the risk of forest fires and also provides an alternative livelihood option to local communities. In upcoming years, climate change will become even more adverse; in that case, our success will depend on how effectively we can mobilize, organize and empower our society to mitigate our forester sources.

14.4 Fire Impacts on Carbon and Air Quality

Forest fires in the Himalayan region are not just a terrestrial disaster, but also an atmospheric crisis. Every time a forest burns, it releases huge amounts of carbon into the atmosphere, disrupting air quality and affecting the climate balance of the region. Using satellite data, researchers assessed the biomass potential and carbon storage of the dominant forest species. The results found that sal (*Shorearobusta*) forests are more tolerant of relatively low-intensity fires, and most of the carbon is preserved despite surface fires. In contrast, pine forests suffer greater carbon loss due to their highly flammable needlelike leaves and frequent fires. Data from the MODIS Burned Area product and the Global Fire Emissions Database can reveal the magnitude of carbon emissions from fires. The eastern Himalaya were the most affected, with the highest CO_2 emissions, followed by Nepal and the western. In Nepal, on average, the loss of biomass was 7.07 million tonnes per year, accounting for the emission of 3.30 million tonnes of carbon per year from 2001 to 2023 (Bhujel et al., 2021). At the same time, carbon balance is also affected as stored carbon in vegetation is being emitted into the atmosphere in the form of gases and above the surface in the form of ash, reversing the forests as carbon sinks to carbon sources (Bargali et al., 2024; Bhujel et al., 2021). Other than the CO_2, other toxic particles and gases are released into the atmosphere, which affect air quality. These aerosols and gases in the high atmosphere give rise to health hazards, reduced visibility and affect the Earth's thermal balance by absorbing and scattering sunlight. When these particles accumulate in snow-covered areas, they reduce the reflectivity of snow, causing faster snow melt.

14.5 Patterns in the Forest Fire and Climate Change Relation

Every year, as winter departs and dry winds start blowing in the Himalayan foothills, a familiar and dangerous threat of forest fires begins on the slopes. From March to June, vast Himalayan forests turn into fire-prone areas, just waiting for a spark. This threat is particularly high in pine forests, the layers of dry needles lying on the ground of these forests are now promoting frequent and more intense fires. The frequency and intensity of forest fires in the Himalayan region has been highlighted in hotspot analysis and their interrelation with climate change performed by Kumar and Kumar (2022) in western Himalayan states of India (Fig. 14.1), where the state of Uttarakhand shows high fire-density grids in comparison to Himachal and Jammu and Kashmir.

The research and management practices on forest fires were initially limited to documentation of fire incidents and post-fire assessment. This now reaches the identification of forest fire contributing factors, prediction modelling, forecasting and real-time monitoring. Key word association analysis (Fig. 14.2) from a recent forest fire research publication suggests that the trend of forest fire research has shifted from assessment of seasonal relationship in 2019 to climate parameter temperature in 2020 to factors, models and climate change in 2021. The thickness of the Keyword association link in the VOSvisualizer-generated map suggested that the forest fire research focuses more on data, impact assessment, factor analysis, temperature importance and model building.

14.6 Predictive Modelling and Technological Interventions for Forest Fire Management

The risk of forest fires has increased multiple-fold in the Himalayan region in recent times, and to tackle this situation, the application of advanced technologies has become essential for effective management in time. Techniques like GIS (Rawat et al., 2022), remote sensing (Kumar et al., 2021) and satellite-based thermal anomaly detection (Babu et al., 2016) are playing a vital role in early detection

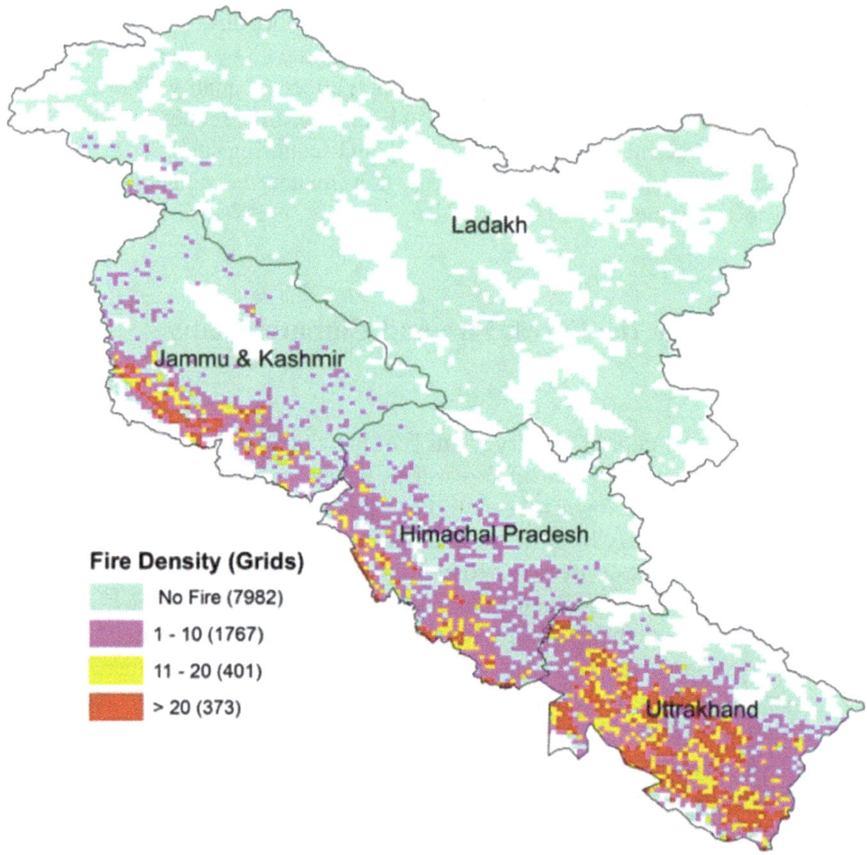

Fig. 14.1 Forest fire density map of western Himalaya (2001–2020); Adopted from Kumar and Kumar (2022)

(Rana et al., 2024), monitoring (Ahmad et al., 2018) and post-fire assessment (Kumar & Kumar, 2022). For example, an integrated decision-making methodology using Multi-Criteria Analysis, Space–Time Cube Analysis and Emerging Hot Spot Analysis has been developed to identify previous forest fires and their distribution, for preparing a fire risk hotspot map for Meghalaya, India. This assessment utilized MODIS thermal anomaly data to analyze the relationship between fire occurrence and triggering factors, including human settlement, Roads, Climate, Elevation, Slope, Aspect, Forest Type, Forest Canopy density, Soil Moisture, Wind Speed and Forest Health (Dhar et al., 2023) (Fig. 14.3). Remote Sensing products such as Landsat, MODIS, Sentinel-2 and VIIRS (Adhikari, 2025; Mamgain et al., 2023; Payra, 2022) provide near-real-time data on fire. Various spectral indices like normalized difference vegetation index, enhanced vegetation index, normalized difference moisture index, topographic wetness index, normalized difference water index and dNBR help in monitoring the condition of vegetation as fuel load, surface moisture and wetness

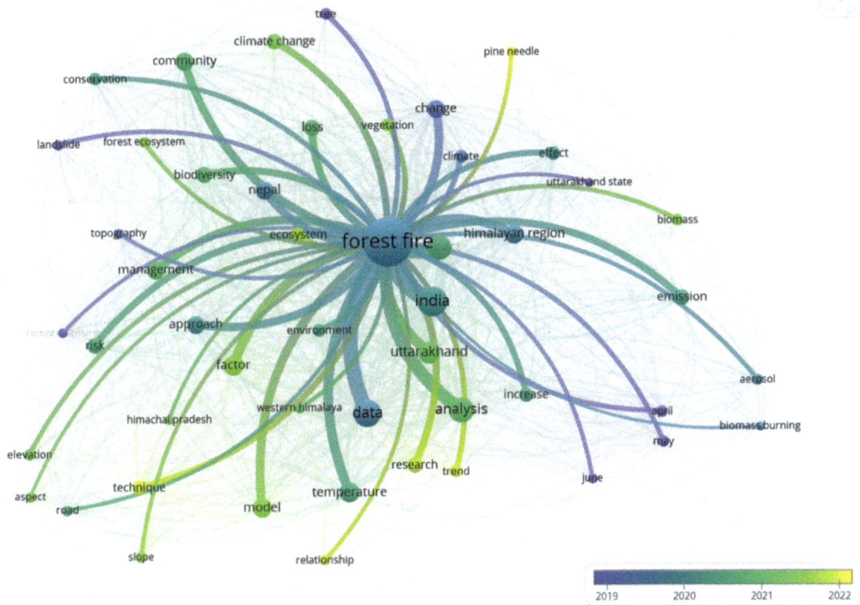

Fig. 14.2 Keyword association network for forest fire research in the Himalayan region

as an indicator of dryness can be used to develop forecasting models and for post-fire impact assessment (Bhattacharya et al., 2025; Mamgain et al., 2023). Predictive Forecast modelling has become more reliable and accurate nowadays (Bar et al., 2023), especially after the incorporation of machine learning (Bar et al., 2023) and deep learning (Rihan et al., 2023). Machine learning algorithms such as Random Forest, Support Vector Machines, Fuzzy Analytical Hierarchy Process, Artificial Neural Network, Boosted Regression Tree, GWBLR Model and Logistic Regression (Banerjee & Sati, 2025; Dhakal et al., 2024; Sharma et al., 2022; Tshering et al., 2020) have been successfully used to identify fire-prone areas.

In addition, multi-criteria techniques such as Analytic Hierarchy Process (KC et al., 2014; Sengar, 2022) and Frequency Ratio (Tshering et al., 2020) have also helped significantly in decision-making. This method analyzes the factors and gives them a weighting to produce fire risk maps. Effective forest fire management requires comprehensive measures, including strong policy support (Bargali et al., 2024), active participation of local communities (Parkash, 2010), early-warning systems (Biswas et al., 2025), well-developed firefighting infrastructure, strict controls on open burning and the integration of forest fire risk maps.

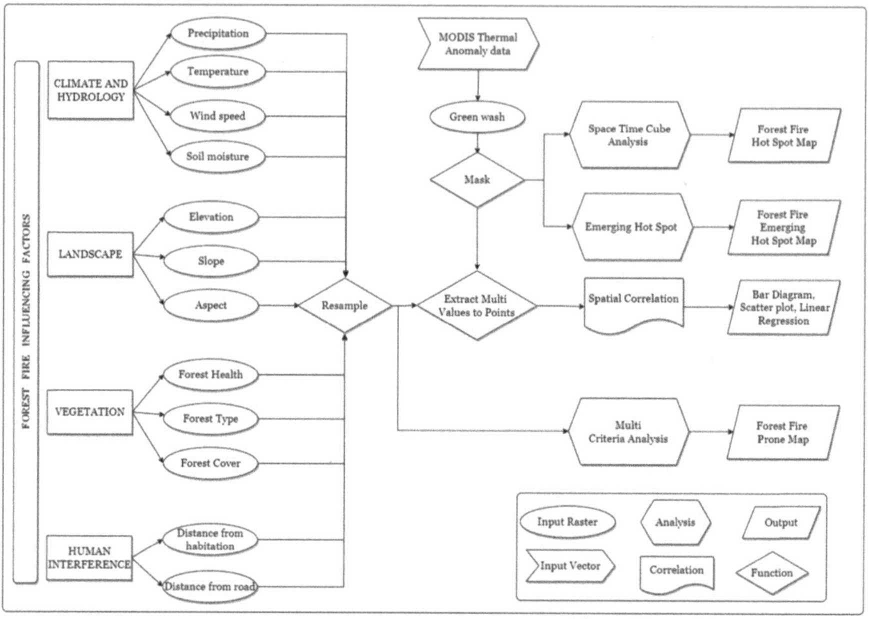

Adopted from Dhar et al. 2023

Fig. 14.3 Flow diagram of the methodology used for forest fire risk mapping and trend analysis in Meghalaya, India

14.7 Technological Advancements and Decision Support for Smart Fire Management Practices

Recent technological advancements transformed the forest fire analysis (Babu et al., 2017) and made it easily accessible (Tiwari et al., 2021), a timesaver and efficient (Murthy et al., 2019) in combating the growing threat of forest fires. Platforms such as Google Earth Engine, ArcGIS and R Studio (Kumari et al., 2025) are now processing large amounts of spatial and temporal data. Technology has emerged in order to train machine learning models like CART, Random Forest and SVM to identify fire-prone areas. Scientists are using indices like NDVI, NBR, SWIR, MNDFI and PMI (Rihan et al., 2023). Satellites like MODIS Terra and Aqua also provide thermal data, which makes it possible to detect active fire hotspots almost instantly (Fig. 14.4). AHP (KC et al., 2014; Sengar, 2022) and FR (Tshering et al., 2020) are two multi-level machine learning algorithms that were used to build fire risk maps for vulnerable areas. It is possible to determine the causes of the fires using these techniques (Sati, 2024). Another important innovation is the fuzzy logic-based approach, which is able to better understand the uncertainty and variability of data. The accuracy of fire maps and early warning systems is significantly increased by integrating data from various sources. In Uttarakhand, a DSS system aids in risk mapping before a fire, rescue operation planning in the event of a fire and post-fire rehabilitation planning (Yadav et al., 2025). Today, researchers are using models such as FARSITE and WRF-SFIRE, which use topography, weather, vegetation and fuel data, to analyze the direction and movement of fires. In order to predict the direction and speed of fires, tools like the WindNinja are used; these models also evaluate carbon emissions from fires. Furthermore, researchers are also using 'ensemble modelling', which integrates several algorithms, including Random Forest, GBM, AdaBoost and DNN, to improve prediction accuracy. While this also increases the complexity of the model, it is being addressed with XAI tools like SHAP and LIME that explain why an area is considered high risk or what factors are responsible for predicting fire (Hang et al., 2024). When machine learning is integrated with Unmanned Aerial Vehicles (UAVs), drone monitoring (Gupta et al., 2024) and crowd-sourced fire reporting, a quicker and more precise response can be achieved (Gupta et al., 2024). All these technological innovations together are writing a new story of fire management in the Himalaya (Table 14.1).

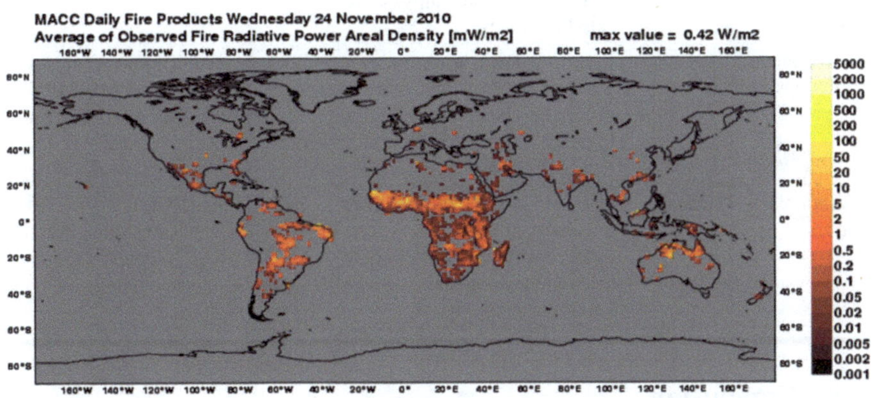

Fig. 14.4 Global map of fire radiative power (FRP) areal density for 24 November 2010, derived by the MACC project under the GMES Atmospheric Service initiative. The map illustrates the daily average FRP from all active fires detected by MODIS sensors (Terra and Aqua), aggregated within 125 km grid cells and expressed in mW/m^2. Adapted from Kaiser et al. (2009)

Table 14.1 Solutions in geoinformatics for forest fire management

Challenge	Emerging solutions	Examples/Tools
Low spatial resolution	High-resolution satellite systems and drones provide finer spatial detail for early detection	PlanetScope (3–5 m), Sentinel-2 (10 m), DJI drones, LiDAR-equipped UAVs
High costs	Open-access platforms, micro-satellites and collaborative data-sharing reduce dependency on costly infrastructure	Google Earth Engine (GEE), FireAlert (India), CubeSats, Copernicus Open Data
Complex data processing	Cloud computing enables large-scale analysis without local infrastructure; AI automates tasks like classification and risk mapping	AWS Earth on Demand, GEE, AI platforms (e.g. TensorFlow, PyTorch), AutoML for satellite imagery
Access limitations (cloud/terrain)	Synthetic Aperture Radar (SAR) can 'see through' clouds and smoke; UAVs and IoT sensors provide low-altitude, ground-based data in inaccessible regions	Sentinel-1 SAR, LoRaWAN sensor networks, FLIR thermal drones, weather-resistant UAVs
Delayed alerts and data gaps	Continuous monitoring via IoT sensor networks and rapid-deployment drone fleets fills temporal data gaps between satellite passes	Smart ForestIoT sensors, AI-triggered UAV swarms, near-real-time VIIRS alerts
Limited integration with ground ops	Real-time Decision Support Systems (DSSs) connect geospatial data with firefighting logistics, evacuation planning and public alerts	Technosylva Wildfire Analyst, FireCast, NASA FIRMS API, Mobile DSS apps for the forest department
Institutional and capacity barriers	National training programmes, university collaborations and open education resources help bridge the geospatial knowledge divide	FSI training (India), UN-SPIDER Knowledge Portal, NASA ARSET, OpenGeoEdu

14.8 Conclusion

The intensity of forest fires in the Himalayan region is increasing rapidly, primarily due to the abnormal conditions of climate change and human interventions. What used to be occasional and limited in area is now occurring more frequently and on a larger scale. This is not only damaging ecosystems but also endangering the livelihoods of millions of people who depend on these forests. As climate change accelerates and traditional forest management is weakened, the incidence of fires and their adverse effects is increasing, including loss of biodiversity, degradation of catchments and increased emissions of greenhouse gases. Especially in a sensitive and biodiverse region like the Himalaya, burning of even a small part of the forest can have extremely severe impacts. A coordinated and multi-layered approach is needed to tackle this serious problem. Emergency fire suppression alone is no longer enough—we must prioritize measures such as predictive modelling, risk mapping and environmental monitoring. Modern technologies such as remote sensing, Geographic Information Systems (GIS) and artificial intelligence have made early warning, real-time monitoring and post-fire assessment possible. If these technologies are combined with traditional local knowledge, people's participation and favourable policies, a robust and effective strategy can be developed. However, solutions cannot be merely technical. For long-term success, it is important to strengthen local institutions, revive traditional forest conservation practices, adapt strategies to local geographical and climatic conditions and promote transboundary cooperation among Himalayan countries. Conservation of these forests is crucial not only from an environmental perspective but also for regional water security, climate balance and the livelihoods of local communities. Therefore, strong political will, grassroots participation and international cooperation, along with technological innovation, are essential to protect Himalayan forests from the increasing threat of fire.

Acnowledgements The authors acknowledge the financial and infrastructure support provided by the Council of Scientific and Industrial Research (CSIR) under CSIR-Infra MLP002619 project entitled 'Ecological Assessment and Sustainable Valorisation of Himalayan Bio-resources for Industrial and Societal Benefit (IHBT-HB)'.

Conflict of Interest The author(s) declare that they have no competing financial interests or personal conflicts that could have appeared to influence the work reported in this article.

References

Adhikari et al. (2025). Spatial dynamics and risk mapping of forest fires in Madhesh Province, Nepal: a multi-criteria decision approach. *Environment and Natural Resources Journal, 23*, 80–94. https://doi.org/10.32526/ennrj/23/20240124

Ahmad, et al. (2018). Himalayan forest fire characterization in relation to topography, socio-economy, and meteorology parameters in Arunachal Pradesh, India. *Spatial Information Research, 26*, 305–315.

Aryalakshmi et al. (2025). Resilience to forest fires: An indicator-based study in a tropical forest of Uttarakhand, India. *IEEE Geoscience and Remote Sensing Letters.*

Asrani et al. (2024). Himalayan chir pine biomass as an adsorbent for the removal of industrial dye. In *Climate crisis and sustainable solutions* (pp. 343–356). Springer.

Babu, K. V. S., Roy, A., & Prasad, P. R. (2016). Forest fire risk modeling in Uttarakhand Himalaya using TERRA satellite datasets. *European Journal of Remote Sensing, 49*, 381–395.

Babu et al. (2017). Assessment of forest fire danger using automatic weather stations and MODIS TERRA satellite datasets for the state Madhya Pradesh, India. In *2017 International conference on advances in computing, communications and informatics (ICACCI)* (pp. 1876–1881). IEEE.

Banerjee, S., & Sati, V. P. (2025). Combining spatial and elementary analysis with qualitative approaches: a novel approach of forest fire risk detection, mitigation, and impact analysis. *Natural Hazards*, 1–24.

Bar, S., Parida, B. R., Pandey, A. C., Shankar, B. U., Kumar, P., Panda, S. K., & Behera, M. D. (2023). Modeling and prediction of fire occurrences along an elevational gradient in Western Himalayas. *Applied Geography, 151*, 102867.

Bargali, H., et al. (2024). Mapping and impact of seasonal forest fires on vegetation, soil, and tree mortality in Uttarakhand Himalaya. *Landscape Online, 99.*

Bhattacharya, O., Sinha, S., Mishra, V. N., Kumari, M., Hasher, F. F. B., Barman, J., & Zhran, M. (2025). Harnessing geospatial tools to map the forest fire: Risk zonation in PauriGarhwal, Uttarakhand. *Results Eng, 25*, 103694.

Bhujel, et al. (2021). Fire-induced carbon emissions from tropical mixed broad-leaved forests of the Terai-Siwalik region, central Nepal. *Journal of Forest Research, 32*, 2557–2565. https://doi.org/10.1007/s11676-020-01256-x

Biswas, et al. (2025). Spatial prediction of forest fires in India: A machine learning approach for improved risk assessment and early warning systems. *Environmental Science and Pollution Research, 32*, 4856–4878. https://doi.org/10.1007/s11356-025-35982-8

Chaudhary et al. (2023). Deforestation in Nepal: Status, causes, consequences, and responses. In *Hazards and disasters series, biological and environmental hazards, risks, and disasters* (2nd edn.) (pp. 277–318). https://doi.org/10.1016/B978-0-12-820509-9.00003-4

Dahal (2024) Increasing incidences of forest fire in Sikkim, India. In *Geo-information for disaster monitoring and management*. Springer, Cham. https://doi.org/10.1007/978-3-031-51053-3_17

Dhakal, et al. (2024). Synergistic approaches in forest fire risk mapping using fuzzy AHP and machine learning models in the ChureTaraiMadhesh Landscape (CTML) of Nepal. *Geomatics Natural Hazards Risk, 15*, 2436540.

Dhar, et al. (2023). Forest fire occurrence, distribution and risk mapping using geoinformation technology: A case study in the sub-tropical forest of the Meghalaya, India. *Remote Sensing Applications: Society and Environment, 29*, 100883.

Dobhal, et al. (2025). Impact of forest fires on vegetation composition and soil properties of *Pinusroxburghii* forests in the Western Himalaya. *Trees for People, 20*, 100794.

Dwivedi, et al. (2024). Co-fermentation of forest pine needle waste biomass hydrolysate into bioethanol. *Biomass Conversion and Biorefinery, 14*(7), 8829–8841.

Gupta, et al. (2024). Sentinel 2 based burn severity mapping and assessing post-fire impacts on forests and buildings in the Mizoram, a north-eastern Himalayan region. *Remote Sensing Applications: Society and Environment, 36*, 101279.

Hang, H. T., Mallick, J., Alqadhi, S., Bindajam, A. A., & Abdo, H. G. (2024). Exploring forest fire susceptibility and management strategies in Western Himalaya: Integrating ensemble machine learning and explainable AI for accurate prediction and comprehensive analysis. *Environmental Technology & Innovation, 35*, 103655.

Immerzeel, W. W., van Beek, L. P. H., & Bierkens, M. F. P. (2010). Climate change will affect the Asian water towers. *Science, 328*, 1382–1385.

Kaiser, J. W., Flemming, J., Schultz, M. G., Suttie, M., & Wooster, M. J. (2009). The MACC global fire assimilation system: First emission products (GFASv0). *ECMWF Technical Memorandum, 596*.

Kaiser, et al. (2025). Long-term soil nutrient and understory plant responses to post-fire rehabilitation in a lodgepole pine forest. *Forest Ecology and Management, 575*, 122359.

Kala L. D., & Subbarao, P. M. V. (2017). Pine needles as potential energy feedstock: Availability in the central Himalayan state of Uttarakhand, India. In *E3S web of conferences* (Vol. 23, p. 04001). EDP Sciences.

Karki, G., et al. (2022). Climate change effects, adaptation and community-based forest management in the Mid-Hills of Tanahu and Kaski Districts, Nepal. *International Forestry Review, 24*(4), 573–593.

Kaushal, S., & Aggarwal, R. K. (2025). A comprehensive review on biomass briquette technology for combating forest fires and environmental conservation. *CABI Reviews, 20*(1), 0038.

KC et al. (2014). Community users' and experts' perspective on community forestry in Nepal: A SWOT–AHP analysis. *Forests, Trees and Livelihoods, 23*(4), 217–231.

Kumar, R., et al. (2021). Identification and characterization of spatio-temporal hotspots of forest fires in South Asia. *Remote Sensing, 13*, 1176. https://doi.org/10.3390/rs13061176

Kumar, S., & Kumar, A. (2022). Hotspot and trend analysis of forest fires and their relation to climatic factors in the western Himalayas. *Natural Hazards, 114*, 3529–3544. https://doi.org/10.1007/s11069-022-05530-5

Kumar, R., Kumari, K., & Thakur, P. K. (2022). Mapping of forest fire burned severity using the Sentinel datasets. *Geocarto International, 37*, 5892–5911.

Kumari, et al. (2025). Earth observation-based characterization of environmental conditions for forest fire risk in western himalayan ecosystems using machine learning approach. *Journal of the Indian Society of Remote Sensing, 53*(1), 307–318.

Malik, et al. (2020). Study of grid-integrated biomass-based hybrid renewable energy systems for the Himalayan terrain. *International Journal of Sustainable Energy Planning Management, 28*, 71–88.

Mamgain, S., Roy, A., & Karnatak, H. C. (2023). Satellite-based long-term spatiotemporal trends of wildfire in the Himalayan vegetation. *Natural Hazards, 116*, 3779–3796.

Mina, et al. (2023). Forest fires and climate attributes interact in central Himalayas: An overview and assessment. *Fire Ecology, 19*, 14. https://doi.org/10.1186/s42408-023-00177-4

Murthy, K. K., Sinha, S. K., Kaul, R., & Srinivas, V. (2019). A fine-scale state-space model to understand drivers of forest fires in the Himalayan foothills. *Forest Ecology and Management, 432*, 902–911.

Parajuli, et al. (2022). A Strategy for involving community forest managers in effective forest fire management in Nepal. *BankoJanakari, 32*(1), 41–51.

Parkash (2010). Methodology used for community-based multi-hazard risk management in Garhwal Himalaya, Uttarakhand State, India. In *Forms of community participation in disaster risk management*. Nova Science Publishers, Inc., New York, USA.

Payra et al. (2022) Application of remote sensing to study forest fires. *Atmos Remote Sens Principles Applications, 239.*

Puri, P. (2024). COVID-19 lockdown and forest fires in the times of climate change for the mountainous states of Himachal Pradesh, Uttarakhand, and Seven Sister States of India: A spatio-temporal interpretation (2012–2022). In M. Mishra, A. J. de Lucena, B. Maharaj (Eds.*) Climate change and regional socio-economic systems in the global south.* Springer, Singapore. https://doi.org/10.1007/978-981-97-3870-0_19

Rana et al. (2024). Development of an algorithm for detection of forest fire in the Himalayan region using Sentinel-2 MSI images. In *2024 15th International conference on computing communication and networking technologies (ICCCNT)* (pp. 1–5). IEEE

Rawat, et al. (2022). GIS-based landscape vulnerability assessment to forest fire susceptibility of Rudraprayag district, Uttarakhand, India. *Spatial Information Research, 30*, 101–113.

Rihan, et al. (2023). Forest fire susceptibility mapping with sensitivity and uncertainty analysis using machine learning and deep learning algorithms. *Advances in Space Research, 72*, 426–443.

Sahana et al. (2025). A mixed-methods approach for identifying high conservation value areas in the high-altitude landscapes of the Indian Himalayan region. *Geo: Geography and Environment.*

Saini et al. (2024). Forest fire burnt area mapping using remote sensing and GIS in PauriGarhwal, Uttarakhand, India. In *2024 IEEE India geoscience and remote sensing symposium (InGARSS), Goa, India* (pp. 1–4). https://doi.org/10.1109/InGARSS61818.2024.10984017

Sapkota, L. M., et al. (2015). Factors affecting collective action for forest fire management: A comparative study of community forest user groups in Central Siwalik, Nepal. *Environment Management, 55*, 171–186. https://doi.org/10.1007/s00267-014-0404-x

Sapkota, R. P., Dhital, N. B., Shrestha, A. K., Khadka, U. R., Pokhrel, C. P., Joshi, G. R., & Rijal, K. (2025). Forest fire impacts plant community assemblage in different ecological zones of central Himalaya, Nepal. *Global Ecology and Conservation, 58*, e03443.

Sati, V. P. (2024). Forest fires in the Indian Central Himalaya: Major drivers, implications, and mitigation measures. *Environmental Hazards, 23*, 390–404. https://doi.org/10.1080/17477891.2024.2304211

Sengar et al. (2022). A fuzzy analytic hierarchy process based analysis for prioritization of enablers to pine briquettes based energy generation in alignment with the United Nations' sustainable development goals: Evidence from India. *Biomass and Bioenergy, 165*(2022), 106580.

Sharma, et al. (2022). Assessing the predictive efficacy of six machine learning algorithms for the susceptibility of Indian forests to fire. *International Journal of Wildland Fire, 31*(8), 735–758.

Sharma, et al. (2023). Spatiotemporal distribution of air pollutants during a heat wave–induced forest fire event in Uttarakhand. *Environmental Science and Pollution Research, 30*(51), 110133–110160.

Sugandha et al. (2025). Scoping of forest resources in light of climate change implications in the Western Himalayan region. In D. Arya, N. Chandra, R. Kumar, M.L. Upadhayay, A.P. Mishra (Eds.) *Climate change impact on Himalayan biodiversity*. Springer, Cham. https://doi.org/10.1007/978-3-031-77149-1_5

Tiwari, A., Shoab, M., & Dixit, A. (2021). GIS-based forest fire susceptibility modeling in Pauri Garhwal, India: A comparative assessment of frequency ratio, analytic hierarchy process and fuzzy modeling techniques. *Natural Hazards, 105*, 1189–1230.

Tshering, et al. (2020). A comparison of the qualitative analytic hierarchy process and the quantitative frequency ratio techniques in predicting forest fire-prone areas in Bhutan using GIS. *Forecasting, 2*, 36–58.

Wang, et al. (2018). Understanding the relationships in nature-human systems to improve social-ecological resilience in the Hindu-Kush Himalayas. *Asian J ConservBiol, 7*, 17–27.

Yadav, et al. (2025). A prototype forest fire decision support system for Uttarakhand, India. *Fire, 8*(4), 149.

Part IV
Early Warning System and Mitigation Strategies

Chapter 15
Seismic Network Monitoring: A Novel Early Warning Framework for Cascaded Hazards Associated with Landslides, Rockfalls, and Avalanches in the Himalaya

Rajesh Rekapalli and N. Purnachandra Rao

Abstract The Himalayan region, often referred to as the "Third Pole," is highly susceptible to cascading geohazards, including landslides, rockfalls, and avalanches, due to its unique geological, climatic, and anthropogenic factors. Developing effective early warning systems (EWS) for these hazards is critical to mitigating their devastating impacts, which have been well-documented in recent literature. This chapter synthesizes findings on the Seismic Network-based Early Warning of Geohazards, particularly the event cascades associated with Landslide, Rockfalls and Avalanches, emphasizing the challenges, advancements, and recommendations for improving EWS in the Himalayan context.

Keywords Himalayan region · landslide · Early warning system · Geohazards

15.1 Introduction

Mountainous regions like the Himalaya are highly susceptible to cascades of mass flow events (Debris flows and Floods) associated with landslides, rockfalls, avalanches, etc. In addition to the complex geology, seismic activity and rapid urbanization increased the degree of such vulnerability in the Himalaya. These events, often triggered by seismic activity, heavy rainfall, or rapid snowmelt, pose significant risks to human life, infrastructure, and ecosystems in mountainous regions (Keefer, 1984; Fischer et al., 2012; Prakash, 2013; Schuster and Wieczorek, 2018; McColl, 2022; Shugar et al., 2021). Globally, Asian countries have faced nearly 59% of the impact due to landslides (Fig. 15.1). Especially, the Indian Himalaya are more prone to cascaded hazards (Sharma et al., 2023). In Sikkim, the Mw 6.92011 earthquake generated many co-seismic landslides within ~100 km of the epicentre,

R. Rekapalli · N. Purnachandra Rao (✉)
CSIR-National Geophysical Research Institute, Hyderabad, India

V. M. Tiwari (ed.), *Himalayan Natural Hazards*, Indian National Science Academy,
https://doi.org/10.1007/978-981-92-3784-5_15

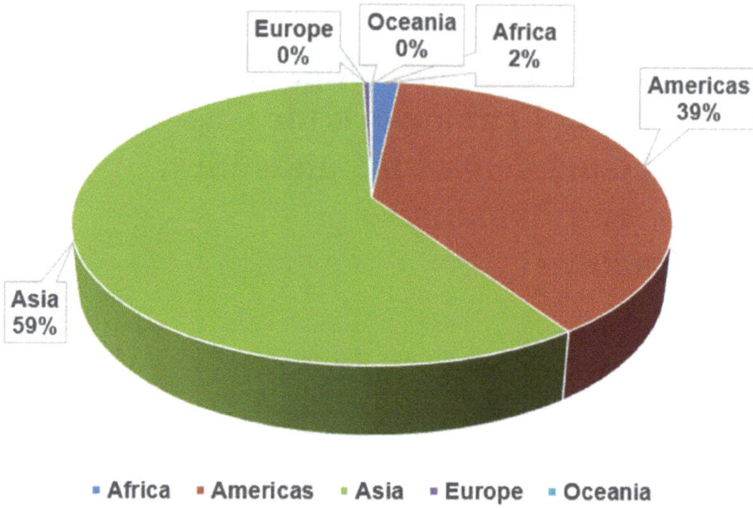

CORRECT THE SPELLING OF PERCENTAGE

Fig. 15.1 Regional impact of landslide disasters (1903–2023): Percentage of total people impacted due to landslides during 1903–2023. *Source* CRED-EMDAT Database

predominantly in the steep slopes near Gangtok and its surroundings. It was argued that post 2011, the reconstruction did not adequately incorporate lessons regarding geotechnical properties of slope soils, exposing the infrastructure to moderate earthquakes and secondary mass movements (Joshi, 2022). A notable recent example is the seventh February 2021 rock-ice avalanche from Ronti Peak in Chamoli that triggered a flash-flood in the Dhauliganga and Alaknanda rivers, causing >200 fatalities and severe damage to dam infrastructure (Rao et al., 2021; Cook et al., 2021; Shugar et al., 2021; Van Wyk de Vries et al., 2022).

The increasing frequency and intensity of these hazards, exacerbated by climate change and anthropogenic activities, underscore the need for robust early warning systems (EWS) to mitigate their impacts (Stahil et al., 2015; Gariano & Guzzetti 2016; Rao et al., 2021; Cook et al., 2021; Capobianco et al., 2025; Alcántara-Ayala, 2025). Traditional early warning frameworks for landslides, rockfalls, and avalanches often rely on localized sensors, such as rain gauges or inclinometers, which may fail to capture the interconnected nature of cascaded hazards in dynamic environments like the Himalaya (Guzzetti et al., 2020). The in-situ sensors may get washed away sometimes during the events. Also, we may miss the events with such in-situ equipment due to the spatial non-stationarity of the events. The spatio-temporal resolutions offered by the present-day satellite monitoring and in-situ observations limit their applicability for early warning. Seismic networks, however, offer a novel solution by leveraging real-time data from distributed seismometers to detect ground vibrations associated with precursory movements or triggering events (Chao et al., 2017). These networks can identify subtle seismic signals, such as micro-seismic activity or low-frequency tremors, that precede large-scale slope failures or snowpack destabilization (Hibert et al., 2019). Thus, the seismic network monitoring has emerged as a promising approach for detecting precursory signals of geohazards, enabling timely warnings and informed decision-making (Dietze et al., 2017a, 2017b; Allen & Melgar, 2019; Rao et al., 2021; Cook et al., 2021).

Another important parameter is the lead time of the Early Warning System. Table 15.1 provides the details on the lead time of different natural hazards in real time. We can notice that earthquakes are the events with the shortest lead time, and the lead time to floods is from hours to days. However, some catastrophic cascades like rockfall-induced floods (as in the case of Feb 07, 2021, Uttarakhand disaster) give us less lead time as the most affected regions are close to the event location and lead times always depend on the velocity of the disaster propagation (seismic waves/ flooded water or debris, etc.) and its velocity. The floods due to intense rainfall can be easily understood and hence provide a good lead time. But the source of floods in mountainous regions are complex (rockfall/landfall in to river flows, breaching landslide dammed lakes and glacial lake outburst, cloud burst events etc.), therefore strategies employed in seismic monitoring offer better solutions as the velocity of the seismic waves generated by the propagating masses are high compared to the velocity of the flowing material (water/debris).

Additionally, studies also suggested that integrating seismic data with machine learning algorithms and geospatial modeling would be a robust option for developing a comprehensive EWS to forecast cascaded hazards with improved accuracy and lead time (Ma et al., 2023).

This chapter discusses the novelty of an early warning framework that harnesses seismic network monitoring to address the challenges of cascaded hazards in the Himalaya. It discusses the merits, demerits, and advancements in the seismic early warning system for real monitoring of cascading events associated with landslides, rockfalls, and avalanches. The following sections outline the seismic EWS framework, methodological developments, success stories, and potential applications, emphasizing its regional scalability and adaptability.

15.2 Seismic Monitoring of Cascaded Hazards Associated with Landslides, Rockfalls, and Avalanches in the Indian Himalaya

In addition to earthquakes that occur in depth ranges of a few kilometers to a few hundred kilometers, shallow subsurface or surface processes, such as floods, landslides, dam breaches, rockfalls, and avalanches, are also recorded by the seismometers in different frequency bands. These seismic observations significantly advanced the understanding of cascaded hazards by enabling the detection and analysis of mass movement events. For instance, seismic data analysis has been pivotal in reconstructing the 2021 Chamoli disaster, where a rock and ice avalanche triggered a

Table 15.1 Lead times of different geo-hazards along with the details of the primary parameter used for detection and protocols used for information dissemination

Name of the geohazard	Parameter used for primary monitoring	Typical early-warning lead time	Warning protocol
Earthquakes	*P*-wave detection, dense seismic networks	Maximum upto ~60 s	Automated alerts issued after *P*-wave detection and before arrival of damaging *S*-and surface waves (Allen et al., 2009; Hoshiba et al., 2008; Wu & Kanamori, 2005)
Landslides	Rainfall thresholds, GNSS, tiltmeters, seismic/acoustic signals	Minutes to days	Forecast-based warnings using rainfall thresholds (hours–days) and short-term alerts during acceleration phases detected by deformation or seismic signals (Intrieri et al., 2019; Guzzetti et al., 2007; Dietze et al., 2017a, 2017b; Roccati et al., 2020; Guzzetti et al., 2020)
Rockfalls	Seismic and acoustic emission, crack monitoring	Seconds to minutes (occasionally hours)	Real-time or near-real-time alerts during fracture propagation or detachment; limited long-term predictability (Helmstetter & Garambois, 2010; Lacroix & Helmstetter, 2011; Cook et al., 2021)
Avalanches	Snowpack stability analysis, meteorology, seismic/infrasound	Hours to days	Primarily forecast-based warnings derived from snowpack and weather evolution; real-time detection possible but limited for spontaneous events (Schweizer et al., 2003; van Herwijnen & Schweizer, 2011)
Floods	Rainfall radar, river gauges, hydrological models	Hours to weeks	Forecast-and monitoring-based warnings using upstream rainfall and river-stage observations, enabling evacuation and preparedness (Alfieri et al., 2012)

deadly debris flow, highlighting the cascading nature of Himalayan hazards. The study enabled a clear demonstration of possible early warning using seismic networks (Rao et al., 2021; Cook et al., 2021).

The integration of seismic records with satellite imagery and numerical modeling has allowed researchers to estimate the volume and trajectory of such events, revealing that approximately 27 million cubic meters of material collapsed from Ronti Peak, causing widespread devastation (Fig. 15.2). The band-pass filtered seismic recording of the Chamoli disaster recorded by the Auli (AUL) station of the CSIR-NGRI seismic network (Srinagesh et al., 2019) situated ~24 km away from the rock-fall location is shown in Fig. 15.3 along with its spectrogram depicting the temporal variation of spectral power. Auli station is equipped with a Reftek 151B-120 broad-band seismic sensor and 24 bit 3 channel Wrangler recorder. The data was acquired

Fig. 15.2 **a** Aerial view shows the Rishi Ganga river with a high-energy debris-laden flood. The river appears choked with sediment, boulders, and muddy water, indicating a sudden upstream mass-movement source rather than rainfall-driven flooding. **b** NTPC hydropower barrage is partially obscured by dust and spray, with visible structural damage and debris accumulation. **c** ITBP Emergency response and rescue operations near the NTPC tunnel and **d** A close view of severely damaged concrete structures of the NTPC hydro power plant during 07 Feb 2021 Chamoli disaster

at 100 samples per second (sps) and down-sampled to 20 sps. It may be noticed that all the cascaded events, i.e., rockfall, debris flow/mass flow and flood events, are clearly recorded by the seismometer with distinct frequency characteristics in the 0.5–5 Hz frequency band.

Further, Fig. 15.4 provides a comparison of spectral power variations at the nearest station AUL (~24 km from the event location) and the farthest station TUN (~190 km away from the rockfall location). The first and second columns of Fig. 15.4 show the filtered signals at these two stations for the 0.5–4 Hz and 0.01–0.5 Hz frequency bands, respectively. It can be seen in Fig. 15.4 that the high-frequency signal corresponding to rockfall is visible at both stations, and the low-frequency signal (0.05–0.01 Hz) is observed only at the nearest station, Auli (AUL). Further, an analysis of signal powers at stations up to 190 km also revealed that the rockfall signal is recorded at all the stations with signal powers above the respective cultural and anthropogenic maximum noise levels observed during daytime and monsoon period in the study area (Cook et al., 2021). Cook et al., have presented the comparison of the three phases of the event (rockfall, debris flow and flood) observed in the seismograms analysis with time of these phases reported from the available photos, videos etc. The inferences from the study of the Rontighad rockfall event and associated mass

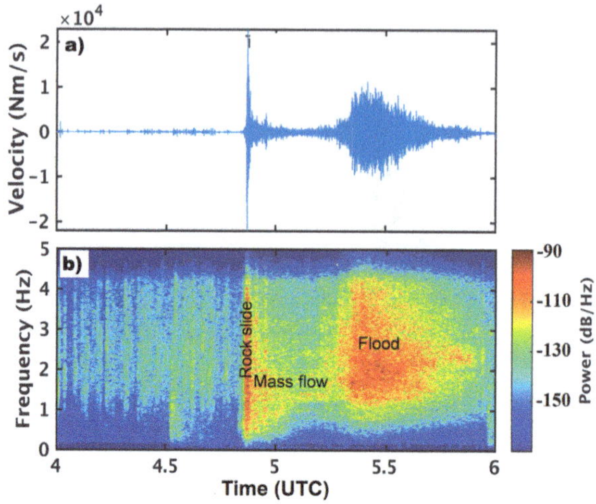

Fig. 15.3 Vertical component of seismic ground velocity during 07 Feb 2021, Chamoli disaster, Uttarakhand, India, at the AUL station of CSIR-NGRI seismic monitoring network

flow and floods using seismic data unveiled the feasibility for early warning of such cascaded hazards using seismic networks.

15.3 Advantages of Seismic Monitoring

Seismic monitoring of rockfalls, landslides, and associated secondary hazards offers distinct advantages over state-of-the-art satellite and in-situ monitoring methods for early warning systems. Below is a comparative analysis highlighting the key advantages of seismic monitoring.

15.3.1 Real-Time Detection and Rapid Response

Seismic Monitoring: Detects ground vibrations caused by rockfalls or landslides in near real-time, enabling rapid identification of events as they occur. Seismic sensors can capture the onset of movement, providing immediate data for early warning systems.

Satellite Monitoring: Relies on periodic imaging (e.g., optical or radar-based systems like InSAR), which may have revisit times of hours to days, delaying detection. Real-time monitoring is limited unless paired with costly, high-frequency satellite systems.

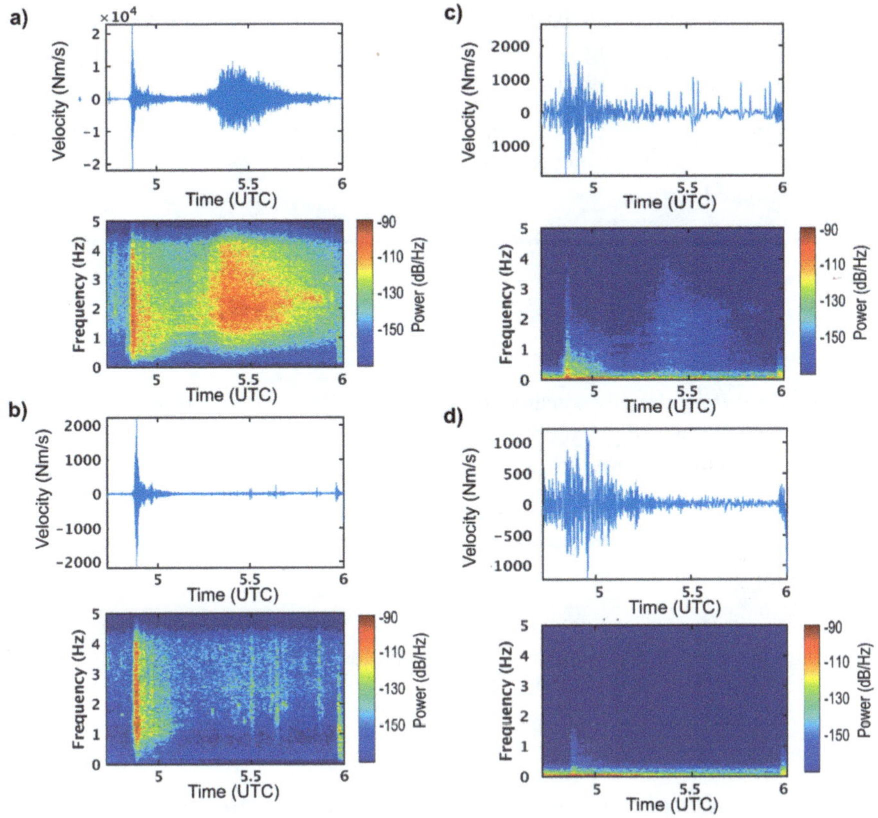

Fig. 15.4 Comparison of vertical velocity of the rock fall, debris/mass flow and flood signals recorded at the nearest station AUL (a and c) and farthest station THU (b and d) during the Chamoli disaster in frequency bands 0.5–4 Hz (a and b) and 0.01–0.5 Hz (c and d) along with the signal spectrogram

In-Situ *Monitoring*: Devices like extensometers or GPS sensors provide real-time data but are localized to specific points, potentially missing broader instability signals until significant movement occurs.

15.3.2 Spatial Coverage

Seismic Monitoring: A single seismic network can monitor large areas (kilometers to tens of kilometers), detecting events across diverse terrains without requiring line-of-sight. It captures signals from remote or inaccessible regions, including secondary hazards like debris flows or avalanches.

Satellite Monitoring: Covers large areas but is limited by cloud cover (for optical satellites), terrain shadowing, or vegetation interference (for radar). Dense vegetation or steep topography can obscure signals, reducing effectiveness.

In-Situ *Monitoring*: Limited to specific instrumented sites, requiring extensive deployment to cover large areas, which is costly and logistically challenging in remote or rugged terrains.

15.3.3 Sensitivity to Small-Scale and Precursory Events

Seismic Monitoring: Highly sensitive to small ground vibrations, allowing detection of precursory signals (e.g., microseismic activity from crack propagation) before major failures. This can provide early warnings for impending rockfalls or landslides.

Satellite Monitoring: Struggles to detect small-scale or precursory movements due to resolution limits (typically meters for InSAR) and temporal gaps between images, missing subtle deformations.

In-Situ *Monitoring*: Can detect localized precursory movements (e.g., crack widening) but may miss broader instability patterns if sensors are not optimally placed.

15.3.4 Continuous Monitoring and All-Time Functionality

Seismic Monitoring: Operates continuously with temporal sampling rates extended up to 1000 samples per second, unaffected by weather, darkness, or vegetation cover, ensuring reliable data collection under all conditions.

Satellite Monitoring: Optical satellites are hindered by cloud cover, fog, or darkness, while radar satellites (e.g., InSAR) are less affected but still face challenges in dense vegetation or complex terrain.

In-Situ *Monitoring*: Functions in all weather conditions but requires regular maintenance (e.g., power supply, sensor calibration), which can be disrupted in harsh environments.

15.3.5 Detection of Secondary Hazards

Seismic Monitoring: Excels at detecting secondary hazards like debris flows, mudflows, or avalanches triggered by initial rockfalls or landslides, as these generate distinct seismic signatures. This allows for comprehensive monitoring of cascading events.

Satellite Monitoring: Limited in detecting secondary hazards in real-time due to temporal resolution and inability to capture dynamic processes like debris flows unless they cause significant surface changes.

In-Situ *Monitoring*: May detect secondary hazards only if sensors are placed in affected areas, which is often impractical for dynamic, unpredictable events.

15.3.6 Cost-Effectiveness for Large-Scale Monitoring

Seismic Monitoring: Relatively cost-effective for regional monitoring, as a network of seismometers can cover vast areas with fewer instruments compared to in-situ methods. Maintenance costs are moderate, and existing seismic networks (e.g., for earthquake monitoring) can often be repurposed.

Satellite Monitoring: High costs for frequent, high-resolution imagery or real-time satellite systems. Free data (e.g., Sentinel-1) is available but often lacks the temporal resolution needed for immediate early warning.

In-Situ *Monitoring*: Expensive to deploy and maintain dense networks of sensors (e.g., GPS, inclinometers) across large or remote areas, with high costs for installation, power, and data transmission.

15.3.7 Robustness in Challenging Environments

Seismic Monitoring: Performs well in rugged, remote, or vegetated areas where access is limited, as seismometers can be placed at a distance from the hazard zone and still detect signals.

Satellite Monitoring: Limited by terrain complexity, vegetation, or atmospheric conditions, which can reduce data quality or availability.

In-Situ *Monitoring*: Vulnerable to damage from rockfalls, landslides, or extreme weather, and installation in hazardous areas increases risk to personnel and equipment.

15.3.8 Integration with Other Systems

Seismic Monitoring: Easily integrated with other geophysical data (e.g., earthquake monitoring) and can be combined with machine learning for automated event detection and classification, enhancing early warning capabilities.

Satellite Monitoring: Can be integrated with ground data but requires complex processing to translate imagery into actionable warnings, often with delays.

In-Situ *Monitoring*: Provides high-precision data for integration but is limited by spatial coverage, making it less scalable for regional early warning systems.

15.3.9 Long-Term Monitoring and Trend Analysis

Seismic Monitoring: Continuous data collection allows for long-term analysis of seismic activity patterns, helping identify areas prone to recurrent rockfalls or landslides.

Satellite Monitoring: Provides long-term deformation trends (e.g., via InSAR) but is less effective for capturing sudden, short-duration events or real-time trends.

In-Situ *Monitoring*: Excellent for long-term monitoring at specific sites but requires extensive maintenance and may not capture regional trends without widespread deployment.

15.4 Challenges in Seismic Monitoring

Despite these advancements, several challenges persist in the application of seismic monitoring for Himalayan hazards. The region's rugged terrain and high altitude limit the deployment of dense seismic networks, restricting data availability for high-altitude mass movement events. The lack of comprehensive hydro-meteorological and seismic data at higher elevations, as noted in studies on hydropower projects, further complicates the development of integrated monitoring and warning systems (IMWS). Additionally, distinguishing seismic signals from landslides, rockfalls, and avalanches amidst background noise and tectonic activity remains a technical challenge.

Anthropogenic factors, such as deforestation, unplanned urbanization, and infrastructure development, exacerbate landslide risks and complicate seismic monitoring efforts. For example, the 2023 Joshimath subsidence event, monitored using PSInSAR and seismic techniques, was attributed to a combination of tectonic activity, paleo-landslide debris, and excessive construction, which overwhelmed the load-bearing capacity of the terrain. This underscores the need for integrated approaches that combine seismic monitoring with geomorphological and hydrological data to account for both natural and human-induced triggers. For example, a map showing the probability of seismic hazard with slope instability index and rivers may enable us to provide the probability of flooding due to landslide/rockfall into a river. Thus, additional integration of the above information alongside the integration of the seismic data with machine learning algorithms and geospatial modeling suggested by Ma et al. (2023) improves accuracy.

In brief, the following are limitations of seismic monitoring that require the attention of the research community.

False Positives: Seismic signals from non-hazardous sources (e.g., traffic, heavy machinery, microtremors etc.) can complicate event detection and need advanced filtering or machine learning algorithms. A recent study by Rekapalli et al. (2024a, 2024b) has provided a global model for discriminating the landslide seismic signals from those of small magnitude earthquakes and background noise using the spectral characteristics of seismic waveforms.

Localization Challenges: Due to a lack of clear phases, as noticed in the case of earthquakes, pinpointing the exact location of a rockfall or landslide from seismic data can sometimes be difficult and requires the development of advanced location methods. Usage of polarization back projection from multiple stations, cross correlation-based travel time computations, etc., may help us to address the phase detection-related issues in estimating the locations. However, the cross-correlation-based tracking methods (Cook et al., 2021) or the multi-station polarization back projection method (Rajesh et al., 2022) can help to overcome limitations to some extent.

Initial Setup Costs: While cost-effective for large areas, installing a seismic network still requires upfront investment in equipment and expertise. Usage of the low-cost sensors may help to reduce the initial cost.

Burtin et al. (2009) highlighted the complexity of analyzing high-frequency seismic noise to characterize Himalayan debris flows, indicating the need for advanced signal processing techniques. However, recent advancements in discrimination of landslide seismic signals from those of earthquakes and noise would be helpful in real time (Rekapalli et al., 2024a, 2024b; Yezarla et al., 2025). These methods use the trend of the power spectral density (PSD) of seismic waveforms and the skewness of power distributions as key features to differentiate landslides from local earthquakes and noise. The studies also showed that the discrimination accuracy is independent of recording station distance and global applicability.

15.5 Discussion

The Indian Himalaya are one of the most seismically active orogenic regions that present a critical natural laboratory for studying cascaded hazards, exemplified by events like the 2011 Sikkim co-seismic landslides and the 2021 Chamoli avalanche-flood disaster. The long-term convergence of the Indian and Eurasian plates continues at ~50 mm/yr., producing complex thrust and extensional structures such as the MCT, MBT, ITSZ, and Spiti–Sutlej Region Fault (SVF) (Kumar et al., 2024). The Spiti–Lahaul region, for example, sits astride the active SVF, where frequent earthquakes yield peak ground accelerations (PGA) of 15–40% g in the most seismic segments (Kaspar & Kieffer, 2024). Historical Kinnaur earthquake, 1975 (Mw 6.8) and Chamoli earthquake, 1999 (Mw 6.6–6.8) triggered widespread landslides, rockfalls and avalanches, damaging roads (e.g., Hindustan-Tibet road) and infrastructure across valleys (Singh et al., 1976; Jain et al., 1999; Prakash, 2013).

The interplay between seismic activity and other triggers, such as intense monsoon rainfall, is a key factor in the Himalaya. Studies suggest that ~15% of global rainfall-induced landslides occur in this region, with 477 of 580 landslides in the Indian Himalaya between 2004 and 2017 being rainfall-triggered (Dikshit et al., 2020; Goswami & Kundu, 2024). Seismic monitoring complements rainfall data by providing insights into subsurface processes, such as fault reactivation or slope destabilization, which are often precursors to landslides.

The 7 February 2021 disaster in the Chamoli district of Uttarakhand, India, was initiated by a massive rock-ice avalanche (~26–27 million m^3) detaching from the north face of Ronti Peak, rapidly transforming into a high-energy debris flow that mixed with river water and evolved into a catastrophic flood downstream in the Ronti Gad, Rishiganga, and Dhauli Ganga valleys (Rao et al., 2021; Cook et al., 2021). The event caused the destruction of infrastructure, including hydropower projects and claimed over ~200 lives. Based on the seismic network, the detachment of the rock chunk occurred at 10:21:14 AM. The study highlighted that the NGRI network would have been sufficient for early warning of the debris flood propagating downstream locations such as Joshimath (~34 km downstream), with ~30-minute lead time based on seismic detection of the initial rockfall. The Chamoli cascade exemplifies the challenge of rapid, compound hazards where tectonic, gravitational, and hydrological processes combine, making traditional hydrometeorology-based flood warning approaches inadequate; instead, seismic network detection is proposed as a basis for real-time debris flow and flood warnings in steep mountain settings.

On the other side, meteorological, hydrological data and modeling frameworks are used in flood early warning systems (EWS) for hydrological events such as river floods and flash floods in other regions of the globe to provide longer lead times, from hours to days from the rainfall, runoff, and river discharge forecast and issuing alerts before peak flows arrive. For example, in the Sirba River Basin in the Middle Niger River system (West Africa), the Global Flood Awareness System (GloFAS) coupled with local hydrometric observations enables forecasting and community alerts for rising flood risk days in advance, providing valuable lead time for preparedness and response activities at national and local levels (Passerotti et al., 2020). Similarly, high-resolution flood inundation forecasting integrated into operational systems in Europe now can offer up to ~17 hours of lead time using numerical weather prediction and hydrodynamic modeling, an approach that has demonstrably enhanced preparedness during events like the 2021 European summer floods (Najafi et al., 2024).

A remote-sensing case study from Central Vietnam illustrates how combining satellite data on convective systems and precipitation can improve the identification of heavy rainfall precursors, thereby enabling earlier detection of flood risk before significant runoff and flood onset occur, highlighting the importance of multisource data fusion in flood EWS (La et al., 2024). These hydrological and meteorological forecasting frameworks generally operate on more predictable precursors—such as rainfall intensity, soil moisture, river stage thresholds, and modeled discharge probabilities—compared to rapid mass wasting-driven floods such as the Chamoli event, which are triggered by sudden gravitational collapses with minimal precursory signals detectable by standard hydro-meteorological tools.

The contrast between the rockfall-triggered flood event in Uttarakhand and conventional rainfall-runoff flood early warning systems underscores the need for multi-hazard integrated warning frameworks in regions prone to compound events. While hydrological EWS can give hours to days of warning for floods driven primarily by meteorological factors, geophysical cascades like the 7 Feb 2021 event may only offer minutes of effective lead time, necessitating dense seismic monitoring and automatic alert dissemination to maximize the utility of any available early signal.

Seismic monitoring has revolutionized the study and management of cascaded hazards, providing critical insights into landslides, rockfalls, and avalanches. While significant advancements have been made, challenges such as data scarcity, signal differentiation, and anthropogenic influences remain. By integrating seismic monitoring with other data sources (infrasound, water level gauges etc.) (Rekapalli et al., 2024a, 2024b) and emerging technologies like machine learning, deep learning, etc. (AI/ML), and fostering collaborative mitigation efforts, we can better prepare for and mitigate the impacts of these devastating hazards in the Himalayan region. Yet, the integration of these tools into an operational early warning system for cascaded hazards, earthquakes causing landslides/rockfalls/avalanches that then trigger downstream disasters, remains an urgent challenge. Future efforts must focus on combining seismic networks, slope deformation sensors, infrasound sensors, water level gauges, satellite data, scenario modelling, and local resilience practices to reduce catastrophe in the Himalaya.

By addressing the region's unique challenges through advanced technologies, standardized frameworks, and community-focused strategies, EWS can significantly reduce the devastating impacts of these cascaded hazards. Continued interdisciplinary research cooperation will be key to building resilient systems that would protect vulnerable Himalayan communities.

15.6 Challenges in Developing Early Warning Systems

The Himalayan region's complex topography, coupled with climate change and human activities, exacerbates the frequency and severity of landslides, rockfalls, and avalanches. Studies indicate that the region experiences a higher incidence of these hazards compared to other global mountainous areas, driven by factors such as steep slopes, heavy rainfall, seismic activity, and permafrost degradation. For instance, the 2021 Chamoli disaster in Uttarakhand, India, highlighted the cascading nature of these events, where a rock and ice avalanche triggered a deadly debris flow, resulting in over 200 fatalities and significant infrastructure damage. The lack of comprehensive historical landslide inventories and the variability in hazard triggers (e.g., rainfall, snowmelt, earthquakes) pose significant challenges to predictive modeling. Additionally, the absence of standardized EWS frameworks tailored to the region's multi-hazard environment limits their effectiveness.

Socioeconomic factors further complicate EWS implementation. Rural Himalayan communities often lack the resources and infrastructure for robust monitoring systems, and low resilience in these areas amplifies vulnerability. High tolerance for landslide risks in developing countries, driven by economic constraints and developmental pressures, often results in inadequate mitigation measures. Moreover, the integration of social science perspectives into EWS design, such as community risk perception and compliance, remains underexplored, hindering effective warning dissemination.

15.7 Conclusions and Future Directions

Seismic monitoring offers significant advantages for early warning of rockfalls, landslides, and secondary hazards due to its real-time detection capability, wide-area coverage, sensitivity to precursory signals, and all-weather and continuous functionality. The approach also helps us to provide quantitative estimates of the energy release, equivalent seismic magnitude, source mechanisms rock or slope failure, and volumes of debris flow and water. Compared to satellite monitoring, it excels in immediacy and robustness, while compared to in-situ methods, it provides broader coverage and lower costs for large-scale applications. For optimal early warning, seismic monitoring is best used in combination with satellite and in-situ methods to leverage their complementary strengths, such as satellite imagery for post-event analysis and in-situ sensors for high-precision local data.

Future research should focus on enhancing the spatial and temporal resolution of seismic monitoring networks in the Himalaya. The deployment of low-cost, portable seismometers could address data gaps in remote areas, improving the detection of small-scale mass movements. Additionally, integrating deep learning techniques with seismic data, as explored in recent studies, offers promise for automated hazard detection and classification. These models can analyze large volumes of Earth observation data, including seismic signals, to predict landslide and avalanche risks with high accuracy.

Collaboration between scientists, policymakers, and local communities is essential to translate seismic monitoring into effective mitigation strategies. The National Disaster Management Authority (NDMA) guidelines on landslides and avalanches emphasize the need for continuous monitoring and early warning systems, which can be enhanced through seismic data integration. Community-based resilience programs, informed by seismic hazard maps, can further reduce vulnerability, particularly in rural areas with low adaptive capacity.

Acknowledgments We thank the Director, CSIR-NGRI, for his permission (NGRI/Lib/2026/ Pub-22) and support to publish this work. We acknowledge Dr. V. M. Tiwari for providing the opportunity to contribute this Chapter.

Funding The authors declare that no funds, grants, or other support were received during the preparation of this manuscript.

Competing Interests The authors have no relevant financial or non-financial interests to disclose."

Author Contributions Both authors contributed to the conception and design, and material preparation, manuscript draft writing, and finalization.

References

Alcántara-Ayala, I. (2025). Landslides in a changing world. *Landslides*, 1–15.

Allen, R. M., Gasparini, P., Kamigaichi, O., & Bose, M. (2009). The status of earthquake early warning around the world: An introductory overview. *Seismological Research Letters, 80*(5), 682–693.

Allen, R. M., & Melgar, D. (2019). Earthquake early warning: Advances, challenges, and future prospects. *Annual Review of Earth and Planetary Sciences, 47*, 361–388. https://doi.org/10.1146/annurev-earth-053018-060457

Alfieri, L., Salamon, P., Pappenberger, F., Wetterhall, F., & Thielen, J. (2012). Operational early warning systems for water-related hazards in Europe. *Environmental Science & Policy, 21*, 35–49.

Burtin, A., Bollinger, L., Cattin, R., Vergne, J., & Nábělek, J. L. (2009). Spatiotemporal sequence of Himalayan debris flow from analysis of high-frequency seismic noise. *Journal of Geophysical Research: Earth Surface, 114*, B04009.

Capobianco, V., Choi, C. E., Crosta, G., Hutchinson, D. J., Jaboyedoff, M., Lacasse, S., Reeves, H., et al. (2025). Effective landslide risk management in era of climate change, demographic change, and evolving societal priorities. *Landslides*, 1–19.

Chao, W. A., Wu, Y. M., Zhao, L., Chen, H., & Peng, H. S. (2017). Seismically triggered landslides: Monitoring and prediction using seismic noise. *Geophysical Research Letters, 44*(12), 6098–6106. https://doi.org/10.1002/2017GL073063

Cook, K. L., Rekapalli, R., Dietze, M., Pilz, M., Cesca, S., Rao, N. P., Hovius, N., et al. (2021). Detection and potential early warning of catastrophic flow events with regional seismic networks. *Science, 374*(6563), 87–92.

Dietze, M., Turowski, J. M., Cook, K., & Hovius, N. (2017a). Spatiotemporal patterns and triggers of seismically detected rockfalls. *Earth Surface Dynamics*. https://doi.org/10.5194/esurf-5-757-2017

Dietze, M., Mohadjer, S., Turowski, J. M., Ehlers, T. A., & Hovius, N. (2017b). Seismic monitoring of small alpine rockfalls–validity, precision and limitations. *Earth Surface Dynamics, 5*(4), 653–668.

Dikshit, A., Sarkar, R., Pradhan, B., & Segoni, S. (2020). Himalayan region: A critical review. In *Recent advances in geographic information system for earth sciences.*

Fischer, L., Purves, R. S., Huggel, C., Noetzli, J., & Haeberli, W. (2012). On the influence of topographic, geological and cryospheric factors on rock avalanches and rockfalls in high-mountain areas. *Natural Hazards and Earth System Sciences, 12*(1), 241–254.

Gariano, S. L., & Guzzetti, F. (2016). Landslides in a changing climate. *Earth-Science Reviews, 162*, 227–252.

Goswami, A., & Kundu, G. (2024). Geological and geomorphological interaction of the landslides in the Himalayan region. In *Landslides in the Himalayan region: Risk assessment and mitigation strategy for sustainable management* (pp. 3–21). Singapore.

Guzzetti, F., Peruccacci, S., Rossi, M., & Stark, C. P. (2007). Rainfall thresholds for the initiation of landslides in central and southern Europe. *Meteorology and Atmospheric Physics, 98*(3), 239–267.

Guzzetti, F., Gariano, S. L., Peruccacci, S., Brunetti, M. T., Marchesini, I., Rossi, M., & Melillo, M. (2020). Geographical landslide early warning systems. *Earth-Science Reviews, 200*, 102973. https://doi.org/10.1016/j.earscirev.2019.102973

Helmstetter, A., & Garambois, S. (2010). Seismic monitoring of Séchilienne rockslide (French Alps): Analysis of seismic signals and their correlation with rainfalls. *Journal of Geophysical Research: Earth Surface, 115*(F3).

Hibert, C., Michéa, D., Provost, F., Malet, J. P., & Geertsema, M. (2019). Exploration of continuous seismic recordings for the detection and characterization of landslide processes. *Journal of Geophysical Research: Earth Surface, 124*(8), 2073–2090. https://doi.org/10.1029/2018JF004876

Hoshiba, M., Kamigaichi, O., Saito, M., Tsukada, S. Y., & Hamada, N. (2008). Earthquake early warning starts nationwide in Japan. *Eos, Transactions American Geophysical Union, 89*(8), 73–74.

Intrieri, E., Carlà, T., & Gigli, G. (2019). Forecasting the time of failure of landslides at slope-scale: A literature review. *Earth-Science Reviews, 193*, 333–349.

Jain, S. K., Murty, C. V. R., Arlekar, J. N., Rajendran, C. P., Rajendran, K., & Sinha, R. (1999). Chamoli (Himalaya, India) earthquake of 29 march 1999. *EERI special earthquake report, EERI newsletter, 33*(7), 1–18.

Joshi, M. (2022). Co-seismic landslides in the Sikkim Himalaya during the 2011 Sikkim earthquake: Lesson learned from the past and inference for the future. *Geological Journal, 57*(12), 5039–5060.

Kaspar, M., & Kieffer, D. S. (2024). The Geomechanics of the Dangkhar landslide, Himachal Pradesh, India. *Geotechnics, 4*(2), 655–672.

Keefer, D. K. (1984). Landslides caused by earthquakes. *Geological Society of America Bulletin, 95*(4), 406–421.

Kumar, P., Kamal, Sharma, M. L., Jakka, R. S., & Pratibha. (2024). Uttarakhand state earthquake early warning system: A case study of the Himalayan environment. *Sensors, 24*(11), 3272.

La, T. V., Nguyen, T. H., Matgen, P., & Chini, M. (2024). Early flood warning using satellite-derived convective system and precipitation data-a retrospective case study of Central Vietnam. In *IGARSS 2024–2024 IEEE international geoscience and remote sensing symposium* (pp. 3752–3756). IEEE.

Lacroix, P., & Helmstetter, A. (2011). Location of seismic signals associated with microearthquakes and rockfalls on the Séchilienne landslide, French Alps. *Bulletin of the Seismological Society of America, 101*(1), 341–353.

Ma, J., Wang, J., & Zhang, X. (2023). Machine learning for landslide early warning systems: A review. *Natural Hazards, 115*(1), 23–45. https://doi.org/10.1007/s11069-022-05512-7

McColl, S. T. (2022). Landslide causes and triggers. In *Landslide hazards, risks, and disasters* (pp. 13–41). Elsevier.

Najafi, H., Shrestha, P. K., Rakovec, O., Apel, H., Vorogushyn, S., Kumar, R., Samaniego, L., et al. (2024). High-resolution impact-based early warning system for riverine flooding. *Nature Communications, 15*(1), 3726.

Parkash, S. (2013). Earthquake related landslides in the Indian Himalaya: Experiences from the past and implications for the future. In *Landslide science and practice: Volume 5: Complex environment* (pp. 327–334). Springer Berlin Heidelberg.

Passerotti, G., Massazza, G., Pezzoli, A., Bigi, V., Zsótér, E., & Rosso, M. (2020). Hydrological model application in the Sirba river: Early warning system and GloFAS improvements. *Water, 12*(3), 620.

Rajesh, R., Purnachandra Rao, N., Paul, H., Venkatesh, V., Sarma, A. N. S., Vijaya Raghavan, R., Suresh, G., & Saha, S. (2022). Landslide location analysis using seismic multi-station polarization back-projection. In *FIGA-2022 held at Wadia Institute of Himalayan Geology, Dehradun.*

Rao, N. P., Rekapalli, R., Srinagesh, D., Tiwari, V. M., Hovius, N., Cook, K. L., & Dietze, M. (2021). Seismological rockslide warnings in the Himalaya. *Science, 372*(6539), 247–247.

Rekapalli, R., Yezarla, M., & Rao, N. P. (2024a). Discriminating landslide waveforms in continuous seismic data using power spectral density analysis. *Geophysical Research Letters, 51*(21), e2024GL110466.

Rekapalli, R., Purnachandra Rao, N., Venkatesh, V., Uma Devi, E., & Sarma, A. N. S. (2024b). Seismic broadband and infrasound data-based river water monitoring. In *ASC2024 held in Belek, Antalya, Turkiye during 3rd to 7th November 2024.*

Roccati, A., Paliaga, G., Luino, F., Faccini, F., & Turconi, L. (2020). Rainfall threshold for shallow landslides initiation and analysis of long-term rainfall trends in a Mediterranean area. *Atmosphere, 11*(12), 1367.

Schweizer, J., Bruce Jamieson, J., & Schneebeli, M. (2003). Snow avalanche formation. *Reviews of Geophysics, 41*(4).

Schuster, R. L., & Wieczorek, G. F. (2018). Landslide triggers and types. In *Landslides* (pp. 59–78). Routledge.

Sharma, S., Talchabhadel, R., Nepal, S., Ghimire, G. R., Rakhal, B., Panthi, J., Kumar, S., et al. (2023). Increasing risk of cascading hazards in the Central Himalayas. *Natural Hazards, 119*(2), 1117–1126.

Shugar, D. H., Jacquemart, M., Shean, D., Bhushan, S., Upadhyay, K., Sattar, A., Westoby, M. J., et al. (2021). A massive rock and ice avalanche caused the 2021 disaster at Chamoli, Indian Himalaya. *Science, 373*(6552), 300–306.

Singh, S., Jain, A. K., Sinha, P., Singh, V. N., & Srivastava, L. S. (1976). The Kinnaur earthquake of January 19, 1975: A field report. *Bulletin of the Seismological Society of America, 66*(3), 887–901.

Srinagesh, D., Mandal, P., Raghavan, R. V., Gupta, S., Suresh, G., Srinivas, D., Tiwari, V. M., et al. (2019). Digital seismic network: To map Himalayan orogen and seismic hazard. *Current Science, 00113891*(4).

Stähli, M., Sättele, M., Huggel, C., McArdell, B. W., Lehmann, P., Van Herwijnen, A., Springman, S. M., et al. (2015). Monitoring and prediction in early warning systems for rapid mass movements. *Natural Hazards and Earth System Science, 15*(4), 905–917.

Van Herwijnen, A., & Schweizer, J. (2011). Monitoring avalanche activity using a seismic sensor. *Cold Regions Science and Technology, 69*(2–3), 165–176.

Wyk, V., de Vries, M., Bhushan, S., Jacquemart, M., Deschamps-Berger, C., Berthier, E., Gascoin, S., Kääb, A., et al. (2022). Pre-collapse motion of the February 2021 Chamoli rock–ice avalanche, Indian Himalaya. *Natural Hazards and Earth System Sciences, 22*(10), 3309–3327.

Wu, Y. M., & Kanamori, H. (2005). Rapid assessment of damage potential of earthquakes in Taiwan from the beginning of P waves. *Bulletin of the Seismological Society of America, 95*(3), 1181–1185.

Yezarla, M., Rekapalli, R., & Gupta, S. (2025). Persistent PSD trends: a tool for seismic landslide detection. *Acta Geophysica, 73*(6), 5399–5408.

Chapter 16
Climate-Induced Disasters and Plausible Remediation: Himalayan Viewpoints

Kalachand Sain

Abstract The Himalayan region is warming much more than the global average, and people living over there are distressed by different kinds of geological, ecological and hydrological disasters with variable intensities. The influence of warming and impact of climate change are visible in terms of excessive rainfall leading to landslides, flash floods, debris flows; weakening of high-altitude rock mass through freezing and thawing; and inconsistent snow precipitation leading to avalanches and flash floods. There are also ice-rock mass avalanches, glacier lakes outburst floods (GLOFs), landslide-lakes outburst floods (LLOFs), fluvial extremes, forest fires, as well as cascading effects caused by earthquakes to landslides to formation of temporary lakes followed by sudden breaching. The hassles are aggravated additionally by anthropogenic loads due to unauthorised civil constructions and some infrastructure without proper impact assessment. All these have been responsible for the disruption not only to the water, energy, food, and health securities but also to the biodiversity, agriculture, forestry, and tourism sectors. The impact will be worsened if temperature rises consistently without any control. Here we synthesise the causes of several climate-induced disasters and provide their plausible remediation to meet the sustainable development goals. The first and foremost measure is the identification of hotspots for hazards through acquisition and analysis of land and satellite data. This will be followed by establishment of state-of-the-art sensors, suited for monitoring and development of Integrated Early Warning Systems (IEWS). The next measure is the stringent implementation of mountain-specific town planning and building codes, and engaging local communities to utilize their traditional knowledge. The final and the most crucial measure is the balance between economy and ecology through a mix of modern scientific approaches and nature-based solution (NbS) for preservation of the Himalayan ecosystem at one side and exploitation of resources for socio-economic growth on the other side.

Keywords Himalaya · Climate-change · Hazards · Remediation · Sustainability

K. Sain (✉)
CSIR-National Geophysical Research Institute, Hyderabad, India

V. M. Tiwari (ed.), *Himalayan Natural Hazards*, Indian National Science Academy,
https://doi.org/10.1007/978-981-92-3784-5_16

16.1 Introduction

The Himalaya is a depot for a variety of natural resources. The high altitude glaciers, snow field, cold springs, river system, lakes, and wetland are lifelines to hundreds of millions of people for irrigation, drinking water, hydropower generation, domestic usages, industrial requirement, even for transportation and entertainment. The pressure-temperature (P-T) regime of the Himalaya has led to the formation of precious minerals and orebodies. The organic rich sediments from Tethys Sea are source of hydrocarbons in thrust-fold belts of the Himalaya. Abundant renewable energy resources like geothermal, wind, and sunshine are also present over there. The roots, flowers, stems, leaves, and barks of medicinal plants are useful sources for curing diverse kinds of ailments. The timbers, gums, resins, and decorative leaves are used for generating economy wealth. The Himalaya is the destination of pilgrimage sites and trekking routes. The ecosystem with different species of animals, birds and trees; striking flower gardens; picturesque landscape; scenic beauty and meadows attract ttourists round the year. All these have made the Himalaya a hub for settlement and agro-economic advancement. Though there lie great opportunities for progress and prosperity with regard to infrastructures development; building hydro-power projects, geothermal or hybrid energy plants; establishment of geoparks or geoheritage sites and green ecotourism; growth of cultivation and livestock farms; expansion of micro-scale industries using local resources, proper management and utilization of resources, and building developmental projects in scientific manner only can meet the aspiration of people living in the Himalayan regions.

However, the ecosystem of the Himalaya is being affected by warming under the influence of present climate change (CC) scenario. The impact is visible in terms of rapid melting and declining surface area of glacier/snow fields; expansion of glacier-lakes and formation of new lakes; uphill swings of snow and tree lines; and elevation-dependent-warming. This has made tremendous influence to the hydrological cycles, agriculture and crop yields, river flows and hydropower generation, plant/animal species distribution, air quality, health, environmental hazards, and socio-economic settings, etc. (Gao et al., 2013; Gupta et al., 2019; Lutz et al., 2016; Sain, 2025; Wang et al., 2016; Zebre et al., 2021). The susceptibility of mountain regions has been aggravated further due to enhanced anthropogenic loads that are being exerted due to unplanned civil construction and several other structures that are built without assessing the load-bearing capacity (Gopirajan et al., 2022).

The communities living in the Himalayan states have been experiencing the worst impact of CC, as the temperature gradient is much higher than that in many other parts of globe (Gentle et al., 2014; Matta et al., 2025; Verma, 2023). The CC has been attributed to the changes in monsoon patterns leading to erratic rainfall and hence floods, as well as extended dry periods; rapid melting of glaciers affecting the water availability and river dynamics for hydropower generation; reduced snowfall and its early melting; environmental hazards—landslides, ice/snow avalanches, floods and debris flow, and enhanced health risks due to air pollution. It can be clarified that glacial-melt increases the water discharge initially for a short period but threatens to water scarcity in the long run, as glacial meltwater will diminish over time. The melt water may form glacial lakes at high-altitude depression, which may breach due to increased volume of water as GLOF. Thawing permafrost and increased rainfall have been destabilizing mountain slopes, causing landslides and potentially blocking stream courses, further exacerbating flood risks due to LLOF. All these have affected properties and people who particularly depend deeply on the agriculture, tourism, and natural resources for their livelihood (Das and Mishra, 2023).

The Himalaya is projected to face further warming with possible temperature rise of 2.6–4.6 °C and disappearances of glaciers by the end of twenty-first century, if the global warming continues without any regulation (Chauhan et al., 2023a; Dimri et al., 2018; Sabin et al., 2020). The impact will be much more rigorous to the river systems, vegetation and traditional farming seasons, rainfall and snowfall patterns, species dispersals, biodiversity, and extreme weather events—heat waves, droughts, cloudbursts or flash floods, and hazards due to landslides, avalanches, debris flow, GLOFs or LLOFs. The future precipitation patterns are also not clear with some models suggesting augmented rainfall, while others expect diminutions, emphasizing the need for localized assessment. Though the CC phenomenon is a world-wide issue and its total mitigation is global, we can take a lot of initiatives in the Himalayan regions, as it has significant challenges, and acts as a natural laboratory for CC research. Thus, it requires a combination of global efforts that can reduce the carbon footprint and make a renaissance in alternate renewable/green energy sectors at one hand; and adaptation to traditional or nature-based strategies through community engagement and adoption to modern tactics at local level on the other hand for evolving effective strategies (Chauhan et al., 2023b; Jeelani et al., 2024; Dhungana et al., 2020; Sain, 2023a, 2024a; Sain et al., 2019, 2023a, 2023b, 2025). Here we present the causes and consequences of some climate-induced environmental hazards and related disasters due to landslides and land subsidence; glacier avalanches and debris flows; GLOFs, cloudburst-triggered flash floods; snow avalanches; and forest fires, and describe numerous strategies for their plausible mitigation with case studies and success stories. Since the youngest Uttarakhand (UK) state in the Indian Himalayan Region (IHR) is ~85% mountainous, and has gone through all types of hazards, we focus our study in the UK Himalaya.

16.2 Landslides

Around one-third of the world's landslides takes place in the Himalaya (Gupta et al., 2022). The friable geology with faults/fractures/fissures, steep slopes, elevation, intense rainfalls, unplanned construction, and sprawling population over the fluvial and un-consolidated materials have been attributed to the susceptibility or vulnerability or risks associated with landslides. Over 1100 landslides have been reported in 2023 compared to 245 in 2022 and 354 in 2021, along with 300 people killed by crumbling mountains between 2015 to 2023 in the UK Himalaya. The brunt of the 1998 Malpa (Pithoragarh), 2003 Varunavrat Parbat (Uttarkashi), and 2012 Okhimath (RudraPrayag) landslides are still afresh (Gupta & Bist, 2004; Martha & Kumar, 2013; Paul et al., 2000). The susceptibility or vulnerability maps for the regions of evolving activities provide useful input in developing green corridors and monitoring the most risk-prone zones against future disaster. The landslide risk maps have been prepared for the Nainital (Kumaun) and Mussoorie (Garhwal) towns (Gupta et al., 2022, 2023) as well as for basins like Goriganga (Kumaun) and Bhagirathi valleys by utilising parameters like slope inclination, slope aspect, gradients, curvatures, direction, elevation, lithology, rock types, rock strength, structures, geomorphology, forest cover, built-up or open area, topographic wetness index (TWI), and distance to drainage, road and thrusts etc. Similar exercises can be made to other areas of the Himalaya.

The risk map of Mussoorie town shows which parts are the highest, medium or least prone to landslides (Fig. 16.1). This also indicates how many buildings are high-, medium- or low-risk prone to landslides. The flowchart for landslide susceptibility mapping and sensitivity analysis, as discussed by Kaur et al. (2024), can be adopted in any area. The landslide in the sparsely vegetated Varunavrat Parbat region, which is situated on right bank of the Bagirathi river with the Uttarkashi township at its foot, was triggered mainly by the rainfall, and partly by the groundwater and developmental activities (Gupta & Bist, 2004). The daily rainfall for July–September 2003 and average annual rainfall for the last 14 years clearly indicated that the rainfall in 2003 was above normal. The debris slide at the top and rock fall in the middle started on the day of rainfall ceased and continued for the next 15 days. The clay-filled joint structures, rock types, and degree of slopes were other factors. The 1500 mm of rainfall, as evidenced in the Varunavrat Parbat, can be considered threshold for landslides in monsoonal months at a similar topographic setting. This urges in developing a stochastic relation between landslides and rainfall intensity duration or total rainfall for assessing the landslides over the risk-prone zones. Since landslides are triggered by rainfall, tectonic disturbances, reservoir drawdown, or human incursions, these can be forewarned by coupling slope instability with the rainfall threshold or other triggering factor(s) by establishing a network of web-based sensors such as rain gauges, piezometers, inclinometers, extensometers, Total Stations, InSAR, and so on in a vulnerable slope region (Fig. 16.2). This can be followed by real-time data assimilation and development of landslides Integrated Early Warning System (IEWS) using IoT-based or AI/ML algorithms (Sain & Kumar, 2022).

16.3 Land Subsidence

Geosciences play an important role in ensuring site-selection, and warranting safety and durability of any project related to strategic advancement and socio-economic growth. The geological, structural, and geomorphological mapping and geotechnical investigation provide valuable information on surface rock types and their strengths; and presence of faults, fractures, and their orientation. Geophysical studies based on

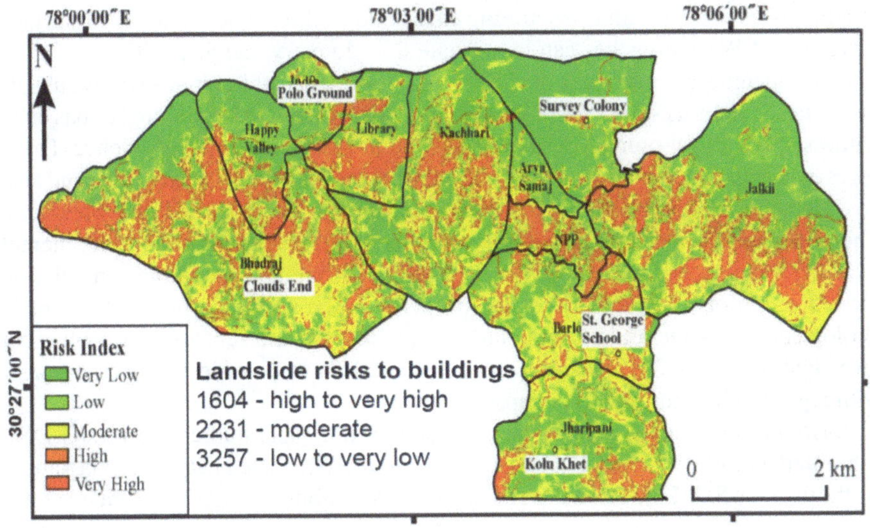

Fig. 16.1 Landslide map of Mussoorie town showing high-, medium- and low-risks with different colors as well as number of buildings possessing different category of risks (Gupta et al., 2023)

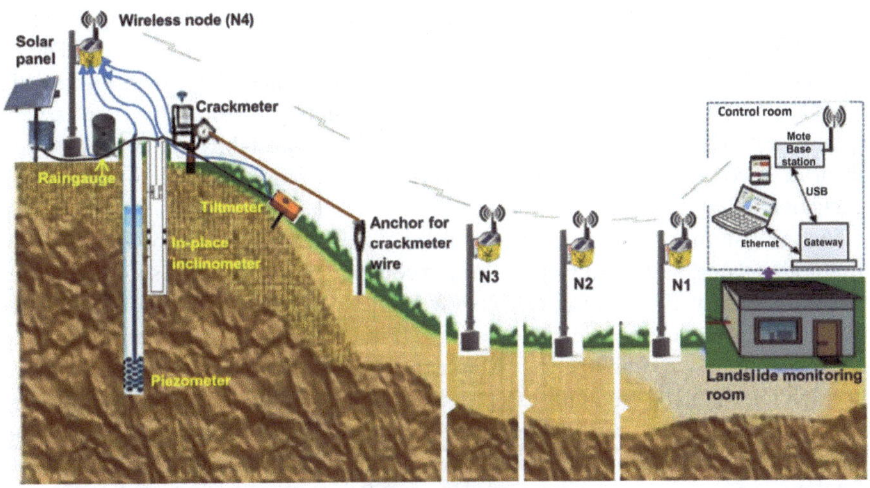

Fig. 16.2 A schematic diagram of monitoring and developing landslides early warning system by deploying sensors at field sites in a hill slope, coupled with wireless device (N4), which subsequently send data through wireless devices (N3, N2 and N1), gateway, and base radio to computer at control room (After Chaulya et al., 2024)

seismic, resistivity, ground penetrating radar and multichannel analysis of surface waves (MASW) approaches can investigate the geometry and disposition of faults/fractures in underground, weak or shear zones or presence of fluids at shallow subsurface. These subsurface features or weak zones can be attributed to sinking, as has been witnessed in Silkyara tunnel-roof collapse and at Joshimath land subsidence. Drill-cores enable to understand in-situ conditions of rocks/soils at different depths and also provide materials for laboratory testing to measure permeability and pore pressures. This information is central for choosing the site and designing the developmental strategy; gauging hotspots for geo-hazards; monitoring and providing remediation for protection of lives, structures, and properties. The surface geological, geomorphological, and structural studies can be married with the subsurface geophysical investigation (Sain, 2019; Sain & Damodara, 2023) to provide better assessment with regard to hazards due to collapse or sinking.

Based on geological, geophysical, and geomorphological studies, the WIHG-DST submitted a report (Vaideswaran et al., 2024) to the UK State Disaster Management Authority (UKSDMA). The geophysical surveys exhibited moisture-saturated zones in shallow subsurface. Such studies have brought out that the land subsidence, surface deformation, and development of surface cracks at Joshimath are due to underground hydrogeological imbalances (Sain, 2023b). Since Joshimath town is situated over a steep slope with dense concrete built-up and partly dense forest covers, it was felt necessary to have a high-resolution topographic map of the region. The Light Detection and Ranging (LiDAR) survey was entrusted to WIHG-DST by the UKSDMA, and this could generate 3D bare-ground digital terrain map along with the mosaic of surface cracks for the Joshimath and the surrounding region. This map is very useful in comprehending land subsidence or landslides by establishing a link between subsurface geophysical features with the surface observations as well as for effective management and mitigation purposes.

The mitigation for landslides and land subsidence can be accomplished by both structural and non-structural means. We can construct guard-walls to safeguard from mass failure, as well as putting retentive wire-mesh, shotcrete, soil nails, rock bolts etc. against the slope movement. Non-structural measures include enforcing building codes, educating communities on "Dos and Don'ts", and prohibiting exposure to high-risk regions. The NbSs in response to landslides include plantation of grasses over the earlier barren areas, Gabion wall construction, avoiding livestock grazing in landslide-prone areas, and prohibition on tillage at risk-prone areas of landslides. Building a concrete drainage system will not allow infiltration of used water into the subsurface haphazardly, which otherwise accelerates soiled rock mass to slide over the subsurface plane of weakness. Other remedial measures include implementation of strict building codes such as prohibition of construction on unstable slopes and flood-prone riverbanks, no settlements and infrastructure projects without any hazards assessment and environment impact assessment, no deforestation and mining in steep regions, hill-cutting with scientific assessment, and no pilgrimage sites and tourism infrastructure in geologically unstable areas.

16.4 Glacier Avalanches

Out of over 10,000 glaciers in the IHR, around 970 glaciers covering an area of ~2857 km^2 exist in the UK Himalaya itself (Kumar et al., 2022; Ravindra et al., 2024; Sain et al., 2021a, b). The rate of melting, thinning, and retreat of glaciers are considered while assessing the glaciers-related hazards as well as for gauging water availability in the nearby regions. Glacial hazards mean glacier detachment and ice avalanche directly that lead to debris flow and flash floods indirectly. A classic example is the 2021 Chamoli disaster caused by a massive ice-rock mass avalanche followed by a huge deluge that killed more than 200 people, destroyed hydropower projects, disconnected many villages by damaging roads and bridges in peak winter (Cook et al., 2021; Mehta et al., 2021; Sain et al., 2021a; Shugar et al., 2021).

The glaciers in different catchment are receding at different rates. The Gongotri and Dokriani Glaciers in the Bhagirathi basin are retreating at 15–20 m per year since 1995, whereas the Chorabari Glacier in the Mandakini sub-basin of Alaknanda basin is retreating at 9–11 m per year during 2003–2017 (Bhambri et al., 2023). Therefore, it is necessary to update the glaciers inventory (Bhambri et al., 2015) at periodic interval. Besides recession, the catastrophes associated with glaciers are produced by several other factors such as sediment/water discharge, lithology, structures, climatic variability, moraine collapse, rainfall, glacial lake outburst, etc. The threats due to glacial hazards are generally examined using the hydrological and meteorological data. Tiwari et al. (2022) have identified seismic precursors of glacial avalanche, first of its kind, for about 3 and a half hours' duration (Fig. 16.3) before the main ice-rock avalanche took place, which caused the 2021 Chamoli disaster (Cook et al., 2021; Sain et al., 2021a; Shugar et al., 2021). This study led for the establishment of a network of seismological stations in addition to the hydro-climatological stations; instant data broadcast through Very Small Aperture Terminal (VSAT) or Global System for Mobile (GSM) communication; automatic analysis and integration of data using suitable AI/ML algorithms; and development of IEWS of glacial hazards. A flow chart for such glacier warning systems has been explained by Sain (2022). Tiwari et al. (2024) also observed continuous peaks of radon anomalies before the event occurred, which decreased suddenly after the event ceased, providing an opportunity for multi-disciplinary precursory study for glacial avalanche.

16.5 Debris Flow

The debris flows in the mountainous regions have been caused by the landslides, glacial/snow avalanches, and floods. The Malpa landslide is one of the worst landslides that wiped away the entire Malpa village, and generated ~one million cubic meters of rock fall and debris flow. The landslide killed ~220 people including 60 pilgrims travelling to Tibet as a part of 'Kailash Manas Sarovar Yatra'. The impact of the 2021 Chamoli ice-rock mass avalanche is still visible. Post-disaster remote

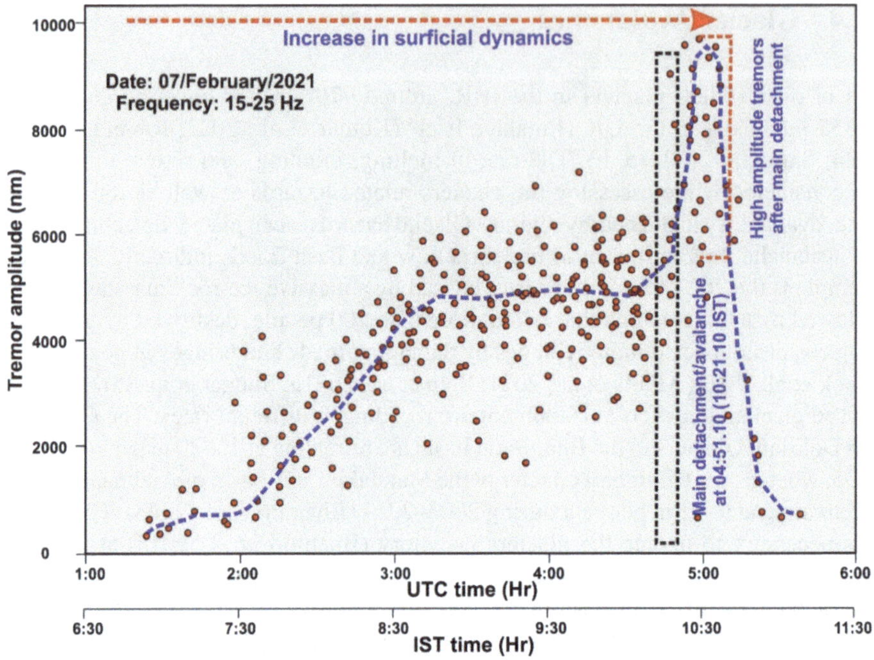

Fig. 16.3 Seismic precursors, observed ~3 and ½ h before the ice-rock mass avalanche occurred at 10:30 IST on seventh February, 2021 in the Rishiganga-Dhauliganga basin (After Tiwari et al., 2022) that led to the development of Avalanche Early warning system by coupling seismological data with the hydrometeorological data

sensing data suggest considerable instability on the landscapes and geomorphology near Raini village lying on the bank of the Rishiganga river, primarily due to toe-cutting by river incision and debris flowing down. The high sediment transfer may impact hydro-projects in the downstream of Chamoli area, possibly by silting reservoir and/or corroding turbine blades. Understanding the sediment dynamics also necessitates mass wasting modelling or debris flow simulation (Westoby et al., 2023). The combined remote sensing, field observations, and numerical modelling show that seven million cubic meters of debris, deposited by the 2021 Chamoli ice-rock mass avalanche in February, was removed along a 30-km-long domain from near its origin by the pre-monsoon and monsoon river flows in 12-months post event. The study also conveyed bed-load waves traveling at 0.1–0.3 km/day and sustained order-of-magnitude increases in suspended sediment concentrations as far as 85 km from the event source. These have wider implications, notably for water quality and downstream hydropower projects that may be disrupted by elevated bed-load and suspended sediment transport. The huge siltation at riverbed also poses threat to inundation of the riverbanks. The stakeholders of hydropower projects must have the IEWS for their respective dams.

A debris flow occurred in the foreland of Gangotri glacier by its former tributary, Meru (Bamak) glacier in July 2017 (Kumar et al., 2019). The Meru Bamak stream was completely dissected that exposed the ice-cored left lateral moraines. The study brought out deposition of ~6.5 million cubic meters of debris in the frontal region of the Gangotri glacier as fan-type feature. A sharp escalation in the concentration of suspended sediments reaching upto 11,000 mg/l at ~3 km downstream was also noticed. The recession of the Gangotri glacier, degraded ice-cored moraine, loose sediments at the front of Meru Bamak, and continuous rainfall have been attributed for such debris flow. Since there were no habitation and structures in nearby area, it did not incur any disaster.

16.6 GLOF

There are more than 20,000 glacier lakes existing in the IHR. Around 1270 glacier lakes covering an area of roughly 7.6 km^2 occur in the UK Himalaya itself (Sain et al., 2021a, 2021b; Kumar et al., 2022; Ravindra et al., 2024). As the dimension of glacier lakes is changing and the new lakes are being formed, the lake inventory (Bhambri et al., 2015) must be upgraded at periodic interval. Since most of the glacier-lakes in UK state are supraglacial, they may not be hazardous. However, the moraine-dammed lakes may pose threat to the downstream. Out of 350 moraine-dammed lakes, only 13 lakes have been identified in 2023 by the National Disaster Management Authority (NDMA) and UK State Disaster Management Authority (UKSDMA) along with several other constituted member organizations. The Bhilangana lake at ~4750 m height is one such lake in the Tehri-Garhwal region of UK Himalaya, which has grown to 0.38 km^2 area in the last 55 years, as per the study made by WIHG-DST (Fig. 16.4). It has been observed that the associated Khatling glaciers have been thinning. The study shows that air temperature gradient (0.79 °C/decade) has been increased considerably in the immediately past decade compared to that (0.28 °C/decade) over the last three decades. Again, downslope of this region is considerable, and thus close monitoring and assessment of this lake is required to avert any disaster due to GLOF, as have been witnessed by complete outburst of Chorabari lake in Kedarnath in 2013 (Dobhal et al., 2013). The Kedarnath GLOF inundated 5500 villages, killed 6000 people, detained more than 1 lakh pilgrims in peak summer, and destroyed huge properties and structures. This also led to more than 10,000 small, medium, and large-size landslides (Martha et al. 2015). Some of these temporarily blocked the stream courses, which were breached subsequently. Thus, the cascading effect—landslides, formation of temporary lakes, outburst of landslides-lake leading to flash floods, debris flow, etc.,—was observed for the Kedarnath GLOF. Assessment and monitoring of glaciers and glacier lakes are very much required with regard to protection from GLOF and related hazards (Kumar et al., 2023). It is also pertinent to evaluate the vulnerability of watersheds due to environmental degradation in the Lesser Himalaya using a series of models, as demonstrated by Chauhan et al. (2024) based on IoT.

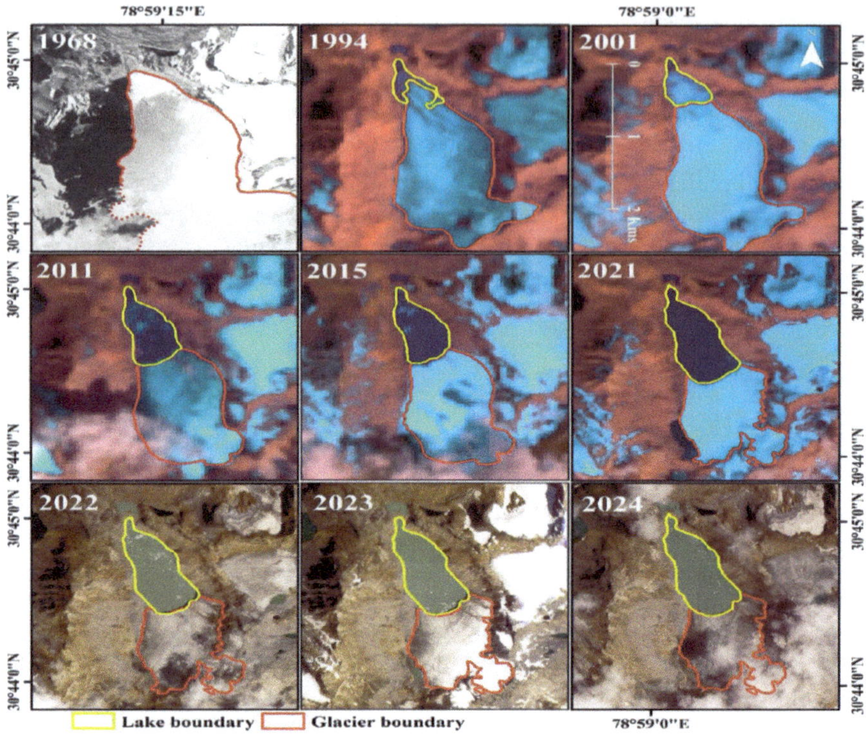

Fig. 16.4 Expansion of proglacial Bhilangana lake between 1968 and 2024 visible at multiple satellite images in the tributary catchment of Bhagirathi river and thinning of associated Khatling glacier (Patidar et al., 2025)

Monitoring of glacier-lakes and associated glaciers cannot be isolated. To comprehend the impact of glacier-related hazards, the retreat, thinning, melting, water and sediment discharges on the upper reach are to be critically appraised. It is necessary if possible to develop droned based or satellite techniques for mapping bathymetry of glacier-lakes and hence volume of lake waters as well as estimating the thickness of glaciers, and thus volume of water stored within the glaciers-ice. The ground penetrating radar and sonar survey, wherever possible, can be utilised for calculating the glaciers' thickness and volume of lake-water respectively for ground truth validation of remote sensing data. The geoengineering, geotechnical or geophysical surveys play an important role to determine the strengths of walls of the moraine-dammed lakes. The area of lakes, rate of water accumulation, amount of downslope, infrastructure and habitation in downstream are considered critical factors for assessing glaciers and glacier-lake related hazards. The hydro-meteorological stations are to be strengthened to observe air temperature, humidity, precipitation, wind speed, wind direction, melt water discharge, sediments discharge etc. and specifically to be deployed at higher elevations, due to elevation-dependent warming and rainfall shifting upward. It is suggested for controlled puncturing or spillways of glacial-lakes, tunnelling, laying pipelines inside, pumping or siphoning off water as remedial measures of GLOF. The effect of black carbon, which might arise from forest fires and/or biomass burning and settle over glaciers is also to be evaluated. This might accelerate glacier melting, as black carbon absorbs more sunlight. Other anthropogenic activities like deforestation, increased urbanization, use of heavy construction materials or urban heat island, land-use-land-cover patterns, influx of tourists, increased load due to vehicles and their vibrations and emission that are attributed to air pollution and warming, impacting the fragility and hydrological cycle. The field campaigns (if possible) must be made at some regular intervals to measure the glacier mass balance using stakes and GPR, which can also validate the observations made from the remote sensing data.

As glacial hazards are not restricted to any geographical or political boundaries, the success of any individual State Disaster Management Authorities on assessing the stresses induced by climatic factor, anthropogenic load, natural phenomena, and land degradation must be disseminated with nearby states to help formulate better strategies for mitigation. Similarly, the impact of any hazard may place downstream countries under risk. The very recent catastrophe due to icerock avalanche flash floods and debris flow that struck Nepal and neighboring Tibet on 26th August 2026 is a typical example of such need. Thus, a Himalayan Council is to be made proactive by involving neighbouring countries. They must share their initiatives with regard to reduction of carbon footprints, slowing down of glacial melt, shielding habitation and infrastructures, and as well as sharing data related to water-induced disasters and GLOF. Need not to mention that the glaciers and glacier-lakes must be categorised into the high-, medium- and least-risk to glacial hazards.

16.7 Flash Floods

The UK Himalaya has experienced flash floods from time to time. On fifth August, 2025, a rainfall in the upper catchment of the Kheer Ganga river triggered landslides and massive mud-water flash floods that destroyed the Dharali village of Uttarkashi district claiming 4 persons dead, 50 feared missing, several goats swept away, and many homes, shops, hotels, homestays and roads destroyed. The village is located on the pilgrim route to Gangotri Dham from Uttarkashi. The brunt of the massacre is also visible at nearby regions located on this pilgrim route. Among a few other surprising incidences, heavy rainfall triggered disaster leading to slope breakdown was observed near Maldevta and surrounding regions of Dehradun in August, 2022 (Sain et al., 2023a). This damaged the channel and bank of the Song and Baldi rivers for a 15 km stretch between Sarkhet and Thano, and completely washed away Sarkhet, Kumalta and the lower portion of Maldevta town. It destroyed and washed away some of huts and temporary stalls built on the banks of the river. A cloudburst, in which more than 100 mm/hour rainfall occurred locally for a short duration, caused double-impact as flash flood in Arakot Nala and landslides downstream in Arakot and nearby villages of Uttarkashi in August, 2019 (Chauhan et al., 2022). This flash flood led to the death of 19 people, 8 were feared dead, and 15 injured, affecting 38 villages, covering an area of 70 km^2, and stranding ~400 people at various places. The Lasko valley in Pithoragarh district of UK state near the Indo-Nepal border area also experienced rainfall-induced disaster in September, 2022 (Sain et al., 2025). This flash flood caused 6 fatalities, 11 reported missing cases, 72 houses were severely damaged and more than 62 families were deserted at multiple places. Other impacts of this catastrophe were roads being washed away, land degradation, occurrence of small landslides, and destruction of minor bridges. Figure 16.5 shows the event due to cloudburst-induced flash floods and related damages in the Lasko valley. The flash flood has been explained as due to constricted drainage systems through which so much rainwater could not be flashed or drained out through the confluence of Kali river and Lasko nala. In fact, the water receded back and inundated the upstream villages on the riverside.

The warming of atmosphere due to CO_2 and other greenhouse gas emission increases the capacity of holding more moisture leading to heavy rainfall concentrated at one place within a short span or defined as the cloudburst. The confluence of SW

Fig. 16.5 Cloudburst-elicited flash flood at Dharchula, Pithoragarh in Indo-Nepal border. **a** Confluence of Kali River and Lasko Khola (Tributary of Kali river) showing bulk of debris deposit from Lasko Khola; **b** Lower level of Khotila Village where 30 houses were completely submerged in flood (Yellow elliptical), whereas red circle indicates water level risen 1 m above the temple-pinnacle; **c** Kali Valley ahead of the confluence in which water level has gone up (Red dotted line); **d** Dungri temple, fully affected by the deluge (yellow dotted line showing water level on the event day (After Sain et al., 2025)

monsoon from the Arabian Sea and weak Westerly Disturbances from the Mediterranean Sea moving eastward intensified the rainfall at the foothills of Himalaya and to an extent over the Himachal and UK Himalaya. This led to torrential rainfall in July–August 2023 in both the Himachal and UK Himalaya.

The landslides in hilly areas and floods in the plains have been attributed mainly to the climatic factor—cloudburst and slow but continuous normal rainfall over a period of time. The other factor is the land degradation—the top most part made up of phyllites, schists, gneisses, quartzite, limestone etc. is highly weathered, fractured, jointed under the influence of cooling and warming phenomena of the environment. The third factor is the anthropogenic activities—deforestation, unauthorized civil construction, slope instability due to toe cutting in an unscientific manner, and improper drainage system that allows random percolation of water from house-hold kitchen and toilet into the subsurface. The rain water makes the top loose soil saturated and heavy, and underground percolation in a chaotic manner accelerates the mass movement over the plane of weakness leading to landslides under gravitational forces.

For remediation, it is suggested for pragmatic check dam policy and rational shifting of people from hazard-prone areas like flood-prone riverbanks to safe and stable areas. Development of flood IEWS through deployment of water level sensors, rain gauge sensors, hydrometric gauges, and cameras that can be radio-linked with the central station can be an effective strategy for damage control. Installation of Doppler radar and lightning sensors would also aid in monitoring. Observations of sediments and water fluxes through the river channel i.e. fluvial dynamics using web-based sensors also play important role in safeguarding the ecology and life of reservoirs and dams, and assessing the water availability and protection from devastating flash floods during the monsoon. Based on geomorphology and all other natural features for a given region, different scenarios of flash floods and damage patterns can be created with different amount of rainfall, which will help provide precautionary and remedial measures, to be undertaken by the local administrative authorities, in such a consequence.

16.8 Snow Avalanches

Repeated snow avalanches were observed in September 2022 in the north of Kedarnath temple town, which is located in the western extremity of the Central Himalaya in Mandakini River valley (Sain et al., 2023b). About 23% of total catchment area of $\sim$67 km^2 up to Rambara is covered by glaciers. From the analysis of aerial data and ground survey, it was established that the snow-avalanches initiated $\sim$6 km uphill of the Kedarnath town from the peaks of Companion glacier in Mahalaya Parvat (5970 masl), adjacent to the Chorabari glacier. The Companion glacier receives accumulation mostly from the snow avalanches. The run-out and deposition zones are located $\sim$5 km uphill of Kedarnath town, thus posing no threat from avalanche hazards to the temple township. The study from slope morphology, grass clearing, snow drifts, and debris cones has identified 15 avalanche sites in the area. The climate-related other snow-avalanche includes the Atalakodi avalanche in Hemkund Sahib in April, 2022 (Sain and Mehta, 2022), which destroyed a section of the trail between Ghangria and Hemkund Sahib. The Atalakodi route is considered a potential area for snow avalanches. The historical data also shows fatalities due to past avalanches in this area. One of these was in 1998 that buried 27 pilgrims, and another was in 2008 that resulted in 8 deaths and 12 injuries. The Atalakodi avalanche affected an area of approximately 4 km long and 150 m width. The steep slopes, structures and lithology have been attributed for the snow avalanches in Kedarnath and Hemkund Sahib areas. The ropeway construction and environment impact assessment study have been recommended towards the mitigation of such disasters. Based on geomorphology and all other natural features for a given region, different scenario of disasters can be created with different amount of snow precipitation, which will provide a precautionary measure, to be undertaken by the local administrative authorities, in such a situation. It is to be mentioned that the Rishiganga-Dhauliganga deluge in February, 2021 was led by heavy snow precipitation over the ice in Raunthi Peak (Cook et al., 2021; Mehta et al., 2021; Sain et al., 2021b; Shugar et al., 2021).

16.9 Forest Fires

The Himalayan region is mostly covered with forests, and forest fire is one of the major disasters in the Himalaya. The dry spell, scanty winter rainfall and reduced snow precipitation have accelerated the spreading of fire (Sain, 2024b). Negligent discard of burning cigarette-buts, matchsticks, garbage-waste or any burning residues, and garbage waste, may be responsible for forest fires. The fire might result from rubbing of branches or from friction in rocks in dry and warm conditions. Lightning also strikes power cables producing sparks and hence forest fire. Around 70% area of the UK Himalaya is forest-cover, ~28% of the forest area has Chir pine trees that have inflammable resin-rich needles, which along with other dry leaves of deciduous trees, weeds for agriculture, grass, wood catch fire by natural processes and/or by anthropogenic activities. Since November 2023 to summer of 2024, there have been ~400 forest fires in the UK state that has affected ~1300 hectares of forest area. Hundreds of forest fires since November 2023 till 2024 summer affecting much more than thousand hectares of forest area in the UK state has necessitated safeguarding against such disaster. The forest fires have caused not only the loss of rare flora and fauna species and biodiversity but are also been accountable for soil erosion and changes in landscapes. The forest fires also impact the economy of many who depend on forests for food, fodder and fuel. Therefore, it is essential to pay attention for the remediation of forest fires.

The forest fires can be alleviated through (i) education or sensitization to local or tribal people for management and efficiently distinguishing the fires; (ii) immediate reporting of fire to local administrators; (iii) implementation of forest fire alert system based on early detection using radio acoustic sound system; (iv) construction of fire lines with variable width built through the leaf litter of a forest floor down to the soils to prevent spread of fire past the line; and (v) arrangement for buying pine needles and other inflammable materials at reasonable prices from locals by stakeholders who can convert these into usable products like (a) biomass pellets and briquettes for cooking and heating purposes or electricity generation, (b) eco-friendly handicraft items, packing materials and home decorating items, and (c) check dams for soil and moisture conservation in an economically-affordable, socially acceptable and environmentally safe manner.

16.10 Conclusions and Way Forward

The Himalaya is stressed under the influence of CC resulting in the melting of glaciers and snow fields impacting water availability and agricultural practices; altering weather patterns; enlarging geological and hydrological disasters; and disrupting to ecosystems. All these pose immense threats to the region's communities. To address these complex issues necessitates a multi-dimensional approach, comprehending hill-specific challenges and technological barriers for sustainability, integration of scientific understanding with local knowledge, strengthening adaptation and mitigation efforts including the NbS, engagement of native communities, and promoting local and global partnership to build resilience and warrant sustainable future for the people living in the Himalaya.

As the prime reason of environmental hazards in the Himalaya is the land degradation under the influences of CC and additional stresses built by the anthropogenic activities, we need to understand these phenomena in a comprehensive manner. In that context, first of all we have to identify the most vulnerable zones based on geoscientific investigation. Secondly, a network of web-based sensors is to be established at the hotspots for transmission of data from field in enabling long-term monitoring through analysis and assimilation of data using AI/ML or IoT. The next step is to develop an IEWS for alert and to build a response system to safeguard the people against the disasters. Once the warning and alert system is fully operational at a basin or in a catchment, a similar network can be deployed at other hazard-prone basins or areas of the Himalaya. The system would be designed by deploying different kinds of sensors for diverse types of hazards related to landslides, glaciers/glacier lakes related threats, flash floods, and forest fires. It is to be clarified that the investment for development and deployment of IEWS is much more economic than the cost involved in rehabilitation and restructuring, and damages to homes and infrastructure. The loss of invaluable lives cannot be equated in terms of monetary loss.

The Himalaya possesses plethora of resources: glaciers and snow field; springs and riverine system; precious minerals and ores; geothermal, hydrocarbon and hydropower energy resources; medicinal plants, tourist attraction etc. that can be harnessed for the socio-economic growth, agricultural development, and strategic advancement but by following certain guidelines and mountain-specific building codes. There must be a balance between the exploitation of these resources for economy and optimal conservation or non-disturbance of the Himalayan fragile ecosystem. Adaptation to the NbS like (i) plantation, vegetation and restoration of forests for protection of slopes and reduction of soil erosion that can diminish the probability of risks due to landslides; (ii) refurbishment of wetland and degraded riverbanks, and maintenance of natural drainage system to control the floods; (iii) rejuvenation of springs and groundwater recharge areas for mitigation of water scarcity; (iv) implementation of sustainable agroforestry and integrated farming for resilience of ecosystem; and (v) empowering local communities for effective implementation and ensuring long-term success of NbS.

The traditional knowledge like usage of light construction materials; implementation of limited storied buildings and satellite townships instead of having populace around one business town; strict implementation of heavy penalty against unauthorised construction and deforestation; impervious drainage system that will not allow infiltration of kitchen or toilet water haphazardly into the subsurface; building retaining walls, wire-mesh, insertion of bolts and shotcrete for arresting slope failure; slope stability assessment before any construction or hill-cutting in a slope region; stringent codes on muck disposal after rock cutting; non-exposure to highly vulnerable zone; disaster-resilient building or retrofitting old buildings; periodic assessment of load bearing capacity at different scales; environment impact assessment, etc. are some of the other measures, which can be considered for mitigation and way forward. We have to sensitize the locals on how to live with the disasters; inculcate fundamental education on the causes and consequences of disasters; build trained manpower for disaster management—preparedness, response, restoration; draw attention of stakeholders to scopes and opportunities for alleviation of challenges; build climate-adaptable and disaster resilient future; and figure out the sustainable growth. The traditional practices, adaptation capacity for resilience, and invaluable knowledge of local communities must be fostered for firming preparedness mechanisms and addressing inherent vulnerabilities, which can better prepare for and recover from increasing risks of the Himalayan states posed by disasters and climate change. It is mentioned here that the major hindrances to the adaptation and adoption of several strategies and way forward are the dearth of funds, deficiency of knowledge, absence of technology, want of time, and lack of mandatory policy.

Acknowledgement Director, CSIR-NGRI is acknowledged for permission to publish this work. The author acknowledges the ANRF-DST for providing him with the J.C. Bose National Fellowship.

References

Bhambri, R., Mehta, M., Dobhal, D. P., & Gupta, A. K. (2015). *Glacier lake inventory of Uttarakhand* (p. 88). Sp. Publication, Himalayan Geology, Wadia Institute of Himalayan Geology.

Bhambri, R., Sain, K., Chand, P., Srivastava, D., Tiwari, S. K., & Yadav, J. S. (2023). Frontal changes of Gangotri glacier, Garhwal Himalaya between 1935 and 2022. *Journal of the Geological Society of India, 99*, 169–172.

Chauhan, P., Sain, K., Mehta, M., & Singh, S. K. (2022). An investigation of cloudburst triggered landslides and flash floods in Arakot region, Uttarkashi district, Uttarakhand, India. *Journal of the Geological Society of India, 98*, 1685–1690.

Chauhan, P., Singh, O., Sharma, J., Bhardwaj, P., Mehta, M., Shah, R. A., & Sain, K. (2023a). Comparative analysis of discharge and sediment flux from two contiguous Glacierized basins of central Himalaya, India, Env. *Monitoring & Assessment, 195*(729), 1–22.

Chauhan, H. K., Gallacher, D., Bhatt, A., & Bisht, A. K. (2023b). The Himalayas: A climate change laboratory. *Environmental Development, 45*(100814), 1–6.

Chauhan, P., Akiner, M. E., & Sain, K. (2024). Forecast future disasters from hydro-meteorological datasets in the Yamuna river basin, western Himalaya using Markov chain and LSTM approach. *Artificial Intelligence in Geosciences, 5*(100069), 1–23.

Chaulya, S. K., Mishra, P. K., Kumar, N., Kumar, V., & Rawani, V. K. (2024). Landslide monitoring and prediction system using geosensors and wireless sensor network. *Discover Geoscience, 2*(6), 1–18.

Cook, K. L., Rekapalli, R., Dietze, M., Pilz, M., Cesca, S., Purnachandra Rao, N., Srinagesh, D., Paul, H., Metz, M., Mandal, P., Suresh, G., Cotton, F., Tiwari, V. M., & Hovius, N. (2021). Detection and potential early warning of catastrophic flow events with regional seismic networks. *Science, 374*, 87–92.

Das, S., & Mishra, A. J. (2023). Climate change and the Western Himalayan community: Exploring the local perspective through food choices. *Ambio, 52*(3), 534–545.

Dobhal, D. P., Gupta, A. K., Mehta, M., & Khandelwal, D. D. (2013). Kedarnath disaster: Facts and plausible causes. *Current Science, 105*(2), 171–174.

Dhungana, N., Silwal, N., Upadhaya, S., Khadka, C., Regmi, S. K., Joshi, D., & Adhikari, S. (2020). Rural coping and adaptation strategies for climate change by Himalayan communities in Nepal. *Journal of Mountain Science, 17*(6), 1462–1474.

Dimri, A. P., Kumar, D., Choudhary, A., & Maharana, P. (2018). Future changes over the Himalaya: Maximum and minimum temperature. *Global and Planetary Change, 162*, 212–234.

Gao, H., Ouyang, Z., Chen, S., & Van Koppen, C. S. A. (2013). Role of culturally protected forests in biodiversity conservation in Southeast China. *Biodiversity and Conservation, 22*, 531–544.

Gentle, P., Thwaites, R., Race, D., & Alexander, K. (2014). Differential impacts of climate change on communities in the middle hills region of Nepal. *Natural Hazards, 74*, 815–836.

Gopirajan, S. A. T., Kumar, P., & Joshi, P. K. (2022). Unraveling the complex and dynamic Himalayan socio-ecological systems: A systematic review. *Environment, Development and Sustainability, 24*, 1532–1559.

Gupta, V., & Bist, K. S. (2004). The 23 September 2003 Varunavat Parvat landslide in Uttarkashi township, Uttaranchal. *Current Science, 87*(11), 1600–1605.

Gupta, A. K., Negi, M., Nandy, S., Alatalo, J. M., Singh, V., & Pandey, R. (2019). Assessing the vulnerability of socio-environmental systems to climate change along an altitude gradient in the Indian Himalayas. *Ecological Indicators, 106*, 105512.

Gupta, V., Sain, K., & Tandon, R. S. (2022). Landslides and slope instability in Mussoorie and Nainital townships (Uttarakhand) in present climate—Change scenario. In A. S. Unnikrishnan, F. T. Tangang, & R. Durrheim (Eds.), *Understanding & Managing Extreme Natural Events* (pp. 391–411). Challenges for Developing Countries, Springer.

Gupta, V., Ram, P., Tandon, R. S., & Vishwakarma, N. (2023). Efficacy of landslide susceptibility maps prepared using different bivariate methods: Case study from Mussoorie township, Garhwal Himalaya. *Journal of the Geological Society of India, 99*, 370–376.

Jeelani, G., Absar, A., Agnihotri, V., Ahmed, S., Alam, A., Azam, M. F., Bhat, M. S., Deshpande, R. D., Dimri, A. P., Jain, S., Juyal, N., Lone, S. A., Mal, S., Maharana, P., Maurya, A. S., Mukherjee, A., Muddu, S., Pottakkal, J., Romshoo, S. A., Sarin, M. M., Sain, K., Sharma, P., & Shrestha, A. B. (2024). Policy framework to combat the challenges of climate change in the upper Indus Basin. *Current Science, 127*(6), 1–5.

Kaur, R., Gupta, V., & Chaudhary, B. S. (2024). Landslide susceptibility mapping and sensitivity analysis using various machine learning models: A case study of Beas valley, Indian Himalaya. *Bulletin of Engineering Geology and the Environment, 83*(228), 1–23.

Kumar, A., Bhambri, R., Tiwari, S. K., Verma, A., Gupta, A. K., & Kawishwar, P. (2019). Evolution of debris flow and moraine failure in the Gangotri glacier region. *Garhwal Himalaya: Hydrogeomorphological aspects, Geomorphology, 333*, 152–166.

Kumar, A., Verma, A., & Sain, K. (2022). Decadal response of Dokriani glacier using high-resolution hydrological data, Indian Himalaya. *Journal of the Geological Society of India, 98*, 62–68.

Kumar, A., Verma, A., Tiwari, S. K., Bhambri, R., Sain, K., Rai, S. K., & Patidar, P. (2023). Heterogeneity in glacio-hydrological processes and estimation of different components in streamflow from central Himalayan glaciers. *Journal of Hydrology: Regional Studies, 49.*:101495, 1–25.

Lutz, A. F., Immerzeel, W. W., Kraaijenbrink, P. D., Shrestha, A. B., & Bierkens, M. F. (2016). Climate change impacts on the upper Indus hydrology: Sources, shifts and extremes. *PLoS One, 11*, e0165630.

Martha, T. R., & Kumar, V. K. (2013). September, 2012 landslide events in Okhimath, India - an assessment of landslide consequences using very high resolution satellite data. *Landslides, 10*, 469–479.

Martha, T. R., Roy, P., Govindharaj, K. B., Kumar, K. V., Diwakar, P. G., & Dadhwal, V. K. (2015). Landslides triggered by the June 2013 extreme rainfall event in parts of Uttarakhand state, India. *Landslides, 12*(1), 135–146.

Matta, G., Kumar, A., Tomar, D. S., & Kumar, R. (2025). Climate change in the Himalayan region: Susceptible impacts on environment and human settlements. *Frontiers in Environmental Science, 13.*:1550843, 1–7.

Mehta, M., Kumar, V., Sain, K., Tiwari, S. K., Kumar, A., & Verma, A. (2021). Causes and consequences of rishi ganga flash flood, Nanda Devi Biospheere reserve, central Himalaya, India. *Current Science, 121*, 1483–1487.

Patidar, P., Kumar, A., Reza, A., Sain, K., Haritshaya, U. K., Kumar, N., Mishra, A., Verma, A., & Anand, S. (2025). *Evolution of glacial lake in the Bhilangana river basin Garhwal Himalaya* (Vol. 39.: e70308, pp. 1–20).

Paul, S. K., Bartarya, S. K., Rautela, P., & Mahajan, A. K. (2000). Catastrophic mass movement of 1998 monsoons at Malpa in Kali Valley, Kumaun Himalaya (India). *Geomorphology, 35*, 169–180.

Ravindra, R., Kulkarni, A. V., Dimri, A. P., Sain, K., Sharma, M. C., Banerjee, A., Sharma, P., Meloth, T., Rashid, I., & Pant, N. C. (2024). Recent Indian studies in Himalayan cryosphere. *Proceedings of the Indian National Science Academy, 90*, 415–425.

Sabin, T. P., Krishnan, R., Vellore, R., Priya, P., Borgaonkar, H. P., Singh, B. B., & Sagar, A. (2020). Climate change over the Himalaya. In R. Krishnan, J. Sanjay, C. Gnanaseelan, M. Mujumdar, & A. Kulkarni (Eds.), *Assessment of climate change over the Indian region—A report of the Ministry of Earth Sciences, Govt. of India* (pp. 207–223). Chakraborty, Springer Open.

Sain, K. (2019). State-of-the-art seismic tools for subsurface imaging, editorial. *Journal of the Geological Society of India, 94*, 337–340.

Sain, K., Dobhal, D. P., Gupta, V., & Sharma, R. (2019). Geo-resources and impact assessment from geological studies, pp.1-27. In A. Gupta (Ed.), *Status of ecosystem health in the Indian Himalaya region, a report under National Mission for sustaining the Himalayan ecosystem* (p. 172). Dept. of Sci. & Tech.

Sain, K., Kumar, A., Mehta, M., Verma, A., Tiwari, S. K., Garg, P. K., Kumar, V., Rai, S. K., Srivastava, P., & Sen, K. (2021a). Perspective on Rishiganga-Dhauliganga flash flood in Nanda Devi biosphere reserve, Garwal Himalaya, India. *Journal of the Geological Society of India, 97*, 335–338.

Sain, K., Kumar, A., & Verma, A. (2021b). Glaciers in Indian Himalaya and major concerns, K. S. Valdiya commemorative volume. *Journal of the Indian Geological Congress, 12*(2) 13(1&2), 39–44.

Sain, K. (2022). Need for development of AI-based integrated warning system (IWS) for mitigation of glaciers/glacial-lakes related hazards with special reference to UK Himalaya. *Journal of the Geological Society of India, 98*, 1012–1014.
Sain, K., & Kumar, P. C. (2022). *Meta-attributes and artificial networking: A new tool for seismic interpretation* (p. 262). AGU & Wiley, Sp. Publication Series.
Sain, K., & Mehta, M. (2022). Atalakodi route of Hemkund sahib: A potential area of snow avalanche. *Journal of the Geological Society of India, 98*, 863–864.
Sain, K. (2023a). Climate change and fossil fuels: Impacts, challenges and plausible mitigation. *Journal of the Geological Society of India, 99*, 454–458.
Sain, K. (2023b). Geo-hazards and remedial measures: Some observations in the light of recent developments at Joshimath. *Guest Editorial, Current Science, 124*(6), 659–660.
Sain, K., & Damodara, N. (2023). *Active seismic tomography: Theory and applications* (p. 124). Wiley.
Sain, K., Mehta, M., Kumar, V., Gupta, V., & Chauhan, P. (2023a). A climatic surprise—Slope instability processes triggered by heavy rain in Maldevta region, Dehradun, UK state of Uttarakhand, India, on 20 august, 2022. *Journal of the Geological Society of India, 99*, 317–320.
Sain, K., Mehta, M., & Kumar, V. (2023b). Avalanche hazards in Kedarnath temple town, Mandakini river valley, UK state of India—A case study. *Journal of the Geological Society of India, 99*, 173–176.
Sain, K. (2024a). Georesources and Geohazards in the Himalaya: A way forward for economy and ecology. *Current Science, 126*(10), 1191–1192.
Sain, K. (2024b). Glacial hazards and forest fires in Uttarakhand Himalaya, and their plausible mitigation. *Current Science, 127*(4), 383–384.
Sain, K. (2025). Himalaya and its bearings—Progress and sustainability: A synoptic view. *Himalayan Geology, 46*(2), 245–247.
Sain, K., Chauhan, P., Bisht, P., & Vaideswaran, S. C. (2025). Cloudburst-induced debris flows and flash floods at Lasko valley in Pithoragarh district, India-Nepal border. *Himalayan Geology, 46*(1), 1–10.
Shugar, D. H., Jacquemart, M., Shean, D., Bhushan, S., Upadhyay, K., Sattar, A., Schwanghart, W., McBride, Kumar, A., Sain, K., et al. (2021). A massive rock and ice avalanche caused the 2021 disaster at Chamoli, Indian Himalaya. *Science, 373*, 300–306.
Tiwari, A., Sain, K., Kumar, A., Tiwari, J., Paul, A., Kumar, N., Haldar, C., Kumar, S., & Pandey, C. (2022). Potential seismic precursors and surficial dynamics of a deadly Himalayan disaster: An early warning approach. *Scientific Reports, 12*(3733), 1–13.
Tiwari, A., Sain, K., Kumar, N., Paul, A., Kumar, A., & Shukla, V. (2024). Seismic and radon signatures: A multi-disciplinary approach to monitor surface dynamics of a hazardous 2021 rock-ice avalanche, Chamoli Himalaya. *Earth Surface Processes & Landforms, 49*, 2965–2979.
Vaideswaran, S., Rawat, G., Kumar, N., Gautam, P.K.R., Yadav, D., Hazarika, D., Kumar Parveen, Anwar, T., Mishra, R.L., & Bose, S. (2024). Study of ground subsidence in Joshimath following the events of January 2023, WIHG-DST report (unpublished) submitted to USDMA, Dehradun.
Verma, S. (2023). Climate change in Himalayas: Impacts on lives and livelihoods, Internat. *Journal of Science and Research Archive, 09*(01), 419–425.
Wang, L., Young, S. S., Wang, W., Ren, G., Xiao, W., Long, Y., Li, J., & Zhu, J. (2016). Conservation priorities of forest ecosystems with evaluations of connectivity and future threats: Implications in the eastern Himalaya of China. *Biological Conservation, 195*, 128–135.
Westoby, M. J., Dunning, S. A., Carrivick, J. L., Coulthard, T. J., Sain, K., Kumar, A., Berthier, E., Haritashya, U. K., Shean, D. E., Azam, M. F., Koppes, M., McCourt, H. R., Upadhyay, K., & Shugar, D. H. (2023). Rapid fluvial remobilization of sediments deposited by a complex Himalayan mass movement process chain. *Geology, 51*(10), 924–928.
Zebre, M., Colucci, R. R., Giorgi, F., Glasser, N. F., Racoviteanu, A. E., & Del Gobbo, C. (2021). 200 years of equilibrium-line altitude variability across the European Alps (1901–2100). *Climate Dynamics, 56*, 1183–1201.

Chapter 17
Risk Assessment and Resilience to Buildings and Infrastructure in Indian Himalayan Region

R. Pradeep Kumar, Ajay P. Chourasia, and D. P. Kanungo

Abstract The Indian Himalayan Region (IHR), one of the world's most fragile and hazard-prone mountain systems, faces increasing risks from earthquakes, landslides, debris flows, glacial lake outburst floods (GLOFs) and cascading disasters due to active tectonics, climate change, rapid urbanization and unplanned infrastructure growth (Bilham et al., 2001; NDMA, 2019; IPCC, 2023). These interacting hazards pose significant threats to settlements, lifeline systems and critical infrastructure. This paper presents an integrated framework for risk assessment and resilience enhancement of buildings and infrastructure in the IHR. The study combines district-level risk prioritization, urban-scale microzonation and site-specific performance-based assessment of critical facilities using probabilistic and deterministic hazard approaches, vulnerability analysis, exposure characterization and multi-hazard loss estimation. The paper highlights the importance of advanced instrumentation systems, structural health monitoring, techno-legal mechanisms, retrofit strategies, resilience financing and emergency response systems for strengthening disaster resilience. It emphasizes that sustainable resilience in the Himalaya requires convergence of hazard science, engineering, governance and community preparedness rather than isolated interventions. The study concludes that adoption of integrated multi-hazard risk assessment and resilience-oriented infrastructure planning is essential for safeguarding lives, ensuring continuity of critical infrastructure and enabling sustainable mountain development in the Indian Himalayan Region.

Keywords Indian Himalayan region · Multi-hazard risk assessment · Seismic resilience · Disaster risk reduction · Performance-based earthquake engineering · Infrastructure and building vulnerability

R. P. Kumar · A. P. Chourasia (✉) · D. P. Kanungo
CSIR–Central Building Research Institute (CSIR-CBRI), Roorkee, India
e-mail: ajayc.cbri@csir.res.in

R. P. Kumar
e-mail: ramancharla.cbri@csir.res.in

D. P. Kanungo
e-mail: dpkanungo.cbri@csir.res.in

V. M. Tiwari (ed.), *Himalayan Natural Hazards*, Indian National Science Academy,
https://doi.org/10.1007/978-981-92-3784-5_17

17.1 Introduction

The Indian Himalayan Region (IHR), stretching about 2500 km long and 250–400 km wide, spanning across 13 Indian states/UTs and shares international boundaries with 4 neighbouring countries, occupying 18% of Indian geographical area. The region is young fold mountains which are tectonically active. The region is often called the 'Water Tower of the Earth' which sustains over 1.4 billion people through its glacier-fed rivers (IPCC, 2023). IHR faces challenges from unsustainable development, climate-change-induced glacier retreat, exemplified by frequent disasters such as earthquakes, landslides, floods, subsidence, glacial lake outburst floods (GLOFs) and avalanche events, posing risk to lives and infrastructure. The overarching goal of disaster risk reduction efforts in IHR is to minimize life loss, reduce downtime of essential services and mitigate long-term social and economic losses (NDMA, 2007, 2019; UNDRR, 2015).

In the fragile mountain ecosystems of IHR, the resilience of buildings, social infrastructure, critical facilities and specialized lifelines such as roads, bridges, ropeways and retaining structures is essential to reduce disaster vulnerability (Bruneau et al., 2003). Enhancing the robustness of these assets against multi-hazard scenario, while systematically addressing their exposure and risk, is fundamental to safeguarding communities and advancing sustainable development. This chapter highlights the need for comprehensive risk assessments and explores pathways to strengthen resilience in a holistic and context-specific manner.

17.2 Hazard Landscape

17.2.1 Seismic

About 61% of India's land area, home to more than 75% of its population, is exposed to moderate to severe levels of earthquake shaking (Raghukanth et al., 2024). In contrast, the entire Indian Himalayan Region (IHR) lies in the most hazardous seismic zones (Zone V and Zone VI), with design earthquake zone factors of 0.333 and 0.5 g, respectively, corresponding to a 475-year return period (Figs. 17.1 and 17.2). This makes earthquake-resistant design and construction not only essential but mandatory in the IHR (BIS, 2025; Jain et al., 2010).

Globally, earthquakes have accounted for more than 40% of disaster-related deaths over the past century (Peek-Asa et al., 2003). It is noteworthy that while the overall frequency of disasters is similar across developed and developing nations, nearly 90% of disaster-related deaths occur in developing countries. For countries like India, this underscores the urgent need to strengthen disaster risk management frameworks to progressively reduce risks and safeguard lives (UNDRR, 2015).

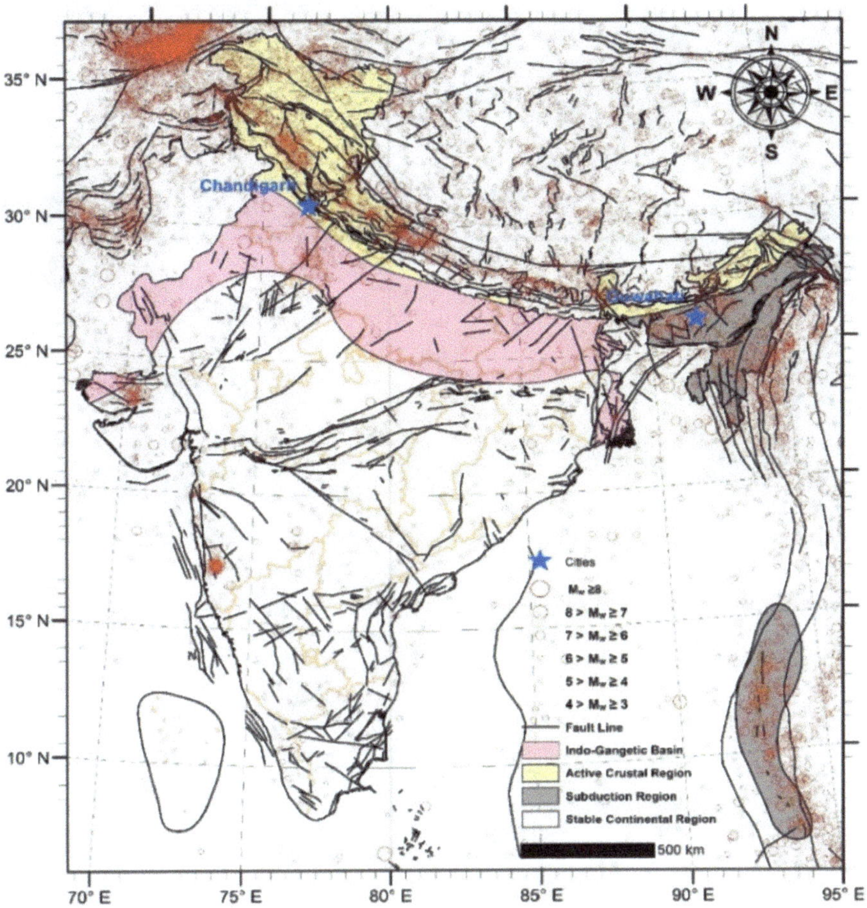

Fig. 17.1 Map of India showing principal tectonic faults and epicentres of past earthquakes considered in earthquake hazard assessment

Approximately 12% of India's total building stock is concentrated in the IHR, demanding focused attention to mitigate the impending earthquake risk. The vulnerability of these buildings is compounded by several factors: most existing structures were constructed before rigorous earthquake-resistant provisions were available; earthquake engineering is not uniformly integrated into undergraduate engineering curricula; untested or non-engineered construction technologies have been widely adopted; maintenance practices remain poor or absent; and municipal bodies lack a robust techno-legal regime to enforce compliance. Addressing these systemic gaps is critical to enhancing seismic resilience in the IHR and, by extension, reducing disaster risk for millions of people dependent on this fragile mountain ecosystem.

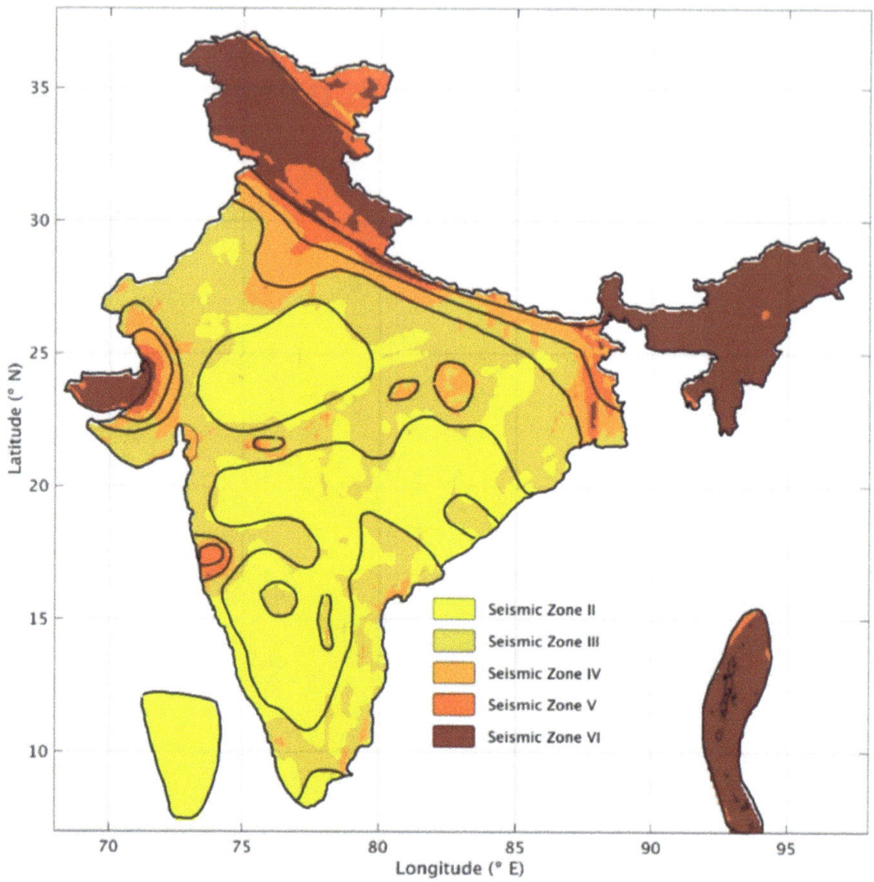

Fig. 17.2 Map of India showing macro-zoning of earthquake hazard as per Draft IS 1893 (Part 1): 2025

17.2.2 Mass Movements

Mountain regions in India, particularly the Himalaya, represent some of the most high-risk environments for large populations of indigenous and economically disadvantaged rural communities. In recent decades, these fragile ecosystems have experienced rapid urbanization and rising population density, which in turn has placed enormous pressure on land resources through the expansion of housing and infrastructure (IPCC, 2023). Unscientific construction practices, mining and quarrying and large-scale infrastructure projects such as roads, dams, reservoirs, telecommunications networks and tourism facilities have often exceeded the natural carrying capacity of these terrains. Such unsustainable development has accelerated environmental degradation, heightening the incidence of natural hazards including landslides, avalanches, glacial lake outburst floods (GLOFs), building collapses, flash floods and cloudbursts. Regrettably, mountain-specific planning and development have received limited attention in policy and governance frameworks.

The Indian Himalayan Region (IHR) is especially vulnerable to landslides and mass movements due to a combination of factors: active seismo-tectonics, rugged topography, fragile geology, prolonged and intense rainfall and human interventions. The dominance of tectonically unstable and geologically young formations, coupled with frequent seismic activity, further amplifies this vulnerability. Among the primary triggers, monsoon rainfall and recurrent seismo-tectonic disturbances play decisive roles in destabilizing slopes. As shown in Fig. 17.3, approximately 13.17% of India's geographical area is susceptible to landslides, with nearly 60% of the IHR falling within the 'very highly susceptible' category, underscoring the urgent need for region-specific strategies to mitigate these cascading risks (GSI, 2022; NDMA, 2019).

17.2.3 Hydroclimatic

Hydroclimatic hazards in the Indian Himalayan Region (IHR) encompass a range of destructive mass-wasting processes, including debris flows, avalanches, glacial lake outburst floods (GLOFs) and landslide lake outburst floods (LLOFs). Among these, debris flows are particularly common in mountainous terrains worldwide and are increasingly prevalent across the Himalaya. Despite their frequency, debris flows remain under-researched in India, even though they often disrupt transportation corridors, block rivers through the formation of temporary landslide dams and cause recurrent damage to lives and property. In many instances, these events escalate into full-scale disasters with devastating socio-economic consequences. With climate change driving shifts in rainfall intensity, snowmelt patterns and glacial dynamics, hydroclimatic hazards are becoming more frequent across almost all Indian Himalayan states (IPCC, 2023). A systematic approach to hazard–vulnerability–risk assessment for debris flows and other hydroclimatic events is urgently required to strengthen disaster risk management and reduce cascading impacts on communities and infrastructure (Fig. 17.4).

17.2.4 Compounding Hazard

Compounding or cascading hazards generally occur when multiple risk events unfold in close succession, amplifying their cumulative impact. These disasters are typically triggered by an initial shock, which sequentially sets off additional hazards. For example, an earthquake may initiate a landslide, which then transforms into a debris flow and ultimately a flash flood. The IHR has witnessed such complex hazard chains with increasing frequency, especially under shifting climatic conditions. A notable case is the 2021 Chamoli disaster, which began with a rock fall that transitioned into a rock–ice avalanche, evolved into a debris flow and culminated in a destructive

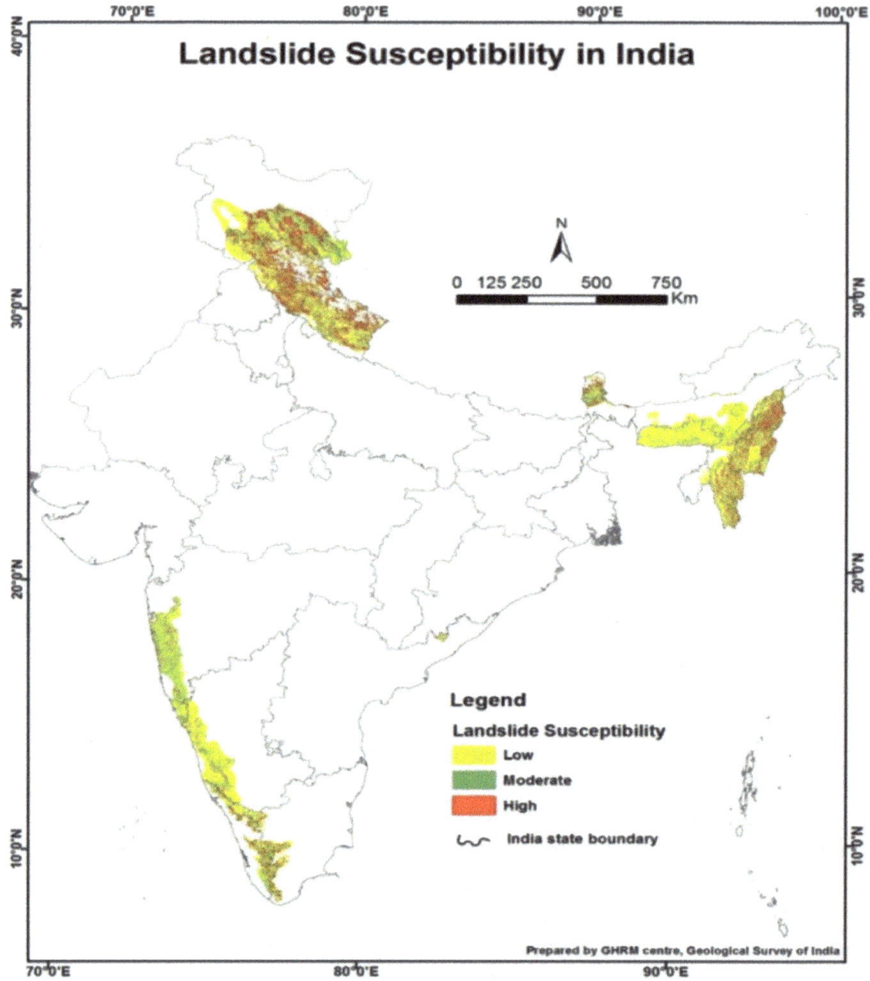

Fig. 17.3 Map of India showing landslide susceptibility (GSI, 2022)

flash flood over several tens of kilometres, devastating downstream communities and hydropower projects. Such events highlight the intensified susceptibility of high-altitude Himalayan environments to multi-hazard interactions, particularly during the monsoon season. These cascading phenomena demand an integrated hazard framework that accounts for interdependencies and feedbacks among geological, hydrological and climatic drivers (Bruneau et al., 2003; UNDRR, 2015).

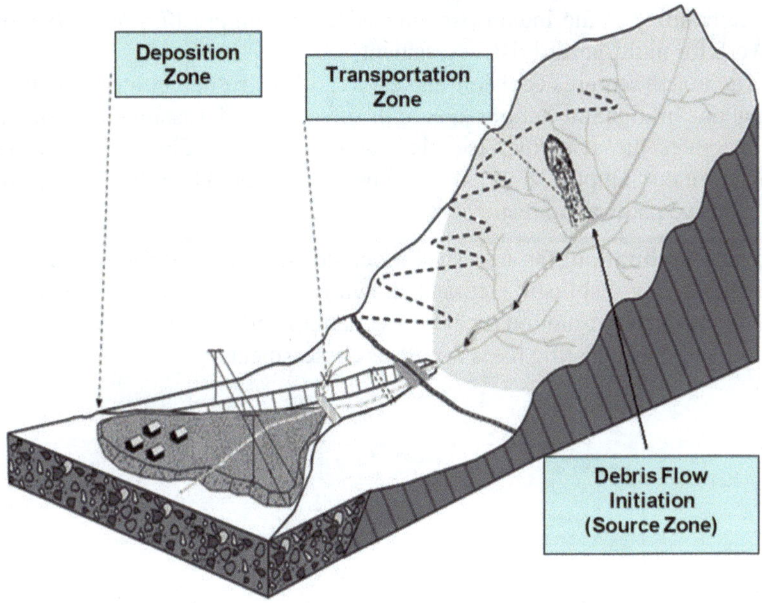

Fig. 17.4 Typical debris flow (hydroclimatic hazard) nomenclature map

17.3 Risk Assessment

Against this backdrop of hydroclimatic and compounding/cascading hazards, risk assessment emerges as the central tool for informed disaster risk reduction (DRR). Risk assessment provides the analytical basis to design strategies ranging from regional development planning to site-specific engineering interventions. It enables the identification of zones with differential levels of hazard exposure, thereby guiding land-use policies, developmental priorities and the formulation of building codes, guidelines and techno-legal regimes. At its core, risk (R) is conceptualized as a function of hazard (H), vulnerability (V) and exposure (E) (Cornell, 1968). While hazard characterizations are increasingly sophisticated, vulnerability and exposure analyses require urgent attention, given rapid urbanization and expanding infrastructure in fragile Himalayan terrain.

At IHR, high strain accumulation, segmented active faults, steep relief and heterogeneous surface conditions converge with unplanned urban growth, making risks deeply complex and spatially variable. A meaningful risk assessment for the region must therefore be multi-scalar (ranging from district-wide evaluations to city/town and asset-level analyses), multi-hazard informed (incorporating ground shaking, liquefaction, landslides, avalanches, GLOFs and cascading chains) and decision-oriented. The framework establishes links between individual Hazard processes such as debris flow, to cascading hazard interactions and ultimately to integrated risk assessment. This calls for the urgent need for a holistic, context-specific approach to resilience building in the Indian Himalayan Region. Figure 17.5 presents a typical framework for multi-hazard risk assessment.

This approach requires decision-makers to allocate resources strategically while ensuring that mitigation actions are both effective and sustainable. The framework integrates the scientific, technical and governance dimensions of seismic risk management, emphasizing five interlinked aspects that collectively provide a comprehensive pathway to resilience:

1. **Hazard Estimation**: Ground shaking could be quantified through both deterministic and probabilistic methods, drawing on data from active faults, seismic catalogues, ground-motion prediction equations and site-specific soil conditions. This forms the scientific basis for all risk assessments.

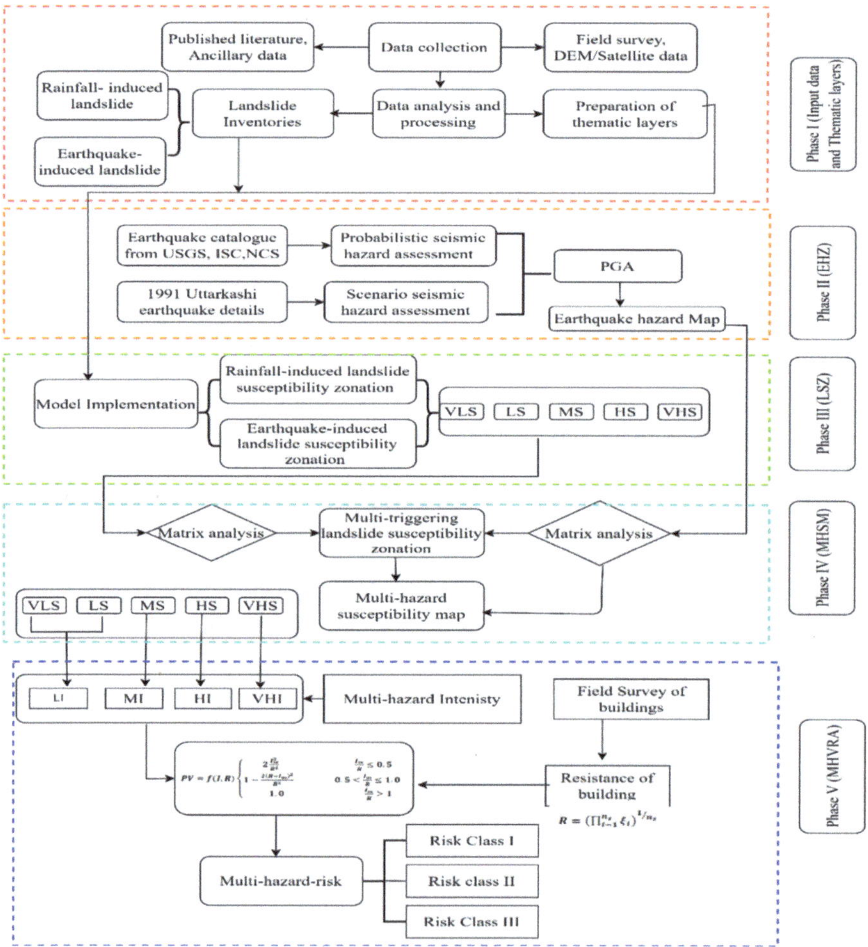

Fig. 17.5 Typical framework for multi-hazard risk assessment

2. **Vulnerability Estimation**: The fragility of the built environment is assessed at three levels, ranging from simplified typology-based approaches for broad overviews to detailed quantitative analyses for critical structures. These assessments capture the diverse performance of building stocks and infrastructure.
3. **Seismic Exposure Estimation**: Human, property and economic assets are mapped using census data, household surveys and extrapolations, providing an inventory of what is at stake. Exposure estimation ensures that risk assessments reflect the scale and distribution of potential losses.
4. **Seismic Risk Estimation**: By integrating hazard, vulnerability and exposure, multi-level risk assessments are developed to quantify expected human, property, financial and business impacts. This step transforms technical assessments into actionable insights for decision-makers.
5. **Risk Communication**: The final and critical stage is the effective dissemination of risk information, tailored to policymakers, institutions and the general public. Communicating risk in accessible yet authoritative formats is essential to mandate, encourage, and enable risk reduction at every level of society.

Above five aspects establish a coherent chain that links scientific knowledge to governance outcomes.

17.3.1 Macro Scale–District Level

At the district level, risk assessment functions as a screening-level tool for prioritization, guiding decisions on where resources should be directed first, which lifelines and assets demand immediate protection and sequencing interventions across administrative boundaries. The analytical approach could be by using probabilistic seismic hazard assessment (PSHA), which incorporates regionally consistent source models that account for both mapped faults and diffuse seismicity (Cornell, 1968; Sreejaya et al., 2022). Ground-motion prediction equations (GMPEs), calibrated for shallow crustal earthquakes of the Himalaya, are combined with site-class information derived from geology, topographic slope and Vs30 maps (Boore & Joyner, 1982; Campbell & Bozorgnia, 2008; Douglas, 2003). Collectively, these inputs generate robust estimates of peak ground acceleration (PGA) and spectral ordinates for return periods of 475 and 2475 years (Wald et al., 1999). To complement this probabilistic approach, deterministic scenario earthquakes in the magnitude range of M7–M8 along major Himalayan fault segments are employed to stress-test logistics, evaluate the resilience of emergency response systems and support the planning of resource pre-positioning strategies.

Parallel to seismic risk, landslide risk assessment at the district level relies on landslide susceptibility zonation (LSZ) mapping. This is achieved by integrating multiple causative factor layers—including slope, lithology, hydrology and land use—with a historic landslide inventory database. The result is the delineation of landslide hazard zones ranging from low to high risk. When superimposed with seismic hazard maps, these outputs generate district-level multi-hazard maps that reflect the complex and compounding hazard environment of the IHR.

A critical dimension of district-scale analysis is exposure modelling, which integrates census-based datasets on population and housing with spatial databases of critical facilities such as hospitals, schools and emergency operations centres (EOCs), along with lifelines including highways, bridges, power networks and water supply systems. High-resolution remote sensing products and cadastral overlays further refine building stock characterization, enabling attribution by typology, height class and plan irregularity.

These hazard and exposure layers are synthesized into composite district risk indices, which not only reflect the concentration of assets and intensity of hazards but also incorporate key modifiers such as structural vulnerability, topographic amplification and access constraints (e.g., dependence on single-route corridors or critical bridge crossings). The resulting outputs take the form of district-scale heat maps, which directly inform the revision of District Disaster Management Plans (DDMPs), guide the preparation of pre-disaster recovery frameworks and underpin risk-sensitive land-use planning.

Importantly, this integrated approach yields prioritized lists of lifeline assets such as bridges, health facilities and water systems that require phased retrofitting and strengthening. By systematically reducing systemic vulnerabilities, district-level assessments not only enhance immediate disaster preparedness but also establish the base for subsequent city-level microzonation and site-specific asset-level evaluations, thereby ensuring a coherent, nested framework of multi-hazard risk reduction across scales.

17.3.2 Micro Scale–City/Town Level

At the town level, risk assessment advances from broad prioritization towards operational planning and regulatory integration, to incorporate likely hazard considerations into microzonation studies, development control rules and emergency response frameworks. Seismic microzonation serves as the scientific basis, integrating geotechnical logging, MASW and SPAC surveys, ambient vibration (H/V) analyses and shear-wave velocity profiling to generate finely resolved maps of spatial amplification (Sánchez-Sesma et al., 2011; Gupta et al., 2023). These maps highlight zones of elevated shaking and are essential inputs for municipal-level codes and land-use planning.

Beyond seismic considerations, dynamic slope instability at the city/town level is evaluated by combining subsurface material variation, geotechnical characteristics, shear-wave velocity profiles and hydro-geological parameters. For hydroclimatic and compounding hazards such as debris flows, flash floods and landslide lake outburst floods (LLOFs), the focus shifts towards the ability to quantify and forecast events, monitor and measure precursors and mitigate their cascading impacts.

Key steps include (a) hazard mapping along natural drainage channels and tributaries within inhabited valleys and (b) scenario-based modelling of potential release zones, including entrainment and phase transitions of mass-movement processes, to estimate runout patterns and downstream impacts. These assessments collectively delineate zones of amplified seismic shaking, liquefaction susceptibility, lateral spreading potential and slope instability, while superimposed fault setbacks capture the compound hazard profile that is characteristic of Himalayan towns. Further, a critical component is the development of ward-level building inventories, systematically mapping the spatial distribution of prevalent typologies such as RC frames (with and without masonry infills), confined and unconfined masonry, timber-laced systems (Dhajji-dewari, Taaq) and random rubble stone in mud mortar. Attributes that critically influence vulnerability, including soft-storey configurations, short-column effects, torsional irregularities and vertical discontinuities, are captured within GIS frameworks. Scenario-based loss modelling builds on these datasets, applying stochastic or physics-based simulations of ground motion to derive fragility-based damage states (Miranda & Taghavi, 2005). The outputs extend beyond structural damage to include estimates of shelter demand, evacuation capacities, accessibility under debris or landslide blockages, hospital surge capacity, timelines for utility restoration and smooth accessibility to critical facilities in a post-disaster scenario. Figure 17.6 shows a typical multi-hazard (rain/landslide, earthquake and combination thereof) map for Kandaghat town in Himachal Pradesh. While, Fig. 17.7 presents a typical risk map of Kandaghat indicating the proposed location of Resilient Centres, as onc of thc mitigation stratcgics.

Ultimately, the ward-level outputs covering damage states, functional downtime and shelter demand directly feed into local retrofitting programmes, school safety initiatives and evacuation route planning with redundancy built-in are important from risk mitigation considerations. By bridging technical microzonation studies with regulatory action and community preparedness, city/town-level assessments provide a vital operational link between broad district-scale prioritization and site-specific, asset-level risk management.

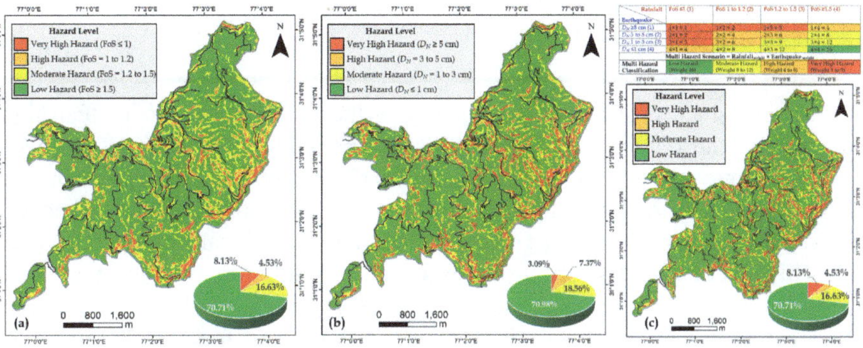

Fig. 17.6 Hazard map due to **a** rainfall/landslide **b** earthquake scenario based **c** multi-hazard analysis for Kandaghat, Himachal Pradesh, India

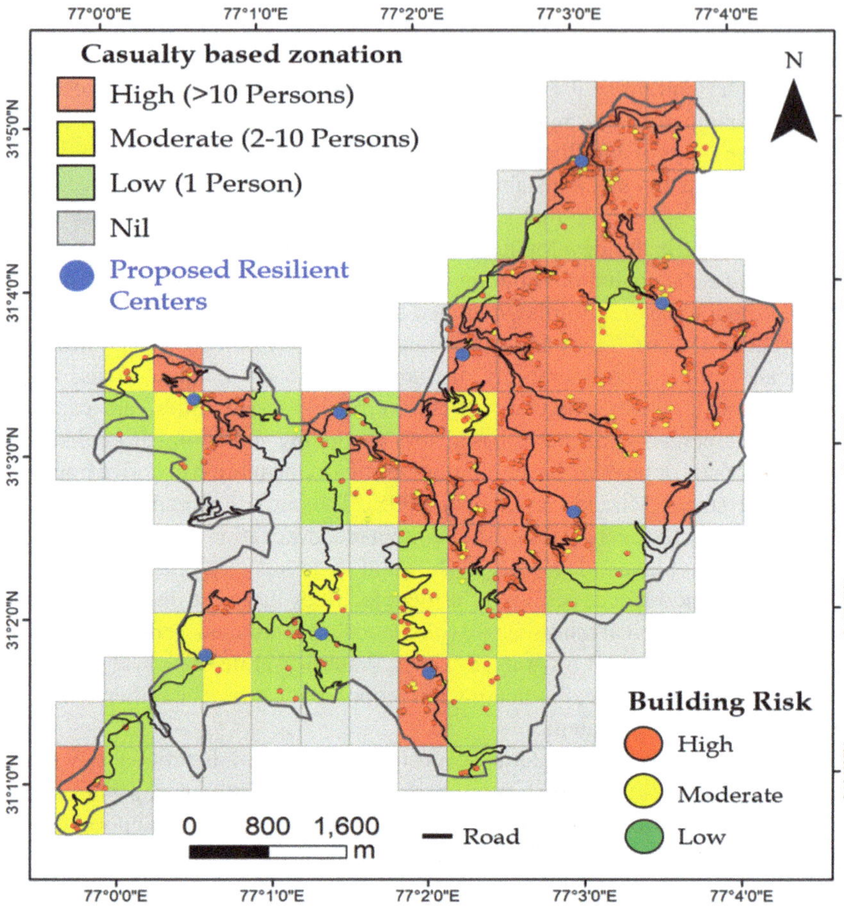

Fig. 17.7 Proposed disaster-resilient centres for the Kandaghat, HP (The dots in map indicate location and vulnerability class of buildings)

17.3.3 Site Specific–Asset Level

The risk assessment of site-specific asset aims on the performance of individual, high-value assets whose functional continuity is indispensable for disaster response and socio-economic stability. Such critical facilities viz. hospitals, emergency operations centres (EOCs), hydropower plants, large dams, bridges, tunnels, fuel depots and data centres, are evaluated from a structural safety and operational resilience perspective. This tier of analysis scrutinizes to the finest resolution, ensuring that assets embedded within broader hazard landscapes are engineered and managed to meet stringent performance objectives. Performance-based design frameworks guide such acceptance criteria (Priestley et al., 2007). Immediate occupancy levels are mandated for hospitals, EOCs and other first-response nodes, while Life Safety or Collapse

Prevention thresholds may be acceptable for transport and utility assets. Functional indicators, residual drift limits, anchorage of non-structural systems, redundancy of utility connections and post-event accessibility are integrated into the evaluation to ensure that survivability is coupled with usability.

Seismic input motions are customized through hazard disaggregation and conditional mean spectra, providing site-specific ground-motion estimates aligned with the fundamental periods of each asset. Carefully selected and spectrally matched time histories are employed in nonlinear response-history analyses, facilitating explicit evaluation of both structural and non-structural performance. Parallel geotechnical characterization through in-situ penetration tests, cyclic triaxial or resonant column laboratory testing and advanced methods for underground structures addresses issues such as liquefaction potential, cyclic mobility and dynamic response of jointed rock masses. For linear infrastructure such as pipelines and transport tunnels, specialized assessments of fault-crossing hazards are in focus.

The integration of structural health monitoring (SHM) technologies, including accelerometers, GNSS arrays and distributed fibre-optic sensing, enables continuous performance tracking and rapid condition assessment during and after seismic events. These data streams ensure operational continuity plans, support trigger-action matrices for cordoning and inspection and facilitate near-real-time decisions on asset reuse. The resulting outputs are asset-specific retrofit packages, adoption of advanced technologies such as base isolation and energy dissipation devices and operational strategies that include resilience into day-to-day functioning (Priestley et al., 2007).

17.3.4 Vulnerability and Fragility

Vulnerability is a fundamental component in the evaluation of risk due to various hazards in IHR. The convergence of vulnerability is looked in the form of 'natural science' (technical, i.e. natural and engineering science) and 'social science'. From social scientists' perspective, vulnerability is represented as the set of socio-economic factors that determine population ability to cope with the stress or changes (Cutter et al., 2003). While, hazard vulnerability is described as the degree of loss to a given element or set of such elements at risk resulting from the occurrence of a natural phenomenon of a given magnitude. United Nations Development Programme (UNDP, 2004) classifies the vulnerability into four major categories: physical vulnerability (PV), i.e. losses due to the damage of physical infrastructure and buildings; environmental vulnerability i.e., loss of natural resources such as water, land resource, etc.; social vulnerability, i.e. loss of population and the social structure; and economic vulnerability, i.e. effect on the economic condition.

Physical vulnerability (PV) determines the degree of loss in terms of structural damage at a given intensity, which can be classified as vulnerability curves, vulnerability matrices and vulnerability indicators. Vulnerability matrices are qualitative methods and the development of matrices are often dependent on expert judgement or empirical data. It is composed of classified intensities and stepwise damage levels.

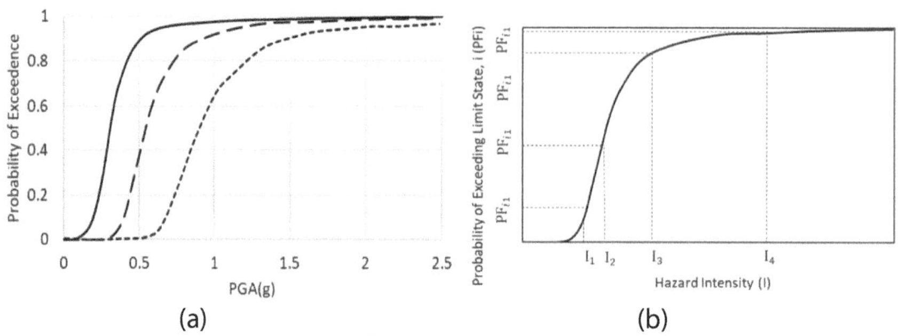

Fig. 17.8 Fragility curve of a building for: **a** single damage state and **b** multiple damage states

While vulnerability curves are usually related to specific building typology damage at different intensity of the hazard, providing quantitative results. Moreover, both the methods neglect the characteristics of the buildings that contribute to their vulnerability. However, the 'vulnerability indicator' is based on the characteristics (structural and non-structural) of the elements at risk. Moreover, indicator-based approach includes the selection of relevant structural and non-structural indicators, assigning weights according to the importance and finally, their sum in terms of a vulnerability index.

Earthquake fragility of a building, an indicator of its expected earthquake performance, is defined as the probability of reaching or exceeding a specific damage level in any element of the building, for a given severity of earthquake shaking (FEMA, 2015; Salvadori & Minnucci, 2017). Today, earthquake shaking is estimated by a probabilistic method because future earthquakes cannot be forecast with certainty. Thus, earthquake fragility of a building is represented realistically, by the probability of earthquake ground motion causing a certain structural damage. For this, a Fragility Curve is prepared (Fig. 17.8), which is a binomial, normal, lognormal or exponential curve with severity of ground motion on the x-axis and probability of exceedance of pre-defined damage state on the y-axis.

Earthquake-induced damage in buildings can be in structural elements or in non-structural elements. Different criteria have been proposed depending on the type of buildings (RCC, Masonry buildings) to define the damage states. The boundary between two damage states is called limit state, which corresponds to the performance that the building is expected to reach after an earthquake. For instance, HAZUS (a hazard description programme of USA) defines the damage states as: no, slight, moderate, extensive and complete damage (Fig. 17.9). Each damage state has a defined qualitative description. This description of damage states is used commonly when deriving Fragility Curves.

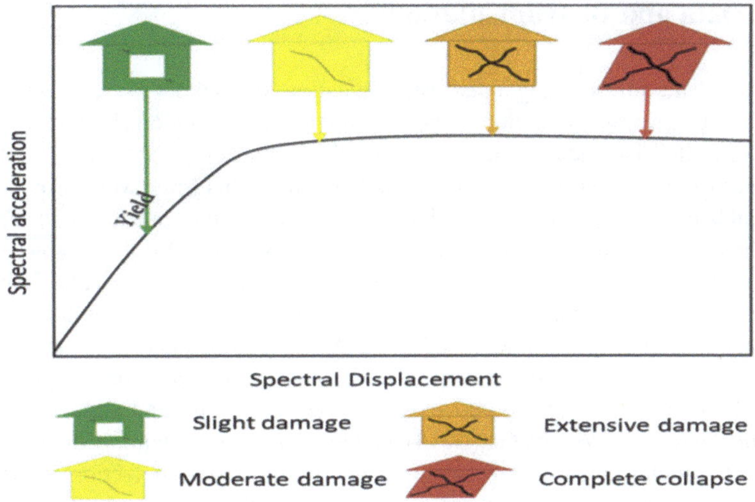

Fig. 17.9 Damage states of a building *Source* HAZUS programme of USA

17.3.5 Loss Assessment

Loss assessment frameworks in the IHR encompass not only direct physical damage but also functional and indirect losses (FEMA, 2015; Miranda & Taghavi, 2005). Direct losses are expressed as repair or replacement costs, disaggregated by building and infrastructure damage states. Functional losses include reductions in hospital bed capacity, bridge throughput and water supply volumes, while indirect losses reflect business interruptions, detour penalties, supply chain disruptions and loss to tourism-dependent economies. Casualty and shelter demand models are explicitly tuned to diurnal and seasonal occupancy, capturing unique hill-town dynamics such as school hours, market congregation and tourist inflows.

Methodologically, nationally tailored tools whether open-source or custom-developed are parameterized with Indian cost libraries, replacement timelines and repair logistics specific to Himalayan conditions. Exposure and fragility modules are explicitly grounded in IHR typologies to maintain contextual validity. Crucially, these risk outputs are linked to financial mechanisms, including contingent credit facilities, parametric insurance covers and pooled risk-sharing instruments. This creates an analytical bridge between technical risk assessments and policy decisions on whether to prioritize retrofitting, incremental strengthening or post-event reconstruction.

17.4 Data and Instrumentation

A resilient and future-ready data-instrumentation framework is fundamental to advancing disaster risk reduction in the Indian Himalayan Region (IHR). Extending the multi-scale risk assessment paradigm, from district prioritization to town-level microzonation and site-specific asset evaluation, this framework provides the empirical backbone for both hazard characterization and real-time decision support.

At the regional scale, the vision of a Himalayan Seismic Observatory is central. This comprises broadband and strong-motion networks deployed at 25–40 km spacing across valleys and corridors, supplemented by rapid-deployment kits for aftershock monitoring. Continuous GNSS arrays complement seismic networks by capturing strain accumulation, interseismic deformation and fault locking processes, while InSAR provides synoptic coverage of line-of-sight ground deformation and coseismic landslides (Wang et al., 2014). Geotechnical arrays, equipped with downhole accelerometers and pore-pressure transducers, monitor liquefaction-prone alluvium, while deep-basin accelerographs help to resolve amplification effects critical for urban settlements.

At the structural scale, instrumentation extends to bridges, hospitals, dams, tunnels, where embedded accelerometers, displacement transducers and fibre-optic sensors are increasingly tied to digital twin models. These enable predictive simulations, near-real-time performance tracking and post-event forensics. Novel data streams from crowdsourced smartphones and UAV-based photogrammetry add further situational granularity, particularly in rapid condition assessments of remote or inaccessible areas.

The data ecosystem is governed through standardized formats, real-time telemetry and automated analytical pipelines, including strong-motion processing, shake map generation and derivation of rapid damage proxies. Event-triggered tasking of satellite constellations and UAV fleets enhances timeliness in post-disaster reconnaissance. Where density and latency permit, Earthquake Early Warning (EEW) systems are envisioned, leveraging near-source arrays to provide precious seconds-to-tens of seconds of lead time for railways, hydropower plants, power grids and hospitals (Wang et al., 2014).

Institutionalization of these capabilities is vital. A 24 × 7 operations cell, staffed by state engineers and scientists trained in both instrumentation and analytics, anchors operational continuity. Long-term sustainability is embedded through systematic O&M contracts, periodic calibration cycles and localization of components to reduce dependence on external supply chains. This ensures that the instrumentation network remains not only technologically robust but also financially and institutionally viable well beyond its initial establishment.

17.5 Mitigation and Retrofit Strategies

17.5.1 System Enablers

Effective mitigation rests on enabling systems of policy, codes, financing and capacity development (Jain et al., 2010; UNDRR, 2015). Enforcement of codes: IS 1893 (seismic design), IS 4326/13827/13828 (masonry and earthen structures), IS 13920 (ductile detailing), IS 13935 (repair/strengthening) and IS 16700 (tall buildings) must be made mandatory, with peer review required for hill-town mid-to-high-rise projects and all critical facilities. Risk-sensitive planning integrates microzonation results into zoning regulations, imposes excavation and slope-cutting limits and enforces drainage and retaining systems, while fault-avoidance zones restrict hazardous siting.

Procurement systems are reoriented to emphasize performance-based specifications and life-cycle cost analysis, while financing mechanisms include retrofit grants, subsidized credit and insurance incentives. Capacity is enhanced through tiered certifications for masons and engineers, standardized hill construction details and e-permitting systems with seismic checkpoints. Community-based programmes such as safer housing pilots, school and hospital safety disclosures and regular citizen drills complement technical measures with public accountability and awareness.

17.5.2 Buildings

For existing masonry buildings, priority retrofits include through-stones, horizontal bands at plinth, lintel and roof levels, corner strengthening, diaphragm anchorage and selective lacing with steel or timber (Basack & Bose, 2016). Replacement of heavy roofs with light materials significantly reduces seismic demand. In unreinforced masonry, confined masonry retrofits represent a practical upgrade pathway.

RC frames require interventions targeting soft-storey vulnerabilities, beam–column joints and column confinement (Mander et al., 1988). Techniques include the addition of steel bracing, RC shear walls, jacketing with RC, steel or FRP and engineered infill management. Non-structural safety measures anchoring parapets, chimneys, MEP systems and hospital equipment are critical for maintaining post-event functionality.

For new builds, hill-appropriate design principles emphasize step-back and step-back-setback configurations, regular and torsion-controlled plans, prefabricated light-weight systems and top-down excavation practices. Rapid, low-cost solutions such as steel/timber banding kits and lintel-band innovations are essential for non-engineered rural housing, particularly when paired with owner outreach and trained local crews.

17.5.3 Bridges and Tunnels

Bridges are among the most critical assets in the IHR, given the reliance on single-route corridors. Seismic isolation bearings, viscous dampers, restrainers and seat length adequacy checks form the first line of retrofit strategies, complemented by abutment strengthening and combined scour seismic checks. Retrofitting priorities are established on the basis of traffic criticality and redundancy.

Tunnels require robust portal stabilization, flexible joints at fault crossings and secondary lining detailing to manage dynamic demands. Provisions for evacuation niches, cross-passages and combined fire–seismic resilience are necessary. Continuous monitoring of overburden slopes and drainage is incorporated into tunnel O&M. At the corridor scale, resilience is enhanced through redundant alignments, stockpiles for rapid replacement of damaged components, pre-contracted emergency works and operational KPIs linked to reopening times.

17.6 Techno-Legal and Techno-Financial Aspects

17.6.1 Targeted Subsidies for Life-Safety Retrofits

Special financial support mechanisms should be directed towards the retrofitting of critical facilities such as schools and hospitals, as well as hill-typology residential houses. Subsidies for essential safety interventions such as seismic bands, roof anchorage and strengthening of non-engineered masonry will directly reduce mortality and injury risks during earthquakes and other hazards. Prioritizing educational and healthcare infrastructure ensures that vulnerable groups remain protected and service continuity is maintained post-disaster.

17.6.2 Resilience Bonds and Parametric Insurance Covers

Innovative financial instruments, including resilience bonds and parametric insurance schemes, should be developed to enable rapid post-disaster recovery (Bruneau et al., 2003). Parametric triggers based on measurable thresholds such as peak ground acceleration (PGA) at seismic stations, rainfall intensity or river stage levels can ensure automatic and timely payouts. These instruments will provide governments and communities with the liquidity required for immediate relief and reconstruction without long delays.

17.6.3 Incentives for Code-Compliant and Retrofitted Buildings

To encourage safer construction practices, premium rebates can be offered for insurance policies covering code-compliant new constructions and certified retrofitted structures. In addition, soft loans through financial institutions can support micro, small, and medium enterprises (MSMEs), particularly guesthouses, homestays, shops and service providers along hill corridors, which are vital for local economies. Such measures will help formalize resilience within both urban and rural construction practices.

17.6.4 Operation and Maintenance (O&M) Annuities

Long-term resilience requires continuous upkeep of critical infrastructure. O&M annuities should be instituted to fund essential activities such as keeping urban drains clear, instrumenting slopes for monitoring and ensuring functionality of emergency routes. By linking financial flows to performance benchmarks, these annuities can ensure that resilience is not a one-time investment but an ongoing practice.

17.7 EOC and Response

17.7.1 Establishment of Emergency Operations Centres (EOCs)

State and district-level EOCs should function as the nerve centres for disaster preparedness, real-time monitoring and coordinated response. These centres must be equipped with GIS-based hazard dashboards, early warning feeds (IMD, CWC, ISRO) and decision-support systems to enable data-driven action. Redundancy in power supply, telecom links and satellite-based backup communication should be built-in to ensure uninterrupted operations during crises.

17.7.2 Integrated Response Mechanisms

EOCs should coordinate with line departments, armed forces, NDRF/SDRF units, health services and community volunteers for rapid mobilization. Pre-designated incident command structures must be activated immediately after an event to ensure clarity of roles and avoid duplication of efforts. Resource inventory management (equipment, shelters, medical supplies, evacuation transport) should be digitized and accessible through the EOC platform.

17.7.3 Community and Local Response Systems

EOCs must be complemented by ward/block-level control rooms to bridge the last-mile response gap. Community Emergency Response Teams (CERTs) should be trained in first aid, light search and rescue and emergency communication, enabling faster lifesaving actions before state machinery arrives. Integration of hill-town-specific systems such as ropeways for evacuation, slope-side escape routes and local shelters should be planned into drills and SOPs.

17.7.4 Drills, Simulations, and Continuous Preparedness

Periodic multi-hazard drills should be conducted involving schools, hospitals, transport agencies and community groups. Simulation-based training and VR-enabled mock drills can be used to strengthen preparedness of responders. Annual review of response time, coordination effectiveness and community participation should feed back into updating contingency plans.

17.8 Conclusions and Way Forward

The seismic risk profile of the Indian Himalayan Region (IHR) is formed by a complex interplay of side effects, the prevalence of vulnerable building typologies and the fragility of access and lifeline networks. Addressing any one of these factors in isolation is inadequate; instead, resilience demands an integrated, multi-scalar approach. A tiered risk assessment framework ranging from district-level screening to city-specific microzonation and building inventories, and finally to asset-level performance-based evaluations, offers actionable resolution that can inform land-use planning, regulatory permitting and operational continuity strategies.

The backbone of such an approach lies in robust data and instrumentation. Dense strong-motion and GNSS networks, coupled with geotechnical arrays and structural health monitoring (SHM), must be integrated into a common analytics platform to enable real-time situational awareness and predictive modelling. Mitigation at scale will only be possible when enforceable building codes are paired with retrofit financing mechanisms, performance-based procurement processes and relentless capacity building of the professional and construction workforce.

The way forward can be envisioned as a phased roadmap. In the *short term (0–2 years)*, the immediate priority is to establish a dedicated Himalayan Seismic Observatory Cell, expand the regional strong-motion network with 300–500 additional stations and publish town-specific design spectra linked to microzonation outcomes. A state-wise School and Hospital Safety Mission should be launched, complete with public scorecards, while hill-town construction checklists are standardized and enforced.

In the *medium term (3–5 years)*, the focus must be on scaling interventions that demonstrate effectiveness. This includes retrofitting the most at-risk 20% of public buildings, piloting Earthquake Early Warning (EEW) systems along two critical transport corridors and within five tertiary hospitals, earthquake alerts/warning to community, accurate earthquake report within 5–10 s. through dense sensor network (Taiwan has sensor density of 30 sqkm area/sensors as against India's 30000 sqkm/sensor) and deploying SHM on at least 100 bridges and 10 tunnels of strategic importance. At the same time, non-structural safety measures should be mandated for all hospitals and public buildings, while performance-based earthquake engineering (PBEE) principles must be embedded into procurement frameworks for essential facilities.

In the *long term (6–10 years)*, the objective should be to institutionalize resilience across the region. By this horizon, all major Himalayan towns should be covered by microzonation maps, performance-based permitting should become standard practice, and redundancy in critical corridors should be strengthened with measurable 'time-to-open' key performance indicators. Financial instruments such as retrofit grants, contingent credit facilities and parametric insurance triggers must be operationalized, explicitly tied to demonstrable risk reduction outcomes.

In closing, the proposition is clear: if India combines science-grade measurement systems with performance-based design, transparent resilience metrics and sustained local capacity, the IHR can make a decisive shift from episodic disaster response to predictable, measurable resilience gains. This transformation would not only safeguard lives and livelihoods but also create an ecosystem for safe and sustainable mountain development in one of the world's most disaster-dynamic regions.

References

Basack, S., & Bose, S. (2016). Seismic vulnerability and strengthening of traditional masonry buildings in Himalayan regions. *International Journal of Disaster Risk Reduction, 19*, 1–12.

Bilham, R., Gaur, V. K., & Molnar, P. (2001). Himalayan seismic hazard. *Science, 293*(5534), 1442–1444.

Boore, D. M., & Joyner, W. B. (1982). The empirical prediction of ground motion. *Bulletin of the Seismological Society of America, 72*(6), S43–S60.

Bruneau, M., Chang, S. E., Eguchi, R. T., Lee, G. C., O'Rourke, T. D., Reinhorn, A. M., Shinozuka, M., Tierney, K., Wallace, W. A., & von Winterfeldt, D. (2003). A framework to quantitatively assess and enhance the seismic resilience of communities. *Earthquake Spectra, 19*(4), 733–752.

Bureau of Indian Standards (BIS). (2025). *Draft IS 1893 (Part 1): 2025–Criteria for earthquake resistant design of structures*. Bureau of Indian Standards, New Delhi.

Campbell, K. W., & Bozorgnia, Y. (2008). NGA ground motion model for the geometric mean horizontal component of PGA, PGV, PGD and 5% damped linear elastic response spectra for periods ranging from 0.01 to 10 s. *Earthquake Spectra, 24*(1), 139–171.

Cornell, C. A. (1968). Engineering seismic risk analysis. *Bulletin of the Seismological Society of America, 58*(5), 1583–1606.

Cutter, S. L., Boruff, B. J., & Shirley, W. L. (2003). Social vulnerability to environmental hazards. *Social Science Quarterly, 84*(2), 242–261.

Douglas, J. (2003). Earthquake ground motion estimation using strong-motion records: A review of equations for the estimation of peak ground acceleration and response spectral ordinates. *Earth-Science Reviews, 61*(1–2), 43–104.

Federal Emergency Management Agency (FEMA). (2015). *HAZUS-MH technical manual*. FEMA.

Geological Survey of India (GSI). (2022). *Landslide susceptibility mapping in India*. Geological Survey of India, Kolkata.

Gupta, I. D., Sreejaya, P., & Sharma, M. L. (2023). Seismic microzonation studies in the Himalayan region: Advances and challenges. *Natural Hazards, 117*(2), 1451–1476.

Intergovernmental Panel on Climate Change (IPCC). (2023). *Climate change 2023: Synthesis report*. IPCC.

Jain, S. K., Mitra, K., Kumar, M., & Shah, M. (2010). A proposed comprehensive model for earthquake safety in India. *Current Science, 98*(10), 1346–1356.

Mander, J. B., Priestley, M. J. N., & Park, R. (1988). Theoretical stress–strain model for confined concrete. *Journal of Structural Engineering (ASCE), 114*(8), 1804–1826.

Miranda, E., & Taghavi, S. (2005). Approximate floor acceleration demands in multistory buildings. I: Formulation. *Journal of Structural Engineering (ASCE)*, **131**(2), 203–211.

National Disaster Management Authority (NDMA). (2007). *National disaster management guidelines: Management of earthquakes*. Government of India, New Delhi.

National Disaster Management Authority (NDMA). (2019). *National disaster management plan*. Government of India, New Delhi.

Peek-Asa, C., Ramirez, M., Seligson, H., & Shoaf, K. (2003). Seismic, structural, and individual factors associated with earthquake related injury. *Injury Prevention, 9*(1), 62–66.

Priestley, M. J. N., Calvi, G. M., & Kowalsky, M. J. (2007). *Displacement-based seismic design of structures*. IUSS Press.

Raghukanth, S. T. G., Iyengar, R. N., & Sharma, M. L. (2024). Updated seismic hazard assessment of India considering recent tectonic and strong motion datasets. *Journal of Earth System Science, 133*(1), 1–18.

Salvadori, L., & Minnucci, G. (2017). Fragility curves and seismic vulnerability assessment of masonry structures. *International Journal of Architectural Heritage, 11*(4), 567–583.

Sánchez-Sesma, F. J., Rodríguez, M., Iturrarán-Viveros, U., Luzón, F., Campillo, M., Margerin, L., García-Jerez, A., Suarez, M., Santoyo, M. A., & Rodríguez-Castellanos, A. (2011). A theory for microtremor H/V spectral ratio: Application for a layered medium. *Geophysical Journal International, 186*(1), 221–225.

Sreejaya, P., Sharma, M. L., & Gupta, I. D. (2022). Probabilistic seismic hazard assessment for the Himalayan region using updated seismotectonic models. *Soil Dynamics and Earthquake Engineering, 156*, Article 107208.

United Nations Development Programme (UNDP). (2004). *Reducing disaster risk: A challenge for development*. UNDP.

United Nations Office for Disaster Risk Reduction (UNDRR). (2015). *Sendai framework for disaster risk reduction 2015–2030*. United Nations.

Wald, D. J., Quitoriano, V., Heaton, T. H., & Kanamori, H. (1999). Relationships between peak ground acceleration, peak ground velocity, and Modified Mercalli Intensity in California. *Earthquake Spectra, 15*(3), 557–564.

Wang, K., Hu, Y., & He, J. (2014). Deformation cycles of subduction earthquakes in a viscoelastic Earth. *Nature, 484*, 327–332.

Chapter 18
Multi Hazard Early Warning Systems: Indian Perspective

M. L. Sharma

Abstract This paper reviews Multi-Hazard Early Warning Systems (MHEWS) within India's disaster risk reduction framework. It traces the evolution of Early Warning Systems (EWS) from traditional observational methods to advanced, people-centered systems that aim to minimize disaster losses through timely forecasting and communication. The paper addresses key natural hazards affecting India, including earthquakes, landslides, floods, cyclones, and others. For each hazard, it discusses characteristics, assessment methodologies, historical case studies, and the role of early warning technologies, stressing the interconnectedness of disasters and the importance of a multi-hazard approach with four key components of effective EWS: risk knowledge, monitoring and warning services, dissemination and communication, and response capability. It highlights India's shift towards Impact-Based Forecasting (IBF) with support from agencies like the India Meteorological Department and the Central Water Commission, alongside innovations like the Common Alerting Protocol and mobile-based alerts. Case studies from India and worldwide illustrate effective EWS, including the Indian Tsunami Early Warning System and various hazard-specific warning systems. The paper advocates for integrated multi-hazard risk frameworks to address interconnections between hazards and enhance climate adaptation and disaster resilience. In conclusion, the paper argues that robust MHEWS are vital for reducing disaster risk and emphasizes the importance of data interoperability, coordination, advanced monitoring, localized communication, and public trust in building resilient communities against increasing climate and disaster risks.

Keywords Multi-hazards · Early warning system · Disaster risks

---

M. L. Sharma (✉)
Head, International Centre of Excellence for Dams and Professor, Earthquake Engineering Department, Indian Institute of Technology Roorkee, Roorkee, India
e-mail: sharmamukat@gmail.com; m.sharma@eq.iitr.ac.in

V. M. Tiwari (ed.), *Himalayan Natural Hazards*, Indian National Science Academy,
https://doi.org/10.1007/978-981-92-3784-5_18

18.1 Introduction

An Early Warning System (EWS) refers to the end-to-end framework that enables detection, forecasting, communication, and preparedness for hazards. The objective of EWS is reducing the effects of natural hazards by providing timely and accurate information. Their main goal is to protect lives, maintain livelihoods, and lessen economic losses, allowing communities and authorities to act appropriately. Since some hazards cannot be entirely prevented or predicted, effective forecasting and monitoring are vital for alerting communities and officials about imminent threats. These warnings facilitate immediate protective actions, safe evacuations, and the protection of vulnerable infrastructure. An effective EWS encompasses risk knowledge, monitoring, forecasting, and communication, which ensure that hazard information reaches at-risk populations in a clear and actionable form. The true effectiveness of EWS depends on the timely execution of the Early Action Plan (EAP), such as evacuation, activation of shelters, protection of critical infrastructure, and mobilization of emergency resources. In practice, the linkage between EWS and EAP ensures that warnings are not merely informational but actionable, enabling communities and authorities to respond promptly and reduce loss of life and damage.

The idea behind EWS has its roots in centuries of traditional knowledge and wisdom, when people relied on natural clues, such as strange animal behavior or changes in the environment, to predict potential hazards. This traditional wisdom has established the foundation of modern EWS driven by progress in meteorological and scientific observation. These developments have enabled earlier risk identification and the distribution of alerts with considerably enhanced precision and reliability (Basher, 2006; UNISDR, 2006). This evolution has been specific to hazards; for example, the India Meteorological Department (IMD) has an established practice of providing weather-related early warnings and agro-meteorological advisories to farmers. Previous disasters in India demonstrated the importance of EWS. The disaster of the 1999 Odisha Super Cyclone, which claimed over 10,000 lives, serves as a clear reminder of how poor communication and the lack of evacuation plans can lead to mass casualties. However, India's disaster management has since evolved; the successful use of modern EWS during Cyclone Phailin (2013) and Cyclone Fani (2019) allowed authorities to move millions of people to safety in record time, successfully reducing the death toll to double digits and demonstrating that timely information is the most effective tool for saving lives. Likewise, the 2004 Indian Ocean tsunami served as a significant boost towards oceanic monitoring and revealed essential deficiencies in preparedness that resulted in the creation of a specialized and efficient tsunami early warning system in the coastal region of India under the Ministry of Earth Sciences (MoES). These advancements highlight a worldwide shift towards international cooperation and strong institutional structures aimed at reducing the human and economic impact of the different disasters (UNESCO-IOC, 2007; UNISDR, 2015).

Over the past fifteen years, the significance of EWS has grown, and they have become a crucial element in climate change adaptation and disaster risk reduction

(DRR) strategies, spanning from global to local levels. Over this period, a paradigm shifts from top-down technical solutions to "people-centered" systems that prioritize the needs and capabilities of those at risk. EWS has consequently received heightened attention within the frameworks of international treaties and action plans. This includes the Hyogo Framework for Action (2005–2015) and its successor, the Sendai Framework for Disaster Risk Reduction (2015–2030), which explicitly targets a substantial increase in the availability of and access to multi-hazard early warning systems by 2030. These frameworks highlight the essential role of EWS in enhancing resilience, more particularly in the developing countries where dense populations and limited infrastructure often exacerbate vulnerability to natural hazards. Integration of scientific forecasting with local knowledge and pre-arranged response mechanisms, an effective EWS, empowers communities to take timely protective actions, thereby minimizing the devastating human and economic impacts of disasters. The EWS is one of the critical components in achieving international goals and frameworks for disaster risk reduction and sustainable development. The United Nations Sustainable Development Goals (SDGs) emphasize building resilient infrastructure (Goal 9), reducing disaster mortality (Goal 11), and strengthening capacity for climate-related hazards (Goal 13), all of which are directly supported by effective EWS. The Sendai Framework for Disaster Risk Reduction (2015–2030) explicitly highlights early warning as a key tool for substantially reducing disaster risk and enhancing preparedness at all levels. Furthermore, the Intergovernmental Panel on Climate Change (IPCC) underscores the role of EWS in adapting to increasing climate-related hazards, enabling early action to protect lives, livelihoods, and ecosystems in the context of a changing climate.

By integrating early warning into broader disaster risk reduction and climate adaptation strategies. EWS significantly enhances the resilience of communities and nations, particularly those in developing regions, and forms an essential tool in protecting lives and livelihoods against both sudden-onset disasters such as tsunamis, flash floods, and earthquakes and slow-onset hazards like prolonged droughts, desertification, and sea-level rise. This chapter reviews multi-hazard early warning systems with a focus on India, covering hazard characteristics, assessment methodologies, and operational EWS case studies.

18.2 Natural Hazards

Natural hazards are extreme environmental events that can pose serious threats to human life, property, and socio-economic stability. These events such as earthquakes, floods, cyclones, landslides, tsunamis, droughts, and volcanic eruptions arise from the complex interplay of geological, meteorological, and hydrological processes. The extent of their impact depends not only on the intensity and frequency of the hazard itself but also on the vulnerability and resilience of the people, infrastructure, and systems exposed to it (Cutter et al., 2003). In other words, two regions experiencing similar events may suffer very different outcomes depending on their level of preparedness and adaptability. To manage these risks effectively, it is crucial to integrate risk assessment, EWS, community-based preparedness, and robust policy frameworks that aim to minimize exposure and build adaptive capacity (UNDRR, 2015). Ultimately, a deep understanding of natural hazards forms the foundation of disaster risk reduction and contributes significantly to achieving sustainable development goals and an effective EWS. Major natural hazards are Earthquake, Landslide, Floods, Cyclone, Tsunami, Forest Fire, Avalanche and GLOF.

18.3 Early Warning Systems

The generic methodology for studying EWS for natural disasters involves a systematic and structured approach aimed at evaluating their effectiveness, identifying operational and infrastructural challenges, and recommending strategies for enhancement. A comprehensive review of existing EWS frameworks, policies, and technological infrastructure is conducted, which includes evaluating hazard detection systems, communication channels, and community response mechanisms. While primary data may include field surveys, stakeholder interviews, and focus group discussions, the secondary data comprise historical disaster records, reports, and relevant scientific publications. This is analyzed to assess the effectiveness of the EWS in terms of timeliness, accuracy, reach of warnings, capacity of communities and authorities to respond promptly and appropriately to alerts. One of the important steps is to integrate technical, social, and institutional findings to formulate actionable recommendations which focus on improving hazard detection, enhancing communication systems, building community preparedness, and strengthening institutional response capabilities.

India's high vulnerability towards various natural hazards, which includes floods in the Indo-Gangetic plains, cyclones along the eastern and western coasts, droughts in arid regions, landslides in the Himalaya and Western Ghats, earthquakes in seismic zones IV and V, heatwaves, and tsunamis necessitates the adoption of a multi-hazard and region-specific EWS framework. The evolution of EWS in the country is reflected in the shift from generalized weather forecasting to more advanced Impact-Based Forecasting (IBF), wherein agencies such as the India Meteorological Department, Indian National Centre for Ocean Information Services, and Central Water Commission provide not only hazard predictions but also actionable insights on potential impacts to lives, livelihoods, and infrastructure at localized scales (Fig. 18.2). This transition has significantly enhanced decision-making for disaster preparedness and response. Furthermore, efforts to strengthen the early warning dissemination mechanisms include the implementation of the Common Alerting Protocol (CAP) for real-time, standardized mobile alerts ensured a rapid and wide-reaching communication across multiple platforms. Concurrently, there is more emphasis on developing specialized monitoring systems tailored for localized and emerging disaster risk, such as glacial lake outburst floods and flash floods in the Himalayan region and urban heatwaves in rapidly expanding cities, thereby reinforcing the need for a dynamic, technology-driven, and region-specific EWS framework in India. Early warning is a strategic approach adopted by societies worldwide to minimize the impacts of disasters. Traditionally, such systems relied on the interconnections between visual observations, collective experience, and community cooperation to anticipate and mitigate losses from impending hazards. However, a modern EWS extends far beyond mere hazard forecasting and information transmission. It represents an integrated framework that ensures communities are informed and prepared to act. For an EWS to be truly effective, it must incorporate and harmonize four essential components: risk knowledge, monitoring and warning services, dissemination and communication, and response capability (Fig. 18.1). Together, these elements form a comprehensive mechanism that transforms hazard information into timely and actionable decisions, thereby reducing vulnerability and enhancing resilience.

Fig. 18.1 Key elements of EWS (UNDP EWS report:2018)

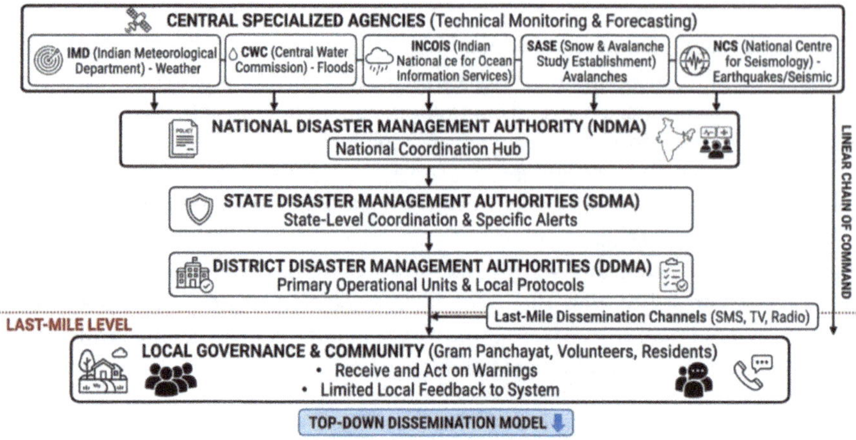

Fig. 18.2 Current state of EWS in India

18.3.1 Risk Knowledge

Risk emerges from the interaction between hazards and vulnerabilities present at a specific location. Consequently, risk knowledge constitutes a fundamental component of an effective EWS. It encompasses the establishment of institutional frameworks, identification of potential natural hazards, assessment of community vulnerabilities, comprehensive risk analysis, and mechanisms for information storage and sharing. This collective understanding forms the baseline for all subsequent EWS activities, ensuring that actions are informed and context-specific. Moreover, when engaging end-users, it is essential that communities clearly understand both the significance and potential benefits of the EWS. Effective risk communication through accessible information on local hazards, communication channels, and emergency response plans will empower communities to make timely decisions and enhance overall resilience.

18.3.2 Monitoring and Warning Services

Monitoring and warning services are core components of an EWS. Monitoring involves the systematic collection of data and proxy indicators related to potential hazards such as rainfall patterns (linked to floods or droughts) or seismic activity (associated with earthquakes). This process may include both instrumental observations and direct human observations, such as visually detecting an approaching wildfire or landslide. However, direct observation alone often provides insufficient lead time for meaningful risk reduction, as the threat is already imminent. In contrast, advanced monitoring technologies such as seismometers, weather radars, and satellite systems enable earlier detection of hazards. While earthquake warnings may only offer seconds of notice, other hazards like storms, floods, and droughts allow for longer lead times (ranging from weeks to months), thereby facilitating proactive, risk-mitigating actions by authorities and communities alike (Mittal et al., 2019).

Broadly speaking, this component is the domain of scientific and technological institutions. For example, in India, the India Meteorological Department (IMD) is the nodal agency for providing cyclone warning services; the Central Water Commission (CWC) has a mandate for providing flood forecasts, and the Indian National Centre for Ocean Information Services (INCOIS) is responsible for providing tsunami warnings.

India's Early Warning System relies on a clear hierarchy of roles from top to bottom, where national agencies like the IMD and CWC provide technical forecasts to NDMA, SDMAs, and local authorities who translate these warnings into district-level actions. To achieve widespread reach, alerts are issued through a multi-channel approach centered on the Common Alerting Protocol (CAP), which pushes geo-targeted SMS and cell broadcasts to mobile users even without internet access. The technical outputs in the form of alerts are converted to easily readable forms for the public through color-coded warnings (Yellow, Orange, Red) and Impact-Based Forecasting, ensuring that scientific data becomes an actionable instruction. The combination of high-tech mobile alerts with traditional methods like community sirens and mock drills bridges the gap between hazard detection and the "last-mile" response necessary to save lives. Furthermore, achieving effective early warning dissemination requires strengthening real-time data integration, improving inter-agency coordination, ensuring multilingual and locally understandable messages. Also, to enhance community awareness, regular drills and capacity-building programs, and leveraging emerging technologies such as GIS, artificial intelligence, and IoT are needed for predictive accuracy and rapid communication. Such effective early warning protocols will ensure that early warnings are not only issued but are also understood, trusted, and acted upon promptly by all stakeholders.

18.3.3 Dissemination and Communication

Dissemination and communication form the critical link between hazard detection and community response within an EWS. Once a threat is identified, timely, accurate, and actionable information must be transmitted through reliable multi-channel communication networks—such as mobile alerts, sirens, broadcast media, and community-based systems—to ensure that at-risk populations can understand and act appropriately. Table 18.1 shows typical warning times for different hazards. This component ensures that clear, timely, and actionable warnings reach all individuals and groups exposed to potential hazards. Effective dissemination depends on robust communication systems capable of delivering understandable and credible information through institutionalized decision-making processes. It involves

Table 18.1 Typical warning times for selected hazards

Hazard type	Warning time	Characteristics	Communication priority
Earthquake	Seconds to ~ 1 min	Near-instantaneous detection	Automated alerts (EEW)
Landslide	Minutes to hours	Rainfall-triggered, localized	Community alerts
Flash flood	Minutes to hours	Rapid onset	Real-time alerts
Cyclone	2–5 days	Trackable systems	Evacuation planning
River flood	Hours to days	Gradual rise	Continuous updates
Tsunami	Minutes to hours	Depends on distance	Immediate evacuation
Fire	Minutes to days	Rapid spread	Local evacuation alerts

Adapted from WMO (2018)

both the physical infrastructure which includes reliable, disaster-resistant communication networks and information technologies and the strategic dimension, which encompasses coordinated interactions among key stakeholders, tailored warning messages, and community-specific communication approaches. The effectiveness of this component ultimately determines whether early warnings translate into informed action, thereby reducing potential losses and saving lives.

In today's world, where technology plays an increasingly pivotal role across all sectors, the tools available for risk communication and information dissemination have expanded significantly in both quality and diversity. Advances in digital communication, mobile technologies, social media platforms, and satellite-based systems have revolutionized the way warnings are transmitted and received. These innovations enable the rapid and widespread distribution of alerts, ensuring that information reaches even remote or vulnerable communities. Moreover, the integration of artificial intelligence (AI), Internet of Things (IoT), and real-time data analytics into early warning communication networks enhances the precision, reliability, and timeliness of messages, thus strengthening community preparedness and response effectiveness.

18.3.4 Response Capability

The final component focuses on the preparedness and response actions taken by individuals, communities, and authorities. Response capability encompasses the centralized knowledge, plans, and operational inputs required to ensure timely and appropriate action by both authorities and communities at risk. This capability can be strengthened through enhanced public and institutional preparedness or by implementing automated emergency responses. For rapid-onset disasters, such as earthquakes, monitoring sensors can trigger immediate safety measures—such as shutting down power and gas supplies—to prevent further harm. Crucially, the effectiveness of response measures depends on community trust in the EWS; repeated false alarms or unnecessary warnings can erode confidence and reduce compliance with future alerts. Therefore, a robust EWS must incorporate strategies to minimize false or unwarranted warnings, ensuring that alerts prompt decisive and proportionate actions when hazards genuinely threaten.

18.3.5 Socioeconomic Aspect of EWS

EWS are not merely technical tools but also social and economic instruments that shape community resilience and disaster management outcomes. Socio-economic factors such as income, education, occupation, and social status significantly influence how communities perceive risks and respond to warnings, with vulnerable groups—including the poor, elderly, and marginalized—often facing limited access to information and resources (UNDRR, 2015). The effectiveness of EWS also depends on awareness and preparedness, which are shaped by literacy, access to communication technologies, and socio-cultural norms that affect how warnings are interpreted and protective actions are taken (Cutter et al., 2003). Economically, EWS can reduce losses by enabling timely evacuation, safeguarding infrastructure, and protecting livelihoods, with investments in such systems providing high cost–benefit returns through avoided damages and continuity of economic activity (World Bank, 2010). Equity and accessibility are critical, requiring multi-channel communication strategies to ensure alerts reach low-income, disabled, and remote populations. Policy and institutional support further influence EWS effectiveness, with social protection measures, insurance mechanisms, and disaster relief programs enhancing resilience for vulnerable populations. Finally, behavioral and social responses are shaped by trust in authorities, past experiences with disasters, and community cohesion, with participatory planning and drills fostering ownership and compliance. Integrating these socio-economic dimensions is therefore essential for the design and implementation of effective and inclusive EWS.

18.4 EWS Case Studies of Different Natural Hazards

18.4.1 Earthquake Early Warning System (EEWS)

Earthquake Early Warning (EEW) systems generally operate through four main steps (Cremen and Galasso, 2020) namely: (i) Event detection and location estimation, (ii) Magnitude estimation, (iii) Ground shaking prediction, and (iv) Alarm (V) Decision-making. Earthquake early warning systems work by detecting the initial, fast-moving P-waves generated by an earthquake using a network of sensors. These systems rapidly analyze the incoming data to estimate the earthquake's origin, magnitude, and the areas likely to experience shaking. With this information, the system predicts when the slower, but more damaging, S-waves will arrive at various locations. Alerts are then sent out via mobile phones, broadcasts, and automated systems, providing people and infrastructure with a brief but potentially life-saving warning before the strong shaking begins (Bhardwaj et al., 2016; Kumar et al., 2024). EEW systems are typically categorized into regional, on-site, and hybrid types, depending on how they address the initial three steps and manage the critical trade-off between warning time and prediction accuracy. A few of the operational EEWS are (i) Japan Meteorological Agency (JMA) EEW System (J-Alert) (Hoshiba et al., 2011; JMA, 2012), (ii) Mexican Seismic Alert System (SASMEX)(Cuéllar et al., 2014; Espinosa-Aranda et al., 2011), (iii) United States-West Coast (ShakeAlert by USGS) (Given et al., 2018; USGS, 2021) (iv) China National EEW Network (Li et al., 2020), (v) Taiwan's Earthquake Early Warning (EEW) System (Wu et. al, 2021), (vi) Israel's Earthquake Early Warning (EEW) System (TRUAA), and Uttarakhand State (India) Earthquake Early Warning System (UEEWS): (Kumar et al., 2024; Rathore et al., 2023).

18.4.2 Cyclone Early Warning System

The accurate and timely detection of cyclones is a critical component of effective early warning systems. A variety of methods and approaches are employed to identify cyclonic activity and support forecasting efforts. Satellites provide continuous high-resolution imagery across visible, infrared, and microwave bands, enabling the detection of cloud formations, wind patterns, and cyclonic structures over oceans. Machine learning techniques now enhance automated identification and parameter estimation of cyclones (Sahana & Mondal, 2023). Ground-based Doppler radars monitor cyclones approaching land by measuring wind velocity and precipitation, helping to locate the cyclone's eye and track its trajectory in real time. Automated weather stations, ocean buoys, and shipborne instruments record atmospheric pressure, wind, temperature, and sea conditions. Abrupt changes in these parameters provide early indicators of cyclonic activity. Numerical Weather Prediction (NWP) models simulate atmospheric processes to forecast cyclone formation, intensity, and trajectory. Ensemble forecasting using multiple simulations improves prediction reliability and early warning lead time. Automated Image Analysis and Machine Learning, such as Convolutional Neural Networks (CNNs), analyze satellite imagery to identify cyclone centers, spiral cloud bands, and intensity, enabling rapid and accurate detection (Sahana & Mondal, 2023). Also, Atmospheric Parameter-Based Techniques detect local pressure minima, wind vorticity, and strong rotational patterns to confirm cyclonic activity and accurately locate cyclone centers.

Integrated cyclone early warning systems (CEWS) provide timely and actionable information to communities, authorities, and emergency responders, helping to reduce the impacts of cyclones through effective preparedness and evacuation measures. Several successful case studies from India and across the world demonstrate the importance of accurate forecasting, rapid communication, and coordinated disaster management. In India, cyclones such as Phailin, 2013 and Fani, 2019 highlighted the effectiveness of advanced warning systems implemented by the India

Meteorological Department (IMD). Timely forecasts issued several days before landfall enabled the evacuation of more than one million people to safer shelters, resulting in comparatively low fatalities despite severe wind speeds and widespread destruction. These events showcased the value of early warning dissemination, modern communication tools, and strong community preparedness.

Similarly, international experiences from Cyclone Nargis in Myanmar (2008), Cyclone Pam in Vanuatu (2015), and Cyclone Winston in Fiji (2016) further demonstrated that effective early warnings and local preparedness significantly reduce cyclone-related casualties and damage. Forecasts, evacuation protocols, public awareness campaigns, and community participation helped protect vulnerable populations and critical infrastructure even during some of the most intense cyclones recorded. Together, these examples emphasize that integrated early warning systems, combined with efficient disaster response coordination and public awareness, play a crucial role in disaster risk reduction and resilience building.

18.4.3 Flood Early Warning System (FEWS)

Flood Early Warning Systems are integrated frameworks designed to detect, forecast, and communicate the risk of impending floods, enabling timely action to minimize loss of life, property, and disruption to society. Drawing from scientific literature, technical reports, and operational case studies, FEWSs are recognized as a cornerstone of modern flood risk management and disaster preparedness. The FEWSs are typically structured around four essential pillars: (i) Meteorological Component, (ii) Hydrological and Hydraulic Modeling, (iii) Threshold-Based Alerts, and Community-Based FEWS.

FEWSs integrate advanced technological approaches such as high-resolution meteorological and hydrological modeling, ensemble forecasting, and automated community alerts with local knowledge and participatory dissemination. A few of the fully operated FEWS include (i) Bangladesh: Flood Forecasting and Warning Centre (FFWC) (Paul et al., 2018), (ii) India: Central Water Commission (CWC) Flood Forecasting (Sharma et al., 2020), (iii) European Flood Awareness System (EFAS) (Pappenberger et al., 2012), (iv) National Water Model, USA, (v) Philippines: Project NOAH (Nationwide Operational Assessment of Hazards), (vi) Mozambique, Flood Early Warning Systems for Limpopo River Basin (CRIDF report, 2017), (vii) Vietnam, Mekong Delta Flood Forecasting (Dang et al., 2021).

18.4.4 Landslide Early Warning System (LEWS)

Modern landslide early warning system development began around 2010 with the operationalization of empirical rainfall thresholds at the territorial scale (Segoni et al., 2018). From 2010 to 2025, LEWS have evolved from rainfall-only, threshold-based models to fused, multi-sensor, community-centered systems. The current operational LEWS in some locations mainly depend upon real-time rainfall monitoring and empirically established thresholds at regional or local levels to forecast landslide events and provide alerts (Martelloni et al., 2012; Segoni et al., 2018). These systems incorporate the monitoring, forecasting, decision-making, and communication elements but frequently encounter problems of false alarms due to only depending on precipitation data (Piciullo et al., 2018). In the recent past, LEWS highlighted the integration of multi-parameter data (e.g., soil moisture and ground deformation), the application of sophisticated models such as machine learning, the facilitation of real-time data integration via IoT, and the enhancement of warning dissemination systems, thereby augmenting the accuracy and reliability of LEWS. Advances include standardized performance metrics, integration of low-cost and satellite sensing, multi-hazard monitoring, cloud-native operations, and explicit coupling of alerts with operational and social frameworks. The trajectory emphasizes technological robustness, spatial representativeness, and human-centered design, enabling more reliable and actionable landslide warnings. The stepwise procedure of LEWS consists of: (i) Hazard Identification and Assessment (ii) Monitoring and Data Acquisition (iii) Trigger Threshold Definition (iv) Data Processing and Analysis (v) Alert Generation and Decision-Making (vi) Communication and Dissemination (vii) Response and Mitigation (vii) Evaluation and Feedback.

Starting from contemporary early warning practices based primarily on rainfall thresholds, LEWS have progressively evolved by integrating multiple cross-cutting technologies. Modern systems now incorporate multi-sensor monitoring, soil and slope displacement measurements, machine learning models, and real-time telemetry, allowing for improved detection and early warning of landslide hazards. These technological advancements enhance the accuracy, lead time, and spatial coverage of warnings, while also supporting community-centered decision-making. Several operational and successful LEWS that illustrate this evolution worth mentioning include (i) Piemonte, Italy (2010–2013) (Tiranti & Rabuffetti, 2010); (ii) Slope-Scale CHASM System, Italy (2013) (Thiebes et al., 2014); (iii) Soil-Moisture Sensor Networks, Indonesia (2016) (Yuliza et al., 2016); (iv) LHASA Satellite Nowcasting, Global (2018) (Krøgli et al., 2018); (v) 5G-IoT Multi-Sensor LEWS, Japan & South Korea (2021) (Li et al., 2021); (vi) Cloud-Native Multi-Hazard LEWS, Philippines & Southeast Asia (2022–2023) (Deng et al., 2023); and (vii) Machine-Learning and Vision-Based LEWS, India (2024–2025). The modern LEWS applied machine-learning models (RF, XGBoost, SVM) and vision-based displacement detection (Dang et al., 2025; Wang and Liang, 2025; Cheng et al., 2025).

18.4.5 Forest Fire Early Warning System (FFEWS)

The development of an Integrated Forest Fire Early Warning System (FFEWS) relies on the combined use of remote sensing, geospatial analysis, meteorological modeling, and machine learning, ensuring accuracy in fire detection, risk prediction, and spatio-temporal monitoring. This integrated approach supports proactive forest fire management and timely decision-making. Many of the techniques used in FFEWS include Satellite-Based Remote Sensing (Csiszar et al., 2014; Giglio et al., 2016), Ground Sensor Networks (Novković et al., 2021), UAV and Drone-Based Surveillance (Sudhakar et al., 2020; Wardihani et al., 2018), Meteorological and Climate Data Integration (Hamadeh et al., 2015; Pandey et al., 2024), Machine Learning and Predictive Modeling, Computer Vision and AI-Based Detection: AI-driven models (Khan & Khan, 2022; Labed et al., 2023) and lastly, citizen science and Crowdsourced Monitoring i.e., Community-based platforms.

To develop a successful FFEWS, the integration of remote sensing, geospatial analytics, IoT-based ground sensors, and machine learning is essential. Such integration enables near-real-time detection, risk prediction, and impact assessment across diverse forested landscapes. As climate change intensifies the frequency and severity of forest fires, future strategies should prioritize multi-sensor data fusion, predictive modeling, and community-based preparedness frameworks to enhance resilience and ensure timely, data-driven decision-making. The successful case studies of operational FFEWS worldwide include Canada—Canadian Wildland Fire Information System (CWFIS (Stocks et al., 2004)), United States—Fire Information for Resource Management System (FIRMS), Australia—Sentinel Hotspots Program (Miller et al., 2017), Europe—European Forest Fire Information System (EFFIS), India—Forest Fire Alert System (FFAS): Developed by the Forest Survey of India, and Indonesia—SiPongi Fire Early Warning and Monitoring System (Herawati & Santoso, 2011).

18.4.6 Tsunami Early Warning System (TEWS)

The development of an effective Tsunami Early Warning System (TEWS) relies on the integration of seismological monitoring, ocean observation networks, satellite-based communication, and numerical modeling. Critically, the effectiveness of these warnings is strongly constrained by source-to-coast travel times, which determine the available lead time between earthquake occurrence and tsunami arrival. In near-field settings where travel times are only a few minutes, warning dissemination and response options are severely limited, whereas in far-field scenarios with travel times of tens of minutes to hours, there is greater scope for evacuation and coordinated Early Warning Actions (EWA). Together, these components enable the rapid detection of tsunamigenic earthquakes, accurate wave propagation modeling, and timely dissemination of warnings to vulnerable coastal regions. The main components are (i) Seismic and Ocean Monitoring, (ii) Numerical Modeling and Forecasting, (iii) Communication and Alert Dissemination and (v) Community Preparedness. A successful TEWS thus combines real-time monitoring, modeling, communication, and community readiness. As oceanic and seismic risks grow due to tectonic activity and coastal development, future systems should emphasize multi-sensor data fusion, AI-based forecasting, and people-centered early action frameworks to strengthen global tsunami resilience. A few of the successfully implemented TEWS include (i) Pacific Tsunami Warning Centre (PTWC), USA (Bernard & Titov, 2015), (ii) Indian National Tsunami Early Warning System (IN-TEWS), India (Rajendran et al., 2013), (iii) Japan Meteorological Agency (JMA) TEWS (Koshimura et al., 2013), (iv) Indonesia Tsunami Early Warning System (InaTEWS) (Liu et al., 2013), (v) Chilean Tsunami Early Warning System (Melnick et al., 2012).

18.4.7 Avalanche/GLOF Early Warning System

Avalanche and GLOF Early Warning Systems (EWS) are designed to detect rapid snow, ice, or glacial lake movements that can threaten downstream communities. These systems integrate remote sensing, ground-based sensors, hydrological and meteorological data, and modeling techniques to provide early alerts, risk mapping, and decision support. Timely warnings help reduce casualties, infrastructure damage, and economic loss in mountainous regions. Similar to other EWS, its components include (i) Remote Sensing and Satellite Monitoring, (ii) Ground-Based Sensors and IoT Networks, (iii) Numerical Modeling and Forecasting, and (iv) Community Preparedness and Response.

Operational Avalanche and GLOF EWS demonstrate that multi-sensor monitoring, numerical modeling, rapid communication, and community engagement are essential for mitigating risks. As glacial retreat and snow instability increase under climate change, future systems should prioritize real-time data integration, AI-based predictions, and people-centered preparedness to enhance resilience in mountain regions. Some of the successful and operational Avalanche/GLOF early warning systems (EWS) include Indian Himalaya—Indian GLOF Early Warning, Nepal—Imja Lake GLOF Monitoring (ICIMOD, 2011), Switzerland—Avalanche Warning System, Bhutan—GLOF and Avalanche Monitoring (Ives et al., 2010), and Peru—Cordillera Blanca GLOF EWS (Carey et al., 2012).

18.5 Interactions of Natural Hazards

Natural Hazard Interactions refer to the phenomenon where one natural hazard triggers, influences, or amplifies another hazard, often creating complex disaster scenarios. Hazards do not always occur in isolation; their interactions can increase the severity, frequency, and spatial extent of disasters. For example, earthquakes may cause tsunamis, droughts can worsen wildfires, and cyclones often lead to floods. Understanding these interactions is key to effective disaster risk management and early warning. Natural hazards rarely occur in isolation, and their interactions can significantly influence disaster outcomes. Based on various worldwide case studies and literature, Bill and Malamud (2014) have categorized the hazard interactions into four types.

18.5.1 Interactions Where a Hazard Is Triggered

A primary hazard can directly trigger one or more secondary hazards which can increase the overall disaster risk and complexity of hazards. For example, earthquakes frequently induce landslides by destabilizing slopes, and these landslides may subsequently dam rivers, generate upstream flooding, or damage critical infrastructure such as roads, bridges, and pipelines. In a few cases, these interactions extend further, meaning the temporary, unstable dams created due to landslide can fail suddenly and produce outburst floods which propagate downstream and exacerbate damage over large areas. Such sequences of interaction of hazards form complex hazard networks, where physical linkages and feedback mechanisms can intensify impacts across space and time, complicating both hazard assessment and disaster risk management (Gill & Malamud, 2014).

18.5.2 Interactions Where the Probability of a Hazard Is Increased

Some hazards do not directly trigger others but change environmental conditions, raising the likelihood of subsequent hazards. For instance, vegetation loss from wildfires can destabilize slopes, making landslides more probable during rainfall. Similarly, subsidence or environmental degradation can increase vulnerability to flooding or other hazards, pushing the system closer to critical tipping points.

18.5.3 Interactions Where the Probability of a Hazard Is Decreased

In some cases, the occurrence of one hazard can reduce the likelihood of another. Heavy rainfall may temporarily lower wildfire risk by saturating soils, and small controlled events, like prescribed burns, can prevent larger disasters. Such interactions highlight that hazards can sometimes have mitigating effects, although these are often context-specific and not easily predictable.

18.5.4 Spatiotemporal Coincidence of Hazards

When multiple hazards occur in the same area and within a short timeframe, their combined impact can exceed or differ from individual effects. Overlapping events can strain infrastructure and communities, hinder recovery, and increase vulnerability to future hazards. However, the exact consequences depend on the timing, magnitude, and nature of the interacting hazards, making planning and mitigation complex.

18.6 From Single-Hazard Assessment to Multi-hazard Risk Frameworks

Traditional hazard and risk assessments often rely on a "single hazard" approach, in which each hazard is analyzed independently. While this method provides insight into individual hazards, it fails to capture the complex interactions, cascading effects, and compounded risks that can occur when multiple hazards affect a region simultaneously or sequentially (UN, 2002; UN-ISDR, 2005).

Multi-hazard approaches are widely promoted by key government and intergovernmental initiatives, but their definition is often unclear. For instance, the Hyogo Framework for Action (2005–2015) advocated for "an integrated multi-hazard approach to disaster risk reduction" (UNISDR, 2005, p. 4), while the Sendai Framework for Disaster Risk Reduction (2015–2030) emphasizes that "disaster risk reduction needs to be multi-hazard" (UNISDR, 2015, p. 10).

Research done by Gill and Malamud (2014) and Kappes et al. (2012) indicates that a comprehensive understanding of hazard potential requires a multi-hazard approach, which explicitly incorporates interactions between hazards, spatial and temporal coincidence, and the dynamic nature of societal vulnerability. Delmonaco et al. (2006) propose documenting both single hazard characteristics and their mutual interrelations, highlighting that hazards should not be treated as independent. Furthermore, Kappes (2012) advocates identifying all relevant hazards in a defined region and constraining their interactions, emphasizing the role of hazard interdependencies in risk assessment.

These studies emphasize that a multi-hazard framework should build on single-hazard approaches by not only identifying multiple hazards but also examining how they interact, cascade, or coincide across space and time. This approach provides a more comprehensive understanding of overall hazard potential and risk. Evidence from case studies in Japan, the USA, the Philippines, and Guatemala highlights that incorporating these dynamics is essential for accurate risk assessment and the development of effective mitigation strategies (Scott et al., 1999; Suleimani et al., 2009; Umbal & Rodolfo, 1996).

A robust multi-hazard methodology should incorporate four key components (Gill & Malamud, 2014, 2016) viz., (i) Hazard Identification and Comparison, (ii)

Hazard Interactions, (iii) Hazard Coincidence, and (iv) Dynamic Vulnerability. Integrating hazard interactions into multi-hazard frameworks significantly improves the accuracy and relevance of risk assessments. By explicitly accounting for interactions between hazards, their spatial and temporal coincidence, and the dynamic nature of vulnerability, multi-hazard approaches overcome the limitations of single-hazard assessments. This holistic perspective enables a deeper understanding of risk, supporting more effective disaster management.

18.7 Multi-hazard Early Warning System as a Mitigation Tool

Warning systems are pivotal in reducing the cascading and compound impacts of interacting natural hazards. By providing timely detection, monitoring, and communication of multiple hazard threats, these systems enable proactive responses and minimize disaster losses. This can be achieved using MHEWS, which can integrate data from various sources—such as seismic, meteorological, hydrological, and volcanic monitoring—to anticipate not only individual hazards but also their potential interactions and cascading effects. This can not only enable proactive, anticipatory responses to complex hazard scenarios but also reduce loss of life, injuries, and property damage.

Effective implementation of MHEWS requires interoperable data systems, institutional collaboration, and local engagement, aligning with the International and National frameworks, which emphasize multi-hazard approaches and integrated risk governance to address complex hazard interactions (Gill & Malamud, 2016). The effectiveness of MHEWS lies in their ability to combine hazard detection, real-time data analysis, communication, and preparedness measures in an integrated framework. By incorporating hazard interaction networks, the MHEWS enhances the accuracy and timeliness of warnings, allowing authorities to anticipate multi-phase disaster events rather than isolated hazards (Gill & Malamud, 2016; Kappes et al., 2012).

This standard structure of early warning system (risk knowledge, monitoring and forecasting, warning dissemination and communication, and preparedness to respond) supports the anticipatory action across different agencies and sectors, ensuring that alerts for one hazard also consider potential secondary or compounding cascading disasters.

18.8 Case Studies for MHEWS

Integrating hazard interactions into MHEWS enhances their effectiveness in mitigating the impacts of cascading and compound hazards. By considering the interdependencies between different hazards and their potential to trigger or exacerbate one another, MHEWS can provide more accurate and timely warnings, enabling communities to take appropriate actions and reduce disaster risks. This approach aligns with the principles outlined in the Sendai Framework for Disaster Risk Reduction (SFDRR), which advocates for comprehensive, multi-hazard risk assessments and the development of inclusive, people-centered early warning systems.

18.8.1 Indian Scenario

India's efforts in implementing MHEWS demonstrate a proactive approach to disaster risk reduction. By enhancing coordination, integrating multiple hazards, and engaging communities, India aims to build a more resilient future in the face of natural hazards.

Indian Tsunami Early Warning System (ITEWS)

Established in 2007 under the Ministry of Earth Sciences and operated by INCOIS, ITEWS is the first multi-hazard real-time system designed to detect and mitigate tsunamis and associated hazards threats along India's coastline. It integrates seismic monitoring, ocean-bottom pressure recorders, coastal tide gauges, and satellite communication to provide rapid alerts. The Tsunami Early Warning Centre (ITEWC) analyzes incoming data and issues bulletins, typically within 10 min of a significant seismic event, allowing timely evacuations (INCOIS, 2021).

Recent upgrades include a state-of-the-art underwater observatory with sensors along a 275 km cable from the Andaman Islands, enhancing detection accuracy and reducing false alarms. India also participates in the UNESCO-IOC Indian Ocean Tsunami Warning and Mitigation System, acting as a Tsunami Service Provider for 25 countries, highlighting regional collaboration in disaster risk reduction.

ITEWS exemplifies an integrated, multi-hazard early warning system combining advanced technology, rapid response, and international cooperation to safeguard coastal populations.

SACHET: India's National Disaster Alert System

SACHET (System for Alerting and Communication of Emergency and Threats) is India's official disaster early warning platform, developed by the National Disaster

Management Authority (NDMA) in collaboration with the Centre for Development of Telematics (C-DOT) and launched in March 2023. It is based on the international Common Alerting Protocol (CAP), ensuring standardized and timely dissemination of alerts across the country. SACHET delivers multi-channel alerts through SMS, mobile app notifications, browser push notifications, and RSS feeds, enabling wide accessibility (INCOIS, 2022). The system provides geo-targeted information, allowing users to receive location-specific alerts, and supports 12 Indian languages, including Hindi, English, Bengali, and Tamil, catering to diverse linguistic communities. Features like "read-out loud" options and "Do's and Don'ts" guidelines enhance accessibility for specially abled individuals. SACHET also integrates real-time weather updates and forecasts from the India Meteorological Department (IMD), Centre Water Commission (CWC) and other alert broadcasting national agencies like ISRO, FSI and NCS. The platform is accessible through a mobile application. SACHET plays a critical role in India's disaster management framework by providing timely, accurate, and inclusive alerts that empower citizens to respond effectively to both natural and man-made hazards, enhancing public safety and resilience.

Odisha Early Warning Dissemination System

Odisha has developed a robust Early Warning Dissemination System (EWDS) to enhance disaster preparedness and response, particularly for coastal communities vulnerable to cyclones and tsunamis. Initiated in 2018, this system was the first of its kind in India, designed to bridge communication gaps during emergencies. The EWDS employs a network of 122 siren towers strategically placed across six coastal districts to provide audible alerts to residents about impending disasters (INCOIS, 2020). It integrates a comprehensive communication network connecting the State Emergency Operation Centre (SEOC) with District and Block Emergency Operation Centres, Cyclone Shelters, and Fish Landing Centres, facilitating real-time information sharing and coordination during emergencies. The system is also aligned with the National Emergency Response Support System (ERSS) and utilizes the Common Alerting Protocol (CAP) for standardized messaging, ensuring clarity and consistency in alerts. The effectiveness of EWDS was demonstrated during Cyclone Fani in 2019, when it enabled the evacuation of approximately 1.2 million people from vulnerable areas, significantly reducing potential loss of life.

Cyclone Multi-hazard Early Warning System (MHEWS)

India's Cyclone MHEWS integrates real-time meteorological, oceanic, and hydrological data to anticipate not only cyclones but also their secondary hazards, such as storm surges, flooding, and landslides. Coordinated by the India Meteorological Department (IMD) alongside the NDMA and state disaster authorities, the system disseminates alerts through multiple channels, including SMS, mobile apps, sirens, social media, and mass media. By linking cyclone forecasts with potential cascading impacts, MHEWS enables timely evacuations, pre-positioning of resources, and coordinated emergency response. Case studies like Cyclone Phailin, 2013 and Cyclone Fani, 2019 show that early, integrated warnings significantly reduce casualties. The system aligns with the Sendai Framework's emphasis on anticipatory action and resilience, making it a vital tool for managing complex, multi-hazard risks in India's coastal regions.

18.8.2 Worldwide Scenario

Globally, MHEWS have emerged as essential tools for reducing disaster risks in complex and interconnected hazard environments. These systems integrate data from meteorological, hydrological, geological, and oceanographic sources to provide comprehensive and timely warnings for multiple hazards and their cascading effects. Countries such as Japan, the United States, Indonesia, and Australia have developed advanced MHEWS that combine real-time monitoring, modeling, and communication networks to anticipate and manage diverse threats such as earthquakes, tsunamis, floods, landslides, and cyclones.

Japan

Japan's MHEWS, led by the Japan Meteorological Agency (JMA), is among the most advanced globally. It integrates seismic, volcanic, meteorological, and oceanic monitoring networks to provide real-time alerts for earthquakes, tsunamis, typhoons, and landslides. The Earthquake Early Warning (EEW) system and Tsunami Warning System operate within seconds of detection, allowing rapid public dissemination through television, mobile alerts, and sirens. This integration of data and communication channels has significantly reduced disaster impacts, as seen during the 2011 Tōhoku earthquake and tsunami.

United States

The U.S. operates a comprehensive MHEWS coordinated by the National Oceanic and Atmospheric Administration (NOAA) and the Federal Emergency Management Agency (FEMA). The system connects weather, flood, drought, wildfire, and tsunami warnings through platforms like the National Weather Service (NWS) and the National Integrated Drought Information System (NIDIS). Advanced satellite monitoring and forecast modeling, coupled with community outreach and impact-based warnings, enhance preparedness and multi-agency coordination across federal and state levels.

Indonesia

Indonesia's MHEWS, led by the Badan Meteorologi, Klimatologi, dan Geofisika (BMKG), integrates the Indonesian Tsunami Early Warning System (InaTEWS) with real-time seismic, buoy, and tide gauge data. After the 2004 Indian Ocean tsunami, Indonesia strengthened its capacity to issue warnings for earthquakes, tsunamis, floods, and volcanic eruptions. Community-based approaches, siren networks, and public education form a critical part of the system, ensuring local action complements technical alerts.

Australia

Australia's Bureau of Meteorology (BoM) manages a national MHEWS that monitors weather, floods, bushfires, cyclones, and tsunamis. The Joint Australian Tsunami Warning Centre (JATWC) combines data from seismic and sea-level networks to issue real-time alerts, while the Australian Warning System (AWS) provides standardized color-coded alerts across hazards. Integration with emergency services and local governments enables coordinated preparedness and response actions, minimizing loss and disruption.

18.9 Multi-hazard Early Warning Challenges and Issues

Multi-hazard early warning systems (MHEWS) face several challenges and issues that can limit their effectiveness despite advances in technology and disaster management frameworks.

Data and Technical Challenges: Effective MHEWS rely on high-quality, real-time data from multiple sources, including seismic, meteorological, hydrological, and volcanic monitoring. In many regions, such data are sparse, outdated, or fragmented (UNDRR & WMO, 2017). Integrating heterogeneous data streams into a coherent system that anticipates interactions and cascading effects is complex; for example, forecasting cyclone-induced floods or earthquake-triggered landslides requires linking multiple models and datasets (Kappes et al., 2012). Current models may not fully capture cascading or compound events, such as GLOFs triggered by earthquakes or floods following landslides, often leading to underestimation of risks (UNDRR, 2015).

Institutional and Governance Challenges: Different agencies often manage different hazards—such as meteorology, seismology, and hydrology—leading to coordination gaps (WMO, 2021). While international frameworks like the Sendai Framework advocate multi-hazard approaches, national or local policies may still focus on single hazards, limiting integrated preparedness (UNDRR, 2015). Limited funding, trained personnel, and maintenance support further hinder the operational sustainability of MHEWS (Kappes et al., 2012).

Communication and Social Challenges: Even when warnings are issued, reaching remote or vulnerable communities in time is challenging. Public understanding and trust in multi-hazard alerts may be low, particularly in areas with linguistic, cultural, or educational barriers (UNDRR & WMO, 2017). Frequent warnings without visible impacts can cause alert fatigue, reducing community responsiveness over time (WMO, 2021).

Multi-hazard Complexity: Predicting interactions among hazards, such as earthquake → landslide → flood sequences, is inherently difficult, and many MHEWS lack the modeling sophistication to account for such cascades (Kappes et al., 2012). Spatial and temporal uncertainties of different hazards complicate integrated forecasting and response planning (UNDRR, 2015).

Technological and Infrastructure Limitations: Sensors, communication networks, and software systems across different hazard domains may not be fully interoperable. Equipment failures, power outages, or cyber threats can compromise the reliability of early warning systems (WMO, 2021).

Addressing these challenges requires investments in integrated monitoring systems, cross-agency coordination, community engagement, risk education, and advanced modeling of cascading and compound hazards. A successful MHEWS combines technical robustness with social inclusiveness, ensuring actionable alerts and effective community preparedness (UNDRR & WMO, 2017; Kappes et al., 2012; WMO, 2021).

18.10 Recommended Multi-hazard Early Warning System

The operation of a MHEWS in India's diverse landscape requires a localized and hierarchical framework that bridges the gap between national-level scientific forecasting and local-level disaster governance. This framework needs to integrate core components of MHEWS viz. risk knowledge, monitoring, forecasting, warning generation, communication, and preparedness into an end-to-end system tailored to regional contexts. Central to this approach is enhanced risk knowledge, combining geophysical datasets (e.g., slope stability, seismicity, and historical GLOF/flood triggers) with granular socio-economic vulnerability profiles to enable impact-based assessments. For real-time monitoring and forecasting, the multi-source data (meteorological, hydrological, and geotechnical) feed to decision-support systems that generate location-specific warnings. To ensure people-centered communication, the system adopts a multi-modal dissemination strategy prioritizing last-mile connectivity through mobile alerts, vernacular channels, and Panchayat-level institutional networks, ensuring that warnings are both accessible and actionable. This MHEWS framework will further institutionalize the preparedness-based early action by linking hazard thresholds to predefined Standard Operating Procedures (SOPs), enabling timely, coordinated responses at the community level. It should align with the Sendai

Framework for Disaster Risk Reduction (UNDRR, 2015) and the World Meteorological Organisation (WMO, 2021) guidelines that emphasize risk knowledge, people-centered communication, and preparedness-based early action.

18.10.1 Risk Knowledge

Comprehensive risk knowledge is the foundation of an effective MHEWS. It requires integrating hazard, exposure, and vulnerability data through multi-hazard maps, exposure inventories, and socio-economic assessments (Kappes et al., 2012). Understanding local contexts, historical events, and climate projections enhances the ability to anticipate cascading risks (UNDRR & WMO, 2017). In India, strengthening the National Database for Emergency Management (NDEM) with GIS and satellite data from ISRO and NRSC can improve multi-hazard analysis. Local assessments should include interactions such as floods triggering landslides or cyclones causing secondary flooding to inform land-use and mitigation planning (UNDRR, 2015; NDMA, 2019).

18.10.2 Monitoring and Detection

Continuous multi-sensor monitoring using meteorological, hydrological, seismic, and remote-sensing technologies enables early detection of diverse hazards. Integrating satellite-based rainfall data with ground observations improves nowcasting for floods and cyclones (WMO, 2021). Automated data validation, sensor redundancy, and open-access data platforms strengthen reliability and transparency (UNDRR & WMO, 2017). In India, hazard-specific monitoring agencies—such as IMD for weather and cyclones, CWC for floods, INCOIS for tsunamis, and SASE for avalanches—should be integrated under a unified national dashboard to enhance real-time coordination and multi-hazard forecasting (NDMA, 2019; WMO, 2021).

18.10.3 Analysis and Forecasting

The analytical strength of a MHEWS depends on interoperable modeling frameworks capable of simulating both individual and cascading hazards. Impact-based forecasting focuses on predicting potential consequences—identifying who and what will be affected—rather than merely forecasting the event (Kappes et al., 2012). Integrating IMD's cyclone models and CWC's flood forecasting tools with remote-sensing data and AI-driven prediction systems enhances anticipatory capabilities (WMO, 2021). Multi-hazard models that assess linked events, such as landslides following heavy rainfall or cyclones causing inland flooding, improve predictive accuracy and support proactive planning (UNDRR, 2015; WMO, 2021).

18.10.4 Warning Dissemination

Timely dissemination of warnings is vital for reducing casualties. Effective MHEWS employs multi-channel communication systems such as SMS alerts, mobile applications, community sirens, television, and local radio networks, to ensure redundancy and reach vulnerable populations. Warnings should be clear, actionable, and linguistically appropriate (UNDRR, 2015). In India, the Common Alerting Protocol (CAP) platform and Cell Broadcast System, managed by NDMA and the Department of Telecommunications, provide coordinated multi-channel alerts, enhancing last-mile connectivity in rural and hilly areas (NDMA, 2019; WMO, 2021). International examples, such as Japan's integrated cell broadcast system, demonstrate the effectiveness of timely, coordinated dissemination. Integration with initiatives like the BharatMAP and Digital India can further expand coverage and accessibility.

18.10.5 Preparedness and Response Linkage

Early warnings are effective only when linked to Standard Operating Procedures (SOPs) and pre-planned response actions. Coordination among response agencies, local governments, and communities is essential to implement clear action protocols for each alert level, reducing confusion during crises and ensuring timely evacuations, particularly in flood- and cyclone-prone areas (UNDRR & WMO, 2017). Embedding early action frameworks into local disaster management plans enhances system responsiveness. In India, State Disaster Management Authorities (SDMAs), district administrations, and local response teams follow predefined SOPs for evacuation, relief, and communication. Regular mock drills and simulation exercises coordinated by NDMA improve operational readiness and community confidence (UNDRR, 2015; NDMA, 2019).

18.10.6 Community Engagement

Community participation is the final and most vital component of MHEWS. Establishing Community Early Warning Committees, Village Disaster Management Committees (VDMCs), and involving local volunteers ensures that warnings translate into timely protective measures (UNDRR, 2015). Integrating local knowledge, gender perspectives, and traditional coping mechanisms enhances system acceptance, credibility, and sustainability. Programs such as Aapda Mitra, alongside engagement with Panchayati Raj Institutions and NGOs, strengthen last-mile connectivity, feedback mechanisms, and overall community ownership, reinforcing resilience against cascading hazards (UNDRR, 2015; NDMA, 2019; WMO, 2021).

18.11 Institutional and Governance Framework

A robust institutional framework is essential to coordinate MHEWS. It clearly defines the roles of meteorological departments, disaster management authorities, communication services, and other stakeholders (WMO, 2021). National coordination mechanisms supported by standardized data-sharing protocols enhance system coherence, interoperability, and accountability (UNDRR & WMO, 2017). In India, the National Disaster Management Authority (NDMA) provides national-level policy guidance, the National Disaster Response Force (NDRF) conducts field operations, and the National Institute of Disaster Management (NIDM) focuses on capacity building and training. Data-sharing agreements between IMD, CWC, INCOIS, and ISRO ensure rapid decision-making and integrated hazard response (WMO, 2021; NDMA, 2019). Regular coordination meetings, cross-sector drills, and adherence to interoperability standards further strengthen governance and system reliability.

18.11.1 Capacity Building and Sustainability

The long-term sustainability of MHEWS relies on continuous capacity development, institutional learning, and research collaboration. Regular training in forecasting, hazard monitoring, communication protocols, and system maintenance ensures technical reliability and operational readiness. Partnerships with academic institutions such as IITs, IIRS, and regional organizations like ICIMOD support innovation in hazard modeling, risk communication, and early warning methodologies. Integrating scientific research into policy-making enables MHEWS to adapt to emerging risks, technological advancements, and changing climatic conditions (Kappes et al., 2012; WMO, 2021).

18.11.2 Monitoring, Evaluation, and Improvement

Regular evaluation of MHEWS performance using indicators such as lead time, forecast accuracy, population coverage, and community response is essential to maintain reliability and efficiency. Post-event reviews, simulation exercises, and drills promote continuous learning and refinement of operational protocols. Institutionalizing such assessments, as recommended by NDMA, ensures transparency, accountability, and system improvement (UNDRR, 2015; UNDRR & WMO, 2017; NDMA, 2019). Lessons learned from major disasters, including Cyclone Amphan (2020) and the Kerala Floods (2018), should inform iterative improvements in hazard forecasting, early-warning dissemination, and community-preparedness strategies.

18.11.3 Implementation Roadmap

Given the vast geographical scale and administrative complexity of the Indian subcontinent, the development of comprehensive MHEWS in a phased implementation approach is essential so that an early warning system is operational and financially viable. This strategy needs to follow the iterative strengthening model advocated by the Sendai Framework for Disaster Risk Reduction (UNDRR, 2015), which prioritizes incremental capacity building over simultaneous, high-risk systemic improvements. Phase I focuses on integrating existing hazard monitoring systems, including IMD, CWC, INCOIS, and SASE, into a unified data-sharing platform. Phase II involves deploying community-based monitoring networks, mobile alert systems, and localized forecasting to enhance last-mile connectivity. Phase III emphasizes national-scale roll-out, cross-border collaboration through platforms such as SAARC and BIMSTEC, and linkages with regional centers like RIMES. Each phase should be supported by sustainable funding, legal frameworks, and public–private partnerships to ensure long-term operational effectiveness and alignment with global best practices (NDMA, 2019; WMO, 2021). Figure 18.3 shows the implementation plan of MHEWS in India.

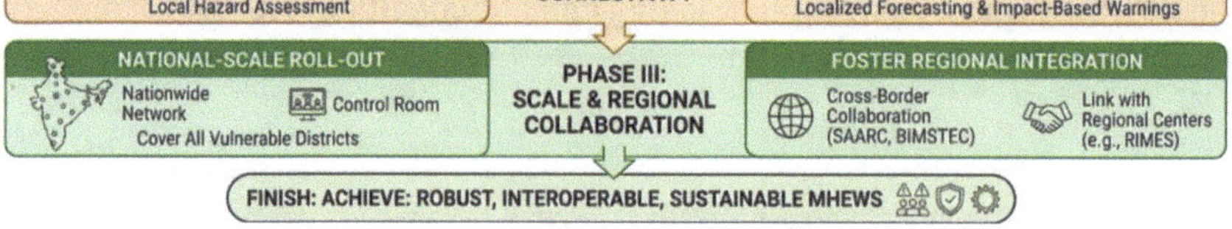

Fig. 18.3 Phased implementation approach for a robust and community-centric MHEWS in India

18.12 Conclusions

EWS are indispensable tools in modern disaster risk reduction, enabling proactive action to safeguard lives and livelihoods across a spectrum of hazards. As hazard environments grow increasingly complex, given increasing anthropogenic activities and cascading risks, the evolution toward MHEWS is essential for comprehensive preparedness and resilience. Effective EWS requires not only robust scientific and technical networks but also equitable communication, community engagement, and integration with local and national policies. As showcased in international and Indian case studies, advancing EWS demands continual innovation, a need for embracing digital technology, inclusivity, and adaptive management. By embedding early warning into the fabric of societies and fostering collaboration across disciplines, geographies, and sectors, we strengthen the global capacity to anticipate, mitigate, and recover from disasters in an era of uncertainty and change.

Acronyms

CAP	Common Alerting Protocol
CEW/CEWS	Cyclone Early Warning/Cyclone Early Warning System
CWC	Central Water Commission
DRR	Disaster Risk Reduction
EAP	Early Action Plan
EEW/EEWS	Earthquake Early Warning/Earthquake Early Warning System
EFFIS	European Forest Fire Information System
EFAS	European Flood Awareness System
ERSS	Emergency Response Support System
EWA	Early Warning Actions
FEWS	Flood Early Warning System
FFEWS	Forest Fire Early Warning System
FFAS	Forest Fire Alert System
FFWC	Flood Forecasting and Warning Centre
FIRMS	Fire Information for Resource Management System
FSI	Forest Survey of India
IBF	Impact-Based Forecasting
IMD	India Meteorological Department
INCOIS	Indian National Centre for Ocean Information Services
IN-TEWS	Indian National Tsunami Early Warning System
IPCC	Intergovernmental Panel on Climate Change
ISRO	Indian Space Research Organisation
IoT	Internet of Things
ITEWC	Indian Tsunami Early Warning Centre
JMA	Japan Meteorological Agency

LEWS	Landslide Early Warning System
MHEWS	Multi-Hazard Early Warning Systems
MoES	Ministry of Earth Sciences
NCS	National Centre for Seismology
NCRMP	National Cyclone Risk Mitigation Project
NDMA	National Disaster Management Authority
NDRRMC	National Disaster Risk Reduction and Management Council
NOAA	National Oceanic and Atmospheric Administration
NWP	Numerical Weather Prediction
PTWC	Pacific Tsunami Warning Centre
SACHET	System for Alerting and Communication of Emergency and Threats
SASMEX	Mexican Seismic Alert System
SDGs	Sustainable Development Goals
SDMA	State Disaster Management Authority
SFDRR	Sendai Framework for Disaster Risk Reduction
SMS	Short Message Service
UEEWS	Uttarakhand Earthquake Early Warning System
UNDRR	United Nations Office for Disaster Risk Reduction
UNISDR	United Nations International Strategy for Disaster Reduction
UNESCO-IOC	Intergovernmental Oceanographic Commission of UNESCO
USGS	United States Geological Survey
WMO	World Meteorological Organization

References

Bernard, E. N., & Titov, V. V. (2015). Evolution of tsunami warning systems and products. *Philosophical Transactions of the Royal Society A, 373*(2053), 20140371. https://doi.org/10.1098/rsta.2014.0371

Bhardwaj, R., Sharma, M. L., & Kumar, A. (2016). Multi-parameter algorithm for earthquake early warning. *Geomatics, Natural Hazards and Risk, 7*, 1242–1264.

Carey, M., Huggel, C., Bury, J., Portocarrero, C., & Haeberli, W. (2012). An integrated socio-environmental framework for glacier hazard management and climate change adaptation: Lessons from Lake 513, Cordillera Blanca, Peru. *Climatic Change, 112*(3–4), 733–767. https://doi.org/10.1007/s10584-011-0215-5

Cheng, P., Sui, H., Wang, J., Hua, L., Liu, J., Zhou, Q., & Zeng, Y. (2025). Forest fire detection based on temporal and spatial correction of background brightness temperature using GF-4 PMI data. ISPRS Annals of the Photogrammetry. *Remote Sensing and Spatial Information Sciences, X-G-2025*, 189–195. https://doi.org/10.5194/isprs-annals-X-G-2025-189-2025

Cremen, G., & Galasso, C. (2020). Earthquake early warning: Recent advances and perspectives. *Earth-Science Reviews, 205*, Article 103184. https://doi.org/10.1016/j.earscirev.2020.103184 ADS+3

CRIDF report. (2017). Case study on flood forecasting systems in the Limpopo River Basin.

Csiszar, I., Schroeder, W., Giglio, L., Ellicott, E., Vadrevu, K. P., Justice, C. O., & Wind, B. (2014). Active fires from the Suomi NPP visible infrared imaging radiometer suite (VIIRS): Product status and first evaluation results. *Journal of Geophysical Research: Atmospheres, 119*(2), 803–816. https://doi.org/10.1002/2013JD020453

Cuéllar, A. D., Espinosa-Aranda, J. M., Suárez, R., Ibarrola, G., Uribe, A., Rodríguez, F. H., Islas, R., Rodríguez, G. M., García, A., & Frontana, B. (2014). The Mexican seismic alert system (SASMEX): Its alert signals, broadcast results and performance during the M 7.4 Punta Maldonado earthquake of March 20th 2012

Cutter, S. L., Boruff, B. J., & Shirley, W. L. (2003). Social vulnerability to environmental hazards. *Social Science Quarterly, 84*(2), 242–261. https://doi.org/10.1111/ssqu.12105

Dang, N., et al. (2021). Flood forecasting and early warning in the Mekong Delta, Vietnam. *Water, 13*(10), 1376. https://doi.org/10.3390/w13101376

Dang, P., Zhu, J., Dang, C., & Zhang, H. (2025). A method for assessing urban flood impacts by combining hydrodynamic models with entity semantic evolution. *International Journal of Digital Earth, 18*, 2515268. https://doi.org/10.1080/17538947.2025.2515268

Delmonaco, G., Margottini, C., & Spizzichino, D. (2006) Report on new methodology for multi-risk assessment and the harmonisation of different natural risk maps, deliverable 3.1, ARMONIA, Rome.

Deng, L., Yuan, H., Chen, J., Zhang, M., Su, G., Zhou, Y., & Chen, Y. (2023). Machine learning-based analysis of landslide deformation using acoustic emission data. SSRN. https://doi.org/10.2139/ssrn.4614293

Espinosa-Aranda, J. M., Cuéllar, A., Rodríguez, F. H., Frontana, B., Ibarrola, G., Islas, R., & García, A. (2011). The seismic alert system of Mexico (SASMEX): Progress and its current applications. *Soil Dynamics and Earthquake Engineering, 31*(2), 154–162. https://doi.org/10.1016/j.soildyn.2010.09.011

Giglio, L., Boschetti, L., Roy, D. P., Humber, M. L., & Justice, C. O. (2016). The collection 6 MODIS burned area mapping algorithm and product. *Remote Sensing of Environment, 217*, 72–85. https://doi.org/10.1016/j.rse.2018.08.005

Gill, J. C., & Malamud, B. D. (2014). *Reviewing and visualizing the interactions of natural hazards.* American Geophysical Union. https://doi.org/10.1002/2014EO240001

Gill, J. C., & Malamud, B. D. (2016). Hazard interactions and interaction networks (cascades) within multi-hazard methodologies. *Earth System Dynamics, 7*, 659–679. https://doi.org/10.5194/esd-7-659-2016

Given, D. D., Cochran, E. S., Heaton, T. H., Hutton, K., Frankel, A., Allen, R. M., Hellweg, M., Vidale, J., & Bodin, P. (2018). Revised technical implementation plan for the ShakeAlert system—An earthquake early warning system for the West Coast of the United States. U.S. Geological Survey Open-File Report 2018–1155, (42 p). https://doi.org/10.3133/ofr20181155

Hamadeh, S., El-Khoury, M., & Habib, M. (2015). Climate change impacts on precipitation and temperature in Lebanon. *Theoretical and Applied Climatology, 120*(3–4), 469–484. https://doi.org/10.1007/s00704-014-1214-2

Herawati, H., & Santoso, H. (2011). Tropical forest susceptibility to and risk of fire under changing climate: A review of fire nature, policy and institutions in Indonesia. *Forest Policy and Economics, 13*(4), 227–233. https://doi.org/10.1016/j.forpol.2011.02.006

Hoshiba, M., et al. (2011). Real-time estimation of ground motion by the Japan meteorological agency earthquake early warning system. *Journal of Seismology, 15*(2), 203–214. https://doi.org/10.1007/s10950-010-9201-6

Indian National Centre for Ocean Information Services (INCOIS). (2020). Early warning dissemination system (EWDS) for coastal communities. Ministry of Earth Sciences, Government of India. https://www.incois.gov.in

Indian National Centre for Ocean Information Services (INCOIS). (2021). Indian Tsunami early warning centre (ITEWC): Operational overview and capabilities. Ministry of Earth Sciences, Government of India. https://www.incois.gov.in

Indian National Centre for Ocean Information Services (INCOIS). (2022). SACHET: A multi-platform dissemination system for ocean and tsunami alerts. Ministry of Earth Sciences, Government of India. https://www.incois.gov.in

International Centre for Integrated Mountain Development (ICIMOD). (2011). *Glacial lake outburst floods in the Hindu Kush Himalaya: Risk assessment and management*. ICIMOD. https://www.icimod.org

Ives, J. D., Shrestha, R. B., & Mool, P. K. (2010). *Formation of glacial lakes in the Hindu Kush–Himalayas and GLOF risk assessment*. International Centre for Integrated Mountain Development (ICIMOD).

Japan Meteorological Agency. (2012). *Earthquake early warning operational overview*. JMA. https://www.jma.go.jp

Kappes, M. S., Keiler, M., von Elverfeldt, K., & Glade, T. (2012). Challenges of analyzing multi-hazard risk: A review. *Natural Hazards, 64*(2), 1925–1958. https://doi.org/10.1007/s11069-012-0294-2

Khan, S., & Khan, A. (2022). FFireNet: Deep learning based forest fire classification and detection in smart cities. *Symmetry, 14*(10), 2155. https://doi.org/10.3390/sym14102155

Krøgli, I. K., et al. (2018). The Norwegian forecasting and warning service. *Natural Hazards and Earth System Sciences, 18*, 1427–1450.

Kumar, P., Kamal, M., Sharma, M. L., Jakka, R. S., & Pratibha. (2024). Uttarakhand state earthquake early warning system: A case study of the Himalayan environment. *Sensors, 24*(11), 3272. https://doi.org/10.3390/s24113272

Labed, S., Touati, H., Herida, A., Kerbab, S., & Sairi, A. (2023). An AI-based image recognition system for early detection of forest and field fires. *European Journal of Forest Engineering, 9*(2), 48–56. https://doi.org/10.33904/ejfe.1322396

Li, X., et al. (2020). Development and application of the ICL earthquake early warning system in China. *Seismological Research Letters, 91*(1), 56–65. https://doi.org/10.1785/0220190174

Li, Y.-R., et al. (2021). IoT-based monitoring and early warning system for the Lianyun Village landslide: 5G and remote sensing integration. *Journal of Wireless Communications and Networking, 2021*, 160.

Liu, P., Titov, V. V., & Satake, K. (2013). Tsunami early warning in the Indian Ocean: System improvements after 2004. *Pure and Applied Geophysics, 170*(6–8), 1183–1198. https://doi.org/10.1007/s00024-012-0540-x

Martelloni, G., Segoni, S., Fanti, R., & Catani, F. (2012). Rainfall thresholds for the forecasting of landslide occurrence at regional scale. *Landslides, 9*(4), 485–495. https://doi.org/10.1007/s10346-011-0308-2

Maurya, A. K., et al. (2022). Chamoli GLOF event: Hydrological and geological perspectives. *Journal of Himalayan Research, 56*(2), 120–136.

Melnick, D., Cisternas, M., Moreno, M., & Norambuena, R. (2012). Estimating coseismic coastal uplift with an intertidal mussel: Calibration for the 2010 Maule, Chile earthquake (M w = 8.8). *Quaternary Science Reviews, 42*, 29–42. https://doi.org/10.1016/j.quascirev.2012.03.022

Miller, M. E., Billmire, M. G., Elliot, W. J., & Robichaud, P. R. (2017). *Rapid response tools and datasets for post-fire modeling in Boreal and Arctic environments*. Presented at the Spring 2017 AFSC Remote Sensing Workshop.

Mittal, H., Yih-Min, W., Sharma, M. L., Yang, B. M., & Gupta, S. (2019). Testing the performance of earthquake early warning system in northern India. *Acta Geophysica, 67*(1), 59–75.

National Disaster Management Authority (NDMA), India. (2019). *National guidelines for the management of cyclones*. Government of India. https://ndma.gov.in

Novković, I., Marković, G. B., Lukić, D., Dragićević, S., Milošević, M., Đurđić, S., Samardžić, I., Ležaić, T., & Tadić, M. (2021). GIS-based forest fire susceptibility zonation with IoT sensor network support, case study—Nature Park Golija, Serbia. *Sensors, 21*(19), 6520. https://doi.org/10.3390/s21196520

Pandey, M., Costache, R., Dash, P., et al. (2024). Optimized ensemble-based flood hazard mapping in low altitude subtropical riverine terrane. *Discover Geoscience, 2*, 56. https://doi.org/10.1007/s44288-024-00037

Pappenberger, F., Cloke, H. L., Balsamo, G., Ngo-Duc, T., Oki, T., et al. (2012). Global runoff routing with the hydrological component of the ECMWF NWP system. *Hydrology and Earth System Sciences, 16*, 4143–4156.

Paul, B. K., Dutta, D., & Rahman, M. (2018). Storm surge modeling and coastal flood risk assessment. *Journal of Coastal Research, 34*(2), 456–469. https://doi.org/10.2112/JCOASTRES-D-17-00128.1

Piciullo, L., Calvello, M., & Cepeda, J. M. (2018). Territorial early warning systems for rainfall-induced landslides. *Earth-Science Reviews, 179*, 228–247. https://doi.org/10.1016/j.earscirev.2018.02.013

Rajendran, C. P., Rajendran, K., Andrade, V., & Srinivasalu, S. (2013). Ages and relative sizes of pre-2004 tsunamis in the Bay of Bengal inferred from geologic evidence in the Andaman and Nicobar Islands. *Journal of Geophysical Research: Solid Earth, 118*(4), 1345–1362. https://doi.org/10.1002/jgrb.50122

Rathore, G., Kumar, A., Jakka, R. S., & Sharma, M. L. (2023). Design and implementation of earthquake early warning dissemination mobile app for Uttarakhand (India). J. Seismol. **27**(1), 203–217. https://doi.org/10.1007/s10950-022-10124-6

Sahana, V., & Mondal, A. (2023). Evolution of multivariate drought hazard, vulnerability and risk in India under climate change. *Natural Hazards and Earth System Sciences, 23*, 623–641. https://doi.org/10.5194/nhess-23-623-2023

Scott, K. M., et al. (1999). Lahar hazards at Mt. Pinatubo, Philippines. *Journal of Volcanology and Geothermal Research, 88*(1–4), 1–15. https://doi.org/10.1016/S0377-0273(98)00101-1

Segoni, S., Piciullo, L., & Gariano, S. L. (2018). A review of the recent literature on rainfall thresholds for landslide occurrence. *Landslides, 15*, 1483–1501. https://doi.org/10.1007/s10346-018-0966-4

Sharma, R., Singh, S., & Kumar, A. (2020). Community-based flood early warning systems in India. *Journal of Flood Risk Management, 13*(3), Article e12664. https://doi.org/10.1111/jfr3.12664

Sudhakar, S., Vijayakumar, V., Sathiya Kumar, C., Priya, V., Ravi, L., & Subramaniyaswamy, V. (2020). Unmanned aerial vehicle (UAV) based forest fire detection and monitoring for reducing false alarms in forest-fires. *Computer Communications, 149*, 1–16. https://doi.org/10.1016/j.comcom.2019.10.007

Suleimani, E., Hansen, R., & Haeussler, P. J. (2009). Numerical study of tsunami generated by multiple submarine slope failures in ResurrectionBay, Alaska, during the Mw 9.2 1964 earthquake. In: P. R. Cummins, L. S. L. Kong, & K. Satake, (Eds.), *Tsunami science four years after the 2004 Indian Ocean Tsunami* (pp. 131–152), Springer.

Thiebes, B., Glade, T., Jäger, S., & Canli, E. (2014, April). Landslide monitoring and early warning systems in Lower Austria—Current situation and new developments [Poster session]. EGU General Assembly 2014, Vienna, Austria.

Tiranti, D., & Rabuffetti, D. (2010). Estimation of rainfall thresholds triggering shallow landslides for an operational warning system implementation. *Landslides, 7*(4), 471–481. https://doi.org/10.1007/s10346-010-0198-8

U.S. Geological Survey. (2021, December 29). 2021 annual overview [Annual report]. U.S. Department of the Interior. https://www.usgs.gov/tools/2021-annual-overview

Umbal, J. V., & Rodolfo, K. S. (1996). The 1991 lahars of southwestern Mount Pinatubo, Philippines, and the evolution of a lahar-dammed lake. In: C. G. Newhall, & R. S. Punongbayan (Eds.), *Fire and mud: Eruptions and lahars of Mt. Pinatubo, Philippines* (pp. 951–970). University of Washington Press.

UNESCO-IOC. (2007). Indian Ocean Tsunami warning and mitigation system (IOTWS)—Implementation plan 2007–2008. Intergovernmental Oceanographic Commission of UNESCO.

UNISDR. (2006). *Developing early warning systems: A checklist*. Third international conference on early warning, Bonn, Germany.

UNISDR. (2015). Sendai framework for disaster risk reduction 2015–2030. United Nations Office for Disaster Risk Reduction.

United Nations International Strategy for Disaster Reduction. (2002). *Living with risk: A global review of disaster reduction initiatives (Draft).*

United Nations International Strategy for Disaster Reduction. (2005, January). Hyogo framework for action 2005–2015: Building the resilience of nations and communities to disasters (A/CONF.206/6). http://www.unisdr.org/wcdr

United Nations Office for Disaster Risk Reduction (UNDRR). (2015). Sendai framework for disaster risk reduction 2015–2030. United Nations. https://www.undrr.org/publication/sendai-framework-disaster-risk-reduction-2015-2030

United Nations Office for Disaster Risk Reduction (UNDRR) & World Meteorological Organization (WMO). (2017). *Multi-hazard early warning systems: A checklist*. UNDRR–WMO. https://www.undrr.org/publication/multi-hazard-early-warning-systems-checklist

United Nations Development Programme. (2018). *Early warning systems for climate resilience: A review of international and regional cooperation.* UNDP.

Wang, S., & Liang, S. (2025). Application of moving target detection in landslide warning based on OpenCV. *Advances in Research, 26*(5), 321–336. https://doi.org/10.9734/air/2025/v26i51488

Wardihani, E. D., Ramdhani, M., Suharjono, A., Setyawan, T. A., Hidajjat, S., Widodo, S., & Saifullah, F. (2018). Real-time forest fire monitoring system using unmanned aerial vehicle. *Journal of Engineering Science and Technology, 13*(6), 1587–1594. https://doi.org/10.1109/JESTEC.2018.9037642

World Bank. (2010). *Natural hazards, unnatural disasters: The economics of effective prevention.* World Bank. https://openknowledge.worldbank.org

World Meteorological Organization. (2018). *Multi-hazard early warning systems: A checklist.* WMO.

World Meteorological Organization (WMO). (2021). *Guidelines on multi-hazard early warning systems: A checklist* (2nd ed., WMO-No. 1150). https://library.wmo.int/doc_num.php?explnum_id=10674

Wu, Y., Mittal, H., Chen, D.–Y., Hsu, T.–Y., & Lin, P.–Y. (2021). Earthquake early warning systems in Taiwan: Current status. *Journal of the Geological Society of India, 97*(12), 1525–1532. https://doi.org/10.1007/s12594-021-1909-6

Yuliza, E., Habil, H., Munir, M., Irsyam, M., et al. (2016). Study of soil moisture sensor for landslide early warning system: Experiment in laboratory scale *Journal of Physics: Conference Series, 739*(1), 012034. https://doi.org/10.1088/1742-6596/739/1/012034

FSC
www.fsc.org
MIX
Papier | Fördert gute Waldnutzung
FSC® C083411

Kolibri 360 GmbH
Ferdinand-Jühlke-Straße 7
99095 Erfurt, Deutschland
produktsicherheit@kolibri360.de